ANNUAL EDITIONS

Microbiology 10/11

First Edition

EDITOR

Brinda Govindan
San Francisco State University

Brinda Govindan is a Lecturer in the Department of Biology at San Francisco State University. Dr. Govindan earned her PhD in Cell Biology at the Yale University School of Medicine. She continued her postdoctoral research as a Damon-Runyon Walter Winchell fellow at University of California at San Francisco before joining SFSU. For the past ten years she has been teaching undergraduate microbiology to allied health students and microbiology majors. She is interested in undergraduate science education research and has been selected as a Research Residency Fellow for the 2009–2010 Biology Scholars Program funded by the National Science Foundation.

Dr. Govindan is the author of many research articles and has also contributed to the development of instructional materials in microbiology. She is a member of the American Society for Microbiology.

Boston   Burr Ridge, IL   Dubuque, IA   New York   San Francisco   St. Louis
Bangkok   Bogotá   Caracas   Kuala Lumpur   Lisbon   London   Madrid   Mexico City
Milan   Montreal   New Delhi   Santiago   Seoul   Singapore   Sydney   Taipei   Toronto

The McGraw-Hill Companies

Higher Education

ANNUAL EDITIONS: MICROBIOLOGY, FIRST EDITION

Annual Editions® is a registered trademark of The McGraw-Hill Companies, Inc.
Annual Editions is published by the **Contemporary Learning Series** group within The McGraw-Hill Higher Education division.

1 2 3 4 5 6 7 8 9 0 QWD/QWD 0 9

ISBN 978–0–07–351552–6
MHID 0–07–351552–3
ISSN 1948–2647

Managing Editor: *Larry Loeppke*
Senior Managing Editor: *Faye Schilling*
Developmental Editor: *Debra Henricks*
Editorial Coordinator: *Mary Foust*
Editorial Assistant: *Cindy Hedley*
Production Service Assistant: *Rita Hingtgen*
Permissions Coordinator: *Lenny J. Behnke*
Senior Marketing Manager: *Julie Keck*
Marketing Communications Specialist: *Mary Klein*
Marketing Coordinator: *Alice Link*
Project Manager: *Joyce Watters*
Design Specialist: *Tara McDermott*
Senior Production Supervisor: *Laura Fuller*
Cover Graphics: *Kristine Jubeck*

Compositor: Laserwords Private Limited
Cover Images: CDC/Janice Haney Carr, The McGraw-Hill Companies, Inc./Don Rubbelke, photographer (inset); © Getty Images/RF (background)

Library in Congress Cataloging-in-Publication Data
Main entry under title: Annual Editions: Microbiology 2010/2011.
 1. Microbiology—Periodicals. I. Govindan, Brinda, *comp*. II. Title: Microbiology.
658'.05

www.mhhe.com

Editors/Academic Advisory Board

Members of the Academic Advisory Board are instrumental in the final selection of articles for each edition of ANNUAL EDITIONS. Their review of articles for content, level, and appropriateness provides critical direction to the editors and staff. We think that you will find their careful consideration well reflected in this volume.

ANNUAL EDITIONS: Microbiology 10/11
1st Edition

EDITOR

Brinda Govindan
San Francisco State University

Editors/Academic Advisory Board continued

Preface

In publishing ANNUAL EDITIONS we recognize the enormous role played by the magazines, newspapers, and journals of the public press in providing current, first-rate educational information in a broad spectrum of interest areas. Many of these articles are appropriate for students, researchers, and professionals seeking accurate, current material to help bridge the gap between principles and theories and the real world. These articles, however, become more useful for study when those of lasting value are carefully collected, organized, indexed, and reproduced in a low-cost format, which provides easy and permanent access when the material is needed. That is the role played by ANNUAL EDITIONS.

Since microbiology is an evolving science, the main goal of *Annual Editions: Microbiology* is to provide the reader with accessible information based on current scientific research in various aspects of this field. Given the large scope of microbiology, every effort has been made to include a broad variety of topics that will be of interest and relevance to the beginning student of microbiology. Introductory microbiology/public health courses often have to cover an enormous amount of material in a short period of time. This anthology will give students the opportunity to further explore topics of interest by learning about current research in microbiology. We hope that the reader will develop critical thinking skills and connect concepts taken from class to applications in the readings. This collection demonstrates how microbes impact our daily lives, both as friend and foe. We hope the reader will gain an appreciation for the many ways in which the lives of humans and microbes are inextricably linked.

We are currently experiencing an explosion in the discovery of new microbes, as well as deeper understanding of known microbes, mainly due to developments in DNA sequencing technologies. While less than 1 percent of all microbes have been isolated by traditional culturing methods, we are learning more about the diversity and nature of the microbial world by examining microbial DNA sequences. The catalog of human microbiota will soon be complete, giving health professionals new insights into preventing and treating human disease. Increasingly, we are also beginning to understand more deeply the complex microbial communities that inhabit our bodies and the consequences of upsetting their delicate balance. The advent of genomic and other "omic" technologies have brought us closer to deciphering the inner workings of pathogenic microbes. These advances promise progress in the development of diagnostic tests and hopefully treatment of infectious diseases. Biotechnology has become a thriving enterprise that uses microbes to generate products of commercial importance, such as vaccines, drugs, and diagnostic tests. Currently, the biotechnology industry is using microbes to address the need for alternative fuel sources in the face of global warming.

Humans are facing another global crisis—the problem of drug-resistant microbes. Since the use of antibiotics began after World War II, microbes have been steadily evolving and adapting to all of the drugs we have been casually using to eradicate them. We are facing increasingly virulent pathogens with a limited arsenal of antimicrobial drugs and virtually no new drugs coming down the pipeline. Many new strategies are being used to combat antibiotic resistance, but perhaps more important than drug discovery is the increased public awareness of this problem and how our own behavior contributes to it. The issue of food safety and microbial contamination is addressed in the wake of public concern over contaminated spinach, peanuts, and a whole host of other products. Bird flu, SARS, tuberculosis, and many other infectious diseases have been splashed across media headlines in recent years. This has prompted growing public concern over the emergence of these "new" diseases. Surveillance databases are being set up to monitor the movement of microbes associated with wild animals; hopefully, this will alert epidemiologists before disease outbreaks occur. Microbes know no borders, and in an age of globalization, this means that microbes are traveling as far and wide as their human hosts.

Annual Editions: Microbiology should be used as a companion to a standard microbiology textbook so that it may update, expand upon, and emphasize certain topics that are covered in the text or present new topics not covered in a standard text. To accomplish this, *Annual Editions: Microbiology* is composed of seven units that cover a broad spectrum of our current knowledge in microbiology. The articles have been carefully selected to cover diverse microbial groups (fungi, bacteria, algae, viruses, protozoa) in each topic area. The order of topics follows the standard content outline of an introductory microbiology textbook. The first unit describes the habitats in which microbes are found—from extremophiles to our own microbiota, and also describes how microbes live together in biofilms. Unit 2 focuses on the deciphering of microbial genomes and the use of genetically engineered microbes to benefit humankind. Units 3, 4, and 5 discuss the problem of drug-resistant microbes, the development of novel antimicrobial agents, and how the immune system responds to pathogens. Unit 6 describes emerging infectious diseases and Unit 7 examines the role of microbes in food production as well as food spoilage and contamination. A topic guide will assist the reader in finding all of the articles on a given subject, and the Internet reference guide will help in further exploring research on a particular topic.

Your input is most valuable to improve this anthology, which we update yearly. We would appreciate your comments and suggestions as you review the current edition.

Brinda Govindan

Dr. Brinda Govindan
Editor

Contents

UNIT 1
Habitats and Humanity: An Introduction to the Microbes Around Us and Inside Us

The concepts in bold italics are developed in the article. For further expansion, please refer to the Topic Guide.

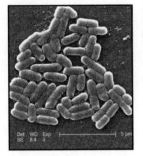

UNIT 2
Genetics and Biotechnology

The concepts in bold italics are developed in the article. For further expansion, please refer to the Topic Guide.

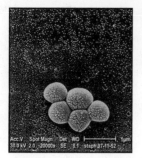

UNIT 3
The Never-Ending Battle: Antimicrobial Drug Resistance

UNIT 4
New Antimicrobial Drug Development

The concepts in bold italics are developed in the article. For further expansion, please refer to the Topic Guide.

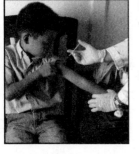

UNIT 5
The Immune Response: Natural and Acquired

The concepts in bold italics are developed in the article. For further expansion, please refer to the Topic Guide.

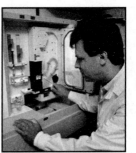

UNIT 6
Emerging Infectious Diseases and Public Health

The concepts in bold italics are developed in the article. For further expansion, please refer to the Topic Guide.

UNIT 7
Microbes and Food: Friend and Foe

The concepts in bold italics are developed in the article. For further expansion, please refer to the Topic Guide.

The concepts in bold italics are developed in the article. For further expansion, please refer to the Topic Guide.

Correlation Guide

The *Annual Editions* series provides students with convenient, inexpensive access to current, carefully selected articles from the public press. **Annual Editions: Microbiology 10/11** is an easy-to-use reader that presents articles on important topics such as *genetics, antimicrobial drugs, immune response, infectious diseases,* and many more. For more information on *Annual Editions* and other *McGraw-Hill Contemporary Learning Series* titles, visit www.mhcls.com.

This convenient guide matches the units in **Annual Editions: Microbiology 10/11** with the corresponding chapters in three of our best-selling McGraw-Hill Microbiology textbooks by Cowan/Talaro, Nester et al., and Talaro.

Annual Editions: Microbiology 10/11	Microbiology: A Systems Approach, 2/e by Cowan/Talaro	Microbiology: A Human Perspective, 6/e by Nester et al.	Foundations in Microbiology, 7/e by Talaro
Unit 1: Habitats and Humanity: An Introduction to the Microbes Around Us and Inside Us	**Chapter 1:** The Main Themes of Microbiology **Chapter 3:** Tools of the Laboratory **Chapter 4:** Prokaryotic Profiles **Chapter 7:** Microbial Nutrition, Ecology, and Growth **Chapter 22:** Infectious Diseases Affecting the Gastrointestinal Tract	**Chapter 1:** Humans and the Microbial World **Chapter 4:** Dynamics of Prokaryotic Growth **Chapter 11:** The Diversity of Prokaryotic Organisms **Chapter 23:** Skin Infections	**Chapter 1:** The Main Themes of Microbiology **Chapter 7:** Elements of Microbial Nutrition, Ecology, and Growth **Chapter 11:** Physical and Chemical Agents for Microbial Control **Chapter 21:** Miscellaneous Bacterial Agents of Disease
Unit 2: Genetics and Biotechnology	**Chapter 9:** Microbial Genetics **Chapter 10:** Genetic Engineering **Chapter 21:** Infectious Diseases Affecting the Respiratory System	**Chapter 5:** Control of Microbial Growth **Chapter 9:** Biotechnology and Recombinant DNA **Chapter 30:** Microbial Ecology	**Chapter 8:** An Introduction to Microbial Metabolism **Chapter 9:** Microbial Genetics **Chapter 10:** Genetic Engineering **Chapter 19:** The Gram-Positive Bacilli of Medical Importance
Unit 3: The Never-Ending Battle: Antimicrobial Drug Resistance	**Chapter 12:** Drugs, Microbes, Host—The Elements of Chemotherapy	**Chapter 8:** Bacterial Genetics **Chapter 21:** Antimicrobial Medications	**Chapter 11:** Physical and Chemical Agents for Microbial Control **Chapter 12:** Drugs, Microbes, Host—The Elements of Chemotherapy
Unit 4: New Antimicrobial Drug Development	**Chapter 6:** An Introduction to the Viruses **Chapter 12:** Drugs, Microbes, Host—The Elements of Chemotherapy	**Chapter 3:** Microscopy and Cell Structure **Chapter 13:** Viruses of Bacteria	**Chapter 6:** An Introduction to Viruses **Chapter 18:** The Cocci of Medical Importance **Chapter 19:** The Gram-Positive Bacilli of Medical Importance
Unit 5: The Immune Response: Natural and Acquired	**Chapter 11:** Physical and Chemical Control of Microbes **Chapter 14:** Host Defenses I: Overview and Nonspecific Defenses **Chapter 15:** Host Defenses II: Specific Immunity and Immunization **Chapter 16:** Disorders in Immunity	**Chapter 16:** The Adaptive Immune Response **Chapter 19:** Applications of Immune Responses	**Chapter 11:** Physical and Chemical Agents for Microbial Control **Chapter 15:** Adaptive, Specific Immunity and Immunization **Chapter 16:** Disorders in Immunity
Unit 6: Emerging Infectious Diseases and Public Health	**Chapter 7:** Microbial Nutrition, Ecology, and Growth **Chapter 13:** Microbe-Human Interactions **Chapter 19:** Infectious Diseases Affecting the Nervous System	**Chapter 17:** Host-Microbe Interactions **Chapter 20:** Epidemiology	**Chapter 13:** Microbe-Human Interactions **Chapter 23:** The Parasites of Medical Importance **Chapter 25:** The RNA Viruses That Infect Humans
Unit 7: Microbes and Food: Friend and Foe	**Chapter 22:** Infectious Diseases Affecting the Gastrointestinal Tract **Chapter 25:** Applied and Industrial Microbiology	**Chapter 27:** Nervous System Infections **Chapter 32:** Food Microbiology	**Chapter 19:** The Gram-Positive Bacilli of Medical Importance **Chapter 20:** The Gram-Negative Bacilli of Medical Importance **Chapter 21:** Miscellaneous Bacterial Agents of Disease **Chapter 22:** The Fungi of Medical Importance

Topic Guide

This topic guide suggests how the selections in this book relate to the subjects covered in your course. You may want to use the topics listed on these pages to search the Web more easily.

On the following pages a number of websites have been gathered specifically for this book. They are arranged to reflect the units of this Annual Editions reader. You can link to these sites by going to *http://www.mhcls.com*.

All the articles that relate to each topic are listed below the bold-faced term.

Internet References

The following Internet sites have been selected to support the articles found in this reader. These sites were available at the time of publication. However, because websites often change their structure and content, the information listed may no longer be available. We invite you to visit http://www.mhcls.com for easy access to these sites.

Annual Editions: Microbiology 10/11

General Sources

World Health Organization: Communicable Disease Surveillance and Response
http://www.who.int/csr/en

The World Health Organization (WHO) maintains this website to provide educational information on communicable disease surveillance and response around the world. This is a great resource for facts, statistics, reports, and educational materials on the global spread of infectious diseases.

National Institutes of Health (NIH)
http://www.nih.gov

Consult this site for links to extensive health information and scientific resources. The NIH is one of eight health agencies of the Public Health Service, which in turn is part of the U.S. Department of Health and Human Services.

The National Institute of Allergy and Infectious Diseases
http://www3.niaid.nih.gov

This is the official website for the branch of the National Institutes of Health dedicated to research on allergies and infectious diseases. The website provides links to research publications, clinical studies, and information on current research projects in infectious diseases.

Centers for Disease Control and Prevention (CDC)
http://www.cdc.gov

Visit the CDC home page to find numerous links to educational information on infectious diseases, immunizations, disease prevention, and current outbreaks.

American Public Health Association
http://www.apha.org

This national organization undertakes advocacy work in the area of public health. This website provides fact sheets and links to information about public health issues in the news.

American Society for Microbiology: MicrobeWorld
http://www.microbeworld.org

This site is sponsored by the American Society for Microbiology and is geared toward educating the general public about issues in microbiology. Find links to news articles written by ASM members as well as the online journal *Microbe.* Summaries of the latest research news in microbiology can be found in concise digests.

UNIT 1: Habitats and Humanity: An Introduction to the Microbes Around Us and Inside Us

The Human Microbiome Project
http://www.nihroadmap.nih.gov/hmp/

The latest research findings of the human microbiome project can be found here. This project was started as a "roadmap initiative" undertaken with funding and support from the National Institutes of Health. Learn all about the microbes that live in and on our bodies.

Astrobiology
http://www.astrobiology.com

This site provides links to information about extremophiles. Discover the extreme environments where microbes can live—from hydrothermal vents to glaciers.

The Center for Biofilm Engineering
http://www.erc.montana.edu

The Center for Biofilm Engineering at Montana State University was established in 1990 as a National Science Foundation Engineering Research Center. This site provides links to current research on the applications of biofilms in industry as well as research on solving problems posed by biofilm formation.

UNIT 2: Genetics and Biotechnology

National Center for Biotechnology Information
http://www.ncbi.nlm.nih.gov

This website is the national resource for information on molecular biology. Whether you are looking for public databases of genome information or news on the latest developments in biotechnology, this site has information for novices and professionals alike. Links to online tutorials can be found that explain how to use many of the tools available.

Fungal Genome Initiative
http://www.broad.mit.edu/annotation/fungi/fgi/

This site provides a list of all current fungal genome sequencing projects. Quick links for students and educators include animations and research publications.

Genome News Network
http://www.genomenewsnetwork.org

Writing for a general audience, the genome news network supplies online reports of the latest research on genes and genomes. A special link to microbial genomes can be found easily. This site lists relevant, current news items concerning public health and disease, bioterrorism, energy and the environment, biotechnology, and drug development.

U.S. Geological Survey
http://water.usgs.gov/wid/html/bioremed.html

This site includes many links to bioremediation research projects sponsored by the U.S. Geological Survey. It provides general information on the subject of bioremediation.

National Renewable Energy Laboratory
http://www.nrel.gov/learning/re_biofuels.html

Learn all about renewable energy, including biofuels, with excellent student resources and links to additional educational sites.

Internet References

UNIT 3: The Never-Ending Battle: Antimicrobial Drug Resistance

Alliance for the Prudent Use of Antibiotics
http://www.tufts.edu/med/apua

The Alliance for the Prudent Use of Antibiotics was founded in 1981 as a nonprofit global organization committed to improving clinical practice worldwide, promoting the appropriate use of antibiotics, and preventing the spread of drug resistance. This site contains fact sheets, reports, and information useful to both the clinician as well as the general public.

World Health Organization
http://www.who.int/drugresistance/en/

Visit the drug resistance section of the World Health Organization's home page to learn more about raising awareness about the spread of drug-resistant microbes that cause infectious diseases like malaria, tuberculosis, and HIV/AIDS.

The Centers for Disease Control and Prevention
http://www.cdc.gov/drugresistance/

The CDC website contains a Q&A about drug resistance as well as a fact sheet on MRSA and links to excellent articles about drug resistance and how to prevent it by changing our behavior.

UNIT 4: New Antimicrobial Drug Development

U.S. Food and Drug Administration
http://www.fda.gov

By exploring the FDA's website you will gain knowledge about new drugs that have been approved for various diseases. You can also find out information about drugs that are currently being tested in clinical trials.

Agricultural Research Service: U.S. Department of Agriculture
http://www.ars.usda.gov

Search within this site for links to articles about alternatives to antibiotics and the development of new antibacterial agents. This site connects you to all of the USDA-sponsored research in microbiology, including the use of antibiotics in animal feed.

Institute for One World Health
http://www.oneworldhealth.org/story

The first nonprofit pharmaceutical company has already made a huge difference in the lives of millions around the world suffering from infectious diseases. Find out the latest on its drug development projects for parasitic and diarrheal diseases.

Office of AIDS Research
http://www.oar.nih.gov

This is the official government website from the National Institutes of Health's Office of AIDS Research. You can search within this site for information on clinical trials related to HIV, research on antimicrobial agents, and new treatments for HIV. The OAR coordinates the scientific, legislative, budgetary, and policy aspects of the NIH AIDS program.

UNIT 5: The Immune Response: Natural and Acquired

Immunization Action Coalition
http://www.immunize.org

The goal of this nonprofit organization is to increase immunization rates and prevent disease by providing educational materials about vaccines to healthcare providers and the general public. The work of this organization is funded in part by the Centers for Disease Control and Prevention in order to disseminate information about its recommended immunization guidelines.

Laboratory of Immunology (NIAID)
http://www3.niaid.nih.gov/labs/aboutlabs/li/

The Laboratory of Immunology is a division of the National Institute of Allergy and Infectious Disease. This website provides links to information about research conducted by this division on all aspects of immunology.

The Institute for Vaccine Safety
http://www.vaccinesafety.edu

The Institute for Vaccine Safety was established at the Johns Hopkins Bloomberg School of Public Health in 1997 with the goal of providing an independent assessment of vaccine safety in order to educate health professionals, citizens, and the media about vaccine safety.

UNIT 6: Emerging Infectious Diseases and Public Health

World Health Organization: Emerging Infectious Diseases
http://www.who.int/topics/infectious_diseases/en

The World Health Organization (WHO) maintains this website to provide educational information on the organization's activities in the area of preventing the spread of infectious diseases. Fact sheets as well as Q&A discussions are posted here. WHO programs around the world can be researched from this site.

National Center for Preparedness, Detection and Control of Infectious Disease
http://www.cdc.gov/ncpdcid/

The NCPDCID conducts infectious disease surveillance and prevention. You can find numerous links to information about emerging infectious diseases.

Infectious Diseases Society of America
http://www.idsociety.org

This website for professionals in the field of infectious diseases contains excellent fact sheets for a general audience as well as links to policy statements and advocacy efforts.

The Johns Hopkins University HIV Guide Q&A
http://www.hopkins-hivguide.org/q_a/index.html?categoryId=9352&siteId=7151#patient_forum

This is a Q&A forum for patients and clinicians run by Johns Hopkins University's Professor Joel Gallant, an internationally recognized expert on HIV disease and Editor in Chief of the *HIV Guide*.

Pandemic Flu
http://www.pandemicflu.gov

This site provides one-stop access to all information about pandemic flu provided by the U.S. government.

Doctors Without Borders
http://www.doctorswithoutborders.org

From treating meningitis to malaria, the charitable organization Doctors Without Borders brings direct relief to millions of people around the world. This website highlights their efforts and also provides links to specific information about prevention and treatment of infectious diseases.

Internet References

UNIT 7: Microbes and Food: Friend and Foe

Center for Food Safety and Applied Nutrition
http://www.foodsafety.gov/list.html

This website contains everything you need to know about microbial contamination of food, product recalls, and the government's food safety program.

American Dietetics Association: Home Food Safety
http://www.homefoodsafety.org/index.jsp

The American Dietetics Association's consumer education outreach program provides this useful site for the general public. Learn about foodborne illness and what you can do to prevent it in your home by handling food safely.

Government Food Safety Information
http://www.foodsafety.gov

This site has links connecting you to the USDA, FDA, CDC, and many other government agencies that deal with food safety issues. Easy to navigate, this site has a wealth of information on foodborne pathogens. It also contains links to product safety alerts and consumer advice.

USDA Food Safety and Inspection Service (FSIS)
http://www.fsis.usda.gov

The FSIS, part of the U.S. Department of Agriculture, is the government agency responsible for inspection of the nation's food supply. Find links here to information on home food safety tips, fact sheets, recalls and Q&A postings.

Fankhauser's Cheese Page
http://biology.clc.uc.edu/fankhauser/cheese/CHEESE.html

David Fankhauser is a biology professor whose website is a perfect introduction to the art of cheesemaking. With detailed instructions, even a novice can enjoy making homemade cheese.

UNIT 1

Habitats and Humanity: An Introduction to the Microbes Around Us and Inside Us

Unit Selections

Learning Objectives

- What adaptations do microbes have in order to live in such diverse environments?

- How would life on Earth be different if there were no microbes?

- Based on the readings, do you think that probiotics are useful? Under what circumstances would you recommend taking probiotics? Explain.

- What do you think of the "bugs in your gut" now? Are they more helpful or harmful? Explain your reasons.

- Why are biofilms dangerous in a healthcare setting? Why do microbes prefer living in a biofilm?

Student Website
www.mhcls.com

Internet References

The Human Microbiome Project
   http://www.nihroadmap.nih.gov/hmp/
Astrobiology
   http://www.astrobiology.com
The Center for Biofilm Engineering
   http://www.erc.montana.edu

Microbes are everywhere on this planet—from the air we breathe to the soil under our feet and even in extreme environments where life is unimaginable. From deep ocean trenches to acidic hot springs, microbes stretch the limits of our imagination when it comes to thinking about where life can exist. Closer to home, microbes are part and parcel of every human being and our bodies could not function without them. New research from the Human Microbiome Project is setting out to catalog microbial populations from individuals, with the hope of being able to pinpoint how these fellow travelers of ours are not merely passengers, but may contribute to both health and disease. This unit provides a brief snapshot into the world of our microbial flora and how our combined behaviors can impact us for better or worse. It also depicts the diverse environments in which microbes are found and how they live. Acting both as friend and foe, microbes are essential to life as we know it.

© PhotoLink/Photodisc/Getty Images

Advances in molecular biology have revolutionized the discipline of microbiology, as previously "unculturable" organisms can now be characterized on the basis of their nucleic acid sequences. The first article in this unit describes how DNA technology has been used by scientists to identify novel groups of microbes in unexpected environments. The take-home message from these studies is that the diversity of microbial life is much higher than what was expected. It seems that we are just beginning to explore the tip of the iceberg when it comes to the microbial universe. This journey will surely challenge how we think about the biosphere and ourselves.

In "Extreme Microbes," the author introduces us to the world of salt-loving archaea and takes us on a trip to some of the extreme environments where microbial life thrives. This article gives an excellent introduction to the evolution of microbes, microbial metabolism, and genomics. Genomic studies of the haloarchaea provide insights into how microbes can survive under such harsh conditions. This work also provides a framework for understanding the differences between archaea and bacteria and how lateral gene transfer may have led to the development of an organism with traits similar to those of bacteria, archaea, and even eukaryotic organisms. A different sort of microbe is introduced in "Pondering a Parasite," which gives the reader an idea of the scope of microbial size, form, and function. The world of fungi is one of symbiosis with plants and humans, but living entwined with these organisms can bring both harm and benefit.

The recurring theme of microbes as both friend and foe emerges in several articles that discuss those organisms that live within (and on) our bodies and the surprising facts about who they are and what they do for us.

Our exploration of normal flora begins in the gut, where microbes begin colonizing us from the day we are born. Recent work by David Relman's group at Stanford University estimates that there are over 400 species of microbes in the human gut. Taking a bacterial "census" for the first time, researchers are discovering surprising facts about our microbial partners. A *Newsweek* article discusses recent work by Jeffrey Gordon's research group that indicates striking variations in how our gut microbes extract energy from the food we eat. Studies on obese mice vs skinny mice showed a marked difference in the predominant types of gut bacteria and how efficient these microbes were at releasing calories from food. Humans who underwent a weight-loss program also exhibited a similar shift in the predominance of different gut bacteria! It's too soon to tell whether these findings will have applications in weight loss or the treatment of malnutrition. In "Nurturing Our Microbes," the question of how to maintain the "good bacteria" in our intestines is addressed by examining the uses of probiotics to replenish our intestinal flora. The intricate balance of microbial flora is key to maintaining our health and well-being, so the disappearance of *H. pylori* from human stomachs in industrialized nations has led to much investigation. Martin J. Blaser's article in *Scientific American* provides an in-depth analysis of the consequences of human behavior on the composition of our microbial flora and how this impacts human health. While *H. pylori* is known to cause gastric ulcers, its absence has led to increased rates of acid-reflux disease and esophageal cancer. This case illustrates the importance of normal microbiota in maintaining distinct environmental niches in the human body. The last article about normal flora looks at the microbes teeming on the surface of our skin. Research from the Human Microbiome Project, a "roadmap initiative" put forth by the National Institutes of Health (NIH), is still underway to characterize all of the microbiota present on humans. This landmark project is like the first mapmakers making sense of our world and its geographic outlines. We are finally going to get the first look at our most intimate landscape. It is hoped that by taking a census of all the microbes inside and on our bodies, we will get a better picture of who we really are (since our cells are outnumbered by microbes 10:1), and thus be better equipped to handle the consequences when our own microbes get out of balance. It is an exciting time to get to know these fellow travelers!

Over the past twenty years, research has shown that microbes do not live alone. Not surprisingly, most microbes prefer to spend their time in tight-knit communities, with all the advantages of communal living including protection from external assaults, namely antibiotics and disinfectants. Most microbes spend their time living as biofilms—forms that we are familiar with when we encounter rocks slippery with algae, bacteria, and fungi, or when we feel the filmy residue of plaque on our teeth. While living cooperatively has tremendous benefits for individual organisms, the collective force of these microbes can be deadly. Unfortunately for humans, biofilms can also develop on plastic catheter tubing and surgical implants, and they are extremely resistant to drugs and disinfectants. The last article of this unit reviews our understanding of biofilm physiology and structure and enumerates how these insights can improve our treatment of bacterial infections.

Microbial Diversity Unbound

What DNA-based techniques are revealing about the planet's hidden biodiversity.

CHRISTINE MLOT

"To see a World in a Grain of Sand"—so begins William Blake's "Auguries of Innocence." What the poet 200 years ago intended to he metaphoric has turned out to be more realistic than be could have realized. The soils, sands, sediments, and waters of Earth harbor a microscopic world of astonishing biological diversity that for most of the 20th century remained unknown. Only in the last 25 years has the hidden diversity started to emerge and its great scope become apparent.

"The microbial world is vastly more complex and vastly more diverse than we thought it was going to be not so long ago," says Norman Pace, a molecular biologist and pioneer of the new environmental microbiology, at the University of Colorado. "We're just getting started, frankly, in describing that diversity."

This new view has come about through use of a suite of molecular techniques that have literally broken open microbial cells and the microbial world to reveal whole new categories of bacteria and archaea in great abundance. The techniques allow researchers to get around the difficulty of isolating and growing most microorganisms in a lab. Less than 1 percent of all microbes have been so cultured. The remaining 99-plus percent are the dark matter of the biosphere: They might be glimpsed under a microscope, but they have never been cultured or characterized.

Microbiologists are now bypassing the culturing and going straight to microbes' DNA, to quantify it, visualize it, sequence it, and insert it into other microbes to study its function. What they are finding "is like an entirely new universe," says microbiologist Jo Handelsman, of the University of Wisconsin. A raft of recent papers in *Science* and *Nature,* among other journals, describes novel microbial genes, proteins, species, and whole communities in unexpected places.

The work promises not only to yield new antibiotics and other useful compounds but to rewrite understanding of the planet's energy and nutrient flows. "I don't know how to find the words to capture the power of that knowledge," Handelsman says. "[It] will transform our understanding of how biogeochemical cycles work, how the biosphere is balanced in every way, how microbial communities work, and how all communities work."

Ribosomal RNA: Biological Bar Code

The first and still the main molecular tool used to identify diversity in the microbial world is a gene that codes for a strand of RNA buried inside the ribosome, the cellular organelle that translates genetic code into amino acids and assembles them into proteins. In the late 1970s biophysicist-turned-evolutionary-biologist Carl R. Woese, of the University of Illinois, first applied this tool and revealed that life ultimately divides into not five kingdoms but what are now known as three domains: bacteria, archaea, and eukarya. Since then the ribosomal RNA gene bas become the most frequently sequenced gene in the public database GenBank, with 160,000 sequences. (By comparison, the most common non-RNA gene, for an HIV protein, has 50,000 sequences.)

The ribosomal gene is the first thing microbiologists check when sampling a new habitat. When genome researcher J. Craig Venter and others recently went fishing for microbial diversity in the Sargasso Sea, for example, they used it as genetic bait for pulling out and assessing novel DNA. "It's almost become the gold standard bar code for organisms," says Woese.

The gene's use as a probe is so standard not only in microbiology but in all phylogenetic work that papers often omit citations to the original research that made it so. It came about, says Woese, from his interest in evolution and his effort to build a universal tree of life. He needed a molecule that would gauge the relatedness of all organisms. "I was particularly interested in the translational apparatus because that is absolutely essential to the cell. . . . Central to it all is the ribosomal RNA. It seemed the logical choice to use."

Ribosomal RNA was also easy to obtain from growing cells, which have thousands of ribosomes churning out protein. RNA, or ribonucleic acid, is the ribosome's main constituent and the root of the organelle's name. Recent work on the structure of the ribosome has demonstrated that it is very much an RNA machine. The ribosomal RNA is distinctive in that none of it is translated into protein; its job is to interact with transfer RNAs as well as the messenger RNA of the protein-to-be.

Three separate single strands of RNA studded with several dozen proteins make up the ribosome's large and small subunit. The large subunit of the ribosome has two strings of RNA, one long and one short; the long string functions as a ribozyme, providing enzyme-like activity that forms the peptide bond between amino acids.

The small subunit of the ribosome contains the medium-length RNA, whose gene has become key for microbial identification. It's commonly referred to as 16S RNA, a measure of the velocity at which it spins out upon centrifugation. The 16S RNA, says Pace, "is the business end of the ribosome." It serves a matchmaking role by lining up the anticodon of the transfer RNA with its corresponding messenger RNA codon. This allows the other end of the transfer RNA, which carries an amino acid, to interact with the ribozyme in the large subunit.

Apart from its abundance, there was another important reason for choosing to work with ribosomal RNA, says Woese. Nucleic acid hybridization studies indicated that ribosomal RNA has a relatively slow rate of mutation. "It is spectacularly highly conserved," says Pace. Ribosomal RNA is like a rock climber contorting to maintain contact with a cliff face. Some of the topography can change without affecting the operation. But change a crucial handhold with the transfer RNA, for example, and the system will fall apart. Strong selection over billions of years has maintained those handhold sequences across all forms of life.

"The ribosomal RNA is so integrated with other components of the translational apparatus, particularly the ribosomal proteins and elongation factors, that if you change one you've got to change so many other things that you're not likely to change at all," says Woese. The translational apparatus, he adds, turns out to be more highly conserved than either DNA transcription or replication.

Compare the 16S ribosomal gene from *Escherichia coli* (bacteria), a methanogen (archaea), and a person (eukarya), and identical runs of about 20 nucleotide bases will line up, like a winning pull on a slot machine. "I love looking at ribosomal RNA sequence alignments," says Pace. "They are invariant across life."

At the same time, parts of the 16S structure are less crucial for the transfer reaction, and over time those nucleotide sequences have mutated and become more variable, allowing assessment of evolutionary change over several different scales. "Ribosomal RNA turned out to be very good" says Woese. "There are parts of the molecule that change in sequence very slowly, and parts that change somewhat faster." Woese likens it to a clock with a sweep-second hand, as well as a minute and hour hand.

By sequencing all or part of the 1500-base 16S gene, microbiologists can measure how different two sequences are, representing how long since their lineages have diverged. In addition to the parts that identify an organism's domain, other parts of the gene are key to genera and even more closely related types.

In the presequencing era, Woese used ribonucleases to cut the RNA into very short pieces from which he was able to build not only a universal tree of life but the first phylogeny of the microbial world. "The consequence was a way to relate organisms

to one another that goes beyond the subjective," Pace explains, "Woese, in articulating the ribosomal RNA phylogenetic tree, put a measuring stick on the table whereby it was now possible to say *how* different two organisms are."

The World According to DNA

It took more than a decade and the advent of whole-genome sequencing for Woese's tree to be confirmed and widely accepted. But the power of the information contained in the ribosomal RNA more quickly opened up the microbial world in a way that Woese says he did not foresee.

In 1980 Pace, a self-described RNA jock, was working on deciphering the sequence and structure of the short string of ribosomal RNA from thermophiles but having trouble getting enough material from these difficult-to-grow microbes. At the same time, University of Wisconsin environmental microbiologist Thomas Brock was reporting finding "kilogram quantities" of microbes growing at near-boiling point in the hot springs of Yellowstone National Park. This, Pace realized, was the place to obtain all the thermophile RNA he needed. And with Woese's work pointing out the bar-code quality of ribosomal RNA, Pace realized something else: It should be possible to pull nucleic acid straight from the environment and identify which organism it came from. Says Pace, "It was very much a eureka moment for me. It still is. The concept that you didn't have to culture an organism, that you could sequence it. Wow."

Around the same time, another experiment on soil from a cold environment opened a window into just how much microbial variety was out there, by examining DNA directly. Vigdis Torsvik, at the University of Bergen, was studying carbon flow in a beech forest in Norway. She found that respiration rates correlated with the DNA content of the soil and with the numbers of bacteria she saw under a microscope. But these measures didn't jibe with the counts of bacteria she was able to grow from the soil sample—they were off by three orders of magnitude, suggesting the presence of many more microorganisms than could be cultured. So she set out to purify the soil DNA and measure how fast the double strands reannealed after heating. The more homogenous the DNA, the faster complementary strands would find each other and pair up. Torsvik found the soil DNA was slow to reanneal, indicating the microbial community it came from was much more genetically diverse than expected. The total soil community diversity, Torsvik calculated, was 200 times higher than what could be obtained by culturing. "It was a very big surprise," she says.

Soon thereafter the invention of PCR (polymerase chain reaction) and automated DNA sequencing gave researchers the means to easily identify the 16S gene and, hence, the organism. Microbiologists were off and sampling everything from Antarctic sediments to ant guts to the Paleolithic paintings of the Altamira caves. The end of the microbial novelty has yet to come into sight.

When Woese put together the first molecular-based phylogeny of bacteria in 1987, he counted about 12 phyla (sometimes called divisions, based on about a 30 percent difference in the sequences of 16S genes), all of which had isolates that could be

4

grown in culture. Today there are published data on 53 phyla, 27 of which have no cultured representatives: They are known only by their DNA signatures. Pace and his colleagues are analyzing data from microbes growing in hypersaline mats that he says could add about another 30 uncultured phyla to the tree.

In some habitats, these elusive microbes are abundant or dominant members of the microbial community. DNA analyses show, for example, that bacteria from the Acidobacterium phylum make up 20 percent to 30 percent of a typical soil community. But so far researchers have been able to isolate only three strains from the group. Another phylum, named OP11 after Yellowstone's Obsidian Pool, where it was first discovered, is very abundant in extremely hot or cold anaerobic habitats but has so far never been cultured.

In the oceans, the ubiquitous phylum known as SAR11, first discovered in the Sargasso Sea, constitutes about a third of the microbial community in sea surface water. Oregon State University microbiologist Stephen Giovannoni and his colleagues describe the group as "among the most successful organisms on Earth." They've been known only by their distinctive 16S sequence, but recently Giovannoni's group isolated the first *Pelagibacter* cells from the phylum by growing them in very dilute seawater. To obtain an organism that's been identified only genetically, microbiologists look to its relatives for clues on how it might be cultured. Sometimes all it takes is time—weeks or months—for the organisms to grow. In still other cases a microbe may require exacting chemical conditions, or it may require interactions with other microbes, and thus cannot be grown in pure culture.

The "uncultured majority," as it's been called, is revealing whole new channels of energy and nutrient flows, as illustrated by the discovery of rhodopsin in the oceans. Until 2000, this light-sensitive molecule was only known to exist in animal eyes and in a group of archaea that live in very saline environments. Then marine microbiologist Ed DeLong, now at the Massachusetts Institute of Technology, and his colleagues reported finding a rhodopsin gene in an uncultured bacterium living in Monterey Bay. By inserting the gene into *E. coli,* they demonstrated that the molecule absorbs a certain wavelength of light that changes the molecule's shape, turning it into a proton pump that drives formation of ATP (adenosine triphosphate), the basic molecular fuel of cells. Like the battery in a hybrid car, rhodopsin seems to provide the bacteria with a supplementary energy-generating system. "We think that they are goosing their metabolism by extracting extra energy from light," says DeLong.

The strategy is turning out to be widespread in marine microbes. "It's probably the rule, not the exception," says DeLong. Venter and colleagues in 2004 reported finding genes for hundreds of rhodopsins from uncultured microbes in the Sargasso Sea, and the genes are turning up in other unexpected organisms as well. Similarly, genes for nitrogen-fixing and nitrite-oxidizing enzymes have been detected in uncultured microbes, indicating hidden avenues for this important nutrient's flow through the biosphere.

As the oldest biological lineages on the planet, microbes are on the front lines of channeling all biogeochemicals through the system. With so many new gatekeepers being discovered,

understanding of the carbon and nitrogen cycles, as well as other basic cycles—and the impacts of globally perturbing them—will change. But first, as DeLong says of the oceans, "we need a parts list before we can understand the metabolism."

The census taking of the microbial world is also changing the profile of archaea, the third domain of life. In addition to the two lineages of archaea that have cultivated representatives, researchers have found a third lineage in the last few years, which is known only from DNA sequences retrieved from high temperature habitats. Extreme habitats such as Yellowstone's hot springs and the Dead Sea were the original source of methanogens and other archaea. But they are also turning up in more mundane habitats, including soil and animal guts, although archaea are still generally less abundant than bacteria in these sites.

Methanogens even show up in the colons of some, but not all, people. Some researchers suspect archaea may turn out to play a role in certain pathologies, such as periodontal disease. The mouth, or "oral microbiome," is an especially rich target of microbial DNA analysis, where only about half of the hundreds of resident microbial types have been cultured.

DNA-based techniques are revealing the full microbial flora that humans carry, which outnumber even our own cells, and identifying the microbes that are involved in sickness and in health. Researchers are finding types that are either new to science or new to the diagnosis in everything from the sputum of cystic fibrosis patients to infected root canals.

Pace predicts that many chronic medical conditions of unknown origin will turn out, through DNA-based studies, to involve bacteria. In what's currently labeled nonbacterial prostitis, for example, "we find suites of organisms," he says. Such conditions will very likely become targets for more effective treatments, as happened when the bacterium *Helicobacter pylori* was determined to be the cause of stomach ulcers.

Mining the Metagenome

The source of those new treatments may very well come from the newly emerging microbial diversity. Bacteria have been the source of scores of antibiotics, all of which have come from the fraction of microbes that can be cultured. Yet this stock of medicine is overused and losing effectiveness as genes for antibiotic resistance proliferate in the microbial world.

There is little doubt that new medicines as well as industrial enzymes and pigments are awaiting discovery in the uncultured majority. "Cultured bacteria and especially soil bacteria are the most prolific producers of all these products, and there's no reason to imagine that the as yet uncultured organisms won't be equally prolific," says Handelsman. "With a thousand-fold more biological diversity in the soil, why won't there be a thousand-fold chemical diversity as well?" The distinctive biochemistry of archaea makes them another rich source of biotech and other applications.

To mine the chemical diversity of the "metagenome," the sum total of the genomes in an environmental sample, researchers have developed techniques to extract large pieces of DNA from the sample and insert them into a cell such as *E. coli.*

The expressed proteins are then screened for a specific function such as antibiotic production. Through this use of metagenomics, also known as environmental or community genomics, "we're hoping to find entirely new classes of genes that have functions that wouldn't be recognized based just on sequence," says Handelsman. Researchers have already found novel antibiotics such as acyl tyrosines, terragine, and turbomycin.

Apart from the practical applications and biogeochemical insights, the greatest contribution of the new microbiology may be in developing a group picture of the most common cells on the planet. Inevitably, there will be refinements to the picture, as lines of horizontal gene transfer are worked out and the usual phylogenetic lumping and splitting take place. But for the first time the full extent of the hidden world underfoot, at the foot of the universal evolutionary tree, is coming into focus. Says Pace, "It is stunningly intellectually satisfying to see this emerging picture of life."

CHRISTINE MLOT (e-mail: cmlot@nasw.org) is a science writer based in Madison, Wisconsin. She has recently been an "embedded reporter" doing bench work in Jo Handelsman's lab at the University of Wisconsin.

Extreme Microbes

Salt-loving microorganisms are helping biologists understand the unifying features of life and molecular secrets of survival under extreme conditions.

Shiladitya DasSarma

Seen from the air, the irregular grid of evaporation ponds at the south end of San Francisco Bay in California is a kaleidoscopic quilt of reds and purples. These ponds take their colors from the single-celled microorganisms that live there, strange beings that thrive in concentrated salt solutions. Such extreme conditions kill almost every form of life on the planet, but not the salt-loving archaea. They are members of an ancient kingdom that existed even before Earth had an oxygen atmosphere. The fact that many archaea live under impossible circumstances—boiling temperatures, lethal radiation, near-complete desiccation—has led scientists to dub them "extremophiles." These organisms may even be capable of hitchhiking through space.

As human beings, our intuitive concept of life is influenced by the visible biosphere—temperate terrestrial, fresh-water and marine environments. But the microscopic world in a drop of near-saturated brine contains a menagerie of bizarre life forms that defy these conventions. Unlike the extremophiles that have adapted to a single extreme condition, haloarchaea are metabolically versatile. They can grow aerobically (with oxygen), anaerobically (in the absence of oxygen) or phototrophically (using light energy), and can adapt to fluctuations in temperature, pH and metal-ion concentration. They are resistant to desiccation, sunlight and ionizing radiation. As a result, they can be found in anoxic salt marshes, hydrothermal vents, perennially frozen Antarctica and pockets of brine deep underground and under the seafloor.

The biggest hurdle to studying most species of archaea is recreating in the laboratory the extreme conditions they need. By contrast, the salt-loving haloarchaea are easily cultivated, and microbiologists have used them in molecular, genetic and physiological experiments. Haloarchaea grow best under hypersaline conditions, from slightly concentrated seawater to near-saturated brine. These attributes, plus the availability of genomic data and tools for molecular manipulation, have elevated them to the status of "model" organisms that shed light on other extremophiles, including other archaea and even higher organisms.

Diversity and Unity

Until the 1970s, scientists believed that all *prokaryotes*—those single-celled microorganisms that lack a nucleus—were "bacteria." The pioneering work of Carl Woese at the University of Illinois at Urbana-Champaign, and his colleagues proved otherwise. Today, prokaryotes are divided into two groups: the "true" bacteria and those evolutionary relics called archaea. Although the two look the same under the microscope, archaea have molecular characteristics that are more similar to nucleus-containing *eukaryotes*—organisms such as yeasts, plants and animals. These traits confirm that archaea are fundamentally distinct from bacteria.

Such distinctions notwithstanding, scientists using halo-archaea have made several landmark findings that have wider implications for microbial and multicellular life forms. For example, evidence from haloarchaea helped H. Gobind Khorana at MIT (with whom I began my studies of haloarchaea) to establish the standard genetic code—the Rosetta Stone of biology that allows the information in genes to be used as a blueprint for proteins. The existence of this code is one of the strongest lines of evidence for the unity of all life on our planet.

Another early discovery first made in haloarchaea was the protein bacteriorhodopsin. First identified by Walther Stoeckenius of the University of California, San Francisco, bacteriorhodopsin uses photons of sunlight to pump hydrogen ions (protons) out of the cell. This action creates a polarized cell membrane. A separate protein complex harnesses the flow of protons trying to reenter the cell to provide energy—similar to the way a mill wheel harnesses the current of a river to do useful work. In archaea, the purple-tinted bacteriorhodopsin proteins cluster in a specialized region of the cell surface called the purple membrane, where they enable the harvest of light energy for growth under conditions where oxygen is scarce. In classic experiments, Stoeckenius and others showed that light could drive the synthesis of adenosine triphosphate or ATP—the cellular currency for energy transactions—in reconstituted spheres that contained bacteriorhodopsin and ATP synthase. The latter is a key enzyme found in mitochondria, those "powerhouse" organelles present in eukaryotic cells. This work provided irrefutable proof of *chemiosmotic coupling,* the mechanism of energy generation used by all cellular life.

Extreme Genomics

Despite its name (more on this later), an organism that was known as "*Halobacterium* species NRC-1" was the first *haloarchaeon*—and one of the first archaea of any kind—to have its genome studied. W. Ford Doolittle at Dalhousie University and my research group (then at the University of Massachusetts Amherst) conducted those early experiments. NRC-1 is in most respects a typical haloarchaeon, widely distributed in hypersaline environments such as the Great Salt Lake. The genome of this species has the unusual property of being spontaneously unstable, such that entire physiological systems, such as the phototrophic purple membrane and buoyant gas-filled vesicles, are sometimes mutated. This curiosity led us to identify a large number of mobile genetic elements—similar to the "jumping genes" described by pioneering geneticist Barbara McClintock in maize. These elements were the first to be discovered in any archaeon. We also found that NRC-1 carries a pair of smaller DNA molecules alongside its chromosome.

The sequence of the NRC-1 genome was completed in the summer of 2000. It was the first complete genome to be sequenced with funds from the U.S. National Science Foundation. The genome consists of a large, circular chromosome (2,014 kilobases) and the two smaller DNA hoops, called plasmids or replicons: pNRC100 (191 kilobases) and pNRC200 (365 kilobases). The pNRC replicons contain many of the DNA repeats that enable genomic rearrangements, including 69 of 91 mobile elements, 33 to 39 kilobases of so-called inverted repeats, which can flip or invert portions of the circles, and 145 kilobases of sequence that are identical in both plasmids.

All this repetition confounded the computer programs tasked with organizing the overlapping fragments of DNA sequence into a seamless genome. The solution to this problem required extensive knowledge of the structure of pNRC100—knowledge that we had painstakingly acquired in the 1980s and '90s by cutting the genome in specific places and analyzing the parts. Later in 2000, we summoned an international consortium of 12 laboratories to meet in Amherst over the winter holidays to identify familiar elements in the genome. Such an analysis was a Herculean task in those days. The computational tools used to analyze genomes were still in their infancy, so we had to write our own computer scripts and manually inspect much of the data.

One of the most exciting discoveries gleaned from that sequence was that the 2,630 predicted proteins were, on average, much more acidic than those of other organisms. The average isoelectric point (a measure of acidity) for predicted *Halobacterium* proteins is only 4.9. By contrast, the values for nearly all non-haloarchaeal species are close to neutral, a 7 on this scale. Acidic proteins carry strong negative charges, which ought to repel other negatively charged molecules in the cell, such as DNA and RNA. But even proteins tasked with binding DNA (itself an acid)—including pieces of the complex machine called the transcription apparatus, which creates RNA from DNA—turned out to be acidic.

Our recent experiments have shown that acidic transcription factors bind DNA with ease in the hypersaline environment inside haloarchaeal cells or isolated in test tubes filled with saline solution. This attraction is remarkable because these proteins and DNA—both negatively charged—ought to repel each other. One possible explanation is that the proteins and DNA work together by sandwiching positively charged ions between nearby, negatively charged side groups. Mutual repulsion between acidic groups helps acidic halophilic proteins to remain in solution under conditions in which neutral, non-halophilic proteins would precipitate or "salt out." Thus, haloarchaea require extremely acidic proteins to maintain cellular functions in a supersaturated-salt environment.

In recent years, the full sequence of the NRC-1 genome has been followed by the publication of five additional haloarchaeal genomes: *Haloarcula marismortui,* a metabolically versatile species from the Dead Sea; *Natronomonas pharaonis,* an alkali (high *p*H)-loving species from the alkaline soda lakes of the Sinai; *Haloferax volcanii,* a moderately halophilic species from Dead Sea mud; *Haloquadratum walsbyi,* a square-shaped organism common in salterns; and *Halorubrum lacusprofundi,* a cold-adapted species from an Antarctic lake. These six genomes currently provide us with an excellent view of haloarchaeal diversity.

Evolutionary Relics

The unveiling of the NRC-1 genome spawned questions about as well as insights into the evolutionary history, or *phylogeny,* of haloarchaea. Although the data confirmed NRC-1 as a true archaeon, the gene that most scientists use as an evolutionary chronometer, the so-called 16S ribosomal RNA, had a unique sequence that prevented easy categorization of the species. This contradiction presented a challenge to the people who name microbes for a living. They had recently renamed several *Halobacterium* species based solely on DNA sequence, a move that wasn't strictly possible for NRC-1 given the ambiguity. Thus, the name "*Halobacterium* species NRC-1" has become controversial, fueling ongoing debate about the meaning and significance of the term "species" among prokaryotic organisms. Some scientists go so far as to suggest that the species concept is inappropriate for prokaryotes and should be ditched in favor of "strain" or "phylotype."

Hoping to find an answer to the phylogenetic puzzle, my coworkers and I compared the entire NRC-1 genome to the few other microbial genomes then available. This preliminary analysis noted interesting similarities between *Halobacterium* NRC-1 and two true bacteria: the gram-positive, spore-forming *Bacillus subtilis* and the radiation-resistant *Deinococcus radiodurans*. But a few years later, after the genomes of dozens of additional species had been added to the databases, the position of *Halobacterium* on the tree of life seemed to have moved, either to a spot near the base of the archaeal branch, or, surprisingly, to a branch of the bacteria.

This relocation disagreed with earlier analyses that had placed *Halobacterium* squarely within the archaea. One possible explanation for the discrepancy is that over time, halophilic archaea may have acquired many genes from unrelated bacterial species in a process called lateral gene transfer. Indeed, some studies concluded that NRC-1 contained nearly as many bacterial genes as archaeal ones, raising the possibility that these unusual microbes began as a kind of prokaryotic jumble.

Despite uncertainties about the precise history of prokaryotic evolution, evidence for the lateral transfer of genes in the haloarchaeal lineage appears convincing. For example, the genes that enable oxygen-based metabolism in *Halobacterium* are arranged the same way as those found in *Escherichia coli* bacteria. The gene sequences are similar to bacteria too. The combination of ribosomal genes that resemble anaerobic archaea and metabolic genes that resemble aerobic bacteria suggests that haloarchaea may have adapted to an oxygen atmosphere by stealing pieces of the respiration machinery through lateral gene transfers from bacteria. Presumably, such events occurred very early in Earth's history when photosynthetic cyanobacteria first began to oxygenate the atmosphere. However, the possibility of multiple transfers, including more recent ones, cannot be ruled out.

A relatively modern example of lateral gene transfer in *Halobacterium* NRC-1 comes from the gene that encodes arginyl-tRNA synthetase (ArgS), an enzyme essential to the manufacture of proteins. The closest cousins to the ArgS protein from NRC-1 are found among ArgS proteins from bacteria, not those from other archaea or haloarchaea—a puzzling situation, given that NRC-1's presumed archaeal ancestors predate bacteria. The most plausible scenario is that following the capture of a bacterial gene that encoded ArgS, the archaeal gene was lost in *Halobacterium* NRC-1.

ArgS isn't the only example of this kind. Based on predictions from the genome sequence, about 40 genes in the two pNRC replicons encode essential proteins, including those that make DNA and RNA, and those that use oxygen for cellular respiration. Since the cell can't do without these genes, the pNRC replicons that contain them may be more accurately labeled minichromosomes than plasmids, which are defined by their dispensability. In NRC-1, the plasmids serve at least two important functions: as reservoirs for captured genes and as vehicles for the lateral transfer of those genes. And when those genes are essential for survival, their transfer may lead a prokaryotic replicon to evolve from a plasmid into a chromosome.

One of the most interesting and still unresolved evolutionary questions about halophiles centers on the origin of bacteriorhodopsin, the protein component of the purple membrane. Bacteriorhodopsin is related to mammalian rhodopsins and like them contains retinal, a substance that is, as the name suggests, important in enabling the eye's retina to perceive light. *Halobacterium* NRC-1 and some other haloarchaea have other retinal-containing proteins too, including halorhodopsin, which pumps chloride ions across the membrane, and photosensory rhodopsin, which can discriminate between useful and harmful sources of light and control the cell's swimming behavior.

Scientists still debate the evolutionary history of these ancient retinal-containing proteins. Although their initial discovery in haloarchaea suggested for a time that such proteins were unique to this group, recent surveys have found similar proteins in many other bacteria and in eukaryotes. Therefore, these proteins may have arisen before archaea, bacteria and eukaryotes first diverged hundreds of millions—perhaps more than a billion—years ago. Alternatively, the genes encoding retinal-containing proteins may have passed by way of lateral gene transfers into planktonic bacteria, some fungi and haloarchaea.

Although the phylogeny of retinal-based pigments remains inconclusive, it's intriguing to think that they co-evolved with chlorophyll-based pigments. Comparing the visible-light absorption patterns of the pigments, it's clear that retinal and chlorophyll have nonoverlapping, complimentary spectra. The co-evolution hypothesis argues that simpler, retinal-based use of sunlight as an energy source evolved in microorganisms that dominated during the anaerobic, "purple phase" of the planet's existence. Later, more complex pigments based on chlorophyll could have evolved to harvest light from regions of the solar spectrum not absorbed by the preexisting species. The success of this strategy led to the "green phase" of Earth's evolution and the attendant oxidation of its atmosphere through photosynthesis—changes that displaced most retinal-based microorganisms. Although speculative, such a scenario suggests that retinal-based nourishment from sunlight is one of the oldest metabolic systems on Earth.

Means of Survival

Haloarchaea endure many stressful conditions that kill other microbes. For instance, *Halobacterium* NRC-1 can withstand intense solar and ionizing radiation that in most organisms causes widespread damage to DNA. This feat is possible because the cells produce additional pigments, such as the red-orange carotenoids, to help shield the genome. NRC-1 also contains a double dose—both the bacterial and eukaryotic versions—of the enzymes used for DNA repair. Photolyase reverses the damage caused by ultraviolet radiation during the day, a process known as "light repair." Exposure to UV light also triggers the production of a recombinase enzyme that aids DNA repair by promoting homologous recombination (using a good copy of the gene to fix the damaged copy). A third enzyme, exinuclease, snips out bits of damaged DNA altogether. Recombinase and exinuclease are critical for "dark repair," which takes place at night when cells can spend their energy fixing sun-damaged genomes. High-energy radiation and desiccation produce a different type of genetic damage in which both strands of the DNA backbone are broken. Such irradiated cells also make copies of the so-called replication protein A, which binds to exposed, single-stranded DNA, as well as to recombinase, and helps repair genetic damage.

Toxic metals and fluctuating ion concentrations also elicit changes in gene expression that promote survival in *Halobacterium* NRC-1. For example, cells resist the toxic effects of arsenic by turning on a cluster of genes encoding enzymes that alter the metal's oxidation state and prepare it for active transport out of the cell. The identity of the pump itself remains a

Haloarchaea in the Classroom

The model halophile *Halobacterium* species NRC-1 is an ideal microorganism for teaching basic microbiology to high school and college students. The primary reason is that the high-salt medium necessary for growing NRC-1 prevents contamination from other types of bacteria or fungi. (The same principle enables salt to act as a food preservative by preventing the growth of microbes that cause spoilage.) As a result, students can conduct their experiments without needing strict adherence to sterile protocols—a difficult enough challenge in a professional research laboratory and one frequently beyond the reach of most high schools. In addition to helping keep their results clean, the hypersaline conditions prevent the growth of potentially pathogenic strains of microorganisms.

When grown in the lab, liquid cultures of NRC-1 have a vibrant pink color that gets students' attention, and the presence of pink, red-orange, or even sectored colonies intrigues them. Students can see and explore the link between genotype and phenotype through the presence or absence of organelles that enable the cells to float when suspended in liquid cultures. These gas-filled organelles are easy to isolate even in a rudimentary laboratory, and students can test their functional properties with a series of quick experiments.

Among its other user-friendly properties, NRC-1 grows in culture media that are inexpensive to make and safe to handle, and its moderate growth rate (a three- to seven-day culture period) can be adapted to a weekly or semiweekly lab schedule. More advanced students can use NRC-1 to learn basic techniques of molecular biology and genetic manipulation. And when the experiments are over, NRC-1 can be stored in salt crystals at room temperature for extended periods of time. Instructors won't need to keep their haloarchaea cultures for thousands of years, but it's interesting to know that they could.

—Priya DasSarma

mystery, which suggests the existence of a novel mechanism in archaea. Another gene near the arsenic cluster seems to be part of a second detoxification system, which transforms arsenite to a volatile form, trimethylarsine.

NRC-1 responds to low oxygen levels in several ways. One is to induce the production of gas-filled vesicles that enable the cells to float to aerobic zones in the water column. But under strictly anaerobic conditions, cells shift to an alternate means of energy conversion that uses dimethyl sulfoxide, produced by other microbes, and trimethylamine N-oxide, produced by fish, to carry out anaerobic respiration. NRC-1 also responds to oxygen starvation by increasing the production of bacterio-opsin, the protein component of the bacteriorhodopsin used for phototrophic growth. The synthesis of bacterio-opsin is linked to levels of the retinal pigment, and the two molecules react to form a single complex. Indeed, the genes that encode the first and last steps of retinal synthesis and the nearby sensor-activator gene are all coordinately induced under low-oxygen conditions.

Scientists don't completely understand how NRC-1 regulates different suites of genes in response to environmental stresses, but my coworkers and I recently proposed a mechanism to explain some of that complexity. Our hypothesis is that two general transcription factors, TBP and TFB—which are sometimes overlooked as rather dull, invariant backdrops to more dynamic processes—are themselves acting as regulators. Although the TBP and TFB proteins are usually encoded by single genes in archaea and eukaryotes, *Halobacterium* NRC-1 carries six genes for TBP and seven for TFB, suggesting the possibility of many different combinations of TBP and TFB during transcription. Specific pairs may activate sets of genes that are intended to work in concert, by way of regulatory sequences found near the genes in question. Other scientists have proposed a similar mechanism to explain the choreography of organ development in some higher organisms. We recently saw evidence of just such a novel regulatory mechanism for haloarchaea, which likely evolved to deal with the stresses and dynamics encountered in their hypersaline environment.

Out of This World

Because haloarchaea tolerate so many forms of environmental stress, I have proposed, along with other scientists, that they are candidate "exophiles"—organisms that might survive on Mars or other planets. The same durability could enable haloarchaea to survive the period of deep-space travel between planets, perhaps encased in salt crystals, which would shield them from some of the damaging radiation. Recently, the European Space Agency found that a strain of *Haloarcula* survived for several weeks in deep space—longer than any other organism that was still capable of dividing. This finding is consistent with the observed capacity of *Halobacterium* NRC-1 to withstand desiccation and radiation in laboratory studies. It also supports the findings of geologists who have isolated haloarchaeal DNA similar to that of NRC-1 from halite (salt) deposits more than 10 million years old.

Some of the chunks of Martian rock that have fallen to Earth as meteorites, including the one that fell on Shergotty, India, and the one that fell on Nakhla, Egypt, have contained halite salt crystals. Thus, it's conceivable that meteorites could act as vehicles for the interplanetary transport of haloarchaea. In this context, it's understandable that the public has been keenly interested in recent reports that halophilic microbes were still alive after being trapped in tiny pockets of brine for hundreds of millions of years—reports that are currently beyond the reach of rigorous scientific validation. Yet the range of haloarchaeal adaptations to daunting conditions suggests that it is still premature to dismiss the idea of Martian life out of hand. There is little data that either support or refute the hypothesis that Earth has exchanged forms of microbial life with other celestial bodies at some point in its history. For now, the fact that Earthly forms of life *could* exist on other planets is remarkable enough all by itself.

References

Bayley, S. T., and R. A. Morton. 1978. Recent development in the molecular biology of extremely halophilic bacteria. *CRC Critical Reviews in Microbiology* 6:151–205.

DasSarma, S., F. T. Robb, A. R. Place, K. R. Sowers, H. J. Schreier and E. M. Fleischmann. 1995. *Archaea: A Laboratory Manual-Halophiles.* Cold Spring Harbor, New York: Cold Spring Harbor Laboratory Press.

DasSarma, S., and P. DasSarma. 2006. Halophiles. In *Encyclopedia of Life Sciences.* Chichester: John Wiley & Sons, Ltd.

DasSarma, S. 2004. Genome sequence of an extremely halophilic archaeon. In *Microbial Genomes,* ed. C. M. Fraser, T. Read and K. E. Nelson. Totowa, New Jersey: Humana Press.

DasSarma, S. 2006. Extreme halophiles are models for astrobiology. *Microbe* 1:120–127.

DasSarma, S., B. R. Berquist, J. A. Coker, P. DasSarma and J. A. Müller. 2006. Post-genomics of the model haloarchaeon *Halobacterium* sp. NRC-1. *Saline Systems* 2:3.

Stoeckenius, W., and R. A. Bogomolni. 1982. Bacteriorhodopsin and related pigments of halobacteria. *Annual Review of Biochemistry* 51:587–616.

The HaloEd Project. http://halo.umbi.umd.edu/~haloed

Woese, C. R., O. Kandler and M. L. Wheelis. 1990. Towards a natural system of organisms: Proposal for the domains Archaea, Bacteria, and Eucarya. *Proceedings of the National Academy of Sciences of the U.S.A.* 87:4576–4579.

SHILADITYA DASSARMA received his PhD in biochemistry in 1984 from the Massachusetts Institute of Technology and then took up a two-year postdoctoral fellowship at Massachusetts General Hospital and Harvard Medical School. He served on the faculty of the University of Massachusetts-Amherst from 1986 to 2001. In 2001, he moved his laboratory to the Center of Marine Biotechnology at the University of Maryland Biotechnology Institute, where he is a full professor. He also holds a teaching appointment in the Graduate Program in Life Sciences at the University of Maryland-Baltimore. His laboratory group includes his wife, Priya DasSarma, who earned her MS in microbiology from the University of Massachusetts win a certification to teach biology. She currently specializes in biotechnology education and science outreach for students of all ages. She discusses the advantages of *Halobacterium NRC-1* as a classroom tool.

Pondering a Parasite

Slimy, repellent, and even deadly, fungi are strange invaders—and, it turns out, our kinfolk.

GORDON GRICE

The lingering death of a tree in my backyard has me thinking. The Eastern red cedar looked bad—shaggy, wasted, a dandruff of gray decline mixed in with the healthier bluish-green needles. We had seen the cause already, in the form of little spiky galls hanging here and there like drab Christmas ornaments. Each of these ornaments was smaller than a golf ball and seemingly made of wood, which might make you think it was some healthy part of the tree. I picked one off. Each small spike was a spout: a hole surrounded by a sharp woody projection. The ball resisted pressure only as much as a fruit might. Carrying it to the cement walk, I crushed it under my heel. It wasn't difficult. Inside, the thing was pulpy and fibrous, its strands of vegetable matter radiating from a tight core. Its wet texture was like that of the new wood you can find under a tree's bark.

I knew this, from books and such, as cedar-apple rust, a parasite with a provocative life cycle that requires it to jump between juniper and apple hosts. (The Eastern red cedar, despite its name, is actually a juniper.) On the leaves and fruit of apple trees, this parasite manifests itself as leathery spots of discoloration. On junipers it appears as these woody galls. The apple and juniper forms of the parasite are different stages of a single life cycle that involves both sexual and asexual spores.

To look at it, though, I wouldn't have recognized this sphere as alien to the tree, made as it was of the tree's own tissues. The tree itself makes the gall, acting on instructions from the fungus, like an animal whose rogue cells produce a tumorous mass.

Through each spiky spout a vivid orange tentacle projected. I tried its texture: wet gummy worms.

Rain revealed more. It rained for two or three days—a steady, soaking rain that had my sons whooping barefoot on the lawn. They were all excitement. "The parasite opened up!" they told me. Indeed. All over the 50-foot tree, the galls had effloresced.

Through each spiky spout a vivid orange tentacle projected. It seemed as if the tree had collided with a swarm of sea anemones. I bent a bough down so the boys could take a closer look. My oldest son poked at one and proclaimed it slimy. I tried its texture myself: wet gummy worms. Our gentlest touches marred them. I almost expected them to recoil.

We usually think of fungus as an unhealthy thing, a sign of disease. That is a slander, for parasitism is only one of the possibilities of fungi.

In the woods behind my house I find uncountable lichens. These are, as every high school student learns, symbionts, a fungus paired with an alga or a similar life form. They are the surface I touch when I lay my hand on a fallen tree; they are the first flaky layer my handsaw bites through when I cut dead wood. When my sons climb to prospect for higher views, half their footholds are ledges of lichen or simple fungus.

I call them uncountable. This is only partly because they are numerous. The other reason I can't quantify them is that they lack integrity. The whole leeward side of a box elder tree is crusted with green: Where does one lichen end and another begin?

Fungi are often colonial rather than singular. A million fungal filaments in a patch of soil may be in communication of a sort, all of them sending strands toward a food source when one detects it. There is, of course, no brain, no central command, merely a shared purpose. If we consider one such aggregation an individual, then the largest life forms we know are fungal, stretching for miles within the soil, their total mass rivaling that of the largest animals on earth—blue whales.

If the individuality we tend to think of as fundamental is lacking in the fungi, then so are the species boundaries. Lichens are only one kind of symbiont; the fungi have many. One style, for example, is the mycorrhiza, a combination of fungi with the roots of a plant. The fungi reach where the plant cannot, bringing in minerals and water the plant needs; the plant in turn shares the food it makes from light and water. Though most people are perhaps not familiar with this arrangement, it is the basis of life as we know it. At least 80 percent of plants cohabit in such a manner, and some estimates go as high as 95 percent. The boundaries are not exactly where we are accustomed to draw them because, practically speaking, the average tree or weed or grass is not merely a plant but a combination of plant and fungus.

It is not easy to grasp this or to see it without benefit of excavation, dissection, and microscopy. But the signs are visible, if you look. Sometimes in wet weather I will find an arc of mushrooms in my front yard. It is this distribution that reveals hidden relationships, for the focus of the arc is an oak tree. The mushrooms are the genitals of fungi intimate with the tree. It is possible to trace thicker roots, barely concealed in the dirt, to aggregations of mushrooms.

A moody morning after rain. The rusts poured forth their tentacles again, but after a few hours the tentacles had withered to brown scraps. My oldest son and I decided to retrieve one of them for study. He held a slender branch taut while I cut. The shear blades met in the wet wood with a squeak and a snap; the branch sprang upward, leaving its severed extremity in my son's hands. He plucked the rust off it avidly.

We watered it in a jar and it revived. Overnight, new tendrils of orange slime pushed their way out through the colandered sphere. It did not need rain, then; any drenching would do.

The next stage of our inquiry took place in a white plastic cup. We had learned that the rust spreads orange spores while it is active. Looking over its gelatinous tentacles, I had no doubt our specimen was alive, but we made the experiment anyway. Into the little white cup it went with a fresh supply of water. In the morning, the water showed orange against the cup, spores coloring the water like a weak dose of Tang.

For a few days my sons were always bringing in bigger, juicier specimens of the rust, poking them with sticks, marveling at them. Then their interest dwindled, and I would find week-old jars of the things on shelves and windowsills, tentacles half dissolved in water, retaining only their color.

It used to be said that a fungus was a sort of defective or degenerate plant, one that lacked chlorophyll and so could not make its own food. It was thus reduced to feeding off the work of others—grubbing in the soil to devour dead animals, fallen leaves, and damp wood.

Biologists know better now. In the last few years, DNA sequencing has revealed strong evidence for all sorts of things we could hardly have suspected. The method is to look for similar strings of DNA, then to analyze them statistically. It is a probability game. All life on earth comes from a common stock; the more dissimilar two life forms are genetically, the longer it has been since they were one. By applying sophisticated mathematical models, geneticists can estimate how long it takes for certain kinds of differences to arise. Then, by comparing these numbers, they can deduce how closely related different kinds of living things are. For instance, human and chimpanzee diverged

much later than human and orangutan. Therefore, we are more closely related to chimps than to orangs.

Much of the genetic data has supported traditional Linnaean taxonomy. For example, the prevailing view has long been that the order containing bats is monophyletic—that is to say, bats are all more closely related to each other than they are to anything else. A minority view held that the two major groups of bats actually had separate origins and that animals such as the squirrellike colugo might be more closely related to one type of bat. But the DNA analysis supports the first view—that all bats belong together in one order, with nothing else included.

Other data have cast the whole Linnaean structure into doubt (which is why there is now a competing branch of systematics, called cladistics, that tries to chart relationships in a much more precise, if cumbersome, way). Some of it is quite counterintuitive. We always knew that whales must have evolved from some sort of land mammal, but who would have suspected they still belong in the order Artiodactyla—the even-toed hoofed animals?

As a result, our vision of life has altered; the kingdoms, the very fundaments of Linnaean biology, have had to be shifted about. The fungi have made up their own kingdom since the late 1800s, but we understand now, better than we used to, at least, what they are. Certain organisms that we called fungi because they were slimy and repulsive and because we didn't know where else to put them—the slime molds, for example—have been exiled from the kingdom. That is not too hard to take, because few of us encounter slime molds with any regularity.

What is harder to take is that these disreputable organisms are our kinfolk. The fungi are not plants at all; they are closer to the animals—to us. We always suspected it, after all. They don't move the way we do, but some of them, the most visible ones, grow so large and are so complex it almost seems a sign of animal life: the mushroom big as a human fist found on your lawn the day after a rain, for example. I have a vivid childhood memory: something smooth and white nesting in the grass, the size and shape of a chicken egg, hard-boiled and peeled. It was not an egg, however; my dog evinced no interest in it. Something about it made me reluctant to touch it. It had no smell, but somehow it reminded me of dog feces, or perhaps merely the Platonic form of disgust. The weird notion that it was an eyeball crossed my mind, and I went so far as to turn it over with a twig, looking for an iris. It was this operation that revealed its true nature, for it tore open under the stick and revealed, first, the stringy origami I associated with the insides of some mushrooms. The other thing this tearing revealed was the smell, which had hitherto been undetectable: the smell of rotten flesh.

Or maybe their overnight appearances simply seem like a sinister kind of magic. Toadstools, elves.

It is not only plants that live in symbiosis with fungi. We animals do it too. We do not like to think about this, because we have for so long conceived of microbes as unclean things, as invaders. But of course we have always been symbionts, dependent on the microbes in our guts to digest our food. We

13

have colonies of fungus inside us. A so-called yeast infection is really an imbalance. It is not a problem that the yeasts are in the human body—they always are. It's a problem that their numbers have, because of some teetering of the pH level, exceeded their usual bounds. It is a natural thing to share our bodies with them, and with all manner of other organisms. As a tree is not simply a tree, we are not simply what we think we are.

Which is not to exonerate them all. Plenty of fungi are pure parasites, and these have invaded almost every form of life. There are specialized fungal parasites of single-celled diatoms and even of other fungi, and some of these cause serious harm. The rust on my red cedar tree was bad for it; it must have been even worse for the apple trees in the neighborhood, for the leaves of the apple are slowly pierced through by the rust, until spore-shooting orange masts sprout from their undersides. The fruit of the apple, too, may be ruined, its hide marred by soft brown patches.

So, too, with the human form. There are fungi to make the feet and the testicles itch, fungi to discolor and deform the fingernails. And there are neighborly fungi content to live within us unobserved, but which will blossom, in the case of a ruined immune system, into devouring sores inside and out. It is a common enough way for people with AIDS to die.

Two years have passed since the last time our red cedar was dazzling in its orange jewelry. The next year we waited in vain for another blooming of that odd fungus life. We had trimmed the galls off where we could reach, hoping to save the tree, but it wasn't our earthbound efforts that drove them away. We could still see dozens of them higher up, dry and drab, refusing to bloom. This is the way of the rust. It dies on apple trees, but on red cedar it can persist, give its life to the wind, or leave its old self to rot.

Our tree grows shaggier, more patched with vanilla and auburn. Within the greenery that remains I can reach dozens of lifeless branches. A good yank is sure to be rewarded with the crack of dead wood. The whole tree is ugly now, truth to tell. It reminds me of nothing so much as the shaggy head of a neglected old man.

Not that I neglect the tree. I prune; I am doing what I can to save its life.

Say Hello to the Bugs in Your Gut

Your small and large intestines are home to countless microbes that some scientists think may play a major role in determining how fat or skinny you are.

PATRICK J. SKERRETT AND W. ALLAN WALKER, MD

Whenever you eat, even if it's just a bowl of cereal while standing at the kitchen counter, you're feasting with trillions of your closest compadres. Bacteria, fungi and other microbes share the bounty as food churns through the vital inner tube that makes up your gut. It isn't a one-way relationship. These microscopic critters are essential contributors to our good health. They break down toxins, manufacture some vitamins and essential amino acids, educate the immune system and form a barrier against infective invaders. A provocative new avenue of research suggests that the makeup of microbes in the gut may influence our weight, too. If true, this could provide new strategies for weight control.

Which species of microbes live in the gut and what they do in there are just two of the many key questions that scientists are asking about this largely unexplored realm. "The landscape of the human gut is truly terra incognita," says Jeffrey Gordon, a genome scientist at Washington University in St. Louis whose research team is spearheading this effort. "The menagerie of microscopic organisms living there acts like an organ that carries out functions that we humans have not had to evolve."

The early work on our gut microbiota (loosely translated from Latin as "community of tiny living things") is challenging our notion of what it means to be human. From an early age, the human body is home to a huge but ever-changing community of microbes: for each one of our cells, there are 10 microbial cells in or on the body. Most of them live in the intestines; the bulk of the rest inhabit the mouth, esophagus, stomach, upper airway, skin and vagina. No one knows how many different species coexist inside the human gut. In the first comprehensive census of the gut, David Relman and his colleagues at Stanford University quit counting after they hit 395 different species in three healthy subjects. The real cast of characters will almost certainly number in the thousands.

Curiously, we don't start life with such a microbial partnership. A developing baby floats in sterile amniotic fluid, protected from bumps—and bugs. That isolation ends during the baby's trip through the birth canal, which is a haven for bacteria. The baby picks up microbes on his or her skin; some get into the mouth. From then on, helpful microbes somehow convince the immune system that they mean no harm. They settle down in hospitable regions, and crowd out those that can't compete.

At first, the microbes in a baby's body resemble those in the mother. Over time, the community takes on its own identity, nudged this way and that by the child's genes, the environment and the unceasing flow of new microbes from food, beverages and unwashed hands. Eventually, an individual's gut microbiota becomes as unique as a fingerprint.

Babies are beset by microbes as they leave the womb. Ultimately, your bug profile is as unique as a fingerprint.

The gut is composed of the small and large intestine. Stretched out, it's as long as a schoolbus. Flatten out the millions of fingerlike projections that line its sides and it would easily cover a tennis court. The small intestine is where much of the food you eat is broken down into simple sugars, fats and amino acids. These are all small enough to be shuttled across the lining of the intestine and into nearby blood vessels. Fiber from fruits, vegetables and grains, along with other indigestible material, passes largely unchanged into the large intestine, also known as the colon.

What's indigestible to you is a seven-course meal to your gut microbiota. The conditions in the colon—dark, moist and free of oxygen—are just what your gut microbiota needs to ferment indigestible material passed on from the small intestine and produce simple sugars and short, chain-free fatty acids. They do this for their own good, but they also share some of these energy-rich substances with their host—us. Some people get up to 10 percent of their daily calories from substances produced by their gut bugs.

Gordon and his colleagues think that the gut microbes in some people are more efficient at extracting energy than those in other people. This could partly explain why some individuals gain more weight on the same diet that allows others to stay lean. Gordon's hypothesis is supported by a series of elegant experiments.

First, Gordon's team raised generations of mice in sterile conditions. These microbe-free mice downed almost one third more food each day than their ordinary counterparts—yet had 40 percent less body fat. When the researchers took samples of gut microbes from ordinary mice and transplanted them into germ-free mice, the newly inoculated rodents began to gain weight even though they weren't eating any extra food. The team took the work a step further with help from a strain of genetically obese mice. Transplanting gut microbes from these fat mice into lean, germ-free mice led to greater gains in body fat than transplanting gut microbes from normal mice.

To see if fat mice had different gut bugs than lean mice, the Washington University team took genetic snapshots using high-tech DNA sequencers. In ordinary normal-weight mice, bacteria belonging to the group known as Firmicutes accounted for about two thirds of the gut's bacterial community. Members of the Bacteroidetes group made up most of the rest. In contrast, genetically obese mice had even more Firmicutes and many fewer Bacteroidetes. By analyzing the sequence of genes extracted from various microbiota samples, Gordon's team discovered that the bacterial community in obese mice had more genes for breaking down complex starches and fiber. In other words, microbes from obese mice were better at releasing calories from the gut's contents than were the microbes from lean mice.

Think of it this way: the gut community in obese mice is like a fuel-efficient car, extracting more energy from food and passing more along to its host than its gas-guzzler counterpart in lean mice.

Mice are mice. Does any of this apply to humans? To find out, the Washington University team asked a dozen obese men and women to follow a low-fat or low-carbohydrate diet for a year. Before starting these diets, these obese volunteers had more Firmicutes and fewer Bacteroidetes in their guts than did several lean volunteers acting as controls—just as was seen in obese and lean mice. As the volunteers lost weight, their microbial communities underwent a remarkable shift, with an increase in the gas guzzlers (Bacteroidetes) and a decrease in the efficient energy extractors (Firmicutes). The type of diet didn't matter; only significant weight loss sparked the shift.

One implication of this work is that the energy content of food isn't a fixed quantity. Consider the 110 calories per cup listed on a box of Cheerios. Some people may get that much, others may get less, depending on their gut microbiota. A difference of just 25 calories a day—that's half a rice cake or one chocolate kiss—between what you take in and what you burn could mean a gain or loss of more than two pounds in a year and 20 pounds over a decade.

On a more practical note, this work suggests that somehow altering the microbial populations in the gut could be one way to modify weight. If Gordon's work continues to pan out, it may be possible someday to use probiotics—dietary supplements containing potentially beneficial microbes—or other microbe-manipulating strategies to aid weight loss by nudging the gut microbiota to be less efficient at extracting energy.

Probiotics are already on the market, most of them containing some form of *Lactobacillus,* best known for its yogurt-making abilities. They're used to fight allergies, diarrhea and a variety of other conditions, although the evidence for their use remains spotty. Major companies such as General Mills and food-ingredient supplier Danisco are exploring links between probiotics and weight control.

It's a bit early to do such microbial gardening for weight loss, cautions Randy Seeley, associate director of the Obesity Research Center at the University of Cincinnati College of Medicine. Though he's impressed with the work Gordon's team has done, he isn't sure the results make sense from an evolutionary perspective. As the obese volunteers lost weight, their gut microbes shifted toward a community that would extract less energy from its food supply. That doesn't make sense from a survival point of view: "If I saw myself getting leaner, I'd want my body to say to my microbes, 'Guys, help me out here,' and make the extra more calories, not fewer," says Seeley.

Gordon is the first to acknowledge that there's a lot of work to be done before anyone can point to gut microbes as a cause of obesity or start manipulating them as a way to lose weight. Even if this line of research doesn't pan out, the convergence of microbiology, molecular biology and a host of other disciplines will shine new light on how we process what we eat and what causes obesity.

While scientists have known for more than a century that we humans live with a huge community of permanent tiny neighbors, it is only recently that research like Gordon's has suggested that these neighbors may have unexpected effects on our health. In response, the National Institutes of Health has launched the Human Microbiome Project, in order to learn more about our gut bugs—starting with their genes. Sequencing of bacterial genes could also help researchers prospect for hitherto unknown chemicals made by our microbes that protect our health. What is making the Human Microbiome Project feasible is the recent development of superfast gene sequencing technologies.

Taking a census is important for several reasons. "We need to know who's there, especially the good bugs that make up the majority of the microbial community, so we can minimize any harm to them when we go after the bad guys," says George Weinstock, a professor of molecular and human genetics at Baylor College of Medicine who is working on the project.

While obesity gets most of the attention, Gordon has his eyes on another prize: fighting malnutrition. Making microbes in the gut more efficient could be one way to help the millions of people around the globe who don't get enough to eat each day or those who involuntarily lose weight while battling cancer or heart failure.

SKERRETT is editor of the Harvard Heart Letter. **WALKER** is the Conrad Taff Professor of Pediatrics at Harvard Medical School. For more on nutrition and health go to health.harvard.edu.

Nurturing Our Microbes

Stewardship of the life teeming within us can pay health dividends.

Janet Raloff

Each of us is a metropolis. Bustling about in everyone's body are tens of trillions of microbes. Some are descended from starter populations provided by mom during birth. Additional bacteria, yeasts, and other life forms hitchhike in with foods. By age 3, everyone's gut hosts a fairly stable, yet diverse, ecosystem.

Most of the tiny stowaways hide out in the gastrointestinal tract—the gut—stealing a share of everything we eat or drink. But that's only fair, because most of these bugs give as good as they take, explains microbiologist Jeffrey I. Gordon. They not only help us digest food, he says, but they also harvest nutrients, manufacture certain vitamins, kill germs, neutralize bacterial toxins, and modulate the immune system. Sickness, antibiotic therapy, or stress, however, can disrupt the ecological balance among gut dwellers—known as flora—diminishing their benefits.

Because these benefits are vital to health—and to averting disease—drug manufacturers are eyeing gut microbes as potential therapeutic targets. In the future, "pharmaceutical companies might be drugging your bugs, not drugging you," suggests Jeremy Nicholson of Imperial College, London.

In the meantime, over-the-counter therapies exist to bug, not drug, the bugs. Known as probiotics, these yogurts and other foods or dietary supplements introduce or replenish beneficial gut species in the digestive system (*SN: 2/2/02, p. 72*).

Probiotic microbes' role in fighting generic diarrheal disease is old hat, but in the past decade, other influences on human immunity and metabolism have emerged. Certain microbial supplements show the potential to reduce the severity of colds and other infections, temper body weight, and even help the elderly fight osteoporosis.

The rub: Research is showing that a probiotic's benefits can be very specific. In fact, it might be more appropriate to view these microbes as a cornucopia of diet-based, over-the-counter micro-pharmacists—each able to dispense only a few therapies or services.

But for all the promise that probiotics offer, they're no panacea, many researchers caution, and may even exhibit disturbing effects (see box). Within a given species, some strains may confer health benefits, others may not.

Yet when the right bug is ingested for a particular condition, even a small dose can trigger dramatic health benefits.

Dining Partners

"The total number of microbes associated with our adult bodies exceeds the total number of our human cells by a factor of 10," says Gordon, of Washington University in St. Louis. So effectively, "we're sort of a superorganism—one that's 90 percent microbial."

Other animals have evolved a similar symbiosis with—or even dependence on—gut microbes, the scientist notes. Rodents born by cesarean section (so they get none of their moms' intestinal flora) and raised under germfree conditions end up smaller than normal, his group found—despite eating "about 30 percent more food than their microbe-laden counterparts."

Germfree animals not only appear less efficient at harvesting calories, he explains, but also "are prone to certain vitamin deficiencies" because gut microbes synthesize certain nutrients, such as vitamins B_{12} and K.

Gut flora also help the body mine minerals from the diet. "We have measured this for calcium," says Jürgen Schrezenmeir of Germany's Federal Research Center for Nutrition and Food, in Kiel.

His team showed that supplementing rats' diets with a probiotic strain of bacteria, *Lactobacillus acidophilus,* kept the animals from losing bone, a symptom of early osteoporosis.

This probiotic, renowned for its copious production of lactic acid, occurs naturally in some yogurts and other fermented dairy products. Bonus intestinal acid should increase the solubility of several minerals, including calcium, Schrezenmeir explains. Extra lactic acid should also spur the growth of cells lining the gut, he says, creating a bigger cadre to sop up released minerals.

To test these hypotheses, his group removed the ovaries from 6-month-old female rats. The ensuing drop in the rodents' production of estrogen mimicked the hormonal environment of postmenopausal women. Over the next 16 weeks, the rats

Not without Risks

Probiotics Exhibit a Dark Side

By design, probiotics should be helpful at best, benign at worst, notes Jeremy Nicholson of Imperial College, London. Side effects can occur, however, so unless people are battling an illness, he warns against consuming such microbes indiscriminately.

"If it ain't broke," he argues, "don't fix it."

The downside of probiotic therapy usually amounts to unexpected diarrhea. However, infections in the liver, heart, and other organs have also been linked to probiotics, according to a 2006 review by Robert J. Boyle of Royal Children's Hospital in Victoria, Australia and his colleagues. Although the infectious agent in some cases was identical to the probiotic used, Boyle's group notes that an indicted strain of microbe may sometimes also "be found in the internal microbiota of healthy humans, so the source of infection in these cases is not conclusively [due to probiotics]."

Last year, researchers reported in the September *Pediatric Intensive Care Medicine* that they had shut down a pediatric trial with *Lactobacillus rhamnosus* GG (LGG), a widely used probiotic, owing to growing concern that it might actually spawn infections.

Looking to cut the risk of hospital-acquired infections in severely ill children, Travis C.B. Honeycutt of WakeMed Health and Hospitals in Raleigh, N.C., and his team began randomly assigning kids to receive a probiotic or a placebo capsule daily while they were hospitalized in an intensive care unit. However, when three reports of LGG blood-borne infections in children emerged in quick succession from neighboring physicians outside the trial, the North Carolina researchers decided to perform an interim analysis to check whether LGG was as benign as they had told their patients' parents it was.

"That analysis showed no benefit in our patients," Honeycutt recalls, "and a trend—although it was not statistically significant—towards increasing infections in our probiotics group."

But the really big wake-up call came last month, when Dutch researchers published findings of a trial using probiotics in people with acute pancreatitis. Patients provided nutrition laced with six probiotics experienced a death rate nearly triple that of people fed just the nutrients (*SN: 2/23/08, p. 115*).

—J.R.

began losing bone, modeling what happens in many elderly women. However, calcium uptake from the diet was somewhat higher—and bone loss somewhat reduced—in animals given *L. acidophilus*.

Calcium uptake and bone mass improved even more when the researchers simply supplemented the animals' diet with a material on which lactic acid bacteria prefer to feed. That

supplement—known as a *prebiotic*—contained carbohydrates that only bacteria can digest.

Rodents receiving both prebiotics and probiotics retained the most bone and dietary calcium, the German team reported in the March 2007 *Journal of Nutrition*. Indeed, the combination restored bone mineral density and bone structure to about the level in rats with intact ovaries, Schrezenmeir says.

Tuning Immunity

Probiotics are usually promoted as supporting intestinal health—a polite way of hinting that they may reduce the risk of diarrhea or bloating. Far less appreciated is the broad range of immune conditions for which they show promise.

The gut "is the body's largest immune organ," notes Arthur C. Ouwehand of the University of Turku, Finland, and of Danisco Innovation, a company that makes probiotics-enhanced foods. That's why investigators at his and other research centers are exploring probiotics to improve immunity.

A study in 2005 by Schrezenmeir and his colleagues showed that daily treatment with a trio of probiotics didn't reduce the incidence of colds. But the supplementation did reduce the severity and duration of cold symptoms—including fever—compared with a group of people that didn't get probiotics.

"We don't know the mechanism" for the probiotic advantage, Schrezenmeir says. However, in individuals given probiotics, the number of activated helper T cells—white blood cells that fight infection—increased, as did the number of germ-killing cells.

Probiotics may move the immune system in the opposite direction as well. Over the past year, several research teams reported some success with probiotics in treating inflammatory bowel disease. At least one study found they could help control exaggerated inflammation in intensive care patients at high risk for multiple organ dysfunction syndrome—a hyperinflammatory condition. And in a paper last August, Ouwehand recounted how probiotics administered to pregnant women and babies reduced the likelihood that high-risk infants developed food allergies.

In its newest work, Schrezenmeir's team incubated immune cells from the blood of healthy or allergic individuals together with several immune-stimulating substances. Cells from all of the people responded, but only cells from allergic people showed an exaggerated response to allergens.

Adding four probiotic microbes or the naked DNA from probiotic bacteria to the mix substantially ratcheted down the response of immune cells, especially for people with allergies. About half of the immune-dampening effect in probiotic-treated cells was attributed to the live bugs, and half to their DNA—released when the beneficial bugs died. The work will appear in an upcoming *Immunobiology*.

Probiotic benefits are typically attributed to the fact that supplemented microbes were alive. However, receptors on the surfaces of both immune cells and cells lining the gut can bind DNA, Schrezenmeir notes. Probiotic DNA won't be accessible to those cells until the microbe dies. His team's new data suggest that probiotics—dead or alive—can affect systems in the body, perhaps by contributing to the communications among the gut's native microbes.

Weight Modulators

A number of food companies are investigating new health applications for probiotic supplements and fortified foods. Among novel functions being explored at the Nestlé research center in Lausanne, Switzerland, is probiotics' control over calorie use.

Company scientists teamed up with researchers in England and Sweden for rodent experiments using strains of *L. paracasei* and *L. rhamnosus,* probiotics that Nestlé discovered years ago.

To create gut ecosystems in rats that model those of humans, the scientists seeded the guts of newborn mice—animals that were still germfree—with microbes from the digestive tracts of human babies. Beginning 6 weeks later, the researchers doctored the animals' drinking water for 14 days with one or the other of the probiotics.

In the Jan. 15 *Molecular Systems Biology,* Nestlé biochemist Sunil Kochhar and his colleagues report that both strains of tested lactobacilli increased the hosts' breakdown and use of simple carbohydrates. The data suggest that by helping people absorb more of the calories present in carbs, these or related probiotics might one day help fight malnutrition in parts of the world where carbohydrate-based diets are common, Kochhar says.

But probiotics can push this metabolic pendulum the other way.

Bile acids, produced mainly in the liver, play an important role in emulsifying dietary fats, a step that readies such lipids for digestion. The Nestlé probiotics broke down taurocholic acid, an especially efficient emulsifying bile acid. The resulting cholic acid "is not a good fat emulsifier," notes Nicholson, a coauthor of the study—and after the probiotic treatment there was a 50-fold higher ratio of cholic to taurocholic acid in the treated animals' guts.

This change diminished the rodents' uptake of dietary fat and also reduced their synthesis of potentially harmful fatty substances in the blood, such as low-density lipoprotein cholesterol.

Where obesity is a problem, the same bugs might help people limit weight gain by diminishing their absorption of fats. "You only need to take in 20 to 30 more calories a day than you expend to make you fat in 2 or 3 years," observes Nicholson. "What we're interested in is looking for [probiotic] microbes that might help you absorb 50 calories less a day."

These metabolic findings complement observations by Gordon's team. The ecology of guts in lean and obese rodents is dominated by different bacteria, the Washington University researchers reported in 2006 in *Nature (SN: 5/19/07, p. 314).* The same holds for people.

After collectively identifying all of the microbial genes present in the guts of the naturally lean and obese mice, "we found

that genes involved in breaking down otherwise indigestible complex carbohydrates were much better represented in the obese animals' gut communities," Gordon says.

His group then transplanted gut flora from a lean or obese mouse into a germfree animal and fed all treated rodents the same amounts. Animals that had received the gut microbes from obese animals gained more fat than did the animals given flora from a lean mouse.

Such experiments "show that differences in gut ecology influence the efficiency with which the bugs extract energy from foods," Gordon says. However, his team's data also show that gut microbes can alter what share of consumed energy will be stored as body fat.

Identifying the specific microbes responsible for these effects could point to new classes of weight-controlling probiotics, Gordon suspects.

Special Effects

For all of their potential weight-modulating similarities, the two Nestlé probiotics had additional—and very different—actions. While the *L. rhamnosus* treatment dramatically decreased gut populations of potentially lethal bacteria known as *Clostridium difficile (SN: 2/18/06, p. 104),* the *L. paracasei* probiotic offered no defense against these germs.

There may be some direct effect of the probiotic microbes on these germs, or even on food metabolism, Nicholson says. But his new data suggest that many of the probiotics' effects might best be characterized as microbial diplomacy—where small delegations of ingested germs persuade an army of resident microbes to adopt activities that better benefit their host.

"Bacteria talk to each other all of the time," he says. Although there may be billions of local organisms, most "tend to behave like multicellular organisms," he explains. These mega-beings coordinate their activity via microbial chatter. They signal their intent through the production and secretion of specific molecules.

"What we think is happening," Nicholson says, "is that the probiotic bugs enter the gut, producing their chemical signals." Relative to the hordes of microbes living in the gut, the incoming microbes make up only a teensy minority. However, based on the chemical dispatches issued during their transit through the intestines, the gut's longtime residents "start to change what they're doing."

Gut flora might make good targets for medicines.

In the new study, Nicholson's group showed that the messages relayed by each of the Nestlé probiotics seem to hit different families of resident flora, leading to different metabolic effects. One implication, he says, is that depending on which microbes permanently inhabit any particular individual's gut, the probiotic's message may resonate loudly or fall on deaf ears.

So which probiotic is most likely to work for an individual may depend on the precise nature of his or her flora, Nicholson maintains. The challenge, he says, will be to find out which flora are present and in what numbers. In a paper due out soon in the *Proceedings of the National Academy of Sciences,* his group will report the ability to get a rough inventory of those flora by analyzing their metabolic detritus in human urine.

Because of "the significant involvement of the gut microbiota in human health and disease," gut flora might make good targets for medicines, Nicholson and his colleagues argue in the February *Nature Reviews: Drug Discovery.*

Consider that there are only about 3,000 human genes available to target with drug therapy—but "probably 100,000 gene targets in your gut microbiome," Nicholson says.

To succeed, drug companies will need a better picture of the human gut's microbial genome. It so happens that the National Institutes of Health recently established the Human Microbiome Project to nail that down.

An Endangered Species in the Stomach

Is the decline of *Helicobacter pylori,* a bacterium living in the human stomach since time immemorial, good or bad for public health?

MARTIN J. BLASER

*H*elicobacter pylori is one of humanity's oldest and closest companions, and yet it took scientists more than a century to recognize it. As early as 1875, German anatomists found spiral bacteria colonizing the mucus layer of the human stomach, but because the organisms could not be grown in a pure culture, the results were ignored and then forgotten. It was not until 1982 that Australian doctors Barry J. Marshall and J. Robin Warren isolated the bacteria, allowing investigations of *H. pylori*'s role in the stomach to begin in earnest. Over the next decade researchers discovered that people carrying the organisms had an increased risk of developing peptic ulcers—breaks in the lining of the stomach or duodenum—and that *H. pylori* could also trigger the onset of the most common form of stomach cancer [see "The Bacteria behind Ulcers," by Martin J. Blaser; *Scientific American*, February 1996].

The prevalence of *Helicobacter pylori* is much lower in developed countries such as the U.S.

Just as scientists were learning the importance of *H. pylori*, however, they discovered that the bacteria are losing their foothold in the human digestive tract. Whereas nearly all adults in the developing world still carry the organism, its prevalence is much lower in developed countries such as the U.S. Epidemiologists believe that *H. pylori* has been disappearing from developed nations for the past 100 years thanks to improved hygiene, which blocks the transmission of the bacteria, and to the widespread use of antibiotics. As *H. pylori* has retreated, the rates of peptic ulcers and stomach cancer have dropped. But at the same time, diseases of the esophagus—including acid reflux disease and a particularly deadly type of esophageal cancer—have increased dramatically, and a wide body of evidence indicates that the rise of these illnesses is also related to the disappearance of *H. pylori*.

The possibility that this bacterium may actually protect people against diseases of the esophagus has significant implications. For instance, current antibiotic treatments that eradicate *H. pylori* from the stomach may have to be reconsidered to ensure that the benefits are not outweighed by any potential harm. To fully understand *H. pylori*'s effects on health, researchers must investigate the complex web of interactions between this remarkable microbe and its hosts. Ultimately, the study of *H. pylori* may help us understand other bacteria that colonize the human body, as well as the evolutionary processes that allow humans and bacteria to develop such intimate relations with one another.

The study of *H. pylori* may help us understand other bacteria that colonize the human body.

A Diverse Bacterium

As soon as scientists began investigating *H. pylori,* it became clear that strains isolated from different individuals are highly diverse. (A variety of strains can also be found in a single stomach.) Although the strains are identical in appearance, their genetic codes vary greatly. Researchers have determined the complete genomic DNA sequences for two separate *H. pylori* strains; each has a single small chromosome of approximately 1.7 million nucleotides, comprising about 1,550 individual genes. (In comparison, *Escherichia coli*—a bacterium inhabiting the intestines—has about five million nucleotides, and humans have about three billion.) Remarkably, about 6 percent of the *H. pylori* genes are not shared between the two strains, and even the shared genes have a significant amount of variation in their nucleotide sequences.

This level of diversity within a species is extraordinary. The genetic differences between humans and chimpanzees—two distinct species—are tiny compared with the differences among

Overview/A *Microbe's* Effects

- Although *Helicobacter pylori* has long colonized human stomachs, improved sanitation and antibiotics have drastically cut the bacterium's prevalence in developed countries over the past century.
- People carrying *H. pylori* have a higher risk of developing peptic ulcers and stomach cancer but a lower risk of acquiring diseases of the esophagus, including a very deadly type of esophageal cancer.
- Studies of the interactions between *H. pylori* and humans may lead to better treatments for disorders of the digestive tract as well as a greater understanding of other bacteria that colonize the human body.

H. pylori strains: 99 percent of the nucleotide sequences in the human and chimp genomes are identical. The substantial variation in *H. pylori*'s genome suggests that either the bacteria have existed for a very long time as a species or that any particular variant is not so much better adapted to the human stomach as to outcompete all the others. In fact, both statements are true.

My laboratory has identified two particular types of variation. In 1989 we created a library of *H. pylori* genes by inserting selected fragments of the bacterium's DNA into cells of *E. coli*. The *E. coli* cells can then produce the proteins encoded by the *H. pylori* genes. We screened the resulting *E. coli* samples using blood serum from a person (me!) who carried *H. pylori* in his stomach; because my immune system had been exposed to the bacterium, the antibodies in my serum would be able to recognize some of the organism's protein products. The first sample that my antibodies recognized contained a gene that we now call *cagA*, which encodes the CagA protein. This was the first *H. pylori* gene found in some but not all strains of the bacterium. Later research indicated that people infected with *H. pylori* strains bearing the *cagA* gene have a higher risk of acquiring peptic ulcer disease or stomach cancer than people with strains lacking the gene.

We now know that *cagA* is part of a region in the *H. pylori* chromosome that also contains genes encoding proteins that form a type IV secretion system (TFSS). Bacterial cells assemble these systems to export large, complex molecules into host cells; for example, *Bordetella pertussis,* the bacterium that causes whooping cough, uses a TFSS to introduce its toxin into the cells of the human respiratory tract. In 2000 research groups in Germany, Japan, Italy and the U.S. determined that several of the *H. pylori* genes near *cagA* encode TFSS proteins that assemble into a structure analogous to a miniature hypodermic needle. This structure injects the CagA protein into the epithelial cells that line the human stomach, which explains why my body produced antibodies to the protein.

After CagA enters an epithelial cell, enzymes in the host chemically transform the protein, allowing it to interact: with several human proteins. These interactions ultimately affect the cell's shape, secretions and signals to other cells. Strains of

H. pylori bearing the *cagA* gene cause more severe inflammation and tissue injury in the stomach lining than do strains without the gene. These differences may explain the increased disease risk in people carrying the *cagA* strains.

In the late 1980s Timothy Cover, then a postdoctoral fellow working with me, began to study some *H. pylori* strains that caused large holes, called vacuoles, to form in epithelial cells in culture. We showed that the active agent was a toxin, dubbed VacA, encoded by a gene that we named *vacA*. In addition to forming the vacuoles, VacA turns off the infection-fighting white blood cells in the stomach, diminishing the immune response to *H. pylori*. Unlike *cagA, vacA* is present in every *H. pylori* strain, but because the gene's sequence varies substantially, only some of the strains produce a fully functional toxin. John C. Atherton, a visiting postdoctoral fellow from England, found four major variations in *vacA:* two (m1 and m2) in the middle region of the gene and two (s1 and s2) in the region that encodes the protein's signal sequence, which enables the protein to move through cell membranes. Subsequent studies showed that the s1 variation could be divided into at least three subtypes: s1a, s1b and s1c.

H. pylori strains with both the m1 and s1 variations produce the most damaging form of the VacA toxin. Thus, it is not surprising that strains bearing this genotype of *vacA*, combined with the *cagA* gene, are associated with the highest risk of stomach cancer. To make matters even more complicated, some people are more susceptible to these kinds of cancers because of variations in their own genes that enhance the inflammatory response to bacterial agents. The worst-case scenario is a person who carries the pro-inflammatory variations and is colonized with *H. pylori* strains containing the *cagA* gene and the s1/m1 *vacA* genotype. The collision of particularly aggressive *H. pylori* strains with particularly susceptible hosts appears to account for most cases of stomach cancer.

Tracing Migrations

Once scientists had discovered ways to distinguish among the *H. pylori* strains that had been collected from around the world, they began to investigate whether strains circulating in different areas varied from one another. Working with Leen-Jan van Doorn of Delft Diagnostic Laboratory in the Netherlands, we found that variations in the *vacA* gene tended to cluster in certain geographic regions: s1c strains predominated in East Asia, s1a in northern Europe and s1b in the Mediterranean area.

My colleague Guillermo I. Perez-Perez and I were particularly interested in studying the *H. pylori* strains in Latin America because the results there could indicate when and how the bacteria arrived in the New World. We initially found that the Mediterranean strain, s1b, was by far the most common, suggesting that *H. pylori* was brought by Spanish and Portuguese settlers or African slaves. We realized, however, that these studies were conducted in Latin America's coastal cities, where the people have mixed European, African and Amerindian ancestries. Working with Maria Gloria Dominguez Bello of the Venezuelan Institute for Scientific Research, we analyzed stomach samples from a more indigenous Amazonian population—people in Puerto Ayacucho, a market town on the Orinoco River in Venezuela—and found that most of the strains had the s1c genotype that is prevalent in East Asia. This work provided evidence that *H. pylori* had been transported across the Bering Strait by the ancestors of present-day Amerindians and thus has been present in humans for at least 11,000 years.

More recent collaborations with Mark Achtman, Daniel Falush and their colleagues at the Max Planck Institute for Infection Biology in Berlin have shown that all modern *H. pylori* strains can be traced to five ancient populations—two arising in Africa, two in western or central Eurasia and one in East Asia. In fact, the genetic variations of *H. pylori* can be used to trace human settlement and migration patterns over the past 60,000 years. Because *H. pylori* is so much more genetically diverse than *Homo sapiens,* the bacteria can better elucidate the history of population movements than can studies of human mitochondrial DNA (the most commonly used marker for such investigations). As researchers attempt to clock the migrations of our species, the mitochondrial studies may provide the hour hand, but the genetic sequences of *H. pylori* may offer a more accurate minute hand.

The genetic variations of *H. pylori* can be used to trace human settlement and migration patterns.

A Microbial Extinction

Humans are the only hosts for *H. pylori,* and the spread of the bacterium involves mouth-to-mouth or feces-to-mouth transmission. The geographic differences in *H. pylori* infection rates—much lower in the developed world than elsewhere—may be partly the result of improvements in sanitation in the U. S., Europe and other developed countries over the past century. But I believe that the widespread use of antibiotics has also contributed to the gradual elimination of *H. pylori.* Even short courses of antibiotics, given for any purpose, will eradicate the bacteria in some recipients. In developing countries where antibiotics are less commonly used, 70 to 100 percent of children become infected with *H. pylori* by the age of 10, and most remain colonized for life; in contrast, fewer than 10 percent of U.S.-born children now carry the organism. This difference represents a major change in human microecology.

Furthermore, the disappearance of *H. pylori* may be a sentinel event indicating the possibility of other microbial extinctions as well. *H. pylori* is the only bacterium that can persist in the acidic environment of the human stomach, and its presence can be easily determined by tests of blood, stool, breath or stomach tissue. But other body sites, such as the mouth, colon, skin and vagina, have complex populations of indigenous organisms. If another common bacterium were disappearing from these tissues, we would not have the diagnostic tools to detect its decline.

What are the consequences of *H. pylori*'s retreat? As noted, the incidences of both peptic ulcer disease (except those cases caused by aspirin and nonsteroidal anti-inflammatory agents

such as ibuprofen) and stomach cancer are clearly declining in developed countries. Because these illnesses, especially stomach cancer, develop over many years, the drop in disease incidence has lagged several decades behind the decline in *H. pylori* infection, but the falloff is startling nonetheless. In 1900 stomach cancer was the leading cause of cancer death in the U.S.; by 2000 the incidence and mortality rates had fallen by more than 80 percent, putting them well below the rates for colon, prostate, breast and lung cancers. Substantial evidence indicates that the continuing extinction of *H. pylori* has played an important role in this phenomenal change. This is the good news.

At the same time, however, there has been an unexpected rise in the incidence of a new class of diseases involving the esophagus. Since the early 1970s, epidemiologists in the U.S., the U.K., Sweden and Australia have noted an alarming jump in esophageal adenocarcinoma, an aggressive cancer that develops in the inner lining of the esophagus just above the stomach. The incidence of this illness in the U.S. has been climbing by 7 to 9 percent each year, making it the fastest-increasing major cancer in the country. Once diagnosed, the five-year survival rate for esophageal adenocarcinoma is less than 10 percent.

Where are these terrible cancers coming from? We know that the primary risk factor is gastroesophageal reflux disease (GERD), a chronic inflammatory disorder involving the regurgitation of acidic stomach contents into the esophagus. More commonly known as acid reflux disease, GERD was not even described in the medical literature until the 1930s. Since then, however, its incidence has risen dramatically, and now the disorder is quite common in the U.S. and other western countries. GERD can lead to Barrett's esophagus, a premalignant lesion first described in 1950 by English surgeon Norman Barrett. The incidence of Barrett's esophagus is rising in tandem with that of GERD, and patients suffering from the condition have an increased risk of developing esophageal adenocarcinoma. It is becoming clear that GERD may initiate a 20-to 50-year process: in some cases, the disorder slowly progresses to Barrett's esophagus and then to adenocarcinoma, paralleling the gradual changes that lead to cancers in other epithelial tissues. But why are GERD and its follow-on disorders becoming more common?

The rise of these diseases has occurred just as *H. pylori* has been disappearing, and it is tempting to associate the two phenomena. When I began proposing this connection in 1996, I was greeted first by indifference and then by hostility. In recent years, though, a growing number of studies support the hypothesis that *H. pylori* colonization of the stomach actually protects the esophagus against GERD and its consequences. What is more, the strains bearing the *cagA* gene—that is, the bacteria that are most virulent in causing ulcers and stomach cancer—appear to be the most protective of the esophagus! In 1998, working with researchers from the National Cancer Institute, we found that people carrying *cagA* strains of *H. pylori* had a significantly decreased risk of developing adenocarcinomas of the lower esophagus and the part of the stomach closest to the esophagus. Then, in collaboration with investigators from the Cleveland Clinic and the Erasmus Medical Center in the

Netherlands, we showed a similar correlation for both GERD and Barrett's esophagus. Independent confirmations have come from the U.K., Brazil and Sweden. Not all investigators have found this effect, perhaps because of differences in the methods of the studies. Nevertheless, the scientific evidence is now persuasive.

A Theory of Interactions

How can colonization by *H. pylori* increase the risk of stomach diseases but protect against esophageal disorders? A possible explanation lies in the interactions between the bacterium and its human host. *H. pylori* has evolved into a most unusual parasite: it can persist in a stomach for decades despite causing continual damage and despite the host's immune response against it. This persistence requires that virtually all the "up-regulatory" events that cause inflammation in the stomach tissue must be balanced by "down-regulatory" events that prevent the damage from worsening too rapidly. There must be an equilibrium between microbe and host; otherwise, the host would die rather quickly, and the bacteria would lose their home before getting a chance to propagate to another person. But how can two competing forms of life achieve this equilibrium? My hypothesis is that the microbe and host must be sending signals to each other in a negative feedback loop.

Negative feedback loops are common in biology for the regulation of cellular interactions. Consider, for example, the feedback loop involving glucose and the regulatory hormone insulin. After you eat a meal, glucose levels in the bloodstream rise and the pancreas secretes insulin. The insulin causes glucose levels to fall, which signals the pancreas to reduce insulin secretion. By modulating the peaks and valleys in glucose levels, the system maintains a steady state called homeostasis. First described in the 19th century by French physiologist Claude Bernard, this concept has become the basis for understanding hormone regulation.

In essence, I took this idea one step further: the feedback relationship can involve microbial cells as well as host cells. Over the years, working with mathematicians Denise Kirschner of the University of Michigan at Ann Arbor and Glenn Webb of Vanderbilt University, our concepts of feedback have become more complex and encompassing. In our current formulation, the *H. pylori* population in a person's stomach is a group of extremely varied strains cooperating and competing with one another. They compete for nutrients, niches in the stomach and protection from stresses. Over the millennia, the long coevolution of *H. pylori* and *H. sapiens* has put intense selective pressure on both species. To minimize the damage from infection, humans have developed ways to signal to the bacteria, through immune responses and changes in the pressure and acidity in the stomach. And *H. pylori*, in turn, can signal the host cells to alleviate the stresses on the bacteria.

A good example of an important stress on *H. pylori* is the level of acidity in the stomach. Too much acid will kill the bacteria, but an extremely low level is not good either, because it would allow less acid-tolerant organisms such as *E. coli* to invade *H. pylori*'s niche. Therefore, *H. pylori* has evolved the ability to regulate the acidity of its environment. For example, strains bearing the *cagA* gene can use the CagA protein as a signaling molecule. When acidity is high, the *cagA* gene produces a relatively large amount of the protein, which triggers an inflammatory response from the host that lowers acidity by affecting the hormonal regulation of the acid-producing cells in the stomach lining. Low acidity, in contrast, curtails the production of CagA and hence reduces the inflammation.

This negative feedback model helps us understand the health effects of *H. pylori*, which depend in large part on the intensity of the interactions between the bacteria and their hosts. The *cagA* strains substantially increase the risk of stomach cancer because they inject the CagA protein into the stomach's epithelial cells for decades, affecting the longevity of the host cells and their propensity to induce inflammation that promotes cancer. Strains lacking the *cagA* gene are much less interactive, so they do not damage the stomach tissues as severely. On the other hand, *cagA* strains effectively modulate acid production in the stomach, preventing acidity levels from rising too high. People who carry strains lacking the *cagA* gene have a weaker modulation of acidity levels, and people who are not colonized by *H. pylori* have no microbial controls at all. The resulting swings in stomach acidity may be central to the rise in esophageal diseases, which are apparently triggered by the exposure of the tissue to highly acidic stomach contents.

The absence of *H. pylori* may have other physiological effects as well. The stomach produces two hormones that affect eating behavior: leptin, which signals the brain to stop eating, and ghrelin, which stimulates appetite. Eradication of *H. pylori* with antibiotics tends to lower leptin and increase ghrelin; in one study, patients who had undergone treatment to eliminate *H. pylori* gained more weight than the control subjects did. Could changes in human microecology be contributing to the current epidemic of obesity and diabetes mellitus (an obesity-related condition) in developed countries? If this research were confirmed, the implications would be sobering. Doctors might need to reevaluate antibiotic treatments that rid the stomach of *H. pylori* (and remove critical bacteria from other parts of the body as well). Although some of the consequences of eradication may be for the better (for example, a reduced risk of stomach cancer) other effects may be for the worse. The balance between good and bad may well depend on the patient's age, medical history and genetic type.

Probiotics

If researchers conclude that *H. pylori* would actually benefit some individuals, should physicians reintroduce the bacterium to these patients' stomachs? For more than 100 years, both medical scientists and laypersons have been searching for probiotics, microbes that can be ingested to aid human health. The earliest studies focused on the *Lactobacillus* species, the bacteria that make yogurt and many cheeses, but the effects of reintroduction were, at best, of marginal value. Researchers have largely failed to find any effective probiotics despite a century of trying.

One reason for this failure is the complexity and coevolution of the human microbiota, the organisms that share our bodies. Our microbiota are highly evolved for living within us and with each other. How likely is it that a newcomer, an unrelated strain of bacteria from outside the body, can successfully rechannel the pathways of interaction in a beneficial way? The existing organisms have survived strong and continuous selection, and this "home court advantage" usually enables them to reject and eliminate any strangers.

But a new day for probiotics may be coming. The key step will be gathering more knowledge of our indigenous microbiota and how they interact with us. I believe that complex interactions take place wherever microbes colonize our bodies (for example, in the colon, mouth, skin and vagina), but because of the array of competing organisms in those tissues, the relations are difficult to elucidate. *H. pylori,* though, largely excludes other microbes from the stomach. By the paradox of its great adaptation to humans and by the accident of its progressive disappearance during the 20th century, *H. pylori* may become a model organism for investigating human microecology.

Once scientists fully catalogue the myriad strains of *H. pylori* and discover how each affects the host cells of the stomach, this research may give clinicians a whole new arsenal for fighting diseases of the digestive tract. In the future, a physician may be able to analyze a patient's DNA to determine his or her susceptibility to inflammation and genetic risks of acquiring different kinds of cancers. Then the doctor could determine the best mix of *H. pylori* strains for the patient and introduce the microbes to his or her stomach. What is more, researchers may be able to apply their knowledge of *H. pylori* to solve other medical problems. Just as the Botox nerve toxin produced by *Clostridium botulinum,* the bacterium that causes botulism, is now used for cosmetic surgery, the toxin VacA could become the basis for a novel class of drugs that suppress immune function. The study of our longtime bacterial companions offers a new avenue for understanding our own bodies and promises to expand the horizons of medical microbiology.

MARTIN J. BLASER is one of the world's foremost experts on *Helicobacter pylori*. He is Frederick H. King Professor of Internal Medicine, chair of the department of medicine and professor of microbiology at New York University School of Medicine. Blaser previously worked at the University of Colorado, the Centers for Disease Control and Prevention, the Rockefeller University, and Vanderbilt University. Since earning his MD at New York University in 1973, he has written over 400 original scientific articles and edited several books on infectious diseases. He is also president-elect of the Infectious Diseases Society of America.

Bacteria Are Picky about Their Homes on Human Skin

Elizabeth Pennisi

Julie Segre is touring the microbial landscape of our body's biggest organ, the skin. In anticipation of a $115 million, 5-year effort by the U.S. National Institutes of Health (NIH), she's traveling from head to toe, conducting a census of some of the trillions of bacteria that live within and upon human skin. Although their project is just getting off the ground, Segre, a geneticist at the National Human Genome Research Institute (NHGRI) in Bethesda, Maryland, and her colleagues have already uncovered a surprising diversity and distribution among skin bacteria. And a few oddities have emerged, too: Microbes known mostly from soils like healthy human skin, living in harmony with us; and the space between our toes is a bacterial desert compared to the nose and belly button.

Segre's work on what bacteria live where "is cool stuff," says Steven Salzberg, a bio-informaticist at the University of Maryland, College Park. "We need to increase our own and the public's awareness of the diversity and quantity of bacterial species on our own skin. The more people are aware, the more we can do to control infection."

Bacteria and other microbes that colonize our skin and other tissues outnumber the human body's cells 10 to 1, forming dynamic communities that influence our ability to develop, fight infection, and digest nutrients. "We're an amalgamation of the human and microbial genomes," says Segre. Recognizing this, NIH last year designated the Human Microbiome Project as one of its two Roadmap initiatives (*Science*, 2 June 2006, p. 1355). Researchers will sequence the genomes of about 600 bacteria identified as human inhabitants and get a handle on the 99% of bacteria that defy culturing but thrive in the skin, nose, gut, mouth, or vagina. "You have to understand what is the normal flora in the healthy skin to understand the impact of flora on disease," says Kevin Cooper, a dermatologist at Case Western Reserve University in Cleveland, Ohio.

As a first step, Segre, NHGRI postdoctoral fellow Elizabeth Grice, and their colleagues have studied five healthy volunteers, swabbing the insides of their right and left elbows. The site chosen isn't as unusual as it sounds; people with eczema often develop symptoms there. To survey the full thickness of skin, the researchers also used a scalpel to scrape off the top cells.

And to reach even deeper, they took small "punches" of skin, a procedure akin to removing a mole.

From all the samples, Grice, Segre, and colleagues pulled out 5300 16S ribosomal RNA genes, which vary from microbe to microbe. After lumping together the most similar 16S genes, they came up with 113 kinds of bacteria and identified these dermal residents by matching the 16S genes to those of known bacteria. (Segre described the results at a recent meeting at Cold Spring Harbor Laboratory, and they are being published online 23 May in *Genome Research*.) "That's a lot of diversity, a lot of different organisms," says Martin Blaser, a microbiologist at New York University, who has done a similar survey of microbes living on the forearm, also finding a lot of diversity.

Yet just 10 bacteria accounted for more than 90% of the sequences. Almost 60% of the 16S genes came from *Pseudomonas*, Gram-negative bacteria that flourish in soil, water, and decomposing organic debris. The next most common one, accounting for 20%, was another Gram-negative soil and water bug, *Janthinobacterium*. Neither had been considered skin microbes before this census. Although there were some differences among the volunteers in the microbes present, their elbows did share a common core set of microbes, the group reports.

The three sampling methods yielded slightly different results, with "punches" revealing a surprising number of bacteria under the skin—1 million bacteria per square centimeter compared with 10,000 from the scrapes. "I would have thought under the skin there would be fewer," says Salzberg.

Segre and her team have also begun sampling 20 other skin sites, including behind the ear and the armpit, from the bodies of volunteers. Skin varies in acidity, temperature, moisture, oil accumulation, and "different environments select for different microbes," says Blaser. Bacteriawise, reports Segre, "no subsite is identical."

Some researchers suspect that shifts in the makeup of skin microbial communities activate the immune system to cause diseases such as eczema. "If you know what the [healthy] flora is, then one strategy is to recolonize the area with the right flora," says Cooper.

Biofilms

A new understanding of these microbial communities is driving a revolution that may transform the science of microbiology.

JOE J. HARRISON ET AL.

When we think about bacteria, most of us imagine a watery milieu, with single-celled organisms swimming about. We might envision these solitary entities getting together with some of their brethren now and then to cause some disease or spoil some food, but once the job is done they return to their isolated existence. This image of bacterial existence, it turns out, is not only oversimplified but perhaps misleading as well. In nature, the majority of microorganisms live together in large numbers, attached to a surface. Rather than living as lonely hermits in the so-called planktonic form, most bacteria spend much of their lives in the microbial equivalent of a gated community—a biofilm.

A mature biofilm is a fascinating construction: It can form layers, clumps and ridges, or even more complex microcolonies that are arranged into stalks or mushroom-like formations. The residents of the biofilm may be a single species or a diverse group of microorganisms distributed in various neighborhoods. Their common bond is a matrix made of polysaccharides, DNA and proteins, which together form an *extracellular polymeric substance*—what many microbiologists just call slime.

It's becoming increasingly clear that the communal life offers a microorganism considerable advantages. The physical proximity of other cells favors synergistic interactions, even between members of different species. These include the horizontal transfer of genetic material between microbes, the sharing of metabolic by-products, an increased tolerance to antimicrobials, shelter from changes in the environment and protection from the immune system of an infected host or from grazing predators. The formation of a biofilm has even been likened to the program by which cells within a multicellular organism differentiate.

An appreciation of the significance of biofilms is a relatively recent phenomenon. Only within the past 15 to 20 years have biologists begun to examine the physiology of these microbial communities. This is an extraordinary state of affairs, given that the Dutch microscopist Antonie van Leeuwenhoek first described biofilms in the late 1600s. Using acetic acid, he had tried to kill a biofilm—the dental plaque on his dentures—but noted that only the free swimming cells could be destroyed. Despite the early discovery of microbial communities, microbiology departed from these observations to focus primarily on planktonic bacteria.

To be sure, not everyone agrees that biofilms are the predominant form of bacteria in nature. The vast majority of laboratory methods used today examine cultured microorganisms in their planktonic mode. But we believe that microbiology is experiencing a shift in how bacteria are conceptualized. We predict that this new perspective of how microorganisms live will have fundamental consequences for medicine, industry, ecology and agriculture.

Biofilms Are Everywhere

Most people are familiar with the slippery substance covering the rocks in a river or a stream. This particular slime is an aquatic biofilm made up of bacteria, fungi and algae. It begins to form after bacteria colonize the rock's surface. These microbes produce the extracellular polymeric substance, which is electrostatically charged so that it traps food particles and clay and other minerals. The matter trapped in the slime forms microscopic niches, each with a distinct microenvironment, allowing microorganisms that have different needs to come together to form a diverse microbial consortium.

A biofilm matrix is considered to be a *hydrogel,* a complex polymer hydrated with many times its dry weight in water. The hydrogel characteristics of the slime confer fluid and elastic properties that allow the biofilm to withstand changes in fluid shear within its environment. So biofilms often form streamers—gooey assemblages of microbes that are tethered to a surface. As running water passes over the biofilm, some pieces may break free and so spread the microbial community downstream. It is believed that bacteria can colonize the lungs of patients on ventilators in this way, causing often-fatal pneumonia in critically ill patients.

A microorganism's extraordinary ability to spread explains how biofilms show up in the unlikeliest of places. The steel hull of a ship at sea can be coated with biofilms that increase the drag on the vessel and so compromise its speed. Other biofilms wreak havoc in the oil industry by facilitating the microscopic corrosion of metals and limiting the lifespan of pipelines. Some biofilms, made up of the ancient lineage of prokaryotes (organisms lacking a nucleus) called *archaea,* can even survive the hostile hydrothermal environments of hot springs and deep-sea hydrothermal vents. The aptly named archaebacterium *Pyrodictium* thrives at the bottom of the sea, growing in a moldlike layer on sulfur crystals in the dark, anaerobic environment of a hydrothermal vent, where temperatures may exceed 110 degrees Celsius.

Perhaps one of the most extraordinary environments where one can find a biofilm is in the belly of a dairy cow. Biofilms are part of the normal complement of microbes in many healthy animals, but the presence of these microbial communities in ruminants provides a rich example of the interactions within a biofilm.

We begin with the *rumen,* the largest compartment of the bovine stomach, which can hold a liquid volume in excess of 150 liters. It is filled with so many microbes that microbiologists refer to cows as mobile fermenters. Bacteria colonize the digestive tract of a calf two days after it is born. Within three weeks the microorganisms have modified the chemistry inside the rumen, which soon becomes home to a reported 30 species of bacteria, 40 species of protozoa and 5 species of yeast. The cells in this biofilm thrive in the mucous layer of the stomach and grow on the food ingested by the animal. Cows, of course, eat grass, which consists largely of cellulose, a complex carbohydrate that cannot be broken down by mammalian digestive enzymes. But cellulose is a perfect fuel for the bacteria in the biofilm, which convert it into a microbial biomass that in turn supplies the proteins, lipids and carbohydrates needed by the cow.

The heart of this process is a microscopic ecosystem that begins when a pioneering planktonic bacterium in the rumen, a species such as *Ruminococcus flavefaciens,* gains access to the inner parts of a leaf, perhaps one that might have been broken by the cow's chewing. These bacteria attach themselves to the cellulose in the inner layers of the leaf and proliferate to form a rudimentary biofilm. The microbes release cellulose degrading enzymes, which produce simple sugars and metabolic byproducts that attract other bacteria—anaerobic fermenters such as the spiral-shaped *Treponema byrantii,* which ingest the sugars and produce organic acids, including acetic acid and lactic acid.

The acidic metabolites would normally slow the growth of the bacteria by a process of feedback inhibition, but it so happens that other microorganisms join the biofilm community and eat the organic acids. These are the *methanogens,* archaea whose actions accelerate the growth of the bacterial community and prevent the inhibitory feedback. As the name suggests, methanogens produce methane—lots of it. Approximately 15 to 25 percent of the global emission of methane, which totals 7.5 billion kilograms per year, is attributable to the flatulence of ruminants. Because methane traps heat in the atmosphere, the biofilm hidden away in a cow's stomach may play a nontrivial role in global climate change.

Animals aren't the only living things that provide a home to biofilms. Microbial colonies have been recognized on tropical plants and grocery-store produce since the 1960s, but it wasn't until the past decade that the term *biofilm* was used to describe bacterial growth on a plant's surface. In this domain, life in a biofilm confers many advantages to the individual cell, including protection from a number of environmental stresses—ultraviolet radiation, desiccation, rainfall, temperature variations, wind and humidity. The biofilm also enhances a microorganism's resistance to antimicrobial substances produced by competing microorganisms or the host's defenses.

Relations between plants and biofilms can be quite varied. In some instances the plant merely serves as a mechanical support, so the biofilm is simply a harmless epiphyte. In other cases, the plant may provide some nutrients for the microbes, such as the saprophytes that feed on decaying plant matter; these too pose no danger to the plant. But there can be trouble when certain epiphytic populations with the genetic potential to initiate a pathogenic interaction with the host grow large enough to overwhelm the host's defense mechanisms. Then the cells in the biofilm coordinate the release of toxins and enzymes to degrade the plant tissue. What began as an innocuous relationship ends in disease.

Belowground, plants and biofilms may also engage in some fairly elaborate interactions. For example, *Pseudomonas fluorescens* colonizes roots and protects plants from pathogens by producing antibiotics that exclude fungi and other bacterial colonizers. But fungal biofilms can also be beneficial to the plant. Certain mycorrhizal fungi penetrate a plant's root cells while also forming an extensive network in the soil; thus they provide a drastic increase in the surface area that the plant can use for the absorption of water and nutrients.

On the other hand, bacteria of the genus *Rhizobium* fix nitrogen from the atmosphere by converting N_2 gas into ammonia (NH_3). This process can involve some intricate chemical signaling between the plant and the bacteria that results in the formation of nodules within the root where the bacterial aggregates engage in nitrogen fixation. Perhaps the most intricate relation involves an interaction between the rhizobia, the mycorrhizal fungi and a plant host. The bacteria form a biofilm on the surface of the fungus, which in turn makes its connection with the plant, and so creates a tripartite symbiotic system that relies on the formation of biofilms by two microorganisms. (Unless the soil is alkaline, the system requires another player, nitrifying bacteria to oxidize the ammonia; they live not in the nodule but in nearby soil.)

Finally, let us consider the pathogenic interactions of biofilms within the plant's vasculature. Unfortunately, vascular diseases are currently untreatable and tend to be devastating to many economically important crops. A few pathogenic biofilms have been described in the water-carrying xylem of plants, but here we'll merely address *Xylella fastidiosa.* This pathogen causes Pierce's disease in grapevines and citrus variegated chlorosis in sweet oranges—diseases that have had a major impact on the wine industry in California and the citrus industry in Brazil, with economic losses exceeding $14 billion in the past decade. Pierce's disease also limits the development of a wine industry in Florida because the bacterium is endemic in that region.

X. fastidiosa is transmitted by xylem feeding insects, called sharpshooters, that acquire the bacteria while feeding from infected plants. The bacteria form a rudimentary biofilm inside the insect's gut, and this allows them to be sloughed off indefinitely in aggregates sufficient to infect another plant when the insect feeds again. In turn, the biofilms clog the plant's xylem and cause symptoms related to water stress. So the biofilm plays a key role in the colonization of the plant vessels, the propagation of the disease and its pathogenicity.

The appreciation of biofilms' importance in plant disease has only just begun, and it will probably take some time for the idea to be applied in plant microbiology. However, the benefits could be significant. A better understanding of the associations between plants and biofilms may lead to more efficacious and environmentally friendly treatments for disease. It may also lead to the development of commercial applications that could improve the beneficial interactions between plants and microorganisms. Indeed, various rhizobia are now being used on commercial farms as a biotic fertilizer.

United We Stand

The Centers for Disease Control and Prevention estimates that up to 70 percent of the human bacterial infections in the Western world are caused by biofilms. This includes diseases such as prostatitis and kidney infections, as well as illnesses associated with implanted medical devices such as artificial joints and catheters and the dental diseases—both tooth decay and gum disease—that arise from dental plaque, a biofilm. In the lungs of cystic fibrosis patients, *Pseudomonas aeruginosa* frequently forms biofilms that cause potentially lethal pneumonias. There is a long list of biofilm-related ailments, and many scientists believe the list will continue to grow as we learn more about the function of these microbial structures.

In almost all instances, the biofilm plays a central role in helping microbes survive or spread within the host. That's because the slimy matrix acts as a shield, protecting pathogenic bacteria from antibodies and white blood cells, the sentinels of the immune system. Biofilms are also notorious for their ability to withstand extraordinarily high concentrations of antibiotics that are otherwise lethal in smaller doses to their planktonic counterparts. In fact, a biofilm can be 10 to 1,000 times less susceptible to an antimicrobial substance than the same organism in suspension.

This challenge, with its grave implications for the fight against pathogens, has been the focus of our research group's investigations. We have developed and licensed to a Canadian startup company a technology (the Calgary Biofilm Device, now called the MBEC Assay) that can be used to rapidly screen biofilms for their sensitivity to antimicrobials. A pharmaceutical laboratory testing a potential drug to fight pneumonia or catheter-related infection can now find out whether a drug that is effective against free-floating pathogens will be successful in eradicating the same organisms in a biofilm.

During the development of this technology, we have learned some remarkable things about biofilms. We have moved on to exploring some pathogenic "co-biofilms" of unrelated species living together, along with specific mechanisms that may be important in drug development. For example, biofilms' resistance to high metal concentrations makes them useful in

A New Way to Look at Microorganisms

The conventional way to grow bacteria is to inoculate a flask that contains a broth of nutrients. If you stir the broth constantly, the cells will have plenty of oxygen and a homogeneous distribution of food. Under these optimal growth conditions, you'll get a nice batch of planktonic bacteria floating in the solution.

Of course, nature rarely provides such a perfectly uniform environment. Bacteria in a biofilm grow in a matrix of heterogeneous microenvironments that vary in oxygen content, nutrient distribution and countless other chemical vagaries. The bacteria that stick to the sides of the laboratory flask form mature biofilms. Ironically, until recently these were largely ignored or destroyed.

Several new technologies have been explicitly developed to grow and examine biofilms in the laboratory. One method uses a rotating disk inside an inoculated broth. The shear force caused by the rotation encourages the formation of a biofilm on the disk. Our laboratory group has also recently developed a biofilm-based assay for examining the effectiveness of antimicrobials in a high-throughput fashion—that is, the device allows us to create 96 statistically equivalent biofilms, and it can also be used to test various dilutions of antimicrobial compounds with a standard microtiter plate, the MBEC assay. We are currently using this tool to discover new substances that may be effective against biofilms.

Another device, called a flow cell, consists of a chamber and an optically transparent surface, such as a glass coverslip. A growth medium is pumped through the chamber, promoting the formation of a thick biofilm on the glass surface. This method allows scientists to examine microbial communities in a confocal laser-scanning microscope (CLSM). Specialized computer software can be used to assemble images captured by CLSM to create a three-dimensional view of a biofilm.

CLSM might be considered as a complement to scanning electron microscopy (SEM). SEM can achieve magnifications that are 10 times greater than CLSM and so can be used to examine the shape and arrangement of single cells, whereas CLSM provides an overview of the biofilm's structure. SEM also kills the microbial community, whereas CLSM is not as invasive. Sequences of images can be compiled into movies that show how microorganisms live and die in a biofilm.

Finally, new methods in proteomics and transcriptomics allow scientists to examine the distribution and patterns of proteins and gene expression in biofilms. The development of these techniques has opened the door to a new view of how microorganisms live.

removing toxic metals from the environment. But a detailed understanding of how the films handle metal toxicity may also open the door to antimicrobial treatments targeted at biofilms.

We and other investigators have learned that part of the extraordinary resilience of bacteria arises from the remarkable heterogeneity inside the biofilm. Microbes closest to the fluid that surrounds the biofilm have greater access to nutrients and oxygen compared with those in the center of the matrix or near the substratum. As a result, the bacteria in the outer layers of the community grow more quickly than those on the inside. This comes into play as a defense mechanism because many antibiotics are effective only against fast-growing cells, so the slow growers within the biofilm tend to be spared. Moreover, the cells in the center of the community are further protected from the environment because the biofilm matrix is negatively charged. This restricts the entry of positively charged substances, such as metal ions and certain antibiotics.

One of the most intriguing defense mechanisms enabled by the formation of a biofilm involves a kind of intercellular signaling called *quorum sensing*. Some bacteria release a signaling molecule, or inducer. As cell density grows, the concentration of these molecules increases. The inducers interact with specific receptors in each cell to turn on "quorum sensing" genes and initiate a cascade of events, triggering the expression or repression of a number of other genes on the bacterial chromosome. Some bacterial strains seem to rely on quorum sensing more than others, but anywhere from 1 to 10 percent of a microbe's genes may be directly regulated by this process.

Quorum sensing is known to affect the production of enzymes involved in cellular repair and defense. For example, the enzymes superoxide dismutase and catalase are both regulated by quorum sensing in *P. aeruginosa,* which forms mucoidal clusters of bacterial cells embedded in cellular debris from the airway epithelial layer in the cystic fibrosis patient's lung. The first enzyme promotes the destruction of the harmful superoxide radical (O_2^-), whereas the second converts the equally toxic hydrogen peroxide molecule (H_2O_2) into water and molecular oxygen. These enzymes help the biofilm survive assaults not only from disinfectants, but also from the cells of a host's immune system that typically kill bacteria by unleashing antimicrobial agents, including reactive oxygen species.

Quorum sensing may also be involved in the defense against antibiotic drugs. Here the mechanism increases the production of molecular pumps that expel compounds from the cell. These so-called *multidrug efflux pumps* reduce the accumulation of the antibiotics within the bacterium and even allow the microbe to grow in the presence of the drugs.

There is also heterogeneity among the cell types in the biofilm that contributes to antimicrobial tolerance. Specialized survivor cells, called "persisters," are slow-growing variants that exist in every bacterial population. They are genetically programmed to survive environmental stress, including exposure to antibiotics. Although persisters do not grow in the presence of an antibiotic, they also do not die. Persisters are not mutants; even in a genetically uniform population of cells a small portion undergo a spontaneous switch to the persistent form. This past year Kim Lewis of Northeastern University demonstrated that persisters generate a toxin, RelE, that drives the bacterial cell into a dormant state. Once antibiotic therapy has ceased, the persisters give rise to a new bacterial population, resulting in a relapse of the biofilm infection.

The use of persister cells as a defense mechanism may have evolved early in the history of life. In this post-genomics era, scientists have learned that many related genes are present in a variety of distantly related bacteria, suggesting that similar genes were present in the primeval common ancestors. Yet the reduced growth rate of the persisters poses a paradox because slowed cell division decreases the fitness of a population. Edo Kussell and his colleagues at Rockefeller University recently proposed that bacterial persistence may have evolved as an "insurance policy" against rare antibiotic encounters. If so, in attempting to overcome bacterial antibiotic tolerance, scientists are battling an ancient mechanism that may have been refining itself for billions of years. If we are ever to succeed in controlling bacterial infection, more research efforts need to be focused on biofilms rather than the comparatively vulnerable planktonic form.

References

Andrews, J. H., and R. F. Harris. 2000. The ecology and biogeography of microorganisms on plant surfaces. *Annual Review of Phytopathology* 38:145–180.

Barea, J. M., R. Azcón and C. Azcón-Aguilar. 2002. Mycorrhizosphere interactions to improve plant fitness and soil quality. *Antonie Van Leeuwenhoek* 81:343–351.

Beech, I. W., and J. Sunner. 2004. Biocorrosion: towards understanding the interactions between biofilms and metals. *Current Opinion in Biotechnology* 15:181–186.

Bjarnsholt, T., P. Jensen, M. Burmølle, M. Hentzer, J. A. Haagensen, H. P. Hougen, H. Calum, K. G. Madsen, C. Moser, S. Molin, N. Hoiby and M. Giskov. 2005, *Pseudomonas aeruginosa* tolerance to tobramycin, hydrogen peroxide and polymorphonuclear leukocytes is quorum-sensing dependent. *Microbiology* 151:373–383.

Ceri, H., M. E. Olson, C. Stremick, R. R. Read, D. W. Morck and A. G. Buret. 1999. The Calgary Biofilm Device: New technology for rapid determination of antibiotic susceptibilities in bacterial biofilms. *Journal of Clinical Microbiology* 37:1771–1776.

Dinh, H. T., J. Kuever, M. Mubmann, A. W. Hassel, M. Stratmann and F. Widdel. 2004. Iron corrosion by novel anaerobic microorganisms. *Nature* 427:829–833.

Hall-Stoodley, L., J. W. Costerton and P. Stoodley. 2004, Bacterial biofilms: From the natural environment to infectious diseases. *Nature Reviews Microbiology* 2:95–108.

Harrison, J. J., R. J. Turner and H. Ceri. 2005 Metal tolerance in bacterial biofilms. *Recent Research Developments in Microbiology* 9:33–35.

Harrison, J. J., R. J. Turner and H. Ceri. 2005. Persister cells, the biofilm matrix and tolerance to metal cations in biofilm and planktonic. *Pseudamonas aerugmosa. Environmental Microbiology* 7:981–994.

Keren, I., D. Shah, A. Spoering, N. Kaldalu and K. Lewis. 2004. Specialized persister cells and the mechanism of multidrug tolerance in *Escherichia coli. Journal of Bacteriology* 186:8172–8180.

Kletzin, A., T. Urich, F. Muller, T. M. Bandeiras and C. M. Gomes. 2004. Dissimilatory oxidation and reduction of elemental sulfur in thermophilic archaea. *Journal of Bioenergetics and Biomembranes* 36:77–91.

Kirchgessner, M., W. Windisch and H. L. Muller. 1995. Nutritional factors for the quantification of methane production. In: *Ruminant Physiology, Digestion, Metabolism, Growth and Reproduction. Proceedings of the 8th International Symposium on Ruminant Physiology,* ed. W, Engelhardt, S. Leonhardt-Marek, G, Breeves and D. Gieseke. Stuttgart: Ferdinande Enke Verlag, pp. 333–348.

Kussell, E., R. Kishnoy, N. Q. Balaban and S. Leibler. 2005. Bacterial persistence: A model of survival in changing environments, *Genetics* 169:1807–1814.

Marques, L. L. R., H. Ceri, G. P. Manfio, D. M. Reid and M. E. Olson. 2002. Characterization of biofilm formation by *Xylella fastidiosa in vitro. Plant Disease* 86:633–638.

McAllister, T. A., H. D. Bae, G. A. Jones and K. J. Cheng. 1994. Microbial attachment and feed digestion in the rumen, *Journal of Animal Science* 72:3004–3018.

Miron, J., D. Ben-Ghedalla and M. Morrison. 2001. Adhesion mechanisms of rumen cellulolytic bacteria. *Journal of Dairy Science* 84:1294–1309.

Morris, C. E., and J. M. Monier. 2003. The ecological significance of biofilm formation by plant-associated bacteria. *Annual Review of Phytopathology* 41:429–153.

Potera, C, 1999. Forging a link between biofilms and disease. *Science* 283:1837–1839.

Ramey, B. E., M. Koutsoudis, S. B. von Bodman and C. Fuqua. 2004. Biofilm formation in plant-microbe associations. *Current Opinion in Microbiology* 7:602–609.

Redak, R. A., A. H. Purcell, J. R. S. Lopes, M. J. Blua, R. F. Mizell III and P. C. Anderson. 2004. The biology of xylem fluid-feeding insect vectors of *Xylella fastidiosa* and their relation to disease epidemiology. *Annual Review of Entomology* 49:243–270.

Stoodley, P., K. Sauer, D. G. Davies and J. W. Costerton. 2002. Biofilms as complex differentiated communities. *Annual Reviews of Microbiology* 56:187–209.

Joe J. Harrison is a doctoral candidate and Raymond J. Turner an associate professor in the Department of Biological Sciences at the University of Calgary. Lyriam L. R. Marques was a postdoctoral fellow with the university's Biofilm Research Group and now serves as associate research director at MBEC BioProducts, Inc. Howard Ceri is a professor of biological sciences at the university and chairs the Biofilm Research Group. He also serves on the Sigma Xi Committee on Publications.

Address for Ceri: Department of Biological Sciences, 2500 University Drive N. W., Calgary, Alberta, Canada T2N 1N4. Internet: ceri@ucalgary.ca

UNIT 2

Genetics and Biotechnology

Unit Selections

Learning Objectives

- What impact have genomic technologies had on our understanding of microbes?

- Describe three ways in which genetically engineered microbes can be harnessed to benefit mankind. Do you have any concerns about the widespread use of genetically modified organisms in our environment?

- How can the diverse metabolic activities of microbes be used to generate alternative energy sources?

- One day in the future all of the human microbiota will be sequenced. What impact do you think this will have on our treatment of disease? If you were a researcher with this information at your disposal, what questions would you ask?

Student Website
www.mhcls.com

Internet References

National Center for Biotechnology Information
http://www.ncbi.nlm.nih.gov
Fungal Genome Initiative
http://www.broad.mit.edu/annotation/fungi/fgi/
Genome News Network
http://www.genomenewsnetwork.org
U.S. Geological Survey
http://water.usgs.gov/wid/html/bioremed.html
National Renewable Energy Laboratory
http://www.nrel.gov/learning/re_biofuels.html

A genome comprises all of the genetic information that defines a species. Biological research is currently in the midst of a "postgenomic era" in which much of our conceptual progress stems from obtaining the complete genomic sequences of organisms. In fact, since 1998 the complete genomes of over 225 types of microbes have been determined! The development of relatively inexpensive high-throughput DNA sequencing technologies has enabled scientists to obtain metagenomic sequences (all of the genetic information of many microbes living in a distinct environmental niche) of organisms previously unidentified by traditional culturing methods. Simply knowing an organism's DNA sequence is not enough, however. Wading through sequences to gain meaningful insights requires an understanding of the new science of bioinformatics—comparing genes from different species in order to make predictions about their functions in the organism. The result of these advances is a wealth of genetic information about microbes that has led to breakthroughs in the diagnosis and treatment of microbial infections.

The very first eukaryotic genome to be sequenced was that of the baker's yeast, *Saccharomyces cerevisiae,* in 1996. In 2000, the Fungal Genome Research Initiative was launched in order to escalate the sequencing of fungal genomes. To date, 25 fungal genomes have been completed. Paul Thacker's article "Understanding Fungi through Their Genomes" provides an interesting look at the role of fungi throughout the history of science. He describes the applications of fungal genomics in the pharmaceutical industry and gives his perspective on the future of this exciting field. Two articles in *Scientific American* focus on genomes from other microbial groups—namely helminths and bacteria. The parasitic bloodsucking worms called schistosomes infect 200 million people around the world every year and are experts at evading detection by the human immune system. Genomic analysis of schistosomes has given researchers more ammunition against these worms in the development of a vaccine. Finally, the genome of *Mycobacterium tuberculosis* has been determined and in "New Tactics against Tuberculosis" the importance of this genetic information is revealed. A new diagnostic test for tuberculosis, approved by the FDA in 2005, has increased specificity due to new genomic understanding of this bacterium. In addition to genomics, other "omics" technologies (proteomics, transcriptomics, chemical genomics, and metabolomics) are yielding significant new insights into the inner workings of *Mycobacterium tuberculosis* (for example, how it defends itself against drugs). This type of research promises to find new drug targets that can knock out the whole system of this deadly bacterium.

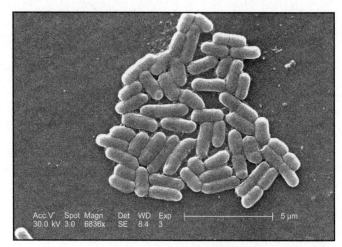

© Centers for Disease Control and Prevention

The field of metabolomics (studying the total chemical output of human metabolism) is still in its infancy, but researcher Jeremy Nicholson is "going with his gut" to find out how the metabolites produced by our normal gut flora can be used to deliver information about the state of health of an individual. According to Nicholson, the metabolome puts genomic information into perspective by looking at how genes interact with the environment. He speculates that personalized medicine based on the metabolome may be in our future and that drugs could be customized to fit a person's bacterial profile. It's too early to tell, but "drugging our bugs" could revolutionize the treatment of disease.

The second part of this unit addresses the role of microbes as our friends in the field of biotechnology. The advent of recombinant DNA technology sparked a revolution in the biological sciences over two decades ago with the commercial production of recombinant insulin. Since then, microbial "workhorses" have been used to churn out industrial scale amounts of various commercial products such as vaccines, interferons, growth factors, and interleukins, to name a few. Recently, the costly anti-malarial drug artemenisin has been mass-produced by recombinant *E. coli.* The use of viruses and bacteria as therapeutic agents has also been studied, and in "Bacterial Therapies: Completing the Cancer Treatment Toolbox," the authors describe how genetically engineered forms of *Salmonella* and *E. coli* can be used to deliver anti-tumor medication in a targeted manner. The advantages of using recombinant bacteria over conventional cancer chemotherapy are numerous; in particular, engineered bacteria do not target noncancerous cells and they are able to penetrate tumor tissue more effectively than conventional drugs.

In addition to improving human health through biotechnology, microbes are also active players in facing the environmental challenges of our time. Historically, microbes have long been used in bioremediation, the process by which microbial metabolism is used to break down toxic waste, sewage, and other pollutants. One organism's trash is a veritable buffet for some microbe on our planet, and we are fortunate to have them around to clean up our messes. Some microbial fuel cells are made from bacteria that generate electricity when they eat organic matter. To make matters even better, the organic matter they eat is sewage and the electricity generated by performing this heroic act is used to power the wastewater treatment plant. If this sounds too good to be true, read "Microbial Moxie" to find out more about how microbes can be used as remote sensors and alternative energy sources. Don't expect them to power a city, but you may be surprised to find out what these tiny fuel cells can do! In the quest for alternative energy sources, microbes are front-runners in the race to develop renewable fuels. Scientists are in the process of engineering *E. coli* to produce molecules similar to gasoline. The aim is to produce biofuels that replace corn-based ethanol. Opponents of corn-based biofuels argue that this method is increasing food costs (especially in poorer countries like Mexico) and they readily welcome an engineered gasoline substitute that would reduce carbon emissions. Regardless of whether they are genetically engineered to produce cheaper drugs or environmentally friendly fuels, microbes are clearly an essential part of our collective arsenal against the scourge of disease and climate change.

Understanding Fungi through Their Genomes

Ranging from microscopic yeast and molds to oversized mushrooms, fungi have been a part of human civilization since before recorded history. As agents of food spoilage, they have long been the bane of societies attempting to store provisions, but fungi can also be vital to the diets of people around the planet. Without them, we would not have mushrooms on our pizza, mold-ripened cheeses, soy sauce, or sliced bread. And let's not forget about cold beer.

PAUL D. THACKER

In the 1800s, fungi led the way to modern microbiology and biochemistry. Peering through a microscope, Pasteur was the first person to observe tiny living creatures. Uncertain about what he was seeing, he classified their activities into "organized ferments" and, when changes occurred without any observable microorganisms, "unorganized ferments." When it became known that the unorganized ferments were the metabolic products of organized ferments, he later suggested the term "enzyme" (*en* = in, *zyme* = yeast).

One of the best studied of these enzymes, diastase, is a product of germinating barley and was used in the malting step of beer production. In 1894 Jokichi Takamine began inoculating thin layers of rice and barley with spores from *Aspergillus oryzae* to optimize production of the enzyme, which converts starch to malt sugar. Much like contemporary scientists, Takamine showed a strong entrepreneurial streak and patented his process, later marketing diastase as a malting enzyme and a digestive aid to treat dyspepsia.

In 1928 medicine vaulted firmly into the modern age with Alexander Fleming's discovery of penicillin, the first "wonder drug." This spurred all sorts of studies into fungal physiology, fermentation technology, and industrial strain development. In 1941 George Beadle and Edward Tatum found the first connection between genes and biochemical function. The one gene-one enzyme hypothesis could not have been formed without the help of *Neurospora crassa.*

When genetics progressed to genomics, *Saccharomyces cerevisiae* became the second species and the first eukaryote to be sequenced. But since this genome sequence became available in 1996, research tools in yeast have bounded forward in huge leaps, while studies of filamental fungi have lagged behind. The gap will soon close.

Over the last two years, a large group of scientists from across the spectrum of mycology and genomics has worked to sequence a broad array of fungal species. In June, the National Human Genome Research Institute (NHGRI), a component of the National Institutes of Health, agreed to fund the costs of the first 7 of 15 proposed species of the Fungal Genome Research Initiative. Researchers in the fungal genomics field anticipate an explosion of information that not only will influence the fields of medicine, industry, and phylogeny but will make mycology one of the hottest fields in science.

"When a young scientist who might not have ever given fungi a serious thought sees the depth and breadth of genomic information that we are going to have in this field," says University of California–Berkeley systematist John Taylor, "this person is going to have to give mycology a serious second look."

What Is Available

Within mycology, only yeast researchers have powerful genomic technologies at their fingertips. The *S. cerevisiae* sequence is well annotated, and the organism can be manipulated with a number of genetic tools such as expression profiling with microarrays, serial analysis of gene expression, protein tagging, two-hybrid interactions, and transposon insertions.

Filamental genetics, however, is much less developed and bits of sequence are flung all across the Internet. A search of 379 genome-project Web sites finds 16 with filamental genetic information representing only 12 species. And some of these

Websites are owned by industry and thus contain proprietary information.

After mining the sequence for information, Cereon/Monsanto released their proprietary, low-resolution sequence of *Aspergillus nidulans*. Access, of course, comes with strings. Researchers using the database have to acknowledge Monsanto in their papers and give the corporation first crack at licensing any agricultural applications stemming from the research. Still, the restrictions have done little to slow the obvious zeal in accessing the genome.

In the last year, however, things have started to improve with the public release of the genome sequences of three other fungi. But for some academics, it was too little, too late. "If you want to be serious about genetic research, then you need the genomic information," says Olen Yoder, who left a tenured faculty position at Cornell University to join Syngenta at their La Jolla, California, research lab. An agricultural research group, Syngenta boasts that it has high-resolution sequences of five fungal genomes. "That just wasn't available to me in academia, so I left. And I know a number of people who did the same thing."

The Squeaky Wheel Gets the Oil

"As a field, mycologists work well together," Taylor says, a statement echoed by many others. This collegial atmosphere was a definite plus in securing funds for what some refer to as mycology's Manhattan Project. According to Taylor, a core filamental fungi group began coalescing only within the last decade, springing out of the *Neurospora* Information Conference, a small biannual meeting that Taylor admits was something of a scientific backwater.

"When I first went to the meeting there were fewer than 100 people," he says. "Still, it was bigger than the previous year's and there was a great deal of excitement." In addition to *Neurospora* researchers, the meeting managed to attract scientists working on other filamental fungi, and things really took off when a charismatic leader, Bill Timberlake, figured out how to introduce genetic material into *Aspergillus*. As more scientists began attending, the conference morphed into the Fungal Genetics Conference. It still convenes every two years at Asilomar, in Pacific Grove, California, and has grown so popular that people have to be turned away.

Another important reason for greater interest in mycology was the release of the *S. cerevisiae* genome sequence in 1996. Combine all the filamental scientists in the world, and you still have fewer people than the number who study this single species of yeast. After the sequence was published, some yeast scientists probably began looking for ways to validate their findings in other organisms, or maybe they saw something more fundamental about their field.

"Suddenly you've got about 10,000 people chasing only 6000 genes," says Rytas Vilgalys, a mycologist at Duke University. "That's been good for mycology, because the yeast people are smart and organized and they're looking for something to do."

In a series of meetings and symposia, various members of the fungal genomics community began working together on a comprehensive initiative for fungal genomics. They chose an initial 15 species with importance in biomedicine, agriculture, industry, eukaryotic biology, and evolution. The Whitehead Institute for Biomedical Research, which has already sequenced *Magnaporthe grisea* and *N. crassa,* was involved from the beginning, and when NHGRI released some funds, seven species were chosen for the initial sequencing. Of course, not everyone got his or her pick.

Researchers complain that the initial seven are heavily weighted toward human pathogens, but because NHGRI is under the National Institutes of Health, this comes as no surprise. They also point out that fungi have become fatal pathogens only in recent years. In the first part of the 20th century, the most important fungal pathogens were both nonfatal: ringworm and thrush. The change from morbidity to mortality is a recent consequence of therapies for patients with AIDS, transplants, or cancer. Although drugs allow people to survive longer with these conditions, fungal infections can quickly undo those gains.

Less than 6 percent of the complete microbial genomes are plant pathogens, and James Sweigard, a senior assistant biologist at DuPont, sees cause for criticism. "The trouble with plant pathogens is that they kill more people than human pathogens, but they do not do this in a very dramatic way," he says. "Famine isn't a problem in the developed world."

Others feel that the heavy focus on human pathogens leaves out species with less obvious importance, such as chytrids, which have been discovered recently as pathogens of amphibians. Joyce Longcore of the University of Maine says that though chytrids are a little-studied group, they have great importance when it comes to phylogeny. "I do wish that a chytrid that was more at the base of the fungal clade had made the list."

Still, despite the moderate criticism of the list, every researcher acknowledges that scientific compromises made to secure funding will matter little over time. "Maybe I didn't get what I wanted, but any sequence that we get is going to inform my research and the research of all of us in this field," Sweigard says.

How to Sequence

Genetically, fungi have slightly larger and more complicated genomes than bacteria. Like us, they have introns, but they are only about 1/300th the size of introns in *Homo sapiens.* To tackle the sequencing, the Whitehead Institute plans to do a deep-shotgun sequence. Specifically, they intend to produce a fine-resolution map by sequencing fragments at "10× coverage," which involves randomly shearing the DNA of multiple clones and then putting the pieces into 4-kilobase (kb) plasmids (90 percent) and much larger 40 kb fosmids (10 percent). For 10× coverage, 10 times the amount of the complete genome is sequenced in a redundant manner. The resulting pieces are strung together by matching overlapping sequences, like copies of the same puzzle jigsawed different ways, producing a high-quality map of the genome.

A deep-shotgun approach leaves some gaps, but the process is quick and cheap. Moreover, the resulting sequence and physical map (that is, where sequences are found on the chromosome) are useful for just about every purpose. A much lower-resolution 3× map of *A. nidulans* proved extremely valuable for years to Monsanto, and since it was published on a Web site only a year ago, the site has received almost 3 million hits.

If a researcher later runs into a gap and needs to fill it in, copies of the various vectors will be available at the Fungal Genetics Stock Center. The FGSC has been around for almost 30 years and maintains a large number of fungal clones, including over 8000 strains of *Neurospora.* "At a 10× coverage, you get several thousand gaps," says Ralph Dean, director of the Fungal Genomics Lab at North Carolina State University. "In 90 percent of these cases, we already have the DNA; we just have to sequence it. But in 10 percent of the cases, we don't have the DNA, so you really end up with a few hundred real gaps."

This filling in of gaps, or closure, can be as time consuming and difficult as the actual sequencing. You can reach a point of diminishing returns, where getting every base pair correct is not very cost-effective. There is also the issue of annotating the genome, that is, locating the individual genes in the sequence. Both closure and annotation can be done piecemeal or systematically in one quick shot. Dean hopes it will be the latter, but that depends on the availability of resources. "What we're hoping for is that we can prove to funding agencies that this is just as important as the sequencing and that funding will be made available," he says.

A Fungus among Us

As lower eukaryotes, fungi are more closely related to humans than are other microorganisms, such as bacteria and viruses. With what little fungal sequence is available, homologues for 30 percent of human proteins can be found, almost twice what is known from *S. cerevisiae* alone. The fungus *A. nidulans* has been the source of much of our knowledge about the genetics of tubulin and microtubules and is also an important model for studying mitosis. Oddly enough, it is this very resemblance to humans that makes fungi a particularly nasty human pathogen, because most therapies designed to kill fungi also harm humans.

"We're building a drug company, and one of our tasks is discovering antifungals," says Peter Hecht, president of Microbia, a three-year-old company." To do this you have to find targets of intervention where you can cripple the pathogen. This can be quite difficult, but what you'd like to find are targets critical to the pathogen, but which are not found in humans."

Also, fungal metabolites have been harnessed for other important pharmaceuticals. The statins were first discovered during antifungal screens in Japan. In high doses these drugs are quite toxic. However, at lower doses they are sold as cholesterol-lowering drugs, a $12 billion worldwide market. Because of the big payoff, pharmaceutical companies have turned all sorts of fungi into minibiofactories, manufacturing statins, antibiotics, fungicides, and other drugs. "At Microbia, we're not just out to find new or safer pharmaceuticals," Hecht says. "We want to manipulate the regulatory controls inside fungal cells to help manufacturers goose up their microfactories."

When Fungi Attack

Although fungi are not important pathogens in humans, they make up 80 to 90 percent of the disease microbes preying on plants. Rice blast caused by *Magnaporthe grisea* leads to losses of crops capable of feeding 60 million people annually. And like another crop pathogen, *Ustilago maydis* (corn smut), it is also an important model for studying host-pathogen interactions.

The genome for *M. grisea* is now complete, as is the genome for rice, and Ralph Dean sees only vast possibilities. "We've now got two genomes, a host and a pathogen," he says. "So now we're going to be studying host and pathogen interactions not just at the molecular level, but also genetically. This really hasn't been done before—understanding how two species interact at the genomic level." He also points out that as other fungal species become available, they will start cross-referencing the *M. grisea* pathways with other fungal pathogens.

"These genomes will certainly impact us," Sweigard says. DuPont has now moved away from the pathogen side to the host side of disease, attempting to modify species such as corn to make them more resistant to pathogens. "We do directive genetics, asking, 'Does the pathogen need this gene to be a pathogen?' If it does need this gene, then we counter the metabolic process by modifying the plant."

"We're interested in why certain fungi are pathogens and how their genomes differ from others, such as saprophytes," says Gillian Turgeon, who collaborates with Olen Yoder at Syngenta. "The ultimate goal is to control the pathogen in the field, and with the species' blueprint, it just shifts the kind of questions you can ask."

During a mutagenesis screen several years ago, Turgeon and colleagues discovered a mutant with reduced pathogenicity. After sequencing the mutated gene, they realized they had discovered a novel gene, and by searching their proprietary genome databases, they discovered the gene had been highly conserved across many different pathogen species.

"There are general pathogen factors, and then there are specific factors for each host," Turgeon explains. "General factors are more interesting, because this allows you to control more fungi." After mutagenizing the gene in other pathogens, she and her colleagues found it reduced fungal disease by around 60 percent in many different plant species. And the gene can also be found in human pathogens. "What we'd like to do, and we have several ongoing collaborations, is to test the gene's virulence in some of the human pathogens. We do not have the data yet, but we are hot on the trail."

The Ongoing Challenge

With every technological advance, from light microscopy to electron microscopy to genetics, the systematics of fungi gets reinvented. Systematic groupings once thought solid are found to be artificial, and species that were thought to be close are discovered to be distant relatives. "They keep changing the names all the time!" says one medical researcher. "I guess this keeps people busy and employed."

Rytas Vilgalys says the confusion is due to the peculiar nature of fungi. "Morphology is misleading, and for fungi we don't really have a fossil record," he says. "Genetics is really the only way to understand these evolutionary relationships."

Take gilled mushrooms, for example. For decades, species with gills were grouped together, even though gills disappear and reappear randomly throughout evolution. There are many examples in which a gilled mushroom made the transition to a puffball-like derivative, and some gilled species can produce both gills and pores. Trying to understand how organisms fit into groups and how these groups are related becomes an act of futility. "It's not [like the situation] in other groups, like birds," Vilgalys says. "There's no way to resolve morphology with phylogeny. You'd go nuts!"

Ecologically, it also becomes useless to understand which morphology leads to greater fitness. "It may not even be useful to think which came first, gills or pores, because species might go back and forth," Vilgalys adds. "There may be some directionality, but some recent rigorous analysis with evolutionary models of how often species have gone from gills to pores or gills to puffballs has found that there is no discernable pattern."

"What genetics did is to take some of the opinion out of systematics," John Taylor comments. "You could go to a meeting of taxonomists and someone would say 'I think this is related to this because I say so.' End of story. It just devolved into 'I've studied these things for 50 years and know more about them than you do.' Now people have to put their data into a matrix and see how it looks."

He does point out, however, that molecular markers are not any better than other characters; it's just that the huge quantity of them evens out opinion and allows statistical analysis. And with genomics, you just get a huge number of data points.

"I'm not really sure what we'll find," he says. "It might be that things are just more complicated than we expected."

Comparisons of fungal divergence from Archaea and bacteria, for instance, were thought to be simple when the only character chosen was a single RNA subunit. "When people started looking at other genes, it became fuzzier. Now the branching order depends on which gene you pick. The hope is that by choosing more genes you get better resolution, but the reality is that you sometimes get confusion."

Magic Mushrooms

What will the future bring for mycology? Taylor is not quite sure. "I'll be perfectly honest," he says, "I really don't know, but I think it's going be more than we can imagine."

Like Ralph Dean, Peter Hecht sees only vast possibilities. At Microbia researchers are examining genomes with the goal of making fungi the engineers of complex biosynthetic pathways leading directly to finished drug products. This saves all the time and energy of modifying a metabolite through traditional chemistry. As Hecht sees it, chemistry becomes irrelevant in the future as pharmaceutical companies move to direct bioprocessing. "The really fun part starts when the information is made available," he says.

And for Yoder, genomic information is the chance at a life-long dream. "Maybe I wasn't visionary enough, but back in grad school I never imagined we would have this sort of information available," he says. "I remember a group of us sitting around and having these late-night discussions about what caused pathogenicity. Was it one gene? Was it many genes? And if so, how many? We didn't have a clue, because there weren't any data. It was just talk. Now I'm going to have a chance at possibly figuring this out, or at least a small part of it."

He pauses for a moment. "It's not philosophy anymore. This is real science."

PAUL D. THACKER (e-mail: pdthacker@yahoo.com) is a freelance science writer based in Jersey City, New Jersey.

Going with His Gut Bacteria

The body and its intestinal flora produce chemicals with hidden health information, Jeremy Nicholson has found. Someday treating disease may mean treating those bacteria.

Melinda Wenner

Jeremy Nicholson was only trying to be thorough. It was 1981, and the young biochemist was using a technique called nuclear magnetic resonance spectroscopy, which can identify chemicals based on the magnetic properties of atomic nuclei. In particular, Nicholson wanted to study how red blood cells absorb cadmium, a metal that causes cancer. Realizing that he would achieve the best results if he could mimic the cells' natural environment, he added a few drops of blood to the cells and ran the test.

"Suddenly there was a huge variety of signals that we hadn't seen before—there were these amazing sets of spectra coming out," Nicholson recalls. A sample of blood or urine contains thousands of metabolites—signatures of all the chemical reactions occurring in the body at a given time. If he could find a way to identify those chemical signatures and their significance, he reasoned, he would be able not only to better understand different diseases—based on chemical reactions that had gone awry—but also to identify early warning signs and potential interventions. That kind of science, he decided, was his kind of science.

Today the 51-year-old Nicholson is one of the world's foremost experts on the so-called metabolome, the collection of chemicals produced by human metabolism. Whereas the genome provides detailed information about a person's genetic makeup, the metabolome is a few steps down the line—it reveals how genes interact with the environment, providing a complete snapshot of a person's physical health. "The genome is really like a telephone directory without any of the names or addresses filled in. On a very basic level, it's got a lot of numbers," explains Nicholson, who now heads the department of biomolecular medicine at Imperial College London. The metabolome "helps to give value to genome information and put it in perspective."

But first it has to be deciphered, and that is no easy task. The job requires the analysis of blood, urine, breath and feces within large populations. For instance, to find potential chemical signatures, or biomarkers, for high blood pressure, Nicholson and his colleagues analyzed the urine of 4,630 individuals from the U.K., the U.S. and Asia and compared the urinary metabolites with blood pressure data to determine if any consistent

Jeremy Nicholson

Bacterial clues: By analyzing the products of intestinal bacteria, Nicholson hopes to fashion new tools for diagnosis and new targets for drugs.

Population boom: The human gut contains some 10 trillion individual bacteria in 1,000 different species.

Father of disciplines: Nicholson's work has spawned two new fields: metabolomics, which studies the metabolites that cellular processes leave behind, and metabonomics, which characterizes the metabolic changes a biological system experiences in response to stressors.

Growing on him: On first noticing metabolic fingerprints that cells leave behind: "I was thinking of them as extremely annoying interferences with mammalian biochemistry. Now I'm almost becoming evangelical about the bloody things."

metabolic differences exist between individuals with hypertension and those without it.

It is kind of like doing science backward: instead of making hypotheses and then devising experiments to test them, he performs experiments first and tries to decipher his results later. He must sift through the range of chemicals produced by the genes people have, the food they eat, the drugs they take, the diseases they suffer from and the intestinal bacteria they harbor.

Those bacteria in particular have become Nicholson's prime focus. They influence how our bodies break down food and drugs and may explain why food affects people differently. For instance, some people cannot derive benefit from one of soy's components because they lack the gut microbes necessary to process it. Although deciphering which metabolites come directly from our gut microbes can be difficult, in some cases it is easy—they are the chemicals that are not produced by cells or ingested in food.

Nicholson focuses on these chemicals both because little is known about them and because they appear to be highly relevant: recent research suggests that gut microbes play a crucial role in human health and disease. They help us absorb nutrients and fight off viruses and "bad" bacteria; disrupting intestinal colonies, such as with a course of antibiotics, often leads to digestive sickness. In fact, Nicholson says, "almost every sort of disease has a gut bug connection somewhere."

Perhaps the most well-known disease-causing gut organism is the bacterium *Helicobacter pylori,* which can trigger peptic ulcer. In the past few years, scientists have linked obesity to the relative abundance of two dominant intestinal bacterial phyla and found that dysfunctional intestinal bacteria are associated with nonalcoholic fatty liver disease, inflammatory bowel disease and some types of cancer. Nicholson even speculates that the organisms could play a role in neurological disorders, such as attention-deficit hyperactivity disorder, Tourette's syndrome and autism. "We have some evidence now that shows that if you mess around with the gut microbes, you mess around with brain chemistry in major ways," Nicholson remarks. He currently collaborates with microbiologists to match metabolites with specific bacteria—there are thought to be 1,000 species and more than 10 trillion bacterial cells inside us at any given time.

This identification process has only recently become possible. Although scientists have been able to extract gut bacteria from fecal samples for many years, it has been next to impossible to culture the samples afterward because they survive only in highly acidic, oxygen-free environments. Thanks to new DNA-sequencing technologies, scientists can now identify gut bacteria fairly easily, and there is growing interest in doing so: the National Institutes of Health launched its Human Microbiome Project last December with the goal of fully characterizing the human gut flora.

Once investigators can correlate metabolites with health, it may one day be possible, Nicholson says, to make urine sticks similar to those used in pregnancy tests to regularly check the fitness of our gut flora. Some companies have already begun selling food products to help keep these populations in line—with live beneficial bacteria (probiotics) or compounds that help these species grow (prebiotics), or combinations of the two (synbiotics). Unfortunately, these medications typically fall into the category of "functional foods," which means they are rarely tested in clinical trials. One exception is VSL #3, a combination of eight bacterial species sold in packet form by the Gaithersburg, Md.–based VSL Pharmaceuticals. In double-blind, placebo-controlled trials, the colonies effectively treated ulcerative colitis and irritable bowel syndrome.

Many possibilities exist for bug-based drugs, and there is a strong need for them, Nicholson maintains. According to a study published by scientists at the pharmaceutical giant Pfizer, the human genome offers only about 3,000 potential drug targets, because just a subset of genes produces proteins that can be bound and modified by drug-like molecules. But "there are 100 times as many genes in the microbial pool," says Nicholson, who regularly works with drug companies to better elucidate how people metabolize medicines. He is "one of a few academics I've met who's interested in the pharmaceutical industry for its problems rather than just for its cash," comments Ian Wilson, a scientist working in England for the pharmaceutical company AstraZeneca. Wilson adds that Nicholson is always full of potential solutions, referring to him as "a bubbling mass of ideas."

Because genes provide only limited information about a person's risk for disease, Nicholson dreams of a time when physicians can provide personalized health care on the metabolome. Simple blood or urine tests would detect the risk of cancer or heart disease early enough to begin preventive therapy; drugs would be tailored to each person's metabolic profile—and in many cases, they would not target our organs but our bacteria. "It opens up visions of a future that we would never have suspected even a few years ago," Nicholson says. "Many microbiologists might argue this is fanciful, but you only make huge progress in science by thinking almost the unthinkable."

MELINDA WENNER is a freelance science writer based in New York City.

New Tactics against Tuberculosis

The pandemic is growing in many places, and strains resistant to all existing drugs are emerging. To fight back, biologists are applying a host of cutting-edge drug development strategies.

CLIFTON E. BARRY III AND MAIJA S. CHEUNG

Bubonic plague, smallpox, polio, HIV—the timeline of history is punctuated with diseases that have shaped the social atmospheres of the eras, defined the scope of science and medicine, and stolen many great minds before their time. But there is one disease that seems to have stalked humanity far longer than any other: tuberculosis. Fossil evidence indicates that TB has haunted humans for more than half a million years. No one is exempt. It affects rich and poor, young and old, risk takers and the abstinent. Simply by coughing, spitting or even talking, an infected individual can spread the bacterium that causes the disease.

Today TB ranks second only to HIV among infectious killers worldwide, claiming nearly two million lives annually, even though existing drugs can actually cure most cases of the disease. The problem is that many people lack access to the medicines, and those who can obtain the drugs often fail to complete the lengthy treatment regimen.

Additionally, TB is evolving faster than our therapies are. In recent years, investigators have observed a worrying rise in the number of cases resistant to more than one of the first-line drugs used to treat the illness. Even more alarming, we have begun to see the emergence of strains that are resistant to every last one of the antibiotic defenses.

The disease is particularly devastating for the developing nations, where some 90 percent of cases and 98 percent of TB deaths occur. Beyond bringing untold suffering and sorrow there, TB harms entire economies. With 75 percent of cases arising in people between the ages of 15 and 54, TB will rob the world's poorest countries of an estimated $1 trillion to $3 trillion over the next 10 years. Furthermore, the disease forces these struggling nations to divert precious resources from other important areas into health care. But the developed world would be mistaken to consider itself safe: although the incidence there is comparatively low, that situation could change if a highly resistant strain were to gain traction.

As bleak as this state of affairs is, we have reason to be hopeful. Cutting-edge biomolecular technologies are enabling researchers to study the complex interactions between the TB

Key Concepts

- Tuberculosis is second only to HIV as the worldwide cause of death from infection, and the pandemic is growing in many places.
- TB is caused by a bacterium. Most cases are treatable, but strains resistant to first- and second-line drugs are on the rise.
- Conventional approaches to developing new antibiotics and vaccines against the disease have mostly failed.
- New tools are enabling scientists to study the TB-causing bacterium in greater detail, offering unprecedented insight into the interactions between pathogen and host. The results are exposing promising new targets for drug therapy.

—The Editors

bacterium and the body in unprecedented detail, generating insights that are informing the development of novel diagnostic tests and drug therapies.

A Short-Lived Success

First identified by German physician Robert Koch in 1882, *Mycobacterium tuberculosis (Mtb)*, the rod-shaped bacterium that causes tuberculosis, exists in both latent and active forms. In a latent infection, the immune system prevents the bacteria from multiplying, thus keeping them from disrupting tissues. Individuals with this form show no symptoms and are not contagious. Latent *Mtb* may persist for months, years or even decades without multiplying or making its host ill. Ninety percent of people infected with *Mtb* never develop active TB disease. But 10 percent of them do develop the active form, particularly those with weakened immune systems, such as young children and individuals who have HIV or are undergoing chemotherapy.

In people with active TB, the bacteria outpace the immune system, rapidly multiplying and spreading out to attack the organs. Primarily an aerobic bacterium, meaning it prefers environments rich in oxygen, *Mtb* has a special affinity for the lungs. Indeed, some 75 percent of patients with active TB exhibit the pulmonary variety of the disease. As the bacteria multiply, they destroy the lung tissue, commonly causing the host to develop such symptoms as a severe cough, chest pain and the coughing up of blood. But other organs are vulnerable, too. In fact, active TB can affect nearly every organ in the body. In children, TB can invade the cerebrospinal column, where it provokes a high fever with systemic shock—a condition known as meningitis. Left untreated, half of people with active TB die of it, most from lung destruction.

A century ago society had no way to combat TB, save for limiting its spread by sequestering affected individuals in sanatoriums. Back then TB, often called "consumption," was widespread even in places that today have a relatively low incidence of the scourge, such as North America and western Europe. Scientists began to gain on the disease in 1921, when a vaccine made by French immunologists Albert Calmette and Camille Guérin, both at the Pasteur Institute in Paris, first entered into public use. (Initially believed to protect against both adult and childhood forms of the disease, the BCG vaccine, as it is known, was later shown through an extensive series of tests to confer consistent protection against only severe childhood forms.)

Twenty-two years later a team led by American microbiologist Selman Waksman developed streptomycin, which despite causing some side effects was the first effective therapy for TB. Waksman's achievement opened the door for the creation in the 1950s of a rapid succession of antibiotics that compensated for streptomycin's weaknesses.

Together these developments brought the era of sanatoriums to a close and significantly lowered the incidence of TB in countries that had the money and infrastructure to tackle the problem. By the 1970s many experts believed that TB had been almost completely eradicated. In reality, however, with international travel on the rise, the largest epidemics were just beginning. To make matters worse, those who would be hit hardest were those who could least afford it: residents of the poorest nations, who would soon also be facing a new and costly killer—HIV.

Today more than half a century after the debut of the first anti-TB drugs, the World Health Organization estimates that fully a third of the world's population (more than two billion people) is infected with *Mtb*. On average, eight million of these carriers a year will develop active TB, and each will infect between 10 and 15 more individuals annually, maintaining the pandemic.

The picture becomes even more frightening when one considers the rising incidence of HIV. People who have latent TB and are HIV-positive are 30 to 50 times more likely than their HIV-negative counterparts to develop active TB, because HIV leaves their immune systems unable to keep TB in check. In fact, TB is the leading cause of death among HIV-positive individuals, claiming the lives of one out of every three worldwide and one out of every two in sub-Saharan Africa, where health care is especially hard to come by. Even if HIV-positive individuals have access to anti-TB drugs, their health will likely deteriorate

Human vs. Microbe

Tuberculosis has dogged humankind for millennia. Here are some key events in this long-standing battle between people and pathogen.

500,000 Years Ago

By that time, TB had begun infecting human ancestors.

1882

Robert Koch identifies *Mycobacterium tuberculosis (Mtb)* as the cause of TB.

1908

Albert Calmette and Camille Guérin develop the BCG vaccine against TB, which is later shown to consistently protect against only severe childhood forms of the disease.

1921

BCG vaccine enters public use.

1943

A team led by Selman Waksman creates the first effective antibiotic against TB: streptomycin.

1960s

Current TB treatment course, which employs four different drugs over a six- to nine-month period, is developed.

1970s

TB is widely believed to have been nearly eradicated.

1981

Scientists identify HIV, which leaves people additionally vulnerable to TB.

1998

TB genome is sequenced.

2005

Improved diagnostic test receives U.S. Food and Drug Administration approval.

2006

Outbreak of extensively drug-resistant TB in KwaZulu-Natal, South Africa.

because dangerous interactions between antiretroviral therapy and first-line TB drugs often force patients to suspend their antiretroviral therapy until the TB is under control.

The Latest Challenge

Perhaps the most disquieting aspect of the present pandemic, however, is the growing problem of the TB bacterium's resistance to antibiotics. To understand how this predicament came

Famous Victims

Tuberculosis has claimed the lives of many luminaries. Among them are:

- All three Brontë sisters
- Anton Chekhov
- Frederic Chopin
- John Keats
- Louis XIII of France
- Molière
- George Orwell
- Cardinal Richelieu
- Jean-Jacques Rousseau
- Erwin Schrödinger
- Henry David Thoreau

Sobering Facts

- A third of the world's population is infected with the TB bacterium; **one in 10** of them will become sick with active TB in their lifetime.
- On average, nearly **four in 10** TB cases are not being correctly detected and treated.
- TB is responsible for a death **every 20 seconds.**
- An estimated **490,000** new cases of TB resistant to first-line drugs and 40,000 cases of TB resistant to second-line drugs occur every year.

to be, consider how TB is treated. The current treatment course, which was developed in the 1960s, is a demanding regimen consisting of four first-line drugs created in the 1950s and 1960s: isoniazid, ethambutol, pyrazinamide and rifampin. Patients who follow the regimen as directed take an average of 130 doses of the drugs, ideally under direct observation by a health care worker. This combination is extremely effective against active, drug-susceptible TB as long as patients are compliant and complete the entire six- to nine-month course.

Drug-resistant strains develop when patients do not complete the full protocol, whether because they start feeling better or because their drug supply is interrupted for some reason. Inconsistent use of antibiotics gives the bacteria time to evolve into a drug-resistant form. Once a drug-resistant strain has developed in one person, that individual can spread the resistant version to others. (For this reason, some authorities argue that it is better to not undergo treatment than to undergo incomplete treatment.)

According to the World Health Organization, nearly 5 percent of the roughly eight million new TB cases that occur every year involve strains of *Mtb* that are resistant to the two most commonly used drugs in the current first-line regimen: isoniazid and rifampin. Most cases of this so-called multidrug-resistant TB (MDR-TB) are treatable, but they require therapy for up to two years with second-line anti-TB drugs that produce severe side effects. Moreover, MDR-TB treatment can cost up to 1,400 times more than regular treatment. Given that most MDR-TB occurs in impoverished countries, this expensive treatment is often not an option. Failure to properly diagnose MDR-TB, along with the high cost of treatment, means that only an estimated 2 percent of MDR-TB cases worldwide are being treated appropriately.

Worst of all, over the past few years health surveys have revealed an even more ominous threat, that of extensively drug-resistant TB (XDR-TB). This type, which made headlines in 2006 following an outbreak in KwaZulu-Natal, South Africa, is resistant to virtually all the highly effective drugs used in second-line therapy. Although XDR-TB is less common than MDR-TB, the possibility that XDR-TB will evolve and spread looms wherever second-line TB drugs are in use. World Health Organization records indicate that 49 countries had confirmed cases as of June 2008. That is a minimum figure, though, because very few countries have laboratories equipped to diagnose XDR-TB.

A Trickling Drug Pipeline

To say that scientists erred in assuming that the first-line drugs from the 1950s would be sufficient to combat TB is a profound understatement. But with the overwhelming majority of TB patients concentrated in some of the world's poorest countries, large pharmaceutical companies have had little incentive since then to invest heavily in research and development for new drugs. And the prevailing wisdom among the greater pharmaceutical conglomerates is still that the cost of drug development— $115 million to $240 million and seven to 10 years per drug— far outweighs the potential global market for such products.

Thanks to government programs and private philanthropic organizations such as the Bill and Melinda Gates Foundation, however, many efforts are under way to create TB antibiotics to both treat drug-resistant cases and reduce the time that it takes to treat normal TB cases.

As a result, a few promising agents are currently in early clinical trials. One such agent, known as SQ109, inhibits cell wall synthesis. It recently completed phase I (safety) clinical trials. Another drug candidate is PA-824, a compound whose ability to attack *Mtb* in both its actively dividing stage and its slow-growing one has generated hopes that the drug could significantly reduce the time needed to treat the disease. PA-824 is in phase II clinical trials, which look at efficacy.

Unfortunately, the odds are against these candidates: historically, fewer than 10 percent of antibiotics that enter early clinical trials garner approval—a success rate that derives in large part from the outmoded logic used to discover these drugs. Fifteen years ago developing new antibiotics was mostly a matter of following a simple formula: identify enzymes that are essential to bacteria survival and that do not have any counterparts in humans; screen libraries of compounds for potent inhibitors of these enzymes; chemically synthesize derivatives of those inhibitors; then optimize the compounds for drug-like properties, such as the ability to get from the stomach to the bloodstream. Yet even the large pharmaceutical companies, masters

Infection Basics
An Ill Wind

Tuberculosis, caused by the bacterium *Mtb,* occurs in both latent and active forms. People can become infected by breathing in even just a few *Mtb* bacteria released into the air when those with active TB cough, spit or talk. *Mtb* causes coughing, the most familiar symptom, because it accumulates abundantly in the lungs, but it can harm other organs as well.

- *Mtb* tends to concentrate in the air sacs, or alveoli, of the lungs because it prefers environments rich in oxygen. In most people, the immune system is able to keep bacterial replication in check, dispatching defensive cells known as macrophages to the site of infection, where they form a shell around the bacteria. But in 10 percent of infected individuals, *Mtb* breaks down the shell, after which it can begin to multiply.
- When unfettered by the immune system, the bacteria destroy the tissue of the lungs; some may also make their way into the bloodstream and infect other parts of the body, including the brain, kidneys and bone. Eventually affected organs may sustain so much damage they cease to function, and the host dies.

Affected Nations
Worldwide Resistance

Tuberculosis occurs in virtually every country in the world, although it is most widespread in developing nations. The incidence of TB caused by strains of *Mtb* resistant to two or more of the first-line drugs for the disease—so-called multidrug-resistant TB (MDR-TB)—has been rising as a result of improper use of antibiotics. Worse still is extensively drug-resistant TB (XDR-TB)—a largely untreatable form identified in 2006; as of June 2008, 49 countries had confirmed cases. Sadly, that figure most likely underestimates XDR-TB's prevalence.

of developing medicines to treat nearly any disease, have been spectacularly unsuccessful in producing new antibiotics using this approach.

For its part, the TB battleground is littered with the corpses of drug candidates that failed. Many of these compounds were highly specific and potent inhibitors of key TB enzymes. In some cases, although they effectively foiled isolated enzymes, they flopped when tested on whole bacterial cells. In others, the compounds thwarted whole bacteria in test tubes (in vitro) but missed their mark when tested in infected animals. TB offers perhaps the most extreme example of the troubling disconnect between the in vitro and in vivo effects of antibiotics. Most of the time investigators have absolutely no idea why drug candidates fail. The crux of the problem is that bacteria are autonomous life-forms, selected throughout evolution for their ability to adapt and respond to external threats. Like modern aircraft, they have all manner of redundancies, bypasses, fail-safes and emergency backup systems. As Jeff Goldblum's character in *Jurassic Park* puts it, life finds a way. Until we truly appreciate the complexities of how TB interacts with humans, new drugs against it will remain elusive. The good news is that we are making progress on that front.

Insights from "Omics"

A key turning point in our TB education came in 1998 with the sequencing of the DNA code "letters" in the *Mtb* genome—a project in which one of us (Barry) participated. That sequence, and those of related organisms, has yielded a trove of insights. Perhaps most importantly, the results showed that of all the

enzymes and chemical reactions that are required for TB to survive in a human, we were considering only a third of them in our in vitro (test tube) tests. We learned, for instance, that *Mtb* devotes a huge amount of its genome to coding for proteins that synthesize and degrade lipids, suggesting that some of those proteins might be worth considering as drug targets. Analysis of the TB genome also hinted that, contrary to conventional wisdom, the bacterium is perfectly capable of living in the absence of air—a suggestion now verified. Under such anaerobic conditions, *Mtb*'s metabolism slows down, making it intrinsically less sensitive to existing antibiotics. Targeting the metabolic elements that remain active under these circumstances is one of the most promising strategies for shortening treatment time.

Translating the information we have gleaned from the genome into discoveries that can help save the lives of people who contract TB has neither been simple nor straightforward. But recently researchers have used those data to make significant advances in diagnostic tests for the disease. Diagnosis can be complicated by the effects of the childhood vaccine, which is given to more than half of all infants born around the world. The vaccine contains a strain of *Mtb* that has lost its virulence yet is still able to induce a child's immune system to react against the TB bacterium. Vexingly, though, the predominant test for TB cannot distinguish between immune responses elicited by virulent *Mtb* and the vaccine form. Hence, the test results for someone who is infected look exactly like the results for someone who has been vaccinated.

While the *Mtb* genome was undergoing sequencing, scientists in Seattle discovered that a large stretch of DNA was missing from the bacterial strain used in the vaccine. Shortly thereafter, independent research teams at the Pasteur Institute, the Albert Einstein College of Medicine and the University of Washington showed that the missing genes were essential to virulence. The deleted region in the vaccine strain thus offered investigators a strategy for improving the specificity of the test. A test that searched only for an immune response directed against the virulence factors absent from the vaccine strain, the researchers reasoned, should be able to distinguish infected individuals from those who had been vaccinated. In fact, just such a test was developed and approved by the U.S. Food and Drug Administration in 2005, and many recent studies have confirmed its accuracy. Unfortunately, so far the cost of the test is high, which restricts its use to the First World.

A Way Forward
Promising Treatment Strategies

The first-line drugs currently used to treat tuberculosis were developed in the 1950s and 1960s. The six-to nine-month regimen is demanding, and failure to comply with it completely has led to the emergence of resistant forms of TB. Developing agents that are easier to administer and cheaper and that hit the *Mtb* bacterium in new ways is critical.

Today

Traditional trial-and-error approaches to identifying drugs against TB have produced some new candidates now in clinical trials.

Future

More recently, scientists have begun trying to understand *Mtb* in far greater detail by studying its genome and other cell components—work that is yielding fresh insights into how the bacterium establishes infection in humans and what its vulnerabilities are. Researchers should be able to inhibit synthesis of ATP much more effectively than drugs now in development can. Likewise, they can probably find compounds that stimulate the bacterium to release even more nitric oxide, which shuts down the cell's ability to breathe, than existing agents do. Blocking synthesis of niacin, the cell's main energy transporter, is another tactic with considerable potential.

Far Future

Ultimately, investigators want to create an in silico model of *Mtb*—a computer model that behaves exactly like its real counterpart does in a human. Such a model would enable researchers to predict the organism's responses to various compounds with far greater precision than is currently feasible.

Drug/Drug Class	How It Works against TB	Stage of Development
Fluoroquinolones (already approved for treating other conditions)	Inhibits DNA replication	Phase III trials (largest tests of efficacy)
Nitroimidazoles (PA-824/OPC67683)	Inhibits cell wall synthesis and cell respiration	Phase II (efficacy)
Diarylquinoline (TMC207)	Inhibits synthesis of the energy storage molecule ATP	Phase II
Oxazolidinones	Inhibits protein synthesis	Phase II
SQ109	Inhibits cell wall synthesis	Phase I (safety)

The *Mtb* genome is not the only new source of data able to provide insight into the TB bacterium's potential vulnerabilities. Scientists can now study all kinds of cell components and processes—from all the proteins in a cell (a discipline known as proteomics) to the amount of messenger RNA (the templates from which proteins are made) made from every gene ("transcriptomics") to the intermediate and final products of cell metabolism ("metabolomics"). These fields are still in their infancy, but already they have borne fruit. Last November, Barry co-authored a paper in *Science* reporting that when TB was treated with PA-824, the bacterial transcriptome reacted exactly as if it had just been poisoned with potassium cyanide. This finding was a vital clue that in metabolizing the drug, *Mtb* releases nitric oxide, a defensive molecule normally made by immune cells in the human body. Armed with this knowledge, we and others are now synthesizing compounds that stimulate the release of larger amounts of nitric oxide than are elicited by PA-824 and so should be even more potent against *Mtb*.

Complementing those approaches, structural genomics seeks to uncover the three-dimensional structure of every protein in *Mtb*—work that can both help identify the still-mysterious functions of many *Mtb* proteins and aid the design and synthesis of drugs targeting particular sites on critical proteins. So promising is this line of inquiry that a global consortium with members from 17 countries is focusing its efforts entirely on the structural genomics of *Mtb*. Thus far the consortium has helped determine the structure of about 10 percent of the organism's proteins.

Another "omics" branch worth noting is chemical genomics, a very recently established field of research that effectively reverses the standard process of drug discovery. Instead of starting with a protein of known function and looking for a compound that inhibits its activity, investigators begin with a compound known to have a desirable trait—such as an ability to inhibit *Mtb* reproduction in cell cultures—and work backward to identify the microbial enzyme impaired by the substance. The compounds can be anything from molecules synthesized in a chemistry lab to products isolated from plants, microbes and even animals. The starting chemical in this case serves strictly to reveal vulnerable enzymes or biological processes, which scientists may then identify as targets for drug development.

What makes this approach so appealing is that it allows us to harness the power of natural selection in our quest to thwart *Mtb*. Before *Mtb* and other mycobacteria found humans to be such appealing hosts, they occupied environmental niches where they had to compete with countless other bacteria for food in a constant arms race. Bacterial ecosystems have therefore undergone multiple rounds of natural selection, and in most cases other bacteria have evolved ways of keeping the mycobacteria in check, as is evident from the diversity of bacteria types in these ecosystems. If researchers could tap into the amazing reservoir of weapons that these competitor bacteria have evolved—applying modern omics tools to identify the defensive molecules, screen them for their anti-TB potential and pinpoint their molecular targets in *Mtb*—we could well uncover entirely new classes of drugs. We could then select those agents that knock out the pathogen's whole system, as opposed to just a single process for which *Mtb* likely has a workaround.

A Model Bacterium

To reap the full benefits of the omics revolution, we need information technology tools capable of making sense of the vast data sets generated by omics experiments. In fact, the development of such tools has become a discipline unto itself, called bioinformatics. And only with these tools can researchers hope to clear another obstacle to drug development: that posed by so-called emergent properties—behaviors of biological systems that cannot be predicted from the basic biochemical properties of their components.

To borrow an example from neuroscience, consciousness is believed to be an emergent property of brain biochemistry. In the case of in vitro *Mtb*, one emergent property is a tendency of the bacteria to form "cords"—serpentine arrays that have a ropelike appearance; these cords result from complex interactions among molecules on the bacterial surface, and their development is not predictable from the properties of the molecules involved. Correspondingly, in a human host, the interactions among such surface molecules and the cells of the immune system result in the formation of a granuloma—a large aggregate of host cells and bacteria that is very difficult for drugs to penetrate. The granuloma, too, is an emergent property of the interaction between *Mtb* and its host.

With the aid of bioinformatics, we hope to ascertain how all 4,000 of *Mtb*'s genes, their corresponding proteins and the bacterium's metabolic by-products react when *Mtb* is treated with a new drug in vitro. Moreover, in the past 10 years we have begun piecing together exactly how the bacterium operates inside of TB patients, as opposed to in vitro. The ultimate goal is to replicate *Mtb* in silico—that is, produce a computer simulation of the bacterium that behaves just like the real thing does in the body. The significance of such an achievement cannot be overstated, because it will enable investigators to accurately predict which bacterial components make the best drug targets and which drug candidates will likely hit those targets most effectively.

To achieve this objective, scientists will need to trace in exquisite detail all of the organism's biochemical pathways (series of reactions) and identify more of the emergent properties that arise from the operation of these pathways. The task is enormous: we still do not know what perhaps a third of *Mtb*'s proteins do in the first place, never mind what their associated pathways are or what emergent properties they spawn. But based on the current rate of progress, we are confident that within the next 20 years we will see a complete in silico bacterium that

---

In the Trenches

New drugs will be critical for combating TB, but public health officials cannot afford to wait until they are available. In the meantime, programs such as the World Health Organization's Stop TB Partnership are working to stem the pandemic by improving quality control at testing facilities, enhancing patient supervision and support, assuring drug supply and educating the public about care, among other tactics. The program aims to reduce the number of deaths from TB by more than half by 2015.

---

acts exactly like its counterpart growing in a test tube in the lab—and maybe even in a human being.

Preventing TB infection in the first place is, of course, better than treating people after they have become sick. To that end, efforts to create a vaccine that confers better protection against the disease than does the BCG vaccine are under way. Some developers are trying to improve the existing vaccine; others are attempting to make entirely new ones. But for the moment, the work is mostly doomed to trial and error because we do not understand why the current vaccine does not work nor how to predict what will work without testing candidates in humans.

In other diseases for which vaccines are available, surviving an initial infection provides immunity to future infection. In TB, however, initial infection does not offer any such protection. A vaccine that is based simply on an attenuated version of TB therefore will not work. And whereas drug development would be greatly accelerated by the development of an in silico bacterium alone, enhanced vaccine development would require both an in silico bacterium and an in silico human to be successful. Such an arrangement would allow us to systematically explore the effects on humans of altering the bacterium.

In his book *The Tipping Point,* Malcolm Gladwell defines said point as "the level at which the momentum for change becomes unstoppable." Never has the need for better diagnostic tests, drug therapies and vaccines against TB been greater. Much work remains to be done, but with the genomes of both *Homo sapiens* and *Mycobacterium tuberculosis* decoded and with an unprecedented amount of brainpower now trained on the problem, the momentum for change truly is unstoppable.

CLIFTON E. BARRY III is chief of the tuberculosis research section at the National Institutes of Health in the Institute of Allergy and Infectious Diseases (NIAID), which he joined in 1991. Barry's research group studies all aspects of tuberculosis drug discovery and genomics and runs a clinical trials program involving patients with highly drug-resistant TB in South Korea. **MAIJA S. CHEUNG** is a fellow at NIAID. A graduate of Middlebury College, Cheung plans to go to medical school to study global public health and infectious disease.

From *Scientific American,* March 2009, pp. 62–69. Copyright © 2009 by Scientific American. Reprinted by permission.

Fighting Killer Worms

Bloodsucking worms called schistosomes are among the world's most worrisome human parasites. A new genome sequence and powerful genetic tools promise to help crack their secrets.

PATRICK SKELLY

Legend has it that vampires create no shadows, cast no reflection and—in more modern versions of the tale—cannot be captured on photographs, film or video. Of course, vampires are only myths. Unfortunately, schistosomes, which behave in some similar ways, are not. These infectious worms dwell in human veins and eat our blood. Among parasitic illnesses, the World Health Organization ranks schistosomiasis, the disease caused by the worms, second only to malaria in terms of the number of people it kills and chronically disables and the drag it imposes on the social and economic development of nations. And, in their own way, schistosomes have achieved invisibility. Cameras can capture these creatures, but our immune system does not.

Investigators have struggled for years against the schistosome's evasiveness. They have been trying to create vaccines able to rally a defense that would pounce on the parasite quickly, thereby preventing disease, or that would help the body to clear existing infections. Vaccines are a necessary and missing component of a global effort to eradicate this illness. So far the results have been disappointing. But schistosome researchers like myself feel we may be at the start of a great leap forward. Genome projects are laying bare the DNA sequence of the parasite, and scientists are beginning to develop powerful new tools to probe its molecular secrets. These weapons may help make it possible to enhance immunity and accelerate vaccine efforts.

Preying on Humans

A vaccine would help avoid an enormous amount of suffering. Some 200 million people, mostly in tropical and subtropical countries, have schistosomiasis, meaning they harbor schistosomes in their blood. In children, persistent infection can retard growth and cause cognitive deficits. And in anyone, it can lead to anemia as well as damage to the intestines, bladder, spleen and liver, resulting in symptoms ranging from bloody diarrhea and cramping to life-threatening internal bleeding and kidney failure. Schistosomiasis can drastically reduce someone's ability to work, crippling both individuals and the economy.

Key Concepts

- Parasitic worms known as schistosomes are a major cause of disability and death in many parts of the world, especially sub-Saharan Africa.
- Although a treatment exists, reinfection is the rule.
- A vaccine would make a world of difference, but none has yet proved effective. Genetic and other tools hold promise for generating new candidates.
  —The Editors

People become infected when they encounter water infested by immature schistosome forms, which, though toothless, easily degrade and penetrate human skin and then enter blood vessels. There the immature parasites develop into adult bloodsucking worms and mate, after which the females begin laying eggs.

Then the eggs make matters worse. As many as half of the hundreds laid daily by each female will lodge in a variety of organs. Unrestrained, they would secrete toxins at a lethal level. The immune system, though usually unable to eliminate the worms, blocks the acute lethality, albeit at the cost of doing damage of its own: it provokes the formation of scar tissue, a major cause of the organ impairment seen in the disease. The immune response to the eggs also apparently helps them to puncture blood vessels, which in the intestinal tract allows them to make their way into feces and thus out of the body to continue development. Eggs that invade the bladder may, alternatively, escape in urine. In water the eggs hatch; then larvae emerge and infect snails. Inside snails the schistosomes replicate asexually before pouring into the water to infect, or reinfect, new human victims. [*For more on the worm's complex life cycle, see box.*]

Good sanitation and snail control have limited the disease in many countries. But in poverty-stricken regions, where clean water is still not available, it thrives. A safe antischistosome drug, praziquantel, was developed in the 1970s. It has few side effects and is now relatively cheap; plus, a single treatment can clear the infection. Reinfection, however, occurs frequently, and the worry looms that schistosomes will gain resistance to this drug. Already cases of schistosomiasis have surfaced that require higher than normal levels of the drug to clear—a possible sign of incipient resistance.

It is because of concern over drug resistance and because prevention is always the best medicine that health officials are eager to add a vaccine to the fight against the parasite—if a practical and effective one can be created. Typical vaccines deliver dead or inactive pathogens or distinctive segments of molecules (often proteins) made by those organisms in a way that induces the immune system to behave as if a true infection has occurred. The system produces cells that specifically recognize molecules present in the vaccine; thereafter some of these cells remain on the alert for the pathogen, ambushing it with antibody molecules directed to the recognized targets with other weapons before the menace can cause illness.

Investigators did not initially expect development of a vaccine against schistosomiasis to be as difficult as it has been. The worms' life cycle suggested the parasites would be a soft target for our mighty immune system. Yet they turn out to be anything but simple to handle.

Swimming with the Enemy

One reason schistosomes initially seemed like they should be an easy target is that they are relatively large and make no effort to find hiding places in the body. The first sight of an adult worm always surprises my graduate students. These biologists are familiar with the microscopic bacteria and viruses that can live in our bodies and that often evade immune attack by hiding inside cells or by outcompeting immune cells through high-speed reproduction: one virus or bacterium can beget millions, indeed billions, of others during the course of an infection.

Schistosomes, on the other hand, are big enough to be viewed by the naked eye. An adult is a centimeter long. Furthermore, the worms that start an infection on day one are the same ones present days, years or even decades later; inside the human body their numbers do not grow, except, of course, by new infections.

And evolution has chosen a hostile home for schistosomes. Lying exposed in the bloodstream would not appear to be an ideal habitat for a parasite. Blood, though nutritious, is a major conduit for all the forces of immunity, which, somehow, the worms avoid.

Beyond being big and brazen, schistosomes possess other features that suggest the immune system could be induced to recognize them if conditions were right. The body's strong reaction to their eggs is one sign of this possibility. What is more, there is nothing intrinsically immunologically invisible about the molecules that make up the worms. R. Alan Wilson and his colleagues at the University of York in England, among other research groups, have shown that if schistosomes are lethally wounded with high doses of radiation and then introduced into experimental animals, the dying parasites do induce strong immunity. Indeed, they serve as effective vaccines, protecting

Formidable Foe

Globally, an estimated 200 million people are infected (20 million severely) and 200,000 die annually.

The schistosome species that cause human disease do not multiply in people but can survive in their blood for 30 to 40 years.

The worms, also known as flukes, once brought down an army. They spread in water, and in 1948 they incapacitated large numbers of soldiers from the People's Republic of China who were preparing for an amphibious assault on Taiwan (formerly Formosa). One historian thus dubbed the worm "the fluke that saved Formosa."

Fast Facts

- Schistosomes probably originated in Asia and then dispersed to India and Africa. They jumped to the Americas in the blood of African slaves.
- The worms lack an anus, so they vomit wastes out the mouth, for the host's bloodstream to whisk away.
- Females do not mature unless they have contact with males; removed from a male's slit, a female will physically regress.
- Schistosomes that sicken humans replicate in aquatic snails. Governments could help limit the worms' spread by eradicating snails from freshwater and by preventing them from colonizing new bodies of water, such as lakes formed when dams are built. Many fear, for example, that construction of the Three Gorges Dam in China will foster new schistosome infections.
- In snails, schistosome larvae often compete with other parasites, some of which like to munch on the larvae. Researchers are considering trying to reduce schistosome populations by seeding ponds with these competitors.
- Schisotosome species that mainly infect water birds can cause a rash known as swimmer's itch in the U.S. and elsewhere. Parasitologist William W. Cort discovered the worm link in 1927, by placing larvae from contaminated water on his own skin and observing the symptoms.

The Global Picture
Where Trouble Lies

Three schistosome species cause most human infections (schistosomiasis). Because the parasite spreads in water contaminated by urine or feces, it is most common in places that lack sanitation systems. Some 85 percent of cases occur in sub-Saharan Africa.

the animal against later challenge by hundreds of healthy schistosomes. Unfortunately, using similarly prepared worms to vaccinate people is impractical.

This animal work has, however, encouraged hope that vaccines can be created inexpensively and in abundance using a single schistosome molecule or a mixture of selected ones as their basis. Three separate species of schistosomes account for the vast majority of human disease—*Schistosoma mansoni, S. haematobium* and *S. japonicum*—and so the ideal vaccine would work against all three. For now, however, researchers are focusing on finding a vaccine that can ward off infection by one species before trying to knock all of them down in one fell swoop.

To date, several schistosome molecules have been explored as vaccines but none has proved strongly effective. One, though, has performed well enough to enter large, phase III clinical trials—the final stage of human testing before a product can be released. This vaccine, developed at the Pasteur Institute of Lille in France, contains the *S. haematobium* version of a protein discovered in 1987: glutathione *S*-transferase. All researchers in the field hope this preparation will succeed, but in the meantime the hunt goes on for other promising vaccine candidates.

Wormy Tactics

Certainly, knowing how schistosomes typically escape immune detection is important if we are to develop vaccines that can overcome that propensity. The parasites have several tricks at their disposal that may explain their seeming invisibility to our defenses. One is that they come armed with a variety of molecules that may allow them to disable or "blind" the immune system. Kalyanasundaram Ramaswamy and his colleagues at the University of Illinois have shown, for instance, that some schistosome molecules can, at least in a test tube, inhibit proliferation of immune cells or induce the cells' death.

In addition, some newly identified schistosome genes look like human ones that are switched on in immune cells. Other genes encode receptors, or docking sites, that are closely related to human receptors that bind small molecules called cytokines (which control the activity of immune cells) or hormones (which convey messages between cells over longer distances). It stands to reason that the parasites would benefit from intercepting signaling molecules that help our bodies to react to infection. The worms presumably use their receptors to essentially spy on intercellular chatter, to gain information about the state of their environment and to prepare counteractive measures before immune cells have a chance to strike.

Schistosomes also possess what seems to be a cloak of invisibility: an unusual covering known as the tegument. Most parasites are covered by a single oily membrane. In addition to that membrane, the outer part of the tegument sports a second, external one that contributes to the parasite's ability to hide. The tegument provides ample protection to the worm as it migrates through our blood, but in the hands of scientists, it is extraordinarily fragile and nebulous. This fragility has made it difficult to answer even basic questions about the tegument's biology,

Infected Again

Reinfection by schistosomes is common even after successful treatment because few individuals develop protective immunity and because in many areas, such as Morogoro, Tanzania, people have little choice but to wash clothes, bathe or cool off in infested water. The high rate of reinfection underscores the urgent need for a preventive vaccine.

Biology Basics
A Complex Life Cycle

The intricate life cycle of the schistosome includes multiplying prodigiously in snails and laying eggs in a person's blood. Those eggs account, by and large, for the long-term effects of infection.

1. **Schistosome eggs** produced in infected individuals enter freshwater in urine or feces.
2. **Snail-invading larvae** called miricadia hatch from the eggs.
3. **Larvae in snails** reproduce and morph repeatedly, ultimately into a human-infecting form.
4. **Released larvae**—cercariae—swim to a new victim, usually emerging in midday to maximize the chance of finding a host.
5. **Cercariae** bore through the skin (despite being toothless), transform into schistosomula and enter veins.
6. **Schistosomula** float to the liver circulation, where they pair up and mature into adults.
7. **Worm** pairs migrate (against the flow of blood) to distant sites to lay eggs.
8. **Eggs** lodge in the intestines or bladder and enter feces or urine, starting the cycle anew.

How Worm Eggs Cause Chronic Disease

Schistosome eggs do harm by working their way into tissues and eliciting destructive immune reactions.

Responses to *S. mansoni* and *S. japonicum* eggs often compromise the liver and intestines and can also lead to bloody diarrhea, lethal internal bleeding and, possibly, colon cancer.

Responses to *S. haematobium* eggs can damage the urinary tract and kidneys and may induce bladder cancer.

such as which proteins reside in it and whether any protrude from its surface. This last question is of keen interest to vaccine designers, because the targets of most successful vaccines are proteins or other molecules that appear on the outside of a pathogen.

We do know, though, that this outer coat can actually acquire *human* molecules from the blood. It is possible to detect, for instance, our own blood-group molecules (which establish the familiar blood types A, B, and so on) attached to the worm's surface. One controversial idea is that these stolen human molecules could act as a disguise, covering the parasite's own molecules and making them invisible to immune surveillance.

Tricks of Our Own

For decades, researchers have tried to pierce this impressive armor of disappearing tricks using the classic tools of molecular biology: isolating schistosome proteins and their genes one by one, then trying to discern the proteins' functions and turn those molecules into effective vaccines. Now this slow and meticulous process may be thrown into higher gear by new technologies and the approaches they make possible.

Overcoming the known and yet undiscovered schistosome evasions would be vastly accelerated by having a catalogue of all the worm's proteins. For that reason, schistosome researchers have been eager to decipher the organism's genome, the complete sequence of DNA codes it uses as a blueprint for constructing every protein it contains.

But like so much else about these creatures, this goal initially proved elusive. For one thing, the schistosome genome—with more than 300 million nucleotide base pairs (the units of DNA)—is the largest parasitic genome that biologists have yet attempted to sequence. (For comparison, the genome sequence of the malarial parasite *Plasmodium* is more than 10 times as small.) Just as daunting was the discovery that almost half the genome is composed of repeated DNA sequences that perform no known function. For researchers, such "junk" DNA makes deriving a completed sequence much more difficult.

Nevertheless, in an international effort spearheaded by Philip T. LoVerde, now at the Southwest Foundation for Biomedical Research, the genome of *S. mansoni* has recently been sequenced, and the sequence is available online for all to analyze. And the Chinese National Human Genome Center in Shanghai is closing in on a listing of all of *S. japonicum*'s active genes.

One great advantage of revealing the full schistosome genome is that every gene can now be seen in context of this organism's entire genetic background. We have learned, for instance, that the parasite has more than one version of some proteins that vaccines could potentially target; this variety might allow schistosomes to function in spite of vaccine-induced immune activity—by using the nontargeted version. Genomic analysis can now identify common structural features shared by such proteins so that those features might be incorporated in a vaccine and thus prevent the worms from escaping immune attack.

Alex Loukas and his colleagues at the Queensland Institute of Medical Research in Australia have taken advantage of the full genome sequence in another way. They screened it for genes whose features suggested the encoded proteins probably protruded from the tegument. The so-called tetraspanin molecules that emerged from the screen have long domains made of greasy

---

A Bright Side?

In experimental animals, schistosomes can prevent or ameliorate a range of debilitating autoimmune disorders, such as Crohn's disease, which causes chronic intestinal inflammation (colitis) in humans. Studies conducted by Joel Weinstock, now at the Tufts University School of Medicine, and his colleagues showed that after mice with colitis were injected with schistosome eggs, they suffered less intestinal swelling and were better protected from lethal inflammation than other mice were.

It turns out that the eggs and Crohn's disease invoke diametrically opposite immune responses. In this immunological tug-of-war, the response elicited by the eggs has the upper hand. Investigators are now hunting for the molecules that elicit these responses, because some might be valuable as therapies for autoimmune diseases.

—P.S.

Schistosome Strategies
How Worms Hide in Plain Sight

Schistosomes have many ways of evading the immune system, some of which are described below. To make successful vaccines, investigators will need to find worm molecules that when delivered to humans will elicit immune responses not defeated by such subterfuges.

Disarm Immune Cells

Larvae release molecules that cripple immune cells needed to clear the larvae.

Don Invisibility Cloak

Adult worms in the blood are covered by an unusual "skin"—the tegument—that displays few parasite proteins on its outer membrane. As a result, the immune system usually takes little notice of the adults.

Dress Up

Human molecules, such as those that determine blood type, can stick to the surface of the worms, possibly helping to further shield the parasites from notice by the immune system.

---

amino acids that would be expected to span the oily surface of the outer membrane, leaving two protein loops exposed on the surface. Recently Loukas's team reported that two of these newly identified proteins, TSP-1 and TSP-2, when used to vaccinate mice, resulted in a substantial reduction in the number of adult worms and eggs in animals; in the case of TSP-2, the reduction was by more than half. The group then showed that in rare cases, people who are putatively resistant to schistosomes—who have avoided infection with the parasite despite years of known exposure—have antibodies against TSP-2 in their blood. In contrast, those who are chronically infected have no detectable level of these antibodies. This finding suggests that recognition of TSP-2 is a component of rare, natural immunity to schistosomes and that the protein might be useful for eliciting protective immunity in a vaccine as well.

The Australian group's work is encouraging for another reason. One might reasonably wonder whether molecules that fail to evoke an immune response in the human body during an infection would be able to do so when delivered as a vaccine. Loukas's team and others, however, have demonstrated in mice that if these molecules are presented to the immune system in the right way, they can indeed, at times, elicit a strong protective response.

In parallel with examining the schistosome genome, researchers are working to understand the functions of the proteins made by the parasites. Such information can help pinpoint which proteins might be the most reasonable to pursue as vaccine candidates. For instance, molecules that the worm always requires to survive or to make eggs in the human body could be useful, because an immune response targeted to them should in principle be deadly to the parasite or limit the destructive egg production.

Playing the Function Card

Several years ago knowledge of protein function led Charles Shoemaker of the Tufts Cummings School of Veterinary Medicine and me to proteins that look promising as vaccine components. These proteins are involved in importing nutrients, such as sugars and amino acids. Schistosomes, as they bathe in blood, not only gobble food through their mouth but also take in many nutrients directly through their tegument, and they require nutrient-importing proteins for this purpose. We also know that to work properly, these proteins must be in direct contact with the host's blood. These molecules are potentially very attractive as vaccine targets because prompting immunity against them could both direct a damaging attack against the parasite (because these proteins are on its surface) and impede its ability to absorb food from the blood.

A focus on function has also raised the possibility of making a vaccine from proteins that the parasites secrete. At first blush, that idea might seem silly: an immune response directed to such molecules would literally miss the target, because these molecules float away from the worm body. But if immune system components bind to these factors and thereby keep the secretions from doing jobs important to the parasite, the vaccine might reduce the worm's survival or its ability to cause disease. An obvious next step would be to shut off secreted genes one at a time, to see which ones are needed the most and would therefore be the best candidate for this approach.

Until recently, standard tools for shutting off genes did not work in schistosomes. But my laboratory and that of Tim Yoshino of the University of Wisconsin-Madison have taken a leaf from the book of 2006 Nobel Prize winners Andrew Z. Fire of Stanford University and Craig C. Mello of the University of Massachusetts Medical School and developed methods for

Vaccine Leads

The vaccine candidate in the most advanced stage of human testing relies on the schistosome protein glutathione *S*-transferase (Sm28GST) to awaken an immune attack targeted to schistosomes. In some trials of this vaccine in animals, fewer worms than usual survived, and those that did produced fewer eggs.

Recent work has identified other schistosome proteins having promise as vaccines. Those called tetraspanins, for instance, peek through the outer surface of adult worms and so can provide clear targets for immune defenses. Tetraspanin Vaccines have provided some protection from infection in animal trials.

Other leads include nutrient transporters (which have to contact the host's blood directly to access nutrients and thus should be accessible to the immune system), as well as molecules secreted by the parasites to maintain infection—such as proteins that degrade host molecules or dampen antiparasite immunity.

—P.S.

silencing specific schistosome genes using a technique called RNA interference [see "Censors of the Genome," by Nelson C. Lau and David P. Bartel; SCIENTIFIC AMERICAN, August 2003]. So it is now possible to silence the genes of secreted proteins and other schistosome proteins to probe their function.

Going forward, vaccine researchers will have other new tools for uncovering the function of schistosome proteins, where they reside and when in the parasite's life cycle they are made. Notably, Paul Brindley of George Washington University, Christoph Grevelding of the University of Düsseldorf in Germany and Edward Pearce of the University of Pennsylvania are developing methods for genetically engineering worms, making it possible to add distinctive tags to a selected parasite protein; such tags will allow scientists to easily track the protein's production and location. Among other advantages, this technique could put to rest the question of which proteins normally reside in the tegument and protrude from its surface. Taking another tack, various groups, including that led by Karl Hoffman of the University of Wales, have created devices called DNA microarrays (commonly called gene chips) that can reveal which mixtures of schistosome genes are switched on at each stage of development.

The many fresh approaches to studying the parasite may yield benefits beyond ideas for vaccines. Knowing this organism's complete genetic makeup, for example, should help pinpoint proteins that are unique to schistosomes and crucial for their survival; novel drugs might then be found that act on those proteins to defeat the worm. Of course, the path from all this new knowledge and know-how to an effective vaccine or treatment is not straightforward or certain. Success will depend on researchers' intellect, intuition, dumb luck, and the level of funding governments and foundations provide. But it is exciting to know that schistosome researchers are moving in directions that were not on the map even a few years ago.

More to Explore

The Immunobiology of Schistosomiasis. Edward J. Pearce and
Andrew S. MacDonald in *Nature Reviews Immunology,* Vol. 2,
No. 7, pages 499–511; July 2002.

Making Sense of the Schistosome Surface. P.J. Skelly and
R. A. Wilson in *Advances in Parasitology,* Vol. 63,
pages 185–284; 2006.

Current Status of Vaccines for Schistosomiasis. D. P. McManus and
A. Loukas in *Clinical Microbiology Reviews,* Vol. 21,
No. 1, pages 225–242; January 2008.

Schistosome life cycle animation: www.wellcome.ac.uk/en/labnotes5/
animation_ popups/schisto.html

Other informative websites: www.cdc.gov/ncidod/dpd/
parasites/schistosomiasis and www.who.int/topics/
schistosomiasis/en

PATRICK SKELLY, who earned his PhD at the Australian National University in Canberra, is assistant professor of biomedical sciences at the Cummings School of Veterinary Medicine at Tufts University and president of the New England Association of Parasitologists.

Bacterial Therapies: Completing the Cancer Treatment Toolbox

ADAM T. ST JEAN, MIAOMIN ZHANG, AND NEIL S. FORBES

C urrent cancer therapies have limited efficacy because they are highly toxic, ineffectively target tumors, and poorly penetrate tumor tissue. Engineered bacteria have the unique potential to overcome these limitations by actively targeting all tumor regions and delivering therapeutic payloads. Examples of transport mechanisms include specific chemotaxis, preferred growth, and hypoxic germination. Deleting the ribose/galactose chemoreceptor has been shown to cause bacterial accumulation in therapeutically resistant tumor regions. Recent advances in engineered therapeutic delivery include temporal control of cytotoxin release, enzymatic activation of pro-drugs, and secretion of physiologically active biomolecules. Bacteria have been engineered to express tumor-necrosis-factor-α, hypoxia-inducible-factor-1-α antibodies, interleukin-2, and cytosine deaminase. Combining these emerging targeting and therapeutic delivery mechanisms will yield a complete treatment toolbox and increase patient survival.

Introduction

To cure cancer completely many of the deficiencies in current therapy must be overcome. Many standard therapies do not effectively target tumors over normal tissue and do not efficiently penetrate tumor tissue.[1-3] Engineered bacteria possess unique abilities to overcome both of these deficiencies.[4,5**] Poor targeting limits the dosage of standard therapies that can be administered because high doses would induce systemic toxicity. Poor penetration reduces the dosage present throughout tumors. Both of these limitations prevent complete eradication of all cancer cells following individual treatments. Between courses of chemotherapy remaining cancer cells repopulate tumors, which leads to loss of local control and recurrence.[6,7] In addition, time between courses of chemotherapy allows individual cancer cell intravasation into blood vessels and increases the chance of metastasis. Tumor recurrence and metastatic disease are the primary cause of mortality from cancer.[8]

To be effective bacterial therapies need to (1) target tumors over normal tissue, (2) be genetically modifiable, (3) be non-toxic, (4) target therapeutically resistant regions of tumors, and (5) deliver an effective anticancer therapeutic.[4] Many different genera of bacteria have been shown to accumulate specifically in tumors over normal tissue in mice including *Clostridium*,[9-13] *Salmonella*,[14,15] *Bifidobacterium*,[16,17] and *Escherichia*.[18] In clinical trials, however, efficacy has been limited by insufficient colonization.[19,20] The next two requirements, toxicity and genetic manipulability, have mostly been addressed.[14,21,22,23*,24] The last two challenges, effective intratumoral targeting and controlled delivery of therapeutics, are the focus of most current research. Enhancing the ability to target all regions of tumors and deliver potent therapeutics would greatly improve efficacy and tumor colonization in human tumors. Many creative approaches are exploiting bacterial processes to tailor bacteria into therapy vectors and cancer cell destroyers. Once these problems have been solved, we believe that bacteria will be a key step toward completing the cancer therapy toolbox.

Genetic Modification and Toxicity

To make bacteria into effective anticancer vectors, considerable research has been performed over the past decade to produce bacteria that are non-toxic and genetically modifiable. Genetic modification is necessary to design bacteria to exhibit desired therapeutic properties. Of the three genera being researched, *Escherichia coli* is the standard organism for genetic engineering, allowing for robust and facile manipulation, and *Salmonella* has similar genetic modification capabilities. Species of *Clostridium* have traditionally posed significant challenges because they are not easily transformable using standard plasmid vectors and electrotransformation techniques.[25] However, a technique utilizing conjugative transfer from *E. coli* was recently developed,[23*,24] which will greatly expand the breadth of genetic modification possible in *Clostridium*.

Bacterial toxicity must be eliminated for many of these strains that are intrinsically pathogenic. A non-toxic (NT) strain of *Clostridium novyi* strain was created by deleting the virulence gene by heat treatment.[21] A non-pathogenic *msbB⁻* and *purI⁻* mutant of *Salmonella typhimurium* has been created[14,22] and found to be non-toxic in clinical trials.[19,20] It has also been shown that non-pathogenic, pro-biotic *E. coli* Nissle 1917 can be utilized in cancer therapies.[26**]

Intratumoral Targeting

From an engineering perspective, the limited penetration of chemotherapeutics into all tumor microenvironments is a provocative problem for which bacteria are ideal solutions. Currently no therapies have been designed to target explicitly the therapeutically resistant regions in tumors. Three of the mechanisms by which bacteria target different intratumoral regions are specific chemotaxis, preferential growth, and hypoxic germination. Specific chemotaxis and preferential growth are employed by facultative anaerobes *Salmonella* and *Escherichia,* while hypoxic germination is employed primarily by obligate anaerobes *Clostridium* and *Bifidobacterium.* These mechanisms, which enable bacteria to target specific tumor regions, also enable bacteria to target tumors over normal tissue. This is the primary reason for the exceptional targeting ability of bacteria: tumors contain unique microenvironments, which are not present in most normal tissue and are attractive to bacteria.

Microenvironments and blood vessel structures are vastly different between normal and tumor tissue. Normal tissue receives oxygen, nutrients, and drug molecules via an organized vasculature system. All cells in normal tissue have access to oxygen and nutrients. In tumors this balance is disrupted by disorganized and variable blood flow,[27] which creates heterogeneous microenvironments, including gradients in cell growth rate and drug concentration, as well as regions of hypoxia and acidity.[28]

Gradients in drug concentration expose some cancer cells to drug levels that are not sufficiently cytotoxic, which prevents complete cancer clearance.[28,29] To overcome these diffusion limitations a method of active drug transport is needed. Because active transport requires energy storage or recruitment from the local environment, not many suitable systems are available. Most cutting-edge therapeutics, including viruses, liposomes, and antibodies rely on passive diffusion and do not actively transport. These modalities can be highly specific to cancer cells but, because of diffusion limitations, cannot penetrate deep into tumor tissue.[30,31] To illustrate the importance of active transport, consider a hypothetical therapeutic that kills cancer cells on contact. This agent would be effective at killing cancer cells lining the tumor vasculature, but would leave surviving cells and not completely eliminate tumors. We call the approach of actively targeting specific therapeutically resistant regions of tumors *targeted intratumoral delivery.*

Salmonella typhimurium, a facultative anaerobe, has been shown to be specifically attracted by chemotaxis toward chemicals excreted by distinct microenvironments in tumors.[5**,32] This ability makes motile bacteria an attractive system because they are able to actively transport, overcome diffusion limitations, and penetrate into therapeutically resistant regions of tumors. To the best of our knowledge, motile bacteria are the only treatment modality that can actively transport and are therefore the only means of effectively treating all regions within a tumor. Experiments with cylindroids, an *in vitro* model that mimics the heterogeneous microenvironment of tumors, showed that *S. typhimurium* bacteria are preferentially attracted to dying cancer cells in therapeutically resistant tumor regions.[32] Further investigations led to a discovery that the aspartate receptor

controls migration toward tumors, the serine receptor initiates penetration, and the ribose/galactose receptor directs *Salmonella* into necrotic regions.[5**] Knocking out the gene for the ribose/galactose receptor causes *S. typhimurium* to accumulate in therapeutic-resistant regions where they induce cancer cell apoptosis.[5**] These findings suggest that *Salmonella* can be directed into any tumor region by manipulating the expression of specific chemoreceptors.

Another mechanism used by facultative anaerobes to target tumors is preferential growth. In cylindroids wild-type *S. typhimurium* was shown to preferentially grow in dying tissue and not in regions of actively growing cells.[32] Targeting can be controlled by creating auxotrophic bacteria that cannot survive without nutrients from specific microenvironments. When injected intravenously, auxotrophs migrate toward regions of tumors that have high concentrations of required nutrients where they preferentially proliferate. Auxotrophs would not be able to grow in normal tissue. To demonstrate this principle an auxotrophic *S. typhimurium* mutant for leucine and arginine has been shown to accumulate throughout tumors in mice bearing metastatic PC-3 human prostate tumors.[33,34,35*]

Salmonella bacteria have also been shown to target non-hypoxic regions and metastases. Because early metastases and viable tumor cells outside necrotic regions are well or partially oxygenated, they are inaccessible to obligate anaerobic bacteria, which cannot tolerate the slightest amount of oxygen.[36] Non-pathogenic strains of *S. typhimurium* have been shown to preferentially accumulate in subcutaneous mouse tumors 2000-fold more than in the liver and spleen, retard tumor growth, and prolong survival.[14,15,22,37,38] It has been shown that, in addition to chemotaxis and preferential growth, facultative anaerobes accumulate in the immune privileged environments of tumors because of limited immune clearance by macrophages and neutrophils.[39,40] In addition to primary tumors, *Salmonella choleraesuis* has been shown to accumulate in lungs metastases and not in the healthy surrounding parenchymal tissue.[41] In orthotopic tumors this strain was shown to increase neutrophil and T cell infiltration and have distinct anti-tumor effects, when administered with cisplatin.[42,43*]

Recently, *Escherichia coli* has been shown to accumulate in tumors.[18, 26**,44] Colonization levels of *E. coli* strains in the spleen and liver were very low compared with tumor tissue.[26**] Because it is a facultative anaerobe, *E. coli* most probably employs the same tumor targeting mechanisms as *Salmonella.* Administration of *E. coli K-12* to mice bearing murine 4T1 breast carcinomas effectively stimulated an anti-tumor immune response and result in major reduction of pulmonary metastatic events.[44]

The best understood mechanism of bacterial tumor targeting is hypoxic germination. When intravenously injected as spores, obligate anaerobes selectively colonize hypoxic tumor regions because they only germinate in poorly oxygenated environments. *Clostridium* is the primary obligate anaerobe investigated as an anticancer agent. In the 1960s several researchers demonstrated that *Clostridium* accumulates in necrotic tumor regions of mice[45–48] and humans.[49] Administration of *C. novyi NT* spores effectively target the avascular/necrotic tumor regions in mice[21] and induced substantial tumor regression in combination with chemotherapeutic drugs and anti-vascular agents.[50] Administration of *C. novyi NT* in combination with radiation therapy greatly increased the efficacy of the radiation.[51] The complete genomic sequence of *C. novyi NT* shows that the spores contain mRNA for redox proteins and that vegetative *C. novyi* produce lipases that enable them to thrive and proliferate in the environments in tumors.[52] Spores of non-pathogenic *Clostridium acetobutylicm* have been tested with the vascular targeting agent combretastatin A-4 phosphate on tumor bearing rats and bacterial growth in tumors was significantly improved.[53] This result indicates that the targeting efficacy of *Clostridium* can be enhanced by anti-vascular agents, which enlarge regions of hypoxia in tumors.

The distinct targeting mechanisms of facultative and obligate anaerobes are complementary. Obligate anaerobes are more selective to tumors because hypoxia is only present in tumors and not in normal tissue. This high specificity makes them preferable for treating large primary tumors. On the contrary, because facultative anaerobes rely on chemotaxis and preferential growth to target tumors, they can target metastases and regions outside tumor necrosis. When assembling the tumor therapy toolbox, these different targeting characteristics will enable personalized therapy directed to the specific characteristics of individual patient's tumors.

Controlled Therapeutic Delivery

Bacteria have the ability to manufacture and secrete proteins, which can be coupled with targeting to apply a specific and focused therapy. Using proteins with different functionality, there are three ways bacterial therapies can be tailored: controlled cytotoxicity, enzymatic drug activation, and biomolecule secretion. These three options, by exploiting bacterial targeting attributes, enable precise control of the time and location of therapeutic delivery in order to maximize efficacy and minimize systemic toxicity. Once the bacteria are located within the tumor environment, they can be triggered to produce compounds that will either directly or indirectly treat the disease.

Temporal control over therapeutic agents, especially for cytotoxic compounds, is necessary because gene expression during transit to the tumor site will distribute products systemically and increase toxicity in healthy tissues. Using gene promoters that are induced by non-toxic, externally applied factors such as radiation or small molecules to control expression can overcome this problem. Radiation exposure is an effective trigger of gene expression because it can easily penetrate human tissue to reach internal tumor sites. Radiation is also a commonly used cancer therapy and infrastructure already exists for its application. This strategy has been demonstrated *in vitro* with a genetically modified strain of *Clostridium* expressing TNFα under the temporal control of the radiation-induced *recA* promoter sequence.[54–56] It was shown that therapeutic radiation exposure levels result in a 44% increase in TNFα expression over non-irradiated samples.[56] This study also noted that basal gene expression under *recA* is significant and compromises the controllability of the method. This problem can be addressed by inserting additional radio-responsive elements into the promoter sequence. One

additional element decreased basal production levels by 30% over the wild type.[55] Coupling these promoters with expression of various therapeutic agents yields highly controllable tools for cancer treatment.

Another mechanism used to trigger gene expression is activation of promoters by small molecules. L-arabinose, which is harmless to mammalian cells and tissues, has been used to activate the P_{BAD} promoter in *E. coli*.[26**] A proof-of-concept test in cell culture using the luminescent *lux* operon showed a 10^5-fold increase over background when 0.02% L-arabinose was added to the growth medium.[26**] *In vivo* mouse models showed that single intravenous and orogastric administrations of L-arabinose could activate luminescence in colonized tumors.[26**] This approach may face challenges if the small molecules used for activation cannot reach the bacterial colonies because of diffusion limitations. However, in preliminary experiments this limitation was not observed, and expression within tumors was detectable within 15–30 min of intravenous administration.[26**]

A second approach for controlling therapeutic delivery is to engineer bacteria to produce enzymes that convert harmless pro-drugs into active agents. *Salmonella* and *Clostridium* have been engineered to express cytosine deaminase (CDase) that cleaves the pro-drug 5-fluorocytosine (5-FC) to chemotherapeutic 5-fluorouricil (5-FU).[38,53,57,58] In a pilot clinical trial utilizing 5-FC and recombinant *Salmonella* expressing CDase, two out of three patients showed tumor colonization. In these patients the average 5-FU concentration showed an intratumoral increase of 300% over blood plasma.[19] This initial trial nicely showed that a pro-drug can be activated selectively in tumor tissues using bacteria. Species of *Clostridium* have also been designed to express a series of nitroreductases for conversion of CB1954 (5-aziridinyl-2,4-dinitrobenzamide) to its 4-hydroxylamine derivative, which is 10 000 times more toxic.[23*] *In vivo* models demonstrated significant tumor regression following combined pro-drug and bacterial treatment. This study employed a repeat dosing structure similar to chemotherapy regimes to show that continued tumor regression could be achieved.[23*]

In a third strategy for therapeutic delivery bacteria could produce biologically active molecules to induce a physiological response. *In vitro* studies have shown that *Clostridum* can be designed to produce therapeutically relevant levels of interleukin-2,[59] a cytokine that causes T-cell-mediated neoplastic death.[60] Hypoxia-inducible factor-1α (HIF-1α) is a transcription factor that triggers genes that allow for cell survival in hypoxic environments.[61] Factors that target this molecule within tumors can inhibit its activity and increase the therapeutic susceptibility of the hypoxic regions. Functional anti-HIF-1α antibody fragments have been expressed by *Clostridium*, which, when combined with hypoxic targeting, may lead to an effective cancer treatment.[62] Bacteria can also be used to deliver DNA, a process referred to as bactofection.[63,64] In this process genes are transferred from bacteria into cancer cells[65,66] that then express the mammalian therapeutic proteins. *Salmonella* containing endostatin[67] and thrombospondin-1[41] genes have been engineered. The products of both of these genes reduce cancer growth by inhibiting angiogenesis. In mouse models, both treatment modalities decreased tumor growth and increased survival times.[41,67]

Conclusions

Cancer is difficult to treat because regions of poor vasculature present transport limitations for oxygen, nutrients, and drugs. Bacterial therapies have an advantage over passive drug molecules because they can actively target intratumoral microenvironments with preferential growth and active motile transport. The cutting edge of bacterial anticancer research is engineering strains to be more than simple vectors and transforming them into potent cancer therapies. The two issues that need to be addressed are intratumoral targeting and controllable therapeutic delivery. Initial efforts with *Clostridium*, *Escherichia*, and *Salmonella* have demonstrated expression of temporally controlled cytotoxins, pro-drug activating enzymes, and biomolecules. The strengths of these targeting and delivery systems are complementary. When combined, they can reach all areas of tumors or metastases and effectively kill all cancer cells. It is possible that bacterial therapies utilizing multiple genera, each expressing different therapeutic modalities, could be administered in unique combinations to individual patients to target and attack cancer in a tailored and highly effective way. Going forward, we are confident that bacterial therapies will provide a powerful tool against tumors in humans, improving efficacy of treatment and resulting in increased patient survival.

References and Recommended Reading

Papers of particular interest, published within the period of review, have been highlighted as:
• of special interest
•• of outstanding interest

1. Jain RK: The next frontier of molecular medicine: delivery of therapeutics. *Nat Med* 1998, **4**:655–657.
2. Brown JM, Giaccia AJ: The unique physiology of solid tumors: opportunities (and problems) for cancer therapy. *Cancer Res* 1998, **58**:1408–1416.
3. Tannock IF, Lee CM, Tunggal JK, Cowan DS, Egorin MJ: Limited penetration of anticancer drugs through tumor tissue: a potential cause of resistance of solid tumors to chemotherapy. *Clin Cancer Res* 2002, **8**:878–884.
4. Forbes NS: Profile of a bacterial tumor killer. *Nat Biotechnol* 2006, **24**:1484–1485.
5. Kasinskas RW, Forbes NS: *Salmonella typhimurium* lacking
•• ribose chemoreceptors localize in tumor quiescence and induce apoptosis. *Cancer Res* 2007, **67**:3201–3209.
   Using cylindroid models, the authors showed how chemotaxis worked in *Salmonella* to target different regions in tumors. Aspartate receptors initiate bacterial migration toward cylindroids, serine receptors launch penetration, and ribose/galactose receptors direct *Salmonella* into necrotic regions. This was a first attempt to explain the tumor targeting mechanism of motile anaerobes. New generic sites that can be engineered to improve tumor targeting efficacy were proposed as well.
6. Davis AJ, Tannock JF: Repopulation of tumour cells between cycles of chemotherapy: a neglected factor. *Lancet Oncol* 2000, **1**:86–93.
7. Davis AJ, Tannock IF: Tumor physiology and resistance to chemotherapy: repopulation and drug penetration. *Cancer Treat Res* 2002, **112**:1–26.
8. Fidler IJ, Singh RK, Yoneda J, Kumar R, Xu L, Dong Z, Bielenberg DR, McCarty M, Ellis LM: Critical determinants of neoplastic angiogenesis. *Cancer J Sci Am* 2000, **6**:S225–236.
9. Fox ME, Lemmon MJ, Mauchline ML, Davis TO, Giaccia AJ, Minton NP, Brown JM: Anaerobic bacteria as a delivery system for cancer gene therapy: *in vitro* activation of 5-fluorocytosine by genetically engineered clostridia, *Gene Ther* 1996, **3**:173–178.
10. Lambin P, Theys J, Landuyt W, Rijken R, van der Kogel A, van der Schueren E, Hodgkiss R, Fowler J, Nuyts S, de Bruijn E *et al.*: Colonisation of *Clostridium* in the body is restricted to hypoxic and necrotic areas of tumours, *Anaerobe* 1998, **4**:183–188.
11. Nuyts S, Van Mellaert L, Theys J, Landuyt W, Lambin P, Anne J: Clostridium spores for tumor-specific drug delivery. *Anticancer Drugs* 2002, **13**:115–125.
12. Theys J, Landuyt AW, Nuyts S, Van Mellaert L, Lambin P, Anne J: Clostridium as a tumor-specific delivery system of therapeutic proteins. *Cancer Detect Prev* 2001, **25**:548–557.
13. Minton NP, Clostridia in cancer therapy. *Nat Rev Microbiol* 2003, **1**:237–242.
14. Low KB, Ittensohn M, Le T, Platt J, Sodi S, Amoss M, Ash O, Carmichael E, Chakraborty A, Fischer J *et al.*: Lipid A mutant Salmonella with suppressed virulence and TNFalpha induction retain tumor-targeting *in vivo, Nat Biotechnol* 1999, **17**:37–41.
15. Pawelek JM, Low KB, Bermudes D: Tumor-targeted Salmonella as a novel anticancer vector. *Cancer Res* 1997, **57**:4537–4544.
16. Fujimori M, Amano J, Taniguchi S: The genus Bifidobacterium for cancer gene therapy. *Curr Opin Drug Discov Dev* 2002, **5**:200–203.
17. Yazawa K, Fujimori M, Nakamura T, Sasaki T, Amano J, Kano Y, Taniguchi S: Bifidobacterium longum as a delivery system for gene therapy of chemically induced rat mammary tumors. *Breast Cancer Res Treat* 2001, **66**:165–170.
18. Yu YA, Shabahang S, Timiryasova TM, Zhang Q, Beltz R, Gentschev I, Goebel R, Szalay AA: Visualization of tumors and metastases in live animals with bacteria and vaccinia virus encoding light-emitting proteins. *Nat Biotechnol* 2004, **22**:313–320.
19. Nemunaitis J, Cunningham C, Senzer N, Kuhn J, Cramm J, Litz C, Cavagnolo R, Cahill A, Clairmont C, Sznol M: Pilot trial of genetically modified, attenuated Salmonella expressing the *E. coli* cytosine deaminase gene in refractory cancer patients. *Cancer Gene Ther* 2003, **10**:737–744.
20. Toso JF, Gill VJ, Hwu P, Marincola FM, Restifo NP, Schwartzentruber DJ, Sherry RM, Topalian SL, Yang JC, Stock F *et al.*: Phase I study of the intravenous administration of attenuated *Salmonella typhimurium* to patients with metastatic melanoma. *J Clin Oncol* 2002, **20**:142–152.
21. Dang LH, Bettegowda C, Huso DL, Kinzler KW, Vogelstein B: Combination bacteriolytic therapy for the treatment of experimental tumors. *Proc Natl Acad Sci U S A* 2001, **98**:15155–15160.
22. Clairmont C, Lee KC, Pike J, Ittensohn M, Low KB, Pawelek J, Bermudes D, Brecher SM, Margitich D, Turnier J *et al.*: Biodistribution and genetic stability of the novel antitumor agent VNP20009, a genetically modified strain of *Salmonella typhimurium*. *J Infect Dis* 2000, **181**:1996–2002.
23. Theys J, Pennington O, Dubois L, Anlezark G, Vaughan T,
• Mengesha A, Landuyt W, Anne J, Blurke PJ, Durre P *et al.*: Repeated cycles of Clostridium-directed enzyme prodrug therapy result in sustained antitumour effects *in vivo*. *Br J Cancer* 2006, **95**:1212–1219.
   Using a novel gene transfer technique, the authors demonstrated the capability to genetically engineer *Clostridium* strains.

The technique was used to produce strains that have superior tumor colonization properties and overexpress nitroreductases for conversion of CB1954 for its toxic derivative.

24. Purdy D, O'Keeffe TA, Elmore M, Herbert M, McLeod A, Bokori-Brown M, Ostrowski A, Minton NP: Conjugative transfer of clostridial shuttle vectors from *Escherichia coli* to Clostridium difficile through circumvention of the restriction barrier. *Mol Microbiol* 2002, **46:**439–452.

25. Nakotte S, Schaffer S, Bohringer M, Durre P, Electroporation of, plasmid isolation from and plasmid conservation in *Clostridium acetobutylicum* DSM 792. *Appl Microbiol Biotechnol* 1998, **50:**564–567.

26. Stritzker J, Weibel S, Hill PJ, Oelschlaeger TA, Goebel W,
•• Szalay AA: Tumor-specific colonization, tissue distribution, and gene induction by probiotic *Escherichia coli* Nissle 1917 in live mice. *Int J Med Microbiol* 2007, **297:**151–162.
Escherichia coli was demonstrated to have significant specialization in targeting tumor tissue over healthy tissues. Additionally, temporal control over protein delivery was shown *in vivo* by using L-arabinose inducible gene promoters.

27. Carmeliet P, Jain RK: Angiogenesis in cancer and other diseases. *Nature* 2000, **407:**249–257.

28. Tredan O, Galmarini CM, Patel K, Tannock IF: Drug resistance and the solid tumor microenvironment. *J Natl Cancer Inst* 2007, **99:**1441–1454.

29. Minchinton AI, Tannock IF: Drug penetration in solid tumours. *Nat Rev Cancer* 2006, **6:**583–592.

30. Graff CP, Wittrup KD: Theoretical analysis of antibody targeting of tumor spheroids: importance of dosage for penetration, and affinity for retention. *Cancer Res* 2003, **63:**1288–1296.

31. Pluen A, Boucher Y, Ramanujan S, McKee TD, Gohongi T, di Tomaso E, Brown EB, Izumi Y, Campbell RB, Berk DA *et al.*: Role of tumor-host interactions in interstitial diffusion of macromolecules: cranial vs. subcutaneous tumors. *Proc Natl Acad Sci U S A* 2001, **98:**4628–4633.

32. Kasinskas RW, Forbes NS: *Salmonella typhimurium* specifically chemotax and proliferate in heterogeneous tumor tissue *in vitro*. *Biotechnol Bioeng* 2006, **94:**710–721.

33. Zhao M, Yang M, Li XM, Jiang P, Baranov E, Li S, Xu M, Penman S, Hoffman RM: Tumor-targeting bacterial therapy with amino acid auxotrophs of GFP-expressing *Salmonella typhimurium*. *Proc Natl Acad Sci U S A* 2005, **102:**755–760.

34. Zhao M, Yang M, Ma H, Li X, Tan X, Li S, Yang Z, Hoffman RM: Targeted therapy with a *Salmonella typhimurium* leucine-arginine auxotroph cures orthotopic human breast tumors in nude mice. *Cancer Res* 2006, **66:**7647–7652.

35. Zhao M, Geller J, Ma H, Yang M, Penman S, Hoffman RM:
• Monotherapy with a tumor-targeting mutant of *Salmonella typhimurium* cures orthotopic metastatic mouse models of human prostate cancer. *Proc Natl Acad Sci U S A* 2007, **104:**10170–10174.
A Leu/Arg auxotrophic mutant of *S. typhimurium A1-R* was demonstrated effective as a mono therapy of metastatic PC-3 human prostate tumor on mice. The auxotrophy limits growth in healthy tissues while allowing for proliferation in all tumor regions.

36. Wei MQ, Mengesha A, Good D, Anne J: Bacterial targeted tumour therapy-dawn of a new era. *Cancer Lett* 2008, **259:**16–27.

37. Forbes NS, Munn LL, Fukumura D, Jain RK: Sparse initial entrapment of systemically injected *Salmonella typhimurium* leads to heterogeneous accumulation within tumors. *Cancer Res* 2003, **63:**5188–5193.

38. Mei S, Theys J, Landuyt W, Anne J, Lambin P: Optimization of tumor-targeted gene delivery by engineered attenuated *Salmonella typhimurium*. *Anticancer Res* 2002, **22:**3261–3266.

39. Liu SC, Minton NP, Giaccia AJ, Brown JM: Anticancer efficacy of systemically delivered anaerobic bacteria as gene therapy vectors targeting tumor hypoxia/necrosis. *Gene Ther* 2002, **9:**291–296.

40. Westphal K, Leschner S, Jablonska J, Loessner H, Weiss S: Containment of tumor-colonizing bacteria by host neutrophils. *Cancer Res* 2008, **68:**2952–2960.

41. Lee CH, Wu CL, Shiau AL: Systemic administration of attenuated *Salmonella choleraesuis* carrying thrombospondin-1 gene leads to tumor-specific transgene expression, delayed tumor growth and prolonged survival in the murine melanoma model. *Cancer Gene Ther* 2005, **12:**175–184.

42. Lee CH, Wu CL, Tai YS, Shiau AL: Systemic administration of attenuated *Salmonella choleraesuis* in combination with cisplatin for cancer therapy. *Mol Ther* 2005, **11:**707–716.

43. Lee CH, Wu CL, Shiau AL: *Salmonella choleraesuis* as
• an anticancer agent in a syngeneic model of orthotopic hepatocellular carcinoma. *Int J Cancer* 2008, **122:**930–935.
Salmonella cholaraesuis was found to preferentially accumulate in both subcutaneous and orthotopic tumor tissues over normal tissues with ratios exceeding 1000 to 1. Additionally, it was observed that intratumoral microvessel density decreased, accumulations of the neutrophils, $CD4^+$ and $CD8^+$ T cells increased, and cell death within the tumor environment was induced.

44. Weibel S, Stritzker J, Eck M, Goebel W, Szalay AA: Colonization of experimental murine breast tumours by *Escherichia coli* K-12 significantly alters the tumour microenvironment. *Cell Microbiol* 2008, **10(6):**1235–1248.

45. Möse JR, Möse G: Oncogenesis by clostridia. I. Activity of *Clostridium butyricum* (M-55) and other nonpathogenic clostridia against the Ehrlich carcinoma. *Cancer Res* 1964, **24:**212–216.

46. Malmgren RA, Flanigan CC: Localization of the vegetative form of *Clostridium tetani* in mouse tumor following intravenous spore administration. *Cancer Res* 1955, **15:**473–478.

47. Parker RC, Plummer HC, Siebenmann CO, Chapman MG: Effect of histolyticus infection and toxin on transplantable mouse tumors. *Proc Soc Exp Biol Med* 1947, **66:**461–467.

48. Thiele EH, Arison RN, Boxer GE: Oncogenesis by clostridia. III. Effects of clostridia and chemotherapeutic agents on rodent tumors. *Cancer Res* 1964, **24:**222–233.

49. Carey RW, Holland JF, Whang HY, Neter E, Bryant B: Clostridial oncolysis in man. *Eur J Cancer* 1967, **3:**37–46.

50. Dang LH, Bettegowda H, Agrawal N, Cheong I, Huso D, Fil P, Loganzos F, Greenbergell L, Barkoczy J, Pettit GR *et al.*: Targeting vascular and avascular compartments of tumors with *C. novyi-NT* and anti-microtubule agents. *Cancer Biol Ther* 2004, **3:**326–337.

51. Bettegowda C, Dang LH, Abrams R, Huso DL, Dillehay L, Cheong I, Agrawal N, Borzillary S, McCaffery JM, Watson EL *et al.*: Overcoming the hypoxic barrier to radiation therapy with anaerobic bacteria. *Proc Natl Acad Sci U S A* 2003, **100:**15083–15088.

52. Bettegowda C, Huang X, Lin J, Cheong I, Kohli M, Szabo SA, Zhang X, Diaz LA, Velculescu Ve, Parmigiani G *et al.*: The genome and transcriptomes of the anti-tumor agent *Clostridium novyi-NT*. *Nat Biotech* 2006, **24:**1573–1580.

53. Theys J, Landuyt W, Nuyts S, Van Mellaert L, van Oosterom A, Lambin P, Anne J: Specific targeting of cytosine deaminase to solid tumors by engineered *Clostridium acetobutylicum*. *Cancer Gene Ther* 2001, **8:**294–297.

54. Nuyts S, Theys J, Landuyt W, van Mellaert L, Lambin P, Anne J: Increasing specificity of anti-tumor therapy: cytotoxic protein delivery by non-pathogenic clostridia under

regulation of radio-induced promoters. *Anticancer Res* 2001, **21**:857–861.

55. Nuyts S, Van Mellaert L, Barbe S, Lammertyn E, Theys J, Landuyt W, Bosmans E, Lambin P, Anne J: Insertion or deletion of the Cheo box modifies radiation inducibility of Clostridium promoters. *Appl Environ Microbiol* 2001, **67**: 4464–4470.

56. Nuyts S, Van Mellaert L, Theys J, Landuyt W, Bosmans E, Anne J, Lambin P: Radio-responsive recA promoter significantly increases TNFalpha production in recombinant clostridia after 2 Gy irradiation. *Gene Ther* 2001, **8**:1197–1201.

57. Dresselaers T, Theys J, Nuyts S, Wouters B, de Bruijn E, Anne J, Lambin P, Van Hecke P, Landuyt W: Non-invasive 19F MR spectroscopy of 5-fluorocytosine to 5-fluorouracil conversion by recombinant Salmonella in tumours. *Br J Cancer* 2003, **89**:1796–1801.

58. Dubois L, Dresselaers T, Landuyt W, Paesmans K, Mengesha A, Wouters BG, Van Hecke P, Theys J, Lambin P: Efficacy of gene therapy-delivered cytosine deaminase is determined by enzymatic activity but not expression. *Br J Cancer* 2007, **96**:758–761.

59. Barbe S, Van Mellaert L, Theys J, Geukens N, Lammertyn E, Lambin P, Anne J: Secretory production of biologically active rat interleukin-2 by *Clostridium acetobutylicum* DSM792 as a tool for anti-tumor treatment. *FEMS Microbiol Lett* 2005, **246**:67–73.

60. Parmiani G, Rivoltini L, Andreola G, Carrabba M: Cytokines in cancer therapy. *Immunol Lett* 2000, **74**:41–44.

61. Semenza GL: HIF-1, O-2, and the 3 PHDs: how animal cells signal hypoxia to the nucleus. *Cell* 2001, **107**:1–3.

62. Groot AJ, Mengesha A, van der Wall E, van Diest PJ, Theys J, Vooijs M: Functional antibodies produced by oncolytic clostridia. *Biochem Biophys Res Commun* 2007, **364**:985–989.

63. Palffy R, Gardlik R, Hodosy J, Behuliak M, Resko P, Radvansky J, Celec P: Bacteria in gene therapy: bactofection versus alternative gene therapy. *Gene Ther* 2006, **13**:101–105.

64. Vassaux G, Nitcheu J, Jezzard S, Lemoine NR: Bacterial gene therapy strategies. *J Pathol* 2006, **208**:290–298.

65. Weiss S, Chakraborty T: Transfer of eukaryotic expression plasmids to mammalian host cells by bacterial carriers. *Curr Opin Biotechnol* 2001, **12**:467–472.

66. Yuhua L, Kunyuan G, Hui C, Yongmei X, Chaoyang S, Xun T, Daming R: Oral cytokine gene therapy against murine tumor using attenuated *Salmonella typhimurium*. *Int J Cancer* 2001, **94**:438–443.

67. Lee CH, Wu CL, Shiau AL: Endostatin gene therapy delivered by *Salmonella choleraesuis* in murine tumor models. *J Gene Med* 2004, **6**:1382–1393.

Acknowledgements—We gratefully acknowledge financial support from the National Institutes of Health (Grant No. 1R01CA120825-01A1), Susan G Komen, For the Cure (Grant No. BCTR0601001), and the Collaborative Biomedical Research Program at the University of Massachusetts, Amherst. Financial support for ATS was provided by the Institute for Cellular Engineering IGERT Program at the University of Massachusetts, Amherst.

Microbial Moxie

Bacteria-based fuel cells provide power.

AIMEE CUNNINGHAM

Anglers casting their lines last September into a Montana creek may not have noticed, but a diminutive power plant was churning away in a shallow spot by the shore. The device generated electricity—with the aid of river-dwelling bacteria—to power a sensor system that wirelessly transmitted data to a receiver about 10 miles away. The underwater device, small enough to fit in a person's hand, was the first attempt to power such a system with a microbial fuel cell.

Microbial fuel cells take advantage of the long-known fact that some microbes produce electricity when they break down organic matter. Only recently, however, have scientists discovered that they could tap into this energy in a practical manner and use it as an alternative energy source.

Today, microbial fuel cells are being explored primarily as a power source for remote sensors and for wastewater treatment, in which the bacteria that break down sewage generate sufficient electricity to run the treatment plant. But the role that the fuel cells could eventually play will depend on whether certain limitations can be overcome.

For starters, the prototype fuel cells don't produce power fast enough to do much more than juice up a clock. "These systems can be very efficient, but they are slow, in terms of the rate that organic matter is converted to electricity," says microbiologist Derek Lovley of the University of Massachusetts at Amherst.

Still, the potential payoff is too good to pass up. "Once the device is constructed, it's basically working for free," says Zbigniew Lewandowski of Montana State University in Bozeman, whose group set up the Montana-stream system.

Electrifying Events

Among the entities zipping along a microbe's metabolic pathways are electrons, which hop from molecule to molecule in the course of various biochemical reactions. When microbes metabolize organic matter in aerobic conditions, they tend to deposit these electrons onto oxygen, an exchange that provides the microbes with chemical energy.

By putting the bacteria in a microbial fuel cell under anaerobic conditions—that is, with no oxygen present—researchers "are stealing some of the energy and harvesting it as electricity," says Lovley.

A basic microbial fuel cell has two chambers, one containing an anode and the other a cathode. The microbes reside under anaerobic conditions on the anode. There, they break down their food, such as glucose, acetate, or the organic compounds in wastewater. Lacking oxygen, these microbes transfer their electrons to the anode. This exchange gives the microbes a small amount of energy to fuel their growth.

A wire connects the anode to the cathode. The cathode chamber harbors oxygen dissolved in water. The electrons travel to the oxygen, generating a current as they move from one chamber to the other.

There is also a selective membrane between the two chambers that enables protons, another product of the microbe's biochemical reactions, to travel from the anode to the cathode. In the cathode chamber, the electrons and protons combine with oxygen to form water.

A different type of fuel cell can operate in a river or a lake. Local microbes colonize an anode stuck into the oxygen-poor sediment. Their electrons travel along a wire connected to a cathode suspended in the overlying water, which contains oxygen (*SN: 7/13/2002, p. 21*).

While electrifying, all this microbial activity doesn't yet translate into much power. There may be room for improvement in both the materials and the microbes.

For the anode, many researchers use a form of carbon called graphite, which conducts electrons. But "that may not be the best material to interact with whatever protein it is that's transferring electrons to the [anode] surface," says Lovley.

At the cathode, meanwhile, the transfer of the electrons to oxygen is slow. Researchers have used various catalysts and electron shuttles to improve cathode performance, but they tend to be expensive or toxic.

The microbes themselves are another limiting factor. Although some microbes, when deprived of oxygen, will ferry their electrons to an anode, says Lovley, they "aren't optimized for electricity production—they've had no evolutionary pressure to do this."

To speed electron transfer, Lovley's group is studying bacteria from the family Geobacteraceae, originally discovered in the sediment of the Potomac River in Washington, D.C. When these common microbes break down organic matter, they transfer their electrons to iron oxides, which makes them adept at using an electrode as their final electron acceptor.

By comparing electrode-dwelling and natural colonies of Geobacteraceae, Lovley's group has identified genes that are more active in the electrode-dwellers and thus likely to be important for transferring electrons to an electrode. In an attempt to increase energy production, the team is now genetically engineering Geobacteraceae to produce more of their electron-transferring proteins.

Wastewater Wattage

Researchers with an eye toward a self-sustaining wastewater-treatment plant are moving forward with prototype microbial fuel cells, says Korneel Rabaey of Ghent University in Belgium.

About 5 percent of U.S. electricity production goes into water and wastewater treatment, says Bruce Logan of Pennsylvania State University in University Park. His team was the first to demonstrate that a microbial fuel cell could produce electricity as it cleans household wastewater (*SN: 3/13/2004, p. 165*).

Along with providing energy savings for developed countries, microbial fuel cells could "revolutionize how we do wastewater treatment" worldwide, Logan says. He points out that 2 billion people live in areas without wastewater-treatment plants.

An inherent difficulty with using domestic wastewater as a raw material, however, is that it contains a lot of matter that biodegrades slowly, says Rabaey. This decreases the rate at which the microbes can pass along their electrons, so it reduces the power they produce. Also, the systems need to deal with a continuous stream of wastewater because it would be impractical to hold wastewater in a tank, notes Largus T. Angenent of Washington University in St. Louis.

In the July 15, 2005 *Environmental Science & Technology* (*ES&T*), Angenent and his colleagues described a microbial fuel cell in which the wastewater flows from the bottom to the top of a tubular system. The team fed the fuel cell continuously with artificial wastewater—a sucrose solution spiked with nutrients and metals—for 5 months.

Rabaey and his colleagues reported another tubular-fuel cell design in the Oct. 15, 2005 *ES&T*. The anode chamber contains porous, graphite granules that are 1.5 to 5 millimeters in diameter and packed to permit liquid to flow through channels among them. Microbial biofilms cover the granules, which serve as the anode surface.

The researchers tested three such fuel cells, each fed continuously for about a year with a solution of a different raw material: acetate, glucose, or household wastewater. The acetate-fed fuel cell generated 90 watts per cubic meter (W/m^3), while the glucose consumer put out 66 W/m^3. The wastewater-fed fuel cell achieved a maximum of 48 W/m^3.

Neither Rabaey's laboratory system nor Angenent's system is self-sustaining. Both rely on a regular supply of a toxic electron shuttle, which gets used up as it acts in the cathode. Logan's systems use an expensive platinum catalyst to speed the final transfer of electrons to oxygen. In the Jan. 1 *ES&T*, his team describes much cheaper metal catalysts containing little or no platinum.

All the researchers report that as they make adjustments to their designs, they continue to increase the power outputs of their laboratory systems. But much work needs to be done to make such systems economically feasible, increase the amount of electricity harvested from the organic matter, and prove that the systems can handle the volumes of a large treatment plant.

Currently, Logan's laboratory system generates 16 W/m^3 as it breaks down wastewater. His goal is 100 W/m^3 in a sustainable system, which he estimates could produce 0.5 megawatt of energy, enough to power a treatment plant for a town of 100,000.

Sensor Success

Powering a remote temperature sensor also requires technical advances to boost the meek wattage now generated by the microbes.

Microbial fuel cells are an ideal solution for remote sensors, says Lewandowski, because they avoid the logistical difficulty of changing batteries in dense wilderness areas or at the bottom of the ocean. Future microbial fuel cells might chug along for years, opening the way to a "drop-it-and-forget-it type of probe," he says.

When Lewandowski's group set up its microbial fuel cell in Hyalite Creek near Bozeman, Mont., the scientists adopted a fuel cell design that differs from the basic approach. The anode is not stuck in the sediment, and the microbes reside on the cathode instead of on the anode. The electrons carrying the current come not from the microbes but directly from the anode, a slab of magnesium alloy, as it slowly corrodes in the water and releases magnesium ions and electrons.

The microbes, which settle on the stainless steel cathode, capture manganese ions present in the water and oxidize them to form encrustations of manganese oxide on the cathode's surface. The electrons then reduce some of the manganese oxides, switching them back to manganese ions, and the cycle continues.

To demonstrate that the fuel cell could power a wireless system, Lewandowski's group housed electric components in a nearby shed. The creek-based fuel cell supplied the entire system with electricity.

By the team's design, energy generated by the fuel cell built up in a capacitor and discharged in short bursts when needed, as a camera flash does. The team also included a component called a DC-DC converter to increase the voltage potential. The converter relayed power to a transmitter, which sent the sensor's water temperature readings wirelessly to a receiver roughly 10 miles away. In the July 1, 2005 *ES&T*, the researchers described the scheme, which ran in the field until September of last year.

Lewandowski's group plans to incorporate into its system a longer-lasting, traditional anode—one that uses electricity-producing microbes on a noncorroding surface—and the researchers say they'd like to scale up the system to power multiple sensors.

Where To?

While researchers say that microbial fuel cells hold great promise, they don't expect them to become major power producers. "You're not going to run an entire city" with them, says Angenent.

But the technology could become a convenient power source for portable electronics, says Lovley. He notes that powering a cell phone during continuous talk mode would require a power-production rate 10 to 100 times as great as that of current microbial fuel cell technology, but he thinks that increase is feasible. With such a system, "10 grams of sugar could theoretically produce power for nearly 2 days of talk time," he says.

Lovley also suggests that the technology could be beneficial for developing countries that don't have well-established power grids.

"There are probably a lot of new opportunities to use this technology—we're still trying to figure out what those are," says Logan. "As it develops and we understand the economics, we will be able to find and define systems. . . . I see a future with lots of [energy] technologies, and there is room for this technology."

Article 15

Eyeing Oil, Synthetic Biologists Mine Microbes for Black Gold

Biotechnology researchers want to reengineer microorganisms to turn agricultural products into gasoline, diesel, and jet fuel.

ROBERT F. SERVICE

What do you do after creating a cheap antimalarial drug that could save millions of lives in the developing world? If you're Jay Keasling, you tackle two equally pressing problems: climate change and the need for abundant renewable fuels.

Keasling, a synthetic biologist at the University of California, Berkeley, made a splash a few years ago when he and his colleagues reengineered two microbes—*Escherichia coli* and baker's yeast—to churn out a costly plant-derived antimalarial compound called artemisinin. Now, Keasling and dozens of colleagues working at the Joint BioEnergy Institute (JBEI) that he directs are trying to create classes of compounds, such as alkanes, that are key components of gasoline and other transportation fuels. "Artemisinin is a hydrocarbon," Keasling says. "We're just trying to engineer organisms to produce different hydrocarbons."

Keasling hopes to leapfrog over a bitter debate among agricultural economists about the value of ethanol, the first-generation biofuel. Keasling and a handful of others are starting from scratch, using synthetic biology to lend *E. coli*, yeast, and other easily grown microorganisms the ability to create mixtures of compounds that can be used to make various things, including gasoline, jet fuel, plastics, and cosmetics. "This is just at the beginning," says Keasling, sitting in his fourth-floor office with a view of the Berkeley campus and the Oakland hills. The combination of rapidly improving bioengineering technology and the massive pull for cheap, low-net carbon fuels has generated enormous excitement, he asserts. "This is just a golden period for this area."

The question is whether it will be black gold. A handful of start-up companies have leapt into the field, some even teaming up with major energy companies such as Chevron and BP. Some of these companies have already begun producing fuels, but none say they can beat the price of conventional petroleum, despite its recent spike to more than $140 a barrel. But with technology improving rapidly, "very likely it can be done," says Vinod Khosla, a venture capitalist with Khosla Ventures in

Menlo Park, California, who has backed several next-generation biofuel start-ups.

For now, most of the new biofuel producers have set their sights on displacing ethanol rather than gasoline and diesel. According to the Renewable Fuels Association, last year the world produced more than 50 billion liters of ethanol fuel. Most of it is made by fermenting food crops—corn kernels in the United States and sugar cane in Brazil.

Critics argue that this method is driving up food costs. The production of corn-based ethanol, they say, requires nearly as much energy from fossil fuels to drive the tractors, produce fertilizer, and run the ethanol plants as what comes out in the alcohol at the end. The end result, they argue, is at best a marginal reduction of carbon dioxide (CO_2). In addition, ethanol can't be shipped through existing oil pipelines because it mixes easily with water, which accumulates in the pipelines.

With so many strikes against ethanol, most synthetic biology groups are working to engineer microbes to produce fuels essentially identical to existing fossil fuels. They want bugs that can grow on the sugars from agricultural waste and other "cellulosic" materials, thereby reducing the need to use scarce agricultural land to grow fuels.

The early results are heartening. Keasling and his colleagues, for example, are having early success at reengineering *E. coli* to produce gasoline-type molecules. The work builds on Keasling's earlier triumph with artemisinin. In that case, Keasling's team focused its efforts on metabolic pathways in *E. coli* and yeast that normally produce small amounts of compounds called isoprenoids, precursors for building many pharmaceuticals such as artemisinin, among other chemicals.

Keasling's team made about 50 genetic changes—adding genes for additional enzymes, promoters, and so on—to the organisms to get them to convert these intermediate compounds to artemisinin. Keasling founded Amyris Biotechnologies, which shares space in the JBEI facility, to commercialize the work. Initially, the bugs produced only tiny amounts of artemisinin.

But by outfitting *E. coli* and yeast with several new genes and turning off the expression of others, Keasling and his colleagues at Berkeley and Amyris boosted a millionfold the production of the antimalarial.

That's good enough to match the average price of artemisinin, about $1 per gram. But Keasling notes that they will have to do even better to make microbial fuel cost-effective, because the production of fuel molecules at $1 per gram translates into a price of about $125 a liter for gasoline. "It has to be cheaper than water," Keasling says. To do that, the team needs to engineer the microbes to increase the flux of the starting material—sugar in this case—through the microbes' fuel-producing pathway.

At a meeting last month in Hong Kong, Keasling reported on a novel strategy to increase the chances that the product of one reaction inside this pathway is properly handed off to the next enzyme in the chain. The team engineered *E. coli* to express a protein scaffold for a trio of enzymes that work in their isoprenoid pathway and bind them in a way that more efficiently transforms starting compounds to the final result. "This increased the flux through the pathway 77-fold," Keasling says. At Amyris, Senior Vice-President of Research Jack Newman says the company is also making significant headway on producing renewable diesel and jet fuel. Amyris has teamed up with Crystalev, one of Brazil's largest ethanol producers, to scale up the company's proprietary technology to make renewable fuels from sugar cane, beginning in 2010.

Across San Francisco Bay in South San Francisco, researchers at LS9 are reengineering *E. coli* and other organisms to make what they refer to as renewable petroleum. Rather than co-opting the microbe's isoprenoid pathway, LS9 researchers are focusing on the pathway that converts sugars to fatty acids, which can then be converted to biodiesel.

Gregory Pal, LS9's senior director for corporate development, notes that fatty-acid biosynthesis is the main route by which organisms convert excess energy into fats. Most organisms have evolved to do this quickly and with high efficiency, as anyone who has tried dieting knows all too well. "We've made on the order of dozens of [genetic] transformations" to maximize that efficiency, Pal says. The company, he adds, has successfully produced a variety of hydrocarbons and is now focused on scaling up the technology. LS9 already has a pilot-scale fermentation facility up and running at its headquarters and plans to open a small-scale production facility by mid-2010.

James Liao and colleagues at the University of California, Los Angeles, are trying a third approach. In the 3 January issue of *Nature,* Liao and his co-authors described how they engineered *E. coli* to produce isobutanol, a longer chain alcohol than ethanol. That increased chain length gives isobutanol a more energetic punch per liter and enables it to be separated from water more easily, Liao says. The molecule can also easily be converted to fuels that can be blended with gasoline as well as transformed into other commodity chemicals.

Like other groups in the field, Liao has his own favored biosynthetic pathway. He and his colleagues have co-opted the metabolic pathway that bugs use to convert common starting materials, called alpha-keto acids, to amino acids. Liao focused on this pathway because it is already adapted to handle large fluxes of hydrocarbons. "Fifty-five percent of the cell's composition is protein, so it needs a lot of amino acids," Liao says of the *E. coli* microbe his team is working with. Liao has teamed up with Gevo, a bioenergy start-up in Pasadena, California, that he says is building a pilot production plant. Liao also notes that his group is making progress on getting photosynthetic bacteria to make isobutanol. The goal is for bacteria to manufacture fuel simply by absorbing sunlight and CO_2.

Getting photosynthetic organisms to produce fuel directly has long been the dream of researchers working with algae that can absorb sunlight and CO_2 to produce plant oils that, in turn, can be converted to biodiesel. But the technology has been plagued by real-world challenges such as the need to harvest algae from large, shallow ponds and separate out the oils.

Another biofuel start-up, Solazyme in South San Francisco, hopes to avoid those pitfalls. Solazyme researchers are working with natural and engineered algal strains to produce what they refer to as renewable biodiesel. Instead of growing their algae outside in the sunlight, Solazyme researchers grow them inside dark stainless steel fermenters, in which the organisms convert sugars to oils. Turning off the algae's photosynthetic apparatus enables the organisms to produce oils more efficiently and makes it less costly to recover, says Harrison Dillon, Solazyme's president and chief technology officer.

The company can already produce biodiesel by the barrel and has had separate versions certified as jet fuel and biodiesel. Solazyme has managed to move quickly, Dillon says, because many of the algal strains start out as efficient oil producers. "They are naturally a long way towards where you want them to be," he says.

Are any of the next-generation biofuels likely to succeed? Khosla and others say the technology has several distinct advantages. First, synthetic biology allows researchers to test hundreds of potential improvements in a short time. "That means we can not only iterate quickly but not stop iterating," Dillon says. And Khosla adds that unlike renewable-energy technologies such as wind and hydropower, the economics of the technology improve as it is scaled up, allowing companies to take advantage of large-scale production efficiencies.

Finally, it's doubtful any biofuel start-up will be beat out by other upstarts. "The demand for fuels is so huge, there will be multiple winners out there," Dillon says. "If you make it at the right price, you can sell as much as you can produce."

Still, for the near term, microbial biofuels will remain dependent on food-based agriculture to generate the sugars needed to feed the bugs. And even if cellulosic wastes can be economically converted to sugars, some wastes will likely need to be left on agricultural and forest land to return needed nutrients to the soil. Despite those drawbacks, Khosla says, next-generation biofuels won't remain a research project for long. "They're a lot closer to market than the time it takes to build a new oil refinery," he says.

From *Science Magazine,* October 24, 2008, pp. 522–523. Copyright © 2008 by American Association for the Advancement of Science. Reprinted by permission.

UNIT 3

The Never-Ending Battle: Antimicrobial Drug Resistance

Unit Selections

Learning Objectives

- What impact has the use of antibiotics had on the evolution of microbes? How do microbes spread drug resistance? What mechanisms to microbes use to render antibiotics ineffective?

- Describe three ways in which you can stop the spread of drug-resistant microbes in our environment.

- If you were writing a letter to Congress to express your concerns about the increasing number of drug-resistant microbes, what two issues would you ask them to address?

- If you were the head of a research group in the pharmaceutical industry, what arguments would you use to convince stock-holders that investing in antibiotic research is not a waste of money?

Student Website
www.mhcls.com

Internet References

Alliance for the Prudent Use of Antibiotics
   http://www.tufts.edu/med/apua
World Health Organization
   http://www.who.int/drugresistance/en/
The Centers for Disease Control and Prevention
   http://www.cdc.gov/drugresistance/

Microbes have been humanity's constant companions and at the same time some of our deadliest enemies. For a brief period in our history, we foolishly thought that we had "won the war" against microbes. This period came after World War II, when antibiotics came onto the scene as the "wonder drugs" that were going to save humanity from infectious diseases. For the first time in history, it seemed that man could combat disease-causing microbes with great ease. Little did we know then that there is no "winning" of this war—at best we can expect to stay in a holding pattern with some gains and all too often too many losses. Microbes, like all other living organisms, are in a race for survival, and our actions can determine the course of their evolution. We are in the midst of a global crisis of epic proportions—the drugs that we thought would save us are now becoming ineffective, as microbes, in their evolutionary struggle for survival, acquire the ability to defeat any new drugs in their environment. This "rise of the superbugs" fills headlines and airwaves and sends people into a panic searching for the next magic drug. However, due to the fact that antibiotic development is not as profitable as developing other more frequently used medications, many drug companies are not investing in the research needed to bring new antibiotics to the market quickly.

Selective pressure from their surrounding environment favors those microbes that acquire the ability to inactivate the chemicals that might cause their destruction. In other words, there were no penicillin-resistant bacteria in our world until penicillin became widely used. Organisms that acquired mutations enabling them to inactivate penicillin gained a competitive edge over those that lacked this ability. Over time, these drug-resistant microbes flourished and spread their genetic advantage to other organisms, ensuring the survival of "their kind" in this new "wonder-drug"-laced environment. In this unit, the rise of drug-resistant microbes is examined, with particular emphasis placed on practical advice as to how individual and collective human behavior can alter the emergence of these resistant "superbugs."

The first two articles in this unit specifically chronicle the emergence of drug-resistant bacteria in hospital environments and in the community at large. In "Germ Warfare," specific strategies for stopping the spread of antibiotic resistant bacteria are outlined. The mechanisms by which bacteria fight back against antibiotics are detailed as well as the methods by which this ability is passed from one bacterium to another. The different symptomatic manifestations of virally caused vs bacterially caused infections are also listed, since people often seek antibiotic treatment when it is not necessary. Antibiotics are specifically directed against bacteria; they have no effect on virally

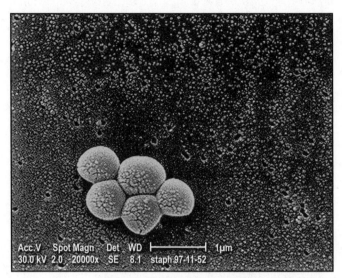

Acc.V Spot Magn Det WD ⊢——————⊣ 1μm
30.0 kV 2.0 20000x SE 8.1 staph 97-11-52

© Centers for Disease Control and Prevention/Janice Carr

caused illnesses. Since the advent of industrialized agriculture, antibiotics have been widely used in livestock to prevent bacterial infections. The consequence of this has been the spread of antibiotic resistance from farms to our food. In the United States, companies don't want to lower their profits, but in Europe, there has been a trend toward only using the minimal amount of antibiotics in agriculture.

"Routes of Resistance" focuses on the precise mechanisms by which microorganisms fight back against antimicrobial agents. The reason why organisms make antibiotics in nature is also explored. Microbial inhabitants of soil naturally produce a wide range of antimicrobial compounds that we have exploited commercially to treat disease. However, since they are surrounded by these antimicrobials, the same organisms have also developed mechanisms to protect themselves against these toxic compounds. Research published in 2008 by George M. Church's laboratory at Harvard has shown the existence of hundreds of bacterial species in soil that can survive solely on a diet of antibiotics. Most striking is the fact that all of the isolated organisms exhibited resistance to multiple classes of antibiotics at concentrations that are used clinically. Although it has not yet been shown, it is possible that these soil-dwelling bacteria could pass on the ability to use antibiotics as their sole carbon source to pathogenic organisms in other environments.

An article from the *American Journal of Nursing* offers practical tips for healthcare workers in dealing with antimicrobial drug resistance. In "Is Your Patient Taking the Right Antimicrobial?" the author tells a very personal story of how her sister's death was caused by inappropriate treatment of a bacterial infection. This poignant and hard-hitting article outlines specific steps that should be taken in making the decision to treat a patient with the most effective antibiotic. Specifically, the review of culture and antibiotic-sensitivity testing reports is urged. By using the most specific antibiotic possible rather than treating patients with broad-spectrum antibiotics, the spread of resistant strains can be kept in check.

An article from the San Francisco Chronicle, "Constant Struggle to Conquer Bacteria" describes the problem of drug development and discovery and examines why there are so few new drugs coming down the pipeline to treat the latest "superbugs." This report points out that since the 1990s pharmaceutical companies that were trying to develop new antibiotics have abandoned research in that field to pursue more profitable drugs.

There are few economic incentives for companies to develop antimicrobials, since these drugs are only used on a "one-time" basis, unlike other drugs that are used by patients with chronic diseases. Unfortunately, if this trend continues, we will be facing drug-resistant microbes with no weapons in our treatment arsenal. It is up to concerned citizens and individuals to make their voices heard on this important issue. The article provides excellent contact information for learning more about bills going through Congress regarding preventing the spread of antibiotic resistance by limiting antibiotic use on farms. "Super Bugged" succinctly summarizes all of the issues and concerns raised in this unit about the rise of "superbugs." This article presents research on DNA pollution and explains how the growing abundance of drug-resistant DNA present in our environment can spread to nonresistant microbes.

The Bacteria Fight Back

In their ongoing war against antibiotics, the bacteria seem to be winning, and the drug pipeline is verging on empty.

GARY TAUBES

Maybe it was just a bad month—an unfortunate statistical fluctuation. Maybe not. As Vance Fowler, an infectious-disease specialist at Duke University Medical Center in Durham, North Carolina, tells it, the first case appeared in early spring 2008: a 13-year-old girl whose bout with the flu evolved into a life-and-death struggle, still ongoing, with necrotizing pneumonia and a particularly pernicious strain of bacteria known as methicillin-resistant *Staphylococcus aureus* (MRSA). Should the girl survive, her life will be "forever changed," says Fowler, from pulmonary disease caused by the death of the lung tissue. The next case, a week or so later, was a research technician from Fowler's laboratory, admitted to the hospital with a facial abscess that showed no signs of healing. Again, MRSA was the cause. A week or so after that, the victims were a husband and wife. "Both were admitted with life-threatening acute MRSA infections out of nowhere," he says. "Multiple surgeries. Life- and limb-threatening infections." Neither one worked in a hospital or a long-term care facility, the kind of environments in which such bacteria might commonly be found. Nor had they visited one recently. So how did they get it? "Bad luck, bad genes, a bad bug, or all three," says Fowler.

The last decade has seen the inexorable proliferation of a host of antibiotic-resistant bacteria, or bad bugs, not just MRSA but other insidious players as well, including *Acinetobacter baumannii, Enterococcus faecium, Klebsiella pneumoniae, Pseudomonas aeruginosa,* and *Enterobacter* species. The problem was predictable—"resistance happens," as Karen Bush, an anti-infectives researcher at Johnson and Johnson (J&J) in Raritan, New Jersey, puts it—but that doesn't make it any easier to deal with. In 2002, the U.S. Centers for Disease Control and Prevention (CDC) estimated that at least 90,000 deaths a year in the United States could be attributed to bacterial infections, more than half caused by bugs resistant to at least one commonly used antibiotic. Last October, CDC reported in the *Journal of the American Medical Association* that the number of serious infections caused by MRSA alone was close to 100,000 a year, with almost 19,000 related fatalities—a number, an accompanying editorial observed, that is larger than the U.S. death toll attributed to HIV/AIDS in the same year.

So far these outbreaks have been concentrated in hospitals, where the environment is particularly conducive to the acquisition and spread of drug-resistant bugs. But the big worry, for Fowler and others, is that they will spread to the wider community—a nightmare scenario, he says. MRSA is particularly worrisome, but so is another class of bacteria, called Gram-negative bacteria, that are even tougher to defeat. These include *A. baumannii,* which has plagued injured soldiers returning from Iraq. For these bacteria, the pipeline of new antibiotics is verging on empty. "What do you do when you're faced with an infection, with a very sick patient, and you get a lab report back and every single drug is listed as resistant?" asks Fred Tenover of CDC. "This is a major blooming public health crisis."

Right Bug, Wrong Place

One of the many misconceptions about bacterial infections is that the bugs involved are not native to the human body or are particularly pernicious to begin with. Virtually all bacteria are capable of causing serious infections, at least in immuno-compromised patients, although most do not. In hospitalized patients, many infections arise from the patient's own bacterial flora, flourishing where they're not supposed to be. Pneumonia, for instance, can be caused when bacteria from the mouth are aspirated to the lungs. Just as *Escherichia coli* is a normal inhabitant of the gut, *S. aureus* colonizes the skin and mucosal surfaces in the nose in 30% of the population. When it sets up shop somewhere else, *S. aureus* can cause a host of infections, including skin abscesses, necrotizing pneumonia, joint infections, and heart valve infections known as endocarditis. Similarly, *S. epidermidis* normally colonizes the human skin, but when it gets into the bloodstream, it can cause sepsis and endocarditis, as well as infections involving prosthetic devices such as pacemakers and artificial joints. The risk of acquiring one of these serious infections is highest in hospitals and health-care facilities simply because these environments offer the greatest opportunities for bacteria to enter the bloodstream or infect open wounds.

Treating a bacterial infection with antibiotics is the obvious first step. But in the 65 years since the first widespread use of penicillin during World War II, infectious-disease specialists have been treated to an ongoing tutorial in the many ways bacteria can acquire and spread resistance to these drugs. Unlike tuberculosis and other bacteria, in which antibiotic therapy simply selects for rare resistance-bestowing mutants, the bacterial strains that are so insidious in hospitals employ far more diverse techniques, says Louis Rice of Case Western Reserve University and the Louis Stokes Cleveland VA Medical Center in Ohio. *S. aureus* and *Enterococcus,* for instance, can acquire resistance by exchanging entire genes or multiple genes with other bacteria, either through plasmids—loops of DNA that are independent of the bacterial chromosomal DNA—or so-called gene cassettes or transposable elements that can be inserted directly into the chromosomal DNA.

Penicillin and all penicillin-like antibiotics are ringlike molecular structures, known technically as β-lactams. They work by attacking a particular cell wall enzyme in the bacteria. The first strains of penicillin-resistant *S. aureus,* which arose within a few years of penicillin's introduction, were strains that have a survival advantage because they naturally produce an enzyme—penicillinase, one of a class of enzymes known as β-lactamases—that destroys the ring structure of penicillin. Within a decade, the effectiveness of penicillin against hospital-acquired staph infections was "virtually annulled," says microbiologist Alexander Tomasz of Rockefeller University in New York City, by "plasmid epidemics" that then spread the penicillinase gene through the entire species of *S. aureus.*

The pharmaceutical industry responded in the 1950s with a host of semisynthetic penicillins designed to be resistant to penicillinases. Methicillin, introduced in 1959, was believed to be the most effective. As Graham Ayliffe, a veteran hospital infection expert at the University of Birmingham in the U.K., recalled, "this was [supposed to be] the end of the resistant staphylococcus." Within 2 years, however, hospitals in Europe were identifying strains of *S. aureus* that were resistant to methicillin: the first MRSA strains.

Researchers later realized that these strains had taken a different route to acquiring resistance. Rather than generating new or different β-lactamases, which could attack the antibiotic directly, they had acquired a new gene entirely, called *mec*A, that coded for a variant of the antibiotic's target: the penicillin-binding protein. When the antibiotics attack the original penicillin-binding protein, explains Tomasz, this "surrogate" binding protein "takes over the task of cell wall synthesis" and works to keep the antibiotic at bay. The *mec*A gene itself, says Tomasz, appears to derive from a common bacterium on the skin of domestic and wild animals known as *S. sciuri.* How *S. aureus* came to acquire the gene is a mystery, but since it did, it passes it on by exchanging entire gene cassettes with the *mec*A gene on them.

Breaking Out

Through the 1970s and 1980s, MRSA remained little more than a nuisance bug, although occasional hospital outbreaks would have to be reined in with strict isolation and control programs. In the mid-1980s, typically only 1% to 5% of all *S. aureus*

isolates were methicillin-resistant, says Henry Chambers, an infectious-disease specialist at the University of California, San Francisco. Around that time, *S. aureus* began to acquire genes that confer resistance to other common antibiotics, apparently from methicillin-resistant *S. epidermidis* and carried on the same mobile cassettes as *mec*A. The result was a bug that was both far more difficult to treat and, as Chambers says, "pretty adaptive to surviving in hospitals." Today, 60% to 70% of all *S. aureus* strains found in hospitals are multidrug-resistant MRSA.

Worries intensified when MRSA appeared a decade ago as a community-acquired infection rather than one exclusive to health-care settings. In 1999, CDC reported on four deaths in Minnesota and North Dakota, all children, all caused by MRSA infections that could not be traced back to hospitalizations by either the patients or family members. Somehow the *mec*A gene had emerged in *S. aureus* strains outside hospitals or health-care facilities. "This was a real biological success story," says CDC's Tenover. "And it all happened off our radar screens." MRSA isolates then began to appear in a range of unexpected community settings: children in day-care centers, army recruits, athletes in contact sports, native Americans living on reservations, prison populations, intravenous drug users, and among men who have sex with men.

The possibility that these MRSA strains were simply hospital strains that had migrated out into the community was refuted by analysis of the gene cassettes carrying the resistance. In hospital strains, these cassettes are relatively large and carry multiple resistance-bestowing genes, explains Tomasz. The "oddball" cassettes carrying the *mec*A gene in the early community-acquired isolates were small and contained only the one gene. In the last few years, however, MRSA strains in the community have begun to acquire multidrug resistance, suggesting that they've been intermingling with the hospital strains.

A half-dozen community-acquired MRSA clones have now spread around the world as their prevalence in the community has continued to increase; in San Francisco, for instance, up to 50% of all *S. aureus* isolates outside health-care settings are now methicillin-resistant. "These methicillin-resistant strains seem to be replacing the susceptible strains of *S. aureus* in the general population," says Mark Enright, an infectious-disease specialist at Imperial College London (ICL), "which means people are carrying strains of MRSA in their nose in the community. Now when they get infections, ones that were formerly treatable are going to be replaced with difficult-to-treat infections."

The public health anxiety increased still further in 2002 with the detection of isolates of MRSA that were fully resistant to the antibiotic vancomycin, traditionally considered the last resort for treating resistant staphylococcal infections. These *S. aureus* isolates seem to have acquired a gene for vancomycin resistance—*van*A—from enterococci, and specifically *E. faecalis,* which are part of the natural flora of the intestinal tract and can cause serious infections in hospitalized patients. When the enterococci developed resistance to common antibiotics in the 1980s, physicians had responded by using vancomycin to treat them. Vancomycin-resistant *Enterococcus* (VRE) was first reported in 1986, and the *van*A gene soon spread throughout the species. Because enterococci readily exchange genetic

1940–2002

1940

Penicillinase, an enzyme capable of destroying penicillin, identified in bacteria

1942

First therapeutic use of penicillin

1943

Penicillin mass-produced

1945

More than 20% of *S. aureus* hospital isolates are penicillin-resistant as penicillinase begins to spread worldwide

1947

Streptomycin approved by FDA

1947

Streptomycin resistance observed

1952

Tetracycline approved by FDA

1956

Tetracycline resistance observed

1958

Vancomycin introduced, although rarely used until the mid-1980s

1959

Methicillin introduced

1961

Methicillin-resistant *S. aureus* (MRSA) observed

1964

Cephalothin, first antibiotic in the cephalosporin class, introduced

1966

Cephalothin resistance observed

1967

Gentamicin approved by FDA

1970

Gentamicin resistance observed

1976

Transferable penicillinase first observed in a gonococcus

1981

Cefotaxime approved by FDA

1983

Cefotaxime resistance observed

1983

First penicillin-resistant *Enterococcus* reported

1987

Vancomycin-resistant *Enterococcus* (VRE) observed

1987

First outbreak of *Klebsiella pneumoniae* resistant to third-generation cephalosporins

1996

S. aureus with intermediate resistance to vancomycin (VISA) reported

1999

Community-acquired MRSA reported

2000

Linezolid, first antibiotic in the oxazolidinone class, approved by FDA

2001

Linezolid-resistant *S. aureus* and VRE observed

2002

S. aureus with complete resistance to vancomycin (VRSA) observed

---

information with other bacterial species, says Tenover, he and other experts assumed that it would soon pass *van*A and vancomycin resistance to MRSA. "Everybody was waiting for the shoe to drop," says Tenover. In 2002 it did, when the Michigan Department of Community Health reported the first isolate of MRSA that had *van*A-mediated resistance to vancomycin. The patient was a 40-year-old diabetic who had recently been given an extended course of vancomycin for a foot ulcer.

Fortunately, vancomycin-resistant MRSA—now known as VRSA—has not developed into the nightmare researchers feared. Only nine isolates have been detected worldwide in

6 years, seven from the same region in Michigan, which suggests that *S. aureus*, unlike *Enterococcus*, loses the ability to compete in the broader environment when it takes on the *van*A gene. That only one of these infections was life-threatening suggests that the vancomycin-resistant bug also loses its virulence.

A Paltry Pipeline

Although MRSA and other Gram-positive bacteria remain a major threat, a half-dozen new antibiotics have either just been approved or are in the pipeline that should work well against them—at least until the bugs evolve more resistance. This is not the case, however, for Gram-negative bacteria, such as *P. aeruginosa, A. baumannii,* and *K. pneumoniae.* These bacteria have both an inner and outer cell membrane, as opposed to the single cell membrane of Gram-positive bugs like MRSA. (The name comes from how these bacteria stain on a Gram stain test.) The pipeline for antibiotics against Gram-negative bacteria, says Bush of J&J, is limited to development programs in a few small companies; only one drug has made it through phase I clinical trials.

Prompted by the emergence of MRSA and VRE in the late 1980s, pharmaceutical companies focused their attention on Gram-positive bugs. Meanwhile, many Gram-negative bugs became resistant to virtually every known antibiotic, or at least every antibiotic that isn't toxic. "These organisms may well start to spread into the community," says Tenover, "and then we really will be in trouble. We have drugs to fall back on for *Staphylococcus.* But when you say, 'Where's the next anti-*Pseudomonas* drug?' I have to scratch my head."

One reason for the dearth of drug candidates is that Gram-negative bacteria are simply harder to kill. First, they have the extra cell membrane the drug has to penetrate. Then they have other defense mechanisms that Gram-positives lack, such as the ability to activate pumps or close down protein channels in the membranes that let these antibiotics in. "They can have three or four mechanisms working at once," says Case Western's Rice. "Even if you develop a new drug entirely, these bacteria may be just as likely to be resistant to new drugs as old ones. It's just really hard."

The problem has been exacerbated by the gradual exodus of pharmaceutical companies from antibiotic development—a trend that began in the 1980s and has accelerated since 2000, in large part because the market is iffy and the chances of success are slim. Of the 15 major pharmaceutical companies that once had flourishing antibiotic discovery programs, eight have left the field entirely, and two others have reduced their efforts significantly. That leaves only five—GlaxoSmithKline, Novartis, AstraZeneca, Merck, and Pfizer—that still have antibiotic discovery efforts commensurate with the size of the problem.

Even though the market for antibiotics is in the neighborhood of $25 billion a year, says Steve Projan, vice president of biological technologies at Wyeth Research in Cambridge, Massachusetts, other drugs, such as antidepressants or antihypertensives, offer a greater bang for the buck because they are often taken for years or decades rather than just a 7- to 14-day course. Resistance only compounds the problem: A drug that takes a decade to develop might be useful clinically for only a handful of years.

What's more, the better the antibiotic, the less health experts want to see it used to avoid the development of resistance. "It's probably the only area of medicine where a drug company can invest all this money to develop a drug, come up with a good one, and then the so-called thought leaders in the field, like myself, will tell people not to use it," says Rice. "We say it's such a good drug that we should save it."

As a result, virtually all the new antibiotics and all those in the pipeline for Gram-positive bacteria are second-generation drugs, that is, incremental improvements on existing classes. The one conspicuous exception—daptomycin, developed for *S. aureus* by the late Frank Tally at Cubist in Lexington, Massachusetts—was first identified 20 years ago by Eli Lilly and Co. and then shelved because it had toxicity problems at high doses.

To the surprise of many, the recent sequencing of more than 650 bacterial genomes has been a "dismal failure" when it comes to drug development, says ICL's Enright. Although genome sequences were expected to yield a "treasure trove of new targets for entirely new classes of antibiotics," as David Pompliano and colleagues at GlaxoSmithKline in Collegeville, Pennsylvania, recently wrote, this simply hasn't panned out. At GlaxoSmithKline, Pompliano and his colleagues spent 7 years and more than $70 million evaluating more than 300 "canonical" bacterial genes that they thought were essential to the viability of the bacteria. The result was just five leads, a success rate, they estimated, that was four-to fivefold lower than for other areas of therapeutics.

Genomics is simply not a good paradigm for discovering new antibiotics, suggests Projan. The genetic approach assumes that a candidate drug can knock out a single gene in the bacterium to render it unfit for survival. But the drugs don't knock out a gene's activity entirely, he says; instead they modulate activity. "As we found out in oncology," says Projan, "sometimes leaving even 5% activity is enough for the tumor to grow. The same thing is true for bacteria." Projan and others suggest that the only route to a new antibiotic—short of pure luck—will be through more fundamental research on the basic biology of the bacteria.

Cutting Back

Barring the discovery of miracle antibiotics to which bacteria cannot evolve resistance—a "laughable" notion, says one expert—the only foreseeable route to curbing antibiotic resistance will be to rein in the use of antibiotics. One obvious way is to lower the risk of acquiring resistant bugs in the hospital. Countries that have mandated rigorous infection control in hospitals, such as Denmark, the Netherlands, and Finland, have been able to keep MRSA infection rates low. These infection-control procedures, however, go far beyond physicians and nurses wearing gloves and protective masks and washing their hands before and after patient contacts, essential as those are. These nations employ a technique known as "active detection and isolation," or "search and destroy," as it's called in the Netherlands. Patients considered at high risk of carrying MRSA and other antibiotic-resistant bugs are cultured when they're admitted to the hospital, and periodic cultures are taken of all patients, particularly those in high-risk wards. The greater the prevalence of pathogens and risk factors, the more frequent this surveillance. Patients who are infected or are carriers are isolated. Health-care workers who are colonized with resistant bacteria can be "decolonized," using skin washes and nasal ointments.

Collateral Damage: The Rise of Resistant *C. difficile*

In April 2002, Mark Miller, an infectious-disease specialist and microbiologist working at Jewish General Hospital in Montreal, Canada, began to suspect that he had an outbreak on his hands. He was used to dealing with the bacteria *Clostridium difficile*, which can cause severe diarrhea in debilitated patients and had been a common problem in hospitals for more than 30 years. But now the number of cases had started to climb, as did their severity. "One of the first indications that we knew we had a problem," says Miller, "was when one of the colorectal surgeons called me and said, 'I just took out my second colon in a month on a *C. difficile* patient.' When we started looking at the numbers, they were absolutely horrendous." At the peak of the outbreak, says Miller, there were 50 new cases of *C. difficile* diarrhea every month in their 600-bed hospital. "Of those, about one in six was dying or going for a colectomy. That's kind of staggering."

Resistance to antibiotics makes for bacteria that are harder to kill, but it can also bestow on a bacterial strain the advantage it needs to spread through the hospital environment and perhaps around the world—a kind of collateral damage in the escalating war between man and microbes. *C. difficile* is an unfortunate case in point. The bacterium has currently been linked to at least 5000 deaths a year in the United States; at the height of the Quebec epidemic it caused more than 7000 serious infections and 1200 deaths in a single year. In many hospitals, *C. difficile* constitutes a greater risk to patients than methicillin-resistant *Staphylococcus aureus* or any other bacteria.

The symptoms of a *C. difficile* infection range from mild diarrhea to severe colitis, and the elderly bear the brunt of the disease. One in four patients will have a recurrence or multiple recurrences. "It's a horrible problem," says Dale Gerding, an infectious-disease specialist at Hines Veterans Administration Hospital and Loyola University in Chicago, Illinois. Patients have to be treated almost constantly with oral vancomycin to prevent recurrences, he says.

C. difficile diarrhea first appeared in the medical literature in the 1970s, mistaken for a side effect of the antibiotic clindamycin. In 1978, physicians realized that the diarrhea was induced by toxins from clindamycin-resistant *C. difficile*, which had colonized the victim's colon after their normal gut flora had been decimated by the clindamycin treatment. *C. difficile* has remained a common hospital infection ever since because the bacteria produce heat-resistant spores that are exceedingly difficult to kill. "They're very resistant to detergents and cleaning agents," says Gerding. "Really, the only thing that destroys them is bleach or hydrogen peroxide."

Through the 1990s, however, *C. difficile* wasn't considered a major threat because the bacteria were susceptible to two antibiotics, vancomycin and metronidazole, the latter of which is inexpensive. As many as 40% of all hospitalized patients are colonized with *C. difficile,* but most tolerate it without symptoms. A series of hospital outbreaks in six U.S. states, beginning around 2000 and capped by the severe Quebec outbreak in a dozen hospitals, suggested that a new, hypervirulent strain of *C. difficile* was circulating.

Since then, the same offending strain has been identified in hospitals in 38 states and has also been linked to outbreaks in Western Europe. What sets it apart from its predecessors, say Gerding and Miller, is its high resistance to the newer fluoroquinolone antibiotics, such as levofloxacin and moxifloxacin. These antibiotics began to be used widely in the late 1980s, and usage has increased steadily ever since. Why this strain induces more severe disease—with a death rate among those infected of 10% compared with 1% in the 1980s—is still a mystery, but one possibility, says Gerding, is a mutation that enables the strain to produce more toxin.

Although Quebec hospitals have reduced the incidence of *C. difficile* infections by two-thirds since the height of the outbreak, through very tight isolation and control and rigorous "housekeeping," says Miller, they have yet to get back to the levels preoutbreak. "*C. difficile* in health-care facilities and hospitals is a very unforgiving organism," he says. It exploits "any lapse in isolation, in housekeeping, in hygiene—whatever it is—and it comes back with a vengeance."

—G.T.

Whether U.S. hospitals should be required to implement active detection and isolation is a long-running controversy. Some specialists—led by University of Virginia epidemiologist Barry Farr, an expert on controlling VRE and MRSA—have argued that it's the only proven method to control hospital MRSA infections. Others have questioned the technique's cost-effectiveness and viability, particularly when rates of MRSA in the community are beginning to rival those in many hospitals.

Vaccines against antibiotic-resistant bacteria would also go a long way to reining in resistance, but only one such vaccine candidate, against *S. aureus,* has ever made it through phase III clinical trials: StaphVAX, licensed by Nabi Biopharmaceuticals. Although patients who received the vaccine had significantly lower rates of *S. aureus* infections at 40 weeks compared to controls, this apparent protection was lost at 54 weeks. A follow-up trial also failed to demonstrate efficacy. Several more vaccines are in development, including a new-generation StaphVAX. Even temporary protection could be useful, argue some experts, either for health-care workers, who could be vaccinated regularly, or for patients who are about to be hospitalized to undergo a procedure.

Ultimately, physicians will have to be persuaded to reduce their use of antibiotics, although that will be a hard sell. One step, for instance, would be to persuade physicians outside hospitals to treat only those patients who are truly infected. A 2001 study from the University of Colorado Health Sciences Center estimated that 55% of all antibiotics prescribed in the United States for upper respiratory infections were unnecessary. This is what Rice describes as the "get-a-little-sniffle-get-a-little-Levaquin" problem. "The patients want it," he says, and "the doctor wants to get the patient out of the office, and the quickest way to do it is to write a prescription." But the societal problem of antibiotic resistance should outweigh whatever personal peace of mind comes from the indiscriminate use of antibiotics, says Tenover.

Similarly, many times physicians prescribe combination "broad-spectrum" antibiotics when a single "narrow-spectrum" antibiotic would do the trick. Understandably, says Rice, physicians are unwilling to wait to treat serious infections until the bug is cultured and they learn to which antibiotics it's still susceptible. But once the crisis is over, usually 1 to 3 days after starting therapy, physicians could switch their patients to the appropriate narrow-spectrum antibiotic.

What the field desperately needs, these experts say, are randomized, controlled trials to establish how long antibiotic therapy should be prescribed for different infections. The data are scarce, and misconceptions abound. The ubiquitous advice in the field—from physicians, patients, and even CDC—is that patients should continue the full course of antibiotics even after they feel better. Because antibiotics tend to have few side effects, physicians consider a longer course to be a no-lose proposition.

But from the perspective of preventing antibiotic resistance, says Rice, "this is totally wrong-headed." In patients with healthy immune systems, he explains, most antibiotics merely stun the bacteria sufficiently to make it easier for the host immune system to do its job. "You can take tetracycline until the cows come home," Rice says, and "all it does is stop most bacteria from growing. It doesn't kill them." Extending the course of the antibiotics unnecessarily increases the likelihood that the patient's normal flora will be inhibited to the point that bacteria resistant to the antibiotic will fill the void.

The few existing studies on the necessary length of therapy have suggested that it is often surprisingly short. Urinary tract infections in young women can be treated with 1- to 3-day courses of antibiotics. The Infectious Diseases Society of America recommends a 3-day treatment for traveler's diarrhea, while acknowledging that 1 day appears to be equally effective. Studies from the 1940s suggested that the "vast majority" of patients with pneumonia get better after 2 or 3 days, says Rice: "Somewhere along the line, that morphed into 7 days, 10 days, 21 days, with no real reason other than making the doctor more comfortable." In May, the U.S. National Institute of Allergy and Infectious Diseases responded to the expert demand and put out a request for proposals for clinical studies that would determine the optimal use of antibiotics, including the optimal duration of therapy. "I think most physicians would respond to compelling data from a well-done trial," says Rice.

One beneficial side effect to curbing antibiotic use is that it may serve to rehabilitate those antibiotics that have lost their effectiveness. "Many of these were wonderful new drugs just 20 years ago, able to treat a wide variety of bugs, both inside and outside the hospital," says Rice. "Now we're at a point where some of them are next to useless, because they've been used for everything."

Germ Warfare

Are we creating a new generation of super bugs?

J. GLENN MORRIS, JR

Merry King was a 48-year-old special education teacher in the suburbs of Washington, DC. She first felt abdominal pain on December 1, 2007, according to *The Washington Post*.

A doctor in a local emergency room, suspecting that ovarian cysts were to blame, sent her home with painkillers. Though she was later hospitalized, King died of methicillin-resistant *Staphylococcus aureus,* also known as MRSA, on December 9.

Invasive MRSA infections strike roughly 94,000 people in the United States each year, according to the Centers for Disease Control and Prevention. An estimated 18,650 of them die, more than the number who die each year of AIDS.

It's not just *Staph.* Physicians are finding resistant strains of *E. coli, E. faecium, Klebsiella, Acinetobacter, Pseudomonas,* and other bacteria. And hospitals are no longer the only place they strike.

The mere we use antibiotics—to treat illness and to raise animals for food—the closer we come to ravaging infections that the "miracle" drugs are powerless to treat.

According to news reports, U.S. Marine Jonathan Gadsden died of a resistant *Acinetobacter baumanii* infection a month after surviving life-threatening injuries in Iraq, and Nicholas Fusco, 57, died after catching methicillin-resistant *Staphylococcus aureus* (MRSA) and vancomycin-resistant *Enterococci* (VRE) while in a New Jersey hospital for minor abdominal surgery. No one knows how 39-year-old Robert Sweitzer of Tucson, Arizona, got the MRSA that killed him.

And antibiotic-resistant bacteria are spreading beyond the hospital. In June, researchers at the University of Iowa found MRSA in 70 percent of the pigs on 10 farms in Iowa and Illinois and in 45 percent of the workers on those farms. Although cooking kills the bacteria, anyone who touches the raw meat could easily end up with an antibiotic-resistant infection. And odds are that it won't be long before most *Staph* is resistant *Staph.*

Here's what you can do to prevent the spread of bacteria that are immune to antibiotics.

Q: What is antibiotic resistance?

A: It's the ability of microorganisms to protect themselves or to fight off the effects of antibiotics, which are compounds that kill bacteria. As a normal part of life, bacteria evolve to sidestep the mechanisms by which antibiotics can kill them.

Q: So medicines no longer work?

A: Right. Bacteria are constantly evolving and changing so as to better survive in their environments. Since the beginning of the antibiotic era, we have used increasing amounts of antibiotics with the intent of killing off harmful bacteria.

But that has driven the evolutionary process so that it's become highly advantageous to the bacteria that have the genes for resistance, because they can live and all the others die.

Q: So antibiotics don't last forever?

A: True. We now recognize that with enough time, antibiotic resistance will emerge in virtually every instance. It's not whether—but when—strains will become resistant to specific antibiotics.

So we have a window during which an antibiotic has the ability to kill certain classes of bacteria. But ultimately, the bacteria are going to mutate and continue to grow despite the antibiotic.

Q: Many people don't remember a time when infections were lethal?

A: Right. Much of modern medicine is based on the assumption that we can quickly and effectively kill any bacteria that are causing infections. That understanding has begun to change as bacteria have become increasingly resistant to antibiotics.

Antibiotics are treasures. We need to use them in a way that minimizes the chance—or the speed—with which resistance will develop.

Beyond MRSA

Q: Which resistant bugs are spreading?

A: Methicillin-resistant *Staphylococcus aureus,* or MRSA, probably has received the most recent attention. It has been a cause of serious infections in hospitals for many years, but only recently has spread into the community, causing infections in people who have not had any contact with hospitals. It is of particular concern because many of the strains being seen in patients outside of hospitals appear to have increased potential for virulence—that is, for causing disease.

Q: And it's not just Staph?

A: Right. Resistance is growing among a number of different organisms in hospitals—like *Enterococcus, Klebsiella, Acinetobacter, Pseudomonas,* and *Enterobacter*—that most people aren't familiar with. It's an ongoing problem in hospitals, where bacteria are developing increasing levels of resistance, because that's where we use the most antibiotics.

But with MRSA, and potentially with other bacteria like *Pneumococci,* we are beginning to see resistant microorganisms outside hospitals.

Q: Why?

A: One reason is the increasing movement of patients out of hospitals and back into the community. It used to be that you'd come in and stay in the hospital for a week. Now you come in and you're there for three hours and they kick you out.

That has increased the antibiotic use in the community and put sicker patients back into the community. And physicians outside of hospitals are prescribing antibiotics more often, so we have more resistance everywhere. It's a major concern.

How Bacteria Resist

Q: How does resistance spread?

A: We are awash in bacteria. We don't see them and aren't aware of them, but our world is swarming with both good and bad bacteria. They are all over everything and, in particular, they are all over us and throughout our bodies. They're in our intestinal tract, and our skin, nose, mouth.

So if you get a strain of bacteria that is antibiotic-resistant, it becomes fairly easy for the genetic material that causes resistance to move to the non-resistant bacteria that have taken up housekeeping right next door.

Q: How?

A: Bacteria are highly promiscuous. It's amazing how often bacteria are swapping off their DNA. Sometimes the genetic material moves on a plasmid, which is a strand of DNA. Sometimes it moves on jumping genes—what we call transposons—or integrons.

So if you bring in a strain of bacteria that has a resistance gene, and you also have a patient who may be taking an antibiotic, there's a high likelihood that those genes are going to move not only to the same species, but to other related species as well.

Q: How can hospitals keep infections from spreading?

A: Traditionally, we've waited to do a culture until patients developed an infection, but they could have spread the organism to a dozen patients by the time we figure it out.

Now there's such a high level of resistance in the community—particularly of MRSA—that there's a raging argument as to whether we should assume that everyone is carrying resistant organisms.

If we made that assumption, everyone who came into the hospital, particularly into an intensive care unit, would be isolated until we proved that they *don't* have resistant organisms.

Q: Does resistance spread because people travel so much?

A: A significant proportion of the resistance that is being seen in foodborne pathogens is associated with overseas travel, but it's hard to know about other pathogens because we're only now starting to collect good international data.

It is clear, however, that we're dealing with a global issue where there is potentially rapid movement of resistant bacteria from one part of the world to another. Bacteria are constantly changing, modifying, and recombining. So if resistant strains are introduced in the United States, we can expect those resistant strains—and genes—to spread.

In the Doctor's Office

Q: How can people avoid unnecessary antibiotics?

A: Many people go to their doctor with a cold or the flu and expect to walk out with a prescription for an antibiotic. If they don't get one, they feel like they've been shortchanged.

Q: So they ask for antibiotics?

A: Yes. But antibiotics are only good for bacterial infections. They don't help for the majority of upper respiratory infections, which are caused by viruses. We've been trying very hard to educate physicians, but we also need to educate patients to minimize antibiotic usage. We need to reduce the evolutionary pressure on bacteria to become resistant.

Q: If you take antibiotics, are you more likely to get a resistant bug later?

A: Yes. For example, in studies we did in Baltimore, we found that the highest rates of antibiotic-resistant *Pneumococci* were not in the inner city, but in the rich suburbs, where kids always get taken to the doctor and always get an antibiotic.

They were more likely to be carrying resistant organisms than a kid in the city who never goes to the doctor and never gets antibiotics. So yes, you should try to minimize antibiotic use.

Q: Unless you really need them?

A: Yes. Clearly there are times when antibiotics are wonder drugs. But sometimes it's a borderline call as to whether or not you need an antibiotic. And in those circumstances, it's better if you don't take that antibiotic because that makes it less likely that the bugs that are on and in you will become resistant.

Q: In what cases do antibiotics not work?

A: There's no advantage to taking antibiotics for most upper respiratory infections—colds, flu. And antibiotics may not do that much for vague bronchitis or for garden variety sinus infections. We don't have strong data that antibiotics are highly effective for a whole series of infections. [See "Bacterial or Viral?" box]

Of course in some instances antibiotics are lifesaving. So it's not that you should never take antibiotics. But ask your physician, "Is this antibiotic necessary?" rather than "What do you mean you're not giving me an antibiotic?"

We need to make that switch. We need to recognize that, to the bacteria that inhabit our bodies, the amount of antibiotics that we use makes a difference.

Q: Are antibacterial soaps a problem?

A: They may be increasing resistance, but it's controversial. Many people have this vision that bacteria are few and far between—like roaches. Let's spray and get rid of them.

The reality is that you're never going to get rid of bacteria, and you don't want to. Our bodies are covered with good bacteria, and it's becoming increasingly apparent that one of our strongest defenses against infection is a group of good bacteria that don't have antibiotic resistance and don't have nasty virulence genes.

Q: Why do we need those bacteria?

A: We want them happily enveloping our entire body—our intestinal tract and our nose and mouth—so when a bad bug does come along, it doesn't have any place to live because the good guys kick it out.

When you take antibiotics that kill off the normal bacteria, there's a vacuum. So you open the door for fairly nasty bacteria that carry resistance genes to move in. That's not a good thing. What you want are normal, happy, friendly, innocuous bacteria everywhere in and on your body.

New Drugs

Q: Can we expect new antibiotics to solve the problem?

A: We don't have many good antibiotics coming down the pipeline. We're moving toward the potential perfect storm, where we have no new antibiotics combined with increasing levels of resistance in microorganisms. We see an inexorable increase in resistance levels, and it's been very difficult to slow it down.

Q: And companies won't work on new drugs if doctors try to use them only rarely?

A: That's the problem. Drug companies have moved away from antibiotics because it's a low-profit area.

It costs a tremendous amount of money to bring an antibiotic to market. Yet if the antibiotic works, a patient may only take it for 3 to 7 days and may never take it again.

Compare that with a drug for hypertension or cholesterol, which a large chunk of the population takes on a daily

basis. There you have a vast market and a lot of money flowing in.

In contrast, people take a much smaller amount of antibiotic. And there are so many antibiotics that the market share for a new one is going to be minimal. From a purely financial standpoint, developing antibiotics makes no sense whatsoever.

Somehow, we need to figure out a way to create the necessary financial incentives so that new antibiotics continue to be produced. If we don't, we're going to be in major trouble.

Q: Are scientists working on new antibiotics to fight bioweapons?

A: Yes, but it's unclear how much impact that's going to have because the research has focused on particularly dangerous pathogens like anthrax or plague or smallpox that hopefully you will never encounter.

Animals

Q: Why are antibiotics used on farms?

A: They are used as a growth enhancer at the subtherapeutic level—that is, at lower doses than you would use to treat an infection. Low doses are particularly good at selecting for resistant strains.

What I mean is that heavy doses kill off *all* the bacteria, including those that have developed weak resistance. In contrast, subtherapeutic doses may not be strong enough to kill off the weakly resistant ones, opening up the opportunity for further evolutionary changes and gene exchange.

Q: How do you know that antibiotic resistance occurs on farms?

A: We looked at a confined animal feeding operation that used tetracycline at subtherapeutic levels to enhance growth. It was an environment where there was constant antibiotic pressure.

When we sampled all across the farm—soil and water and all through the farm environment—we found that the resistance genes for tetracycline had permeated the entire bacterial community. Everything was tetracycline-resistant. And the bacteria carried between four and five genes that made them resistant to tetracycline.

We see the potential for the same thing in humans. When there is long-term use, or even short-term use, of an antibiotic, you see the evolution and transfer of antibiotic resistance genes from one species or strain to another.

Q: How do resistant bacteria from the farm reach people?

A: It works at two levels. You could touch raw chicken and then a raw food like salad or eat some undercooked chicken and became infected with a foodborne pathogen such as a resistant *Campylobacter* or *Salmonella*. If so, the antibiotics used to treat those infections may not work. That's of concern, but it's a straightforward process.

What's much more complicated is that ordinary food may not carry disease-causing bacteria, but it's still covered with *Enterococci* and other harmless bacteria. And those bacteria

Bacterial or Viral?

Should you see a doctor when you get a bad cold, cough, or sore throat? Before you decide, remember that cold symptoms typically last 7 to 11 days. Don't be surprised if a cough or runny nose lasts 2 weeks. Doctors often start to worry about pneumonia if a cough lasts more than 3 weeks (and when patients have other symptoms). Sometimes, you just have to wait it out.

Bacterial Infections Cause

- Some ear infections
- Severe sinus infections
- Strep throat
- Urinary tract infections
- Many wound and skin infections

Viral Infections Cause

- Most ear infections
- Colds
- Influenza (flu)
- Most coughs
- Most sore throats
- Bronchitis
- Stomach flu (viral gastroenteritis)

Source: www.mayoclinic.com/health/antibiotics/FL00075.

may carry resistance genes that they picked up in the farm environment.

Q: Because there are so many antibiotics on the farm?

A: Right. And if those bacteria are ingested or come into contact with humans, their resistance genes may transfer to harmless bacteria that live in our bodies.

Q: So the bugs that cause urinary tract infections may pick up the resistance genes from normal bugs?

A: Exactly. Several years ago, a very interesting study looked at a particularly nasty strain of *E. coli* that causes UTIs. The investigators thought that this strain was spreading so rapidly on a national level because it had a foodborne origin. That would explain how it might spread in multiple communities across the country.

We tend to think about discrete diseases, like a urinary tract infection or a foodborne disease infection. But in reality, UTIs generally arise from bacteria that are present in your intestinal tract. The bacteria in your intestinal tract may come from food or may be influenced by the bacteria that you ingest.

Q: Do farms cause more resistance than hospitals?

A: We don't know. Even though there's a much higher likelihood that you're going to get a resistant organism in the hospital, not that many people go into hospitals. It's a relatively rare event.

However, there is enough food consumed that in some mathematical models we've created, the risk of acquiring resistance from the farm environment works out as a significantly greater risk than the hospital environment.

Off Their Meds?

In recent years, under pressure by health activists and the medical community to protect antibiotics, a growing number of companies have promised to cut back on the drugs. (We say "promise" because no one tests the animals' feed to make sure that it's antibiotic-free.)

Here we list companies that claim—on their labels or Web sites—to use no antibiotics, even to treat illness. Any certified organic meat or poultry also fits into this category ("No antibiotics ever").

Companies in the second list ("No antibiotics, except to treat illness") say that they don't use the drugs to make animals grow faster or to keep illness from spreading.

Foster Farms, for example, says that only roughly 1% of its chickens get treated with antibiotics. Tyson, the nation's largest poultry producer, says that it sells the few chickens that are treated with antibiotics (again, about 1%) to food service operations, not to grocery stores.

However, Tyson—like nearly all poultry producers—does start with eggs that have been injected with antibiotics before they hatch. In fact, even organic chickens often come from antibiotic-injected eggs. Apparently, it's tough to find eggs that aren't injected. (That's one reason why the U.S. Department of Agriculture's definition of organic starts when the chicken is two days old).

And Tyson (and most other chicken producers) sometimes give their animals a class of anti-microbial drugs called ionophores. They are used primarily to kill protozoa and aren't used in people, so there is little worry that they will lead to antibiotic resistance.

Despite those quibbles, chicken companies deserve credit. "Poultry producers have cut back in the last five years or so," says Rebecca Goldburg, senior scientist at the Environmental Defense Fund in New York. "Pork is the biggest problem now."

If you want to know whether your meat or chicken is raised with antibiotics, ask the company. Just listen carefully to the answer.

If the company boasts that it doesn't use fluoroquinolones, for example, don't be impressed. The FDA banned fluoroquinolones—a class of antibiotics that includes Cipro—in chicken feed in 2005. (The FDA proposed a ban in 2000, but Bayer Corporation launched a lengthy battle that delayed the ban by nearly five years.)

What led the FDA to act? Up to 80 percent of broiler chickens in some supermarkets are contaminated with *Campylobacter* bacteria, and fluoroquinolone-resistant strains were making people sick. By 2002, one in five *Campylobacter* infections was resistant.

Portions of the industry have stopped using antibiotics, particularly with the movement toward organic foods. But we still do not have good national data on the amount that is used, so it's difficult to know what proportion is attributable to agriculture vs. human therapeutics.

Breaking the Drug Habit

Here's a sampling of companies that have cut back—or cut out—antibiotics.

No Antibiotics Ever

- Any organic meat or poultry
- Bell & Evans chicken
- Cargill "Good Nature" pork
- Chipotle chicken, beef, pork
- Coleman Natural
- Laura's Lean Beef
- Murray's Chicken
- Nature Select chicken
- Panera chicken (except soups, Chicken Salad Sandwich, and Frontega Chicken Panini)
- Wegman's "Food You Feel Good About" beef, chicken, pork
- Whole Foods
- Wild Oats

No Antibiotics, Except to Treat Illness

- Foster Farms chicken
- Tyson chicken

Q: Some people say that 70 percent goes to agriculture.

A: That's one estimate. The industry says it's closer to 10 or 20 percent. Somewhere in the middle is the truth.

The reality is that the government does not have the authority to get production figures from drug companies. So we have no way of knowing what the usage rates are.

Q: Why doesn't the food industry stop using antibiotics?

A: In many parts of the food industry, profit margins are razor thin. Companies claim that antibiotics can make the difference between bankruptcy and success.

In contrast, Europe has moved away from antibiotics. But as the industry here points out, Europe works on a much smaller scale. The food industry feels that Europe can afford to do it, but we don't have that luxury to stop in such a cut-throat business.

Q: How should it be done?

A: We need a national goal to reduce the overall antibiotic use in hospitals, outside of hospitals, and in agriculture. I would like to see us move toward the European model, where there is minimal subtherapeutic antibiotic use in food agriculture.

Q: Are some companies doing that?

A: A number have dropped back from the subtherapeutic use of antibiotics as a growth enhancer. My impression is that it's still widespread. Until we can get good data, there is no way we can find out what's going on.

J. GLENN MORRIS, Jr., is an Internationally recognized public health scientist and physician who directs the University of Florida's Emerging Pathogens Institute. He began his career at the Centers for Disease Control, has served on four National Academy of Sciences food safety committees, and was part of a U.S. Department of Agriculture team that helped rewrite the nation's food inspection regulations. Morris is a member of *Nutrition Action*'s scientific advisory board.

Routes of Resistance

Our focus on using antibiotics to kill bacteria has blinded us to their diverse functions in the organisms that make these chemicals.

Robert L. Dorit

Last year, some 50 million pounds of antibiotics were used in the United States, an amount that would correspond to roughly 5 tablespoons—or 75 doses—of antibiotics per person. In fact, much of this antibiotic—as much as 70 percent by some estimates—is being used not to treat infections, but instead to promote food production, as antibiotics have become a key ingredient in the American food chain. We are, in short, marinating the living world in antibiotics.

Against this backdrop, the emergence of antibiotic-resistant pathogenic bacteria can hardly come as a surprise. Antibiotics, after all, are used to suppress and kill bacteria, and bacteria, like every other living thing, have no greater evolutionary imperative than to stay alive. Our overuse of antibiotics methodically rewards any bacterium fortunate enough to carry a mutation that confers even slight resistance to the substance in its environment. The less-fortunate bacteria die, leaving behind empty ecological space for resistant strains to fill. Slowly and inexorably, we thus enrich the world with bacteria that now scoff at our attempts to control them.

The Rise of Resistance

It did not take long for antibiotic resistance to emerge once the drugs entered the arsenal of modern medicine. Penicillin, discovered in 1929, came into widespread use on the battlefield during World War II: Some 100 million doses were produced between 1943 and 1945. But even before the end of the war, the first penicillin-resistant strains of bacteria appeared. Over the next five decades, the pattern repeated numerous times. The discovery of every new antibiotic class created excitement by promising to bring bacterial infections to heel. Inevitably, however, the edge of this new tool for the treatment of infections became blunt with overuse as resistant strains grew common. The useful clinical life of new antibiotics, defined as the time between clinical introduction and the rise of resistant strains, is often no more than a matter of three to five years. Over the past decade, moreover, antibiotic resistance, once only a public-health concern, has morphed into a measurable cause of death. Conservatively, untreatable bacterial infections result in some 100,000 deaths per year in the United States. The early promise of the age of antibiotics—an end to infectious diseases—now seems absurdly naive.

But every day, in clinics and hospitals around the world, doctors are forced to administer antibiotics to clear up bacterial infections that threaten the lives of patients. And every day, the bacteria—both the pathogens we are targeting and the innocent bystanders that end up as collateral damage—respond by evolving mechanisms to resist the antibiotics. Each dose is a skirmish in a larger war. Yet, focused as we are on what we see as an epic battle between humans and pathogens, we forget that both antibiotics and antibiotic resistance have been part of the microbial world for the past 3 billion years. Yes, the widespread use of antibiotics to combat infections has had profound implications for the evolution of infectious disease, but it is a mere blip in the history of the microbial world.

Four out of five antibiotics in use today are based on naturally occurring compounds produced by bacteria and fungi. The organisms that produce antibiotics have not, of course, been doing so over billions of years just so that we might discover them and put them to therapeutic use in the 21st century. So what are antibiotics doing for the organisms that produce them? The explanation most obvious to us, of course, is to see these compounds as part of the competitive arsenal of the bacteria that produce them. In this model, antibiotic-producing bacteria kill their competitors and make ecological space available for themselves and their genetic kin. Seen this way, antibiotics in microbial ecosystems and antibiotics in the clinic do one and the same thing: kill bacteria.

To be sure, naturally produced antibiotics in intact microbial ecosystems enable producing strains to kill other bacteria. Some antibiotics, such as a class of proteins called the bacteriocins, are not subtle: A single molecule entering a target cell will kill it. These molecules are designed to kill and nothing more. But the concentrations of many other antibiotics in natural ecosystems seem consistently too low to kill surrounding organisms effectively. Why would bacteria bother to produce antibiotics—which are expensive molecules to synthesize—at concentrations too low to reap the benefits? It now appears that at lower concentrations, antibiotics may well, to the bacteria, mean something utterly different than they do at the massive concentrations encountered in the clinic. In their original context, antibiotics may not be killers at all, but instead messengers enabling cell-to-cell communication both within and across bacterial species boundaries.

Off-Label Meaning

Researchers such as Fernando Baquero and Jose L. Martinez of the National Center for Biotechnology in Madrid, and Julian Davies of the University of British Columbia, among others, have begun to articulate a radically different perspective on the role played by the now-mislabeled "antibiotics." Far from being just stone-cold killers, naturally occurring antibiotics appear to be part of the vocabulary of bacteria, words in the nuanced language of this world unseen by humans. Like language, antibiotics are capable of conveying multiple meanings to different recipients.

At sublethal concentrations, antibiotics can have profound and unexpected effects on surrounding cells. Several different antibiotics, when present at a small fraction (less than 1 percent) of their lethal concentration, coordinate the expression of whole sets of genes in bacteria that sense the antibiotic. These coordinated responses are not simply the molecular expression of panic (the aptly named "SOS response") expected when an antibiotic is present. The signal may, for instance, induce *Pseudomonas aeruginosa* bacteria to develop into a bacterial biofilm—an architecturally complex, surface-bound conglomeration of cells—which, regrettably, makes bacteria much less sensitive to antibiotics in clinical settings. In another irony, low concentrations of antibiotic appear to upregulate the expression of a suite of genes responsible for increased virulence, in effect transforming a benign bacterium into a pathogen. These and other examples, in short, suggest that antibiotics may be far more than just toxic substances designed to kill all bacteria within range. And in the latest of many surprises, George M. Church of Harvard Medical School and his colleagues have isolated hundreds of soil bacteria able to subsist on antibiotics as their sole source of nutrition. This is not resistance but rather downright insolence, as these bacteria think of antibiotics not as threatening molecules, but instead as food.

Our ability to understand the origin and role of antibiotics has, in effect, been hindered by the uses to which we put them. In a fit of intellectual narcissism, we assume that if we use them in clinical settings to kill bacteria, that must be what they evolved for. However, when we recast them in a subtler role as agents of competition, and also as regulators and communicators in the bacterial world, we can view their lethal (and thus, for us, salutary) effects in a different light. It is when we administer doses of antibiotics that are orders of magnitude greater than those encountered in nature that a subtle, modulated signal is transformed into a deafening, and increasingly deadly, roar.

If antibiotics are part of the whispered conversations of the microbial world, what then should we make of antibiotic resistance? The last 65 years of clinical antibiotic use have exposed a dizzying array of mechanisms that bacteria use to survive the presence of toxic antibiotic concentrations. The diversity and complexity of these mechanisms—enzymes that chemically disable antibiotics, shape-shifting targets that prevent antibiotics from binding, complex pumps that eject the antibiotic from the target cell—have raised an obvious evolutionary question: Where do these mechanisms come from? Part of the answer, unsurprisingly, is self-defense. Many of the bacteria that produce antibiotics also harbor mechanisms that make themselves impervious to that antibiotic compound. In the case of the lethal protein antibiotics discussed previously, the bacteriocins, producer bacteria also synthesize a second protein that pairs up with the antibiotic, rendering it inactive until it reaches its target.

More than One Answer

In the classic militaristic interpretation, antibiotics exist to target competitors, and resistance genes evolve to protect producers. But here again, we might do well to remember that massive concentrations of antibiotics used in clinical settings, and the consequences of such massive use, may be obscuring subtler phenomena in the microbial world. A second effect, which I will call the Pete Townshend effect, may also underlie some of the evolution of resistance. In this scenario, antibiotics are playing their signaling role, but the levels of this signal molecule are so high in and around the producing cell that they will cause unintended damage as a byproduct of their high concentration. Resistance evolves in producer strains to protect them from the damaging consequences of overwhelming signal strength. As the antibiotic diffuses away from the producer, its concentration declines. The antibiotic loses its lethality and regains its signaling function. If antibiotics have multiple meanings, resistance too must mean more than one thing.

Our ability to examine the genomes of both culturable and unculturable organisms has changed the way we look at the origin of antibiotic resistance mechanisms. Resistance is everywhere, and we don't always know why. We now realize, for example, that the bacteria in a gram of soil harbor hundreds of different genes that can, in the right setting, contribute to resistance. For instance, work by Gerard D. Wright of McMaster University and others, reveals that bacteria in the soil already harbor the rudiments of resistance to all known antibiotics, including entirely synthetic antibiotics with no counterpart in the natural world. Virtually every bacterial cell in these environmental samples is, on average, resistant to eight antibiotics.

The collection of mechanisms responsible for resistance in these soil bacteria has come to be called the "environmental resistome" (as it rhymes, as everything must nowadays, with "genome"). The samples used to investigate the resistome are not ones where clinical strains selected for resistance might be congregating (although wastewater streams from hospitals and water treatment plants in large cities harbor their own impressive collection of clinic-selected resistant strains). Instead, virtually every sample used has a stable microbial ecosystem that has never been exposed to clinical levels of antibiotics and has not been in contact with resistant clinical bacterial cultures. Nonetheless, every one of the samples harbors a vast repository of resistance genes.

This environmental resistome provides a glimpse of the raw materials that clinical pathogens have accessed in the antibiotic age. What is sobering about the discovery of the resistome is the revelation that the microbial world already possesses a functional arsenal of defenses against the antibiotics we develop for clinical use. The resistome also demands a subtle, but critical, change in our perspective on resistance: Many of the mechanisms involved in so-called resistance may be playing other critical roles in cells coexisting peacefully with their neighbors. If antibiotics at sublethal doses really are signal molecules, perhaps resistance is really a signal modulator, altering the meaning and modifying the effects of the message on the recipient cell. Once again, the context in which we usually first encounter resistance mechanisms—in clinical infections as they become insensitive to therapeutic doses of antibiotics—may be obscuring the many roles these mechanisms play in the microbial world.

Rethinking the Fight

We cannot lose sight of how profoundly 50 million pounds of annual antibiotic use impacts the microbial world. The widespread use of antibiotics has created a black market for antibiotic resistance mechanisms where bacteria poach ready-made shortcuts to resistance: It has enhanced and rewarded the evolution of a vast network that allows for the transfer of resistance along nongenealogical lines. These mechanisms of horizontal transfer now connect the extensive environmental resistome with any pathogen (or commensal bacterium) exposed to clinical concentrations of antibiotic. As a result, bacteria in the presence of potentially lethal concentrations of antibiotics no longer face the daunting evolutionary challenge of remaining alive while they and their descendants cobble together resistance one point mutation at a time. Instead, fully functional resistance located on moveable genetic elements can now be acquired by swapping with other bacteria, or by taking such elements up directly from the environment.

To make matters more challenging, many of these ready-made resistance elements are packaged as a set. Multiple genes encoding resistance to multiple antibiotics now travel as a team (up to 74 different resistances have been found to occur together). The ubiquitous presence of human-administered antibiotics in the environment selects strongly for bacteria that are able to acquire these resistance packs, and may also be selecting for the evolution of increasingly mobile and promiscuous elements equipped with increasingly diverse arrays of resistances. As mobile elements, their Darwinian imperative is to make and move copies of themselves, and that in turn may depend on their ability to keep their host bacteria alive in a sea of antibiotics.

Many forces—social, economic and medical—propel excessive antibiotic use. An expanded perspective on the role of antibiotics and resistance in no way contradicts the urgent need to reduce antibiotic consumption. Our current behavior, moreover, drowns out the subtle melody unfolding in microbial ecosystems and superimposes the deafening roar of therapeutic antibiotics and increasingly resistant pathogens in its place. We can hope that new antibiotics will continue to be developed and broadly prescribed, buying us a few more years. We can keep doing what we have always done, and hope we can reach a stalemate with ever more-aggressive bacterial pathogens. Or we can instead start paying more attention to what microbial ecosystems have to teach us, acknowledging the lessons that have always been embedded in a gram of soil. Our relationship to the unseen world extends far beyond pathogens and our efforts to combat them. We can, if we put our minds to it, begin to see a world in a grain of sand.

ROBERT L. DORIT is an associate professor in the Department of Biological Sciences at Smith College. His work focuses on experimental evolution of molecules and bacteria, and on the design of novel antibiotics. Address: 435 Sabin-Reed Hall, Clark Science Center, 44 College Lane, Northampton, MA 01063. Internet: rdorit@smith.edu.

Is Your Patient Taking the Right Antimicrobial?

The author describes the all-too-common phenomenon of inappropriate antimicrobial prescribing and the role it played in her sister's illness and death. The author details the ways in which bacteria become resistant to antimicrobials, discusses the prevalence and costs of healthcare–associated infections resulting from antimicrobial resistance, and provides practical tips on using culture-and-sensitivity reports to ensure that patients are receiving the appropriate antimicrobial treatments.

Her sister's death informs the author's examination of nosocomial infection, antibiotic resistance, and the importance of culture-and-sensitivity reports.

Mary C. Vrtis, PhD, RN

On February 24, 1997, my youngest sister, Catherine Vrtis, died from severe complications of infection and its treatment. Cathy had been on peritoneal dialysis for chronic renal failure. In the months before she died, she had been hospitalized three times with high fevers and severe abdominal pain. She was 31 years old.

Cathy had developed renal failure at age 19 because of an unspecified autoimmune process. The youngest of nine siblings, she was the second to have developed renal failure: our sister Rosie had done so at age 13. By the time we found our paternal grandfather's autopsy report and discovered that he too had died of autoimmune kidney disease, Rosie had already received two transplanted kidneys—the first from our sister Donna and the second from our sister Jeanne after the autoimmune disorder had damaged the first transplanted kidney. (Our brother Tony also donated a kidney to Rosie at a later date.) Even though it was clear that this was a familial disease, I donated a kidney to Cathy when she was 22. She was 11 years younger than me, and I loved her as if she were my own child. It was not a difficult decision: the disease would likely recur, but my kidney could buy her some time.

Could Cathy's death have been prevented if her clinicians had been paying more careful attention? Perhaps. Despite recurrence of the disease seven years after the transplantation, Cathy's overall health was very good until October 1996. She suddenly developed severe abdominal pain and was hospitalized. A culture grown from a sample of peritoneal fluid on October 18, during Cathy's first hospitalization, showed the presence of *Escherichia coli*. The infection was attributed to contamination during peritoneal dialysis, even though Cathy was meticulous in performing it. She was placed on IV ciprofloxacin and released three days later to continue the antimicrobial therapy at home, through an implanted central venous catheter. Her abdominal pain was attributed to peritonitis, so peritoneal dialysis was stopped and hemodialysis—through a subclavian dialysis catheter, three times a week—was begun.

On November 12 severe abdominal pain recurred, and her fever was 103.8°F. She was hospitalized again and continued receiving IV ciprofloxacin until December 22, when she went home for a Christmas celebration.

After dialysis on December 26, Cathy ran a fever of 103.5°F and was hospitalized a third time, with a diagnosis of sepsis. Extreme abdominal pain radiated to her left shoulder. Ciprofloxacin therapy was resumed. She had gone through several courses of ciprofloxacin (a fluoroquinolone) over a two-month period, and she was still ill. This should have suggested to her clinicians one of two possibilities: that the infecting organism (or organisms) had become resistant to fluoroquinolones or that she had a concurrent infection with anaerobic bacteria, against which fluoroquinolones have "marginal" activity.[1, 2]

(Intraabdominal abscesses are typically caused by a mixture of both aerobic and anaerobic bacteria.[2]) Additional cultures of peritoneal fluid, although indicated, were not taken. A computed tomographic (CT) scan of Cathy's abdomen found nothing abnormal. Because long-term use of intravascular catheters and dependence on hemodialysis are two major risk factors for catheter-related sepsis,[3] her hemodialysis catheter was replaced.

On January 20 or soon after, Cathy's condition deteriorated rapidly. She became hypotensive and experienced acute respiratory distress, requiring mechanical ventilation. According to the physician who intubated her, the cause was an overwhelming Gram-negative sepsis. For the third time since November, the nephrologist refused our requests for an infectious disease consultation because he didn't see the need. Therefore, additional laboratory testing was not performed at that time. A second abdominal CT scan was also negative. No cause for her abdominal pain was identified. About two weeks later, when Cathy's abdominal pain was intractable despite high doses of morphine and hydromorphone, an exploratory laparotomy with lysis of adhesions was performed. The surgeon said the abdominal adhesions were so severe that the bowel was "completely glued together." A second laparotomy was later performed to remove recurrent adhesions; the transplanted kidney, assumed to be infected by then, was also removed.

Infectious Disease Consultation

After the first operation in early February, the high fever returned and Cathy continued to deteriorate. Finally, on *February 21,* at my insistence, the hospital's infectious disease physician was called in. (In retrospect, it's remarkable how difficult it was for us, Cathy's family, to get the nephrologist to take this obvious step. I had spoken to the nephrologist on previous visits and by phone from my home, 600 miles away. But it wasn't until I had come to stay with Cathy in the ICU and made it clear that I would go in person to the hospital's medical director if our demands weren't met, that the nephrologist consented to an infectious disease consultation.)

The consulting physician reviewed Cathy's records from all three hospitalizations and, within an hour, noticed that the final culture report from October 18 showed not only *E. coli* but also the anaerobic organism *Bacteroides fragilis.* The culture-and-sensitivity report had been sent to the nephrologist, who had apparently overlooked or ignored this crucial piece of information. The physician then met with Cathy and us and reported her findings. In addition to the chart review, she had conducted an extensive literature review and had spoken with other infection control physicians to determine whether infection with this organism had ever been

associated with peritoneal dialysis. Her findings suggested that peritonitis caused by *B. fragilis* was always secondary to a bowel rupture, and she concluded that Cathy's severe pain in October had been caused by a ruptured bowel. The primary diagnosis had been incorrect since the first hospitalization.

First-generation fluoroquinolones such as ciprofloxacin, which Cathy had been given, are ineffective against most anaerobic bacteria, such as Bacteroides.[4, 5] Metronidazole, to which *B. fragilis* is susceptible, was initiated by IV infusion. (Note that in 2005 Hermsen and colleagues recommended the use of the newer broad-spectrum quinolones, levofloxacin or moxifloxacin, in addition to metronidazole to treat abdominal anaerobic infection.[2])

The Final Months

Cathy spent the last two months of her life in the hospital. Although she had an occasional good day, her overall condition continued to deteriorate because of complications of infection, including

- systemic sepsis, caused by an unknown organism, possibly *B. fragilis.*
- methicillin-resistant *Staphylococcus aureus* (MRSA) bacteremia, for which she was treated with IV vancomycin.
- systemic candidiasis, for which she needed IV amphotericin B. (Antibiotics suppress protective flora, allowing for an overgrowth of *Candida albicans,* which produces hydrolytic enzymes and proteinases that damage and digest human cells.)
- Stevens-Johnson syndrome, a life-threatening reaction to antimicrobials, featuring massive skin loss and huge, weeping wounds. This adverse drug reaction is associated with a 5% to 15% mortality rate.[6, 7]
- respiratory failure, secondary to acute respiratory distress syndrome, for which she required continuous mechanical ventilation for more than one month.
- disseminated intravascular coagulopathy (DIC), which caused bleeding at IV puncture sites and the gums, lips, and gastrointestinal tract. (DIC occurs when the clotting sequence is activated within the blood vessels because endotoxins are released from the cell walls of dead Gram-negative bacteria.[8] Because circulating clotting factors are reduced, bleeding from damaged tissues such as IV insertion sites cannot be stopped.)

Two months and four operations after Cathy's third admission, she had massive gastrointestinal bleeding. She bled to death, despite many transfusions.

Scope of the Problem

I have no way to know whether Cathy's life could have been saved if a clinician had noticed the report indicating the presence of *B. fragilis*. Would Cathy have experienced the secondary *Candida* and MRSA infections? Would the skin sloughing of Stevens–Johnson syndrome have occurred if she hadn't needed additional antimicrobials to treat the MRSA? Would she have spent the last month of her life on a ventilator if these complications hadn't occurred? I will never know. Still, I can't help but think that maybe I would still have my sister in my life if someone had noticed the culture report and intervened sooner.

Healthcare Associated Infections

Data from the National Nosocomial Infections Surveillance system and other sources suggest that about 1.7 million hospitalized patients in the United States developed a health care-associated infection in 2002; those infections caused or were associated with about 99,000 deaths.[9] Urinary tract infections topped the list, followed by surgical-site infections, pneumonia, and bloodstream infections. Urinary tract infections occurred in an estimated 561,667 cases (33% of the total), and the mortality rate was 2.3% (approximately 13,088 deaths). Surgical-site infections occurred in an estimated 290,485 cases (almost 17% of the total), and the mortality rate was 2.8% (8,205 deaths). While there were fewer cases of pneumonia and bloodstream infections, these conditions are associated with much higher mortality rates: an estimated 250,205 hospitalized patients acquired health care associated pneumonia, and 14.4% (35,967 patients) died as a direct result; bloodstream infections, identified in 248,678 hospitalized patients, resulted in 30,665 deaths, a mortality rate of 12.3%.

The geriatric population is particularly at risk for health care–acquired infections for a variety of reasons (as noted by several commentators in a special issue *Emerging Infectious Diseases* from the Centers for Disease Control and Prevention; go to www.cdc.gov/ncidod/eid/vol7no2/contents.htm). For example, the immune response may be diminished as a result of factors such as chronic diseases, medications, malnutrition, functional impairments, diminished cough reflex, dysphagia, thinning skin, incontinence, an enlarged prostate, or invasive devices (such as urinary catheters and feeding tubes).[10] Also, older adults with diminished functional status are especially at risk for infections related to complications of limited mobility, including respiratory infections, pneumonia, and infected pressure ulcers.[11] As health care delivery moves out of acute care into community-based settings, the challenges in addressing infection control issues are rapidly changing.[12]

Antimicrobial Resistance

With more microorganisms developing greater resistance to antimicrobial agents, the number of deaths resulting from infection is expected to rise. Of particular concern are those organisms with resistance to multiple antimicrobial medications, such as MRSA, vancomycin-resistant enterococci (VRE), and vancomycin-resistant *Staphylococcus aureus* (VRSA), as well as organisms that produce extended spectrum β-lactamases (ESBLs), such as *E. coli, Klebsiella pneumoniae,* and *Enterobacter* species.[13–16] (ESBLs produced by these bacteria can inactivate multiple classes of antibiotics, including penicillins, cephalosporins, and aztreonam [Azactam], by breaking down the β-lactam that interferes with the formation of bacterial cell walls.[15, 17, 18])

Costs

The numerous costs associated with antimicrobial resistance are borne not only by patients but by health care organizations, insurers, the pharmaceutical industry, employers, and the government. They include the costs of treatment, prevention, research, and development of new therapeutic agents, as well as those related to lost productivity and diminished marketability of existing medicines. The annual cost of infection in the United States because of drug-resistant bacteria has been estimated at $4 billion to $5 billion.[19]

In terms of its effects on patients and the costs associated with it, antimicrobial resistance is an enormous problem. (For an interesting discussion of how resistance-related costs are calculated and why better methods are needed, see www.journals.uchicago.edu/doi/pdf/10.1086/323758.) But how does antimicrobial resistance develop and how does it work?

Causes and Mechanisms of Resistance

Antimicrobials target particular microorganisms or groups of organisms; no single antimicrobial targets every type of infectious agent. A patient may be receiving one or more antimicrobials, but that doesn't ensure that the patient's infection is being treated. When a microorganism is *not sensitive* to a particular drug, no amount of the drug—not even a toxic amount—will have the desired effect on the infection. If the antimicrobial isn't appropriate for the specific type of infectious organism, the microorganisms will continue to multiply, and the infection will worsen.

Whenever an antimicrobial medication is administered to treat an infectious organism, the patient's normal bacterial flora—which produce substances that may inhibit or kill pathogens and which compete with them for nutrients and binding sites—are also exposed to the drug. If the normal flora are sensitive to the medication and are killed, pathogens that are not sensitive to that medication or that have developed mechanisms for countering its effects then thrive.[15, 20] Life-threatening superinfections—new infections that are caused by a virus, bacterium, or fungus different from the one that caused the initial infection—can result. For example, systemic candidiasis—which Cathy developed—is often a superinfection. *Clostridium difficile,* which thrives as normal bowel flora are eliminated, causes severe diarrhea and may result in pseudomembranous colitis and other life-threatening complications. The Society for Healthcare Epidemiology of America has named *C. difficile* as the number-one cause of nosocomial infectious diarrhea.[12]

How Resistance Develops

Bacteria that are unaffected by one or more antimicrobials are said to be resistant. Some bacteria are inherently resistant, having evolved this trait over time by spontaneous mutation and natural selection, in response to harmful substances produced by other bacteria and fungi—in effect, naturally occurring antibiotics. In recent decades, many kinds of bacteria that were previously susceptible to antimicrobials have developed resistance as an adaptation to increasing amounts of antimicrobials in their environment (human bodies and the animals we eat). Sometimes this is caused by spontaneous gene mutation, but at other times it results from the transfer of genetic material from a resistant organism to a susceptible one, which is of special concern to clinicians. Either way, once resistance develops, bacteria pass on the trait to subsequent generations.

In the case of mutation, more than one gene may have to mutate for resistance to develop. According to Tenover,

> Although a single mutation in a key bacterial gene may only slightly reduce the susceptibility of the host bacteria to [a specific] antibacterial agent, it may be just enough to allow its initial survival until it acquires additional mutations or additional genetic information resulting in full-fledged resistance to the antibacterial agent. However, in rare cases, a single mutation may be sufficient to confer high-level, clinically significant resistance upon an organism.[15]

As examples of those rare instances, Tenover points to rifampin resistance in *S. aureus* and fluoroquinolone resistance in *Campylobacter jejuni.*

When several different bacteria or strains of bacteria infect a single host, they can exchange genetic material—and antimicrobial resistance—in several ways. This sharing, also known as horizontal gene transfer, occurs in at least three ways: conjugation, transformation, and transduction.

Conjugation. In cell-to-cell conjugation, genetic material in the form of plasmids is transferred from one bacterium to another. A plasmid is a circular piece of genetic material that often carries a resistance gene and is independent of the chromosomes of the host bacterial cell. Plasmids often carry only a few genes and can replicate themselves. A single bacterial cell might have one plasmid or many, and may even have many copies of the same plasmid. When bacterial cells conjugate, one bacterium extends a tube called a sex pilus toward another bacterium and penetrates it. After the plasmid copies itself, one copy goes through the pilus to the other cell, transferring resistance to the formerly susceptible bacterium.

In the laboratory, for example, antimicrobial resistance to the fluoroquinolones has been shown to transfer via plasmids from *K. pneumoniae* to previously sensitive strains of *E. coli.*[21] VRE also become resistant by acquiring *vanA* and *vanB* resistance genes from other bacteria in this manner.[22] The *vanA* gene is also easily transferred through conjugal plasmid transfer to other organisms, such as VRSA.[23] The development of the ability to produce ESBLs can also result from the transfer of bacterial genetic material from one kind of bacteria to another. For example, Heritage and colleagues were able to demonstrate the transfer of an ESBL plasmid from *K. pneumoniae* to a laboratory strain of *E. coli* by cultivating both organisms on the same agar plate overnight. The plasmid conferred resistance to cephalosporins in subsequent *E. coli* clones.[24] All of these organisms are becoming serious problems in health care facilities because of cross contamination among patients, primarily as a result of poor hand hygiene.[16, 25]

Transformation is another way for susceptible bacteria to acquire resistance. In this case, bacteria incorporate into their own chromosomes or plasmids genetic material that's been released into their environment by dead bacteria.[15]

Transduction. Even bacteria can become infected with viruses. When a virus called a bacteriophage infects a bacterial cell, it takes over the reproductive mechanism of the bacterium in order to replicate its own DNA or RNA, producing more virus particles. If the bacterium is resistant, its resistance gene may be incorporated into the viral genetic material in such a way that it becomes transferable to other bacteria that the virus may subsequently infect. According to Tenover, however, "this is now thought to be a relatively rare event."[15]

Mechanisms of Resistance

Not all antimicrobials act in the same way to kill bacteria. In response, bacteria have developed different ways to protect themselves from the specific antimicrobials used to treat them, including[15, 26]

- inactivation or destruction of the antimicrobial agent, usually by enzymes such as β-lactamases.
- active removal (efflux) of the antimicrobial from the microorganism.
- changes in the cell-wall binding site that prevent the antimicrobial from attaching to the bacterium.
- changes in the cell wall, such as thickening, that prevent the organism from absorbing the medication.
- changes in the drug target sites within the bacterial cell, such as ribosomes, that prevent the antimicrobial from binding to them, thus interrupting microbial metabolism.
- development of an alternative metabolic pathway so that internal processes critical to organism survival, such as folate synthesis, are no longer affected by the drug.

When bacteria acquire resistance through exposure to a single antimicrobial, the organism also frequently becomes resistant to other similar drugs. For example, a microorganism that produces β-lactamase enzymes can deactivate many antimicrobials, including almost all cephalosporins, penicillins, and aztreonam. Even though they may have only one mechanism of resistance, β-lactamase producers are multidrug resistant. Similarly, organisms that produce carbapenem-hydrolyzing enzymes will inactivate both imipenem (Primaxin) and meropenem (Merrem). Such widespread resistance creates major obstacles to treating active infection.

Reviewing Culture-and-Sensitivity Reports

When culture-and-sensitivity testing is ordered, the nurse or physician obtains a sample of body fluid (such as blood, sputum, urine, or wound drainage) from the patient and sends it to the hospital's laboratory. There, the sample is incubated to see whether any organisms present in the sample will "culture out," or grow. If growth occurs, tests are then performed to determine which microorganisms are present and whether the organisms will grow in the presence of specific antimicrobials. With automated equipment, it's possible to have results within one day. Sometimes further tests must be conducted manually, in which case it may take up to three days to receive the culture report.

Most nurses are not legally responsible for prescribing the correct medications but are obligated to ensure the appropriateness of the medications patients receive. Nurses should review culture-and-sensitivity results as soon as they are available and act when a medication needs to be changed. Be proactive: make looking at laboratory results a part of your practice when reviewing a chart for the first time, then ensure that the culture-and-sensitivity results are back whenever a new order is written.

You can make the difference between a short-term infection and a long, life-threatening illness by taking the following steps:

- Depending on the system your facility uses, look in the computer system or in the patient's chart for the results after cultures are obtained. Most hospital laboratories will provide a preliminary result as well as a final report. Look for both.
- If multiple tests were performed, final results may be obtained on different dates. Make sure you check both aerobic and anaerobic results, if both were ordered.
- When a culture is positive for a pathogen, verify that the physician or NP has a copy of the report—don't simply fax it to the provider's office. Call to make sure that the report arrived and point out your concerns to the office staff.
- If you're sure that the physician or NP has seen the results and you have expressed your concerns about insensitivity or resistance but the physician or NP fails to act, get further help. If you are in a hospital, contact the infection control practitioner (ICP). Even the smallest hospitals have ICPs. The ICP usually has the knowledge and the clout necessary to address inappropriate care.
- If you don't feel comfortable going to the ICP, notify your supervisor. Your supervisor can notify the organization's infection control department, the infectious disease director, or the facility's medical director. Make sure to clearly document the actions that have been taken.
- Review information on the microorganism causing the infection. If drug-resistant pathogenic bacteria such as MRSA, VRE, or VRSA are involved, follow your facility's isolation guidelines. You may want to contact the ICP at your facility for more advice. Hospital ICPs typically review all positive culture reports, but early intervention can decrease the risk of cross contamination.

- Review the sensitivity data. Laboratories use similar techniques and methods of reporting when performing culture-and-sensitivity testing. If the medication's name has an "R" next to it, that means the organism is resistant; if an "I," an intermediate response is indicated (the organism is inhibited; some, but not all, of the bacteria will be killed); if an "S," the organism is sensitive to it and the medication is likely to cure the infection. The numbers under the heading "MIC [minimum inhibitory concentration] sensitivity" refer to the "lowest concentration of an antimicrobial drug that will inhibit the visible growth of microorganisms after overnight incubation."[27]
- Determine whether the medication prescribed to the patient is appropriate, based on the sensitivity data. If a particular bacterium shows an intermediate response to the antimicrobial, continued exposure to it increases the risk that the infectious organism (and others in the body) will adapt to the hostile environment and become resistant to that type of drug.
- If the patient is not taking a medication to which the bacteria are sensitive, an appropriate antimicrobial must be identified. The NP or physician must be made aware of the problem as soon as it becomes apparent. Finding the right medication is sometimes difficult because the costs of testing are far too high for microbiology laboratories to test every antimicrobial available. The sensitivity test uses drugs representing various classes of antibiotic medication. Antibiotics of a given type tend to cause a similar response in the same bacteria. For example, organisms that are resistant to cefazolin (Ancef, Zolicef) will probably be resistant to *all* cephalosporins.
- If the antibiotic that the patient takes is not on the report list, sensitivity to that particular drug was not tested.
- If the culture shows that multiple types of bacteria are present, the process becomes even more complicated. Ideally, finding a single antimicrobial medication to which all the organisms are sensitive is the best option. More likely, the patient will need more than one medication.
- When yeast (most often *C. albicans*) is also detected, antibacterial medications will not affect it. (*Candida* species often appear in routine cultures while other fungi, such as *Histoplasma capsulatum,* do not; special fungal cultures are needed to detect them.) The first line of defense will have to be antifungal agents, such as fluconazole (Diflucan, Trican) or amphotericin B. Patients with serious, life-threatening bacterial infections often develop secondary systemic fungal infections because of the suppression of normal flora.
- If the infection is viral, the culture report will be negative unless special viral cultures were obtained. The antimicrobials discussed here will not kill viruses. If the patient's immune response is inadequate, it may be necessary to prescribe an appropriate antiviral (assuming that one is available).
- If the patient is taking the wrong medication, make certain that the physician or NP knows this. Point out the medications that the microorganism is sensitive to. You may save a life!

References

1. Hooper DC. New uses for new and old quinolones and the challenge of resistance. *Clin Infect Dis* 2000;30(2):243–54.
2. Hermsen ED, et al. Levofloxacin plus metronidazole administered once daily versus moxifloxacin monotherapy against a mixed infection of *Escherichia coli* and *Bacteroides fragilis* in an in vitro pharmacodynamic model. *Antimicrob Agents Chemother* 2005;49(2):685–9.
3. Fowler VG, Jr., et al. Risk factors for hematogenous complications of intravascular catheter–associated *Staphylococcus aureus* bacteremia. *Clin Infect Dis* 2005;40(5):695–703.
4. Appelbaum PC. Quinolone activity against anaerobes. *Drugs* 1999;58 Suppl 2:60–4.
5. Golan Y, et al. Emergence of fluoroquinolone resistance among Bacteroides species. *J Antimicrob Chemother* 2003;52(2):208–13.
6. Wolf R, et al. Life-threatening acute adverse cutaneous drug reactions. *Clin Dermatol* 2005;23(2):171–81.
7. Ghislain PD, Roujeau JC. Treatment of severe drug reactions: Stevens–Johnson syndrome, toxic epidermal necrolysis and hypersensitivity syndrome. *Dermatol Online J* 2002;8(1):5.
8. ten Cate H. Pathophysiology of disseminated intravascular coagulation in sepsis. *Crit Care Med* 2000;28(9 Suppl):S9–S11.
9. Klevens RM, et al. Estimating health care-associated infections and deaths in U.S. hospitals, 2002. *Public Health Rep* 2007;122(2):160–6.
10. Strausbaugh LJ. Emerging health care-associated infections in the geriatric population. *Emerg Infect Dis* 2001;7(2):268–71.
11. Nicolle LE. Preventing infections in non-hospital settings: long-term care. *Emerg Infect Dis* 2001;7(2):205–7.
12. Simor AE, et al. Clostridium difficile in long-term-care facilities for the elderly. *Infect Control Hosp Epidemiol* 2002;23(11):696–703.
13. Graham PL, 3rd, et al. A U.S. population-based survey of *Staphylococcus aureus* colonization. *Ann Intern Med* 2006;144(5):318–25.
14. Muto CA, et al. SHEA guideline for preventing nosocomial transmission of multidrug-resistant strains of *Staphylococcus aureus* and enterococcus. *Infect Control Hosp Epidemiol* 2003;24(5):362–86.

15. Tenover FC. Mechanisms of antimicrobial resistance in bacteria. *Am J Med* 2006;119(6 Suppl 1):S3–S10; discussion S62–S70.

16. Siegel JD, et al. *Management of multidrug-resistant organisms in healthcare settings, 2006.* Atlanta: Healthcare Infection Control Practices Advisory Committee. Centers for Disease Control and Prevention; 2006. http://www.cdc.gov/ncidod/dhqp/pdf/ar/mdroGuideline2006.pdf.

17. Centers for Disease Control and Prevention. *Laboratory detection of extended-spectrum beta-lactamases (ESBLs).* 1999. http://www.cdc.gov/ncidod/dhqp/ar_lab_esbl.html.

18. Thomson KS. Controversies about extended-spectrum and AmpC beta-lactamases. *Emerg Infect Dis* 2001;7(2):333–6.

19. Institute of Medicine (U.S.). Harrison PF, Lederberg J, editors. *Antimicrobial resistance: issues and options: workshop report.* Washington, DC: National Academy Press; 1998. http://www.nap.edu/catalog.php?record_id=6121.

20. Shlaes DM, et al. Society for Healthcare Epidemiology of America and Infectious Diseases Society of America Joint Committee on the Prevention of Antimicrobial Resistance: guidelines for the prevention of antimicrobial resistance in hospitals. *Infect Control Hosp Epidemiol* 1997;18(4):275–91.

21. Martinez-Martinez L, et al. Quinolone resistance from a transferable plasmid. *Lancet* 1998;351(9105):797–9.

22. Centers for Disease Control and Prevention. *Vancomycin-resistant enterococci (VRE) and the clinical laboratory.* 1999. http://www.cdc.gov/ncidod/dhqp/ar_lab_vre.html.

23. Weigel LM, et al. High-level vancomycin-resistant Staphylococcus aureus isolates associated with a polymicrobial biofilm. *Antimicrob Agents Chemother* 2007;51(1):231–8.

24. Heritage J, et al. SHV-34: an extended-spectrum beta-lactamase encoded by an epidemic plasmid. *J Antimicrob Chemother* 2003;52(6):1015–7.

25. Weinstein RA. Controlling antimicrobial resistance in hospitals: infection control and use of antibiotics. *Emerg Infect Dis* 2001;7(2):188–92.

26. Stratton CW. Dead bugs don't mutate: susceptibility issues in the emergence of bacterial resistance. *Emerg Infect Dis* 2003;9(1):10–6.

27. Andrews JM. Determination of minimum inhibitory concentrations. *J Antimicrob Chemother* 2001;48 Suppl 1:5–16.

MARY C. VRTIS is a legal nurse consultant with Aperio Medical Legal Consulting in Front Royal, VA. Contact author: marycvrtis@yahoo.com. The author of this article has disclosed no significant ties, financial or otherwise, to any company that might have an interest in the publication of this educational activity.

Constant Struggle to Conquer Bacteria

Beating bad bugs: Economics and evolution frustrate effort to eliminate a growing threat.

Sabin Russell

D r. Ian Friedland was sitting at his Mountain View office on a rainy October afternoon when the telephone rang with some long-awaited news: The Food and Drug Administration had just approved doripenem, the powerful antibiotic he had labored over since 2004.

"Shortly after the drug was approved," the 50-year-old Johnson & Johnson researcher said, "we began to hear stories about it being used successfully in patients." It was the kind of outcome that pharmaceutical scientists spend their lives trying to attain.

It is also the kind of story that has become frighteningly rare.

Since the early 1990s, drug companies that had built their businesses on early antibiotic research have been leaving the field. As a consequence, there has been a steady decline in the number of new antibiotics approved by the FDA—even as the existing ones are losing ground to a surge of drug-resistant bacterial strains such as Staphylococcus aureus.

In the five-year period from 1983 through 1987, the FDA approved 16 new antibiotics. During a similar five-year span that ended last year, only five made the cut.

At the same time, the overall level of antibiotic drug research has dropped. In 2006, a survey by the Infectious Diseases Society of America counted only 13 potential new antibiotics in mid- to large-scale clinical trials—a stage of drug development that requires years to complete and offers no guarantee of success.

Medical researchers describe the supply of new drugs under development as "the pipeline." The pipeline for new antibiotics is running dry.

Intravenous doripenem, approved on Oct. 12, is the only important antibiotic licensed since 2005. Distantly related to penicillin, it will be used at first to treat severe and life-threatening abdominal infections in hospitalized patients.

The reasons for the decline in antibiotic research and development are complex, but drug company economics are at the core of it. Drugs that treat chronic conditions such as heart disease, arthritis and diabetes must be taken for a lifetime. A good antibiotic can clear an infection in a week to 10 days.

With the cost of developing new drugs ranging between $110 million and $800 million, cautious investors are putting their money into research that promises the biggest payout.

But scientists also acknowledge that the bugs themselves are proving more difficult to fight. Most antibiotics are derived from toxins created by bacteria to use as weapons against competing bugs. Much of drug development in the second half of the 20th century relied on finding molds that produced these natural toxins, and then tinkering with them to make them easier to manufacture.

More recent efforts focus on the deliberate design of new drugs. Scientists probe disease-causing organisms for weaknesses, and create molecules to attack those weak points. The approach has worked for AIDS drugs, but has yet to hit gold in the effort to find new antibiotics.

Dr. Jack Edwards, chief of infectious diseases at Harbor-UCLA Medical Center, said it was "quite easy" in the 1970s to make new antibiotics such as cephalosporins—a penicillin-like family of drugs that includes well-known products such as Keflex. "It's clear now that it is not so easy to make a new antibiotic," he said.

> **"It's clear now that it is not so easy to make a new antibiotic."**
>
> —Dr. Jack Edwards, chief of infectious diseases
> *at Harbor-UCLA Medical Center*

The struggle to find new antibiotics to replace the old has been going on since the start of the antibiotic revolution.

Scottish researcher Alexander Fleming discovered penicillin in 1928 when a spot of mold contaminated a petri dish of Staphylococcus aureus and produced a distinctive ring where the bacteria would not grow.

It took another 15 years before British and American scientists came up with a practical way to make penicillin on a

Antibiotics on Farm

Critics Seeking to Diminish the Use of Anti-Microbials in Animal Feed

Consumer advocates have been campaigning for years to curb the use of antibiotics in agriculture, citing studies that show that 70 percent of all U.S. antibiotics are administered in low doses—not to treat disease, but to promote the growth of pigs, sheep, chicken and cattle.

Low doses of antibiotics in animal feeds have been shown to boost the speed of food-to-muscle conversion by 5 percent, and can prevent the spread of disease in the tight quarters of modern factory farms.

But as early as 1963, British researchers tied the emergence of drug-resistant strains of salmonella in humans to antibiotics fed to cattle. Among the drugs routinely found in animal feed are erythromycin, penicillin and streptomycin. Critics warn that the use of antibiotics in feed at low dosages helps to breed resistant bacteria in the gut of farm animals—threatening the future of these drugs for use in animals or humans.

Major antibiotic classes such as tetracyclines and the Cipro-like fluoroquinolones have already been compromised, according to Keep Antibiotics Working, a coalition backed by environmental groups and the American Medical Association.

The stakes are high. The Union of Concerned Scientists calculated in 2001 that U.S. farm interests were using 24.6 million pounds of anti-microbials—almost 40 percent higher than industry estimates.

Ron Phillips, vice president of the Animal Health Institute, a Washington trade group for agricultural drugmakers, maintains that growth promotion accounts for only 4.5 percent of antibiotic consumption in agriculture. The rest are used to prevent, treat or control the spread of disease. "Antibiotics," he says, "are a net positive for both animal health and human health."

After antibiotics were banned from animal feed in Europe beginning in 1995, Phillips said, farmers there found they had to use more antibiotics to care for illnesses that cropped up in their livestock.

Keep Antibiotics Working nevertheless is pushing for a federal ban on antibiotics in feed. Introduced by Sens. Edward Kennedy. D-Mass, and Olympia Snow, R-Maine, the "Preservation of Antibiotics for Medical Treatment Act" would phase out in two years antibiotics deemed "important in human medicine."

In response to pressure from consumer groups, McDonald's declared four years ago its intention to phase out the purchase of meats from chicken and livestock fed the drugs to promote growth. The Food and Drug Administration in 2005 banned the use of a Cipro-like drug, Baytril, to treat bacterial infections in poultry, after drug-resistant strains of Campylobacter—a common food-poisoning organism—were found in chicken. Cases of Cipro-resistant Campylobacter were also rising in humans.

The FDA is considering an application for approval of the antibiotic cefquinome, a proposed veterinary drug that is similar to the human drug cefepime. In the fall of 2006, an FDA advisory committee recommended against approval.

"It was surprising what the committee did, because it was stacked with veterinarians and animal science people," said Stephen Roach, director of public health programs for Keep Antibiotics Working.

"The USDA is very reluctant to say that antibiotic use causes a problem, and the FDA has traditionally been in the middle. But I feel that in the last several years, they have been more accommodating to industry," said Roach.

A final decision on approving cefquinome is still pending.

---

commercial scale. Yet almost as soon as penicillin went into common use in the 1940s, a new variety of staph appeared that the antibiotic could not treat.

Hospitals eventually set up special wards to isolate the growing number of patients with penicillin-resistant staph. A new antibiotic, methicillin, came out in 1960, but within a year some staph germs had developed a defense against that drug, too.

Alarmed by the need to keep ahead of rapidly mutating bacterial strains, researchers since then have developed four successive generations of cephalosporins—the first came out in 1964.

These drugs repeatedly raised hopes that the resistance problem could be tamed, but even as scientists developed new generations of cephalosporins, bacteria methodically evolved ways to sidestep each one.

When bacteria began to resist the fourth—and final—generation of cephalosporins, researchers came up with carbapenems, yet another promising family distantly related to

penicillin. Carbapenems are, like thoroughbreds, powerful, expensive and potentially dangerous.

Doripenem, the new Johnson & Johnson antibiotic, is the latest carbapenem.

While germs such as methicillin-resistant Staphylococcus aureus, or MRSA, have been grabbing headlines of late, there are other bugs such as Acinetobacter, Klebsiella and Pseudomonas that are evolving resistance to most existing antibiotics at an alarming pace. Those three, classified as Gram-negative, carry microscopic pumps that drive antibiotics out of their system. They have proved particularly adept at dodging new drugs, and researchers have been largely stumped in their efforts to come up with substitutes.

Lawmakers in Washington have repeatedly offered bills containing packages of incentives to encourage drug companies to develop new antibiotics more rapidly, but specialists such as UCLA's Edwards acknowledge that is a hard sell in Congress.

"There is a general dissatisfaction with the pharmaceuticals industry, based in part on the concept that their profit margins are too large," he said. "But we need to keep industry interested in making new antibiotics."

The latest effort to spur on the drug companies occurred last year, but by the time a large FDA reauthorization bill reached President Bush's desk for signature on Sept. 27, key incentives—such as allowing antibiotic makers to stave off generic competition when they find new uses for their drugs—had been stripped out.

"There are no provisions in the bill that will directly lead to stimulation of antibiotic development," said UCLA infectious disease specialist Dr. Brad Spellberg.

With few incentives to keep big drug companies from bowing out, entrepreneurs such as John and Mike Flavin are trying to fill the gap.

John is president and Mike chief executive of Advanced Life Sciences, a Chicago-area startup that is carrying out late-stage clinical trials of cethromycin. The antibiotic was initially developed, and later dropped, by nearby Abbott Laboratories. The Flavin brothers acquired rights and continued to test it because it shows promise as a pneumonia drug and a treatment for inhaled anthrax.

They have raised $71 million through stock offerings to take the drug through final clinical trials.

"As entrepreneurs, the business case we saw was this lack of a pipeline," said John Flavin. "This product would address a growing need in the face of no real competition. That puts us in the right place, at the right time."

This month, as a result of provisions in the new FDA law signed by the president in September, federal regulators convened a meeting with infectious disease doctors and drugmakers to clarify just what kind of proof is required in clinical trials to win approval of drugs to treat pneumonia. A meeting will be held in April to turn those discussions into formal recommendations.

Looming over those talks is the unfortunate history of Ketek—and the question of how to balance speedy approval against the risk of serious side effects.

When Paris-based Sanofi-Aventis won FDA approval for Ketek as a treatment for respiratory illnesses in April 2004, there was talk of a blockbuster drug. More than 28 million prescriptions have been sold worldwide.

But in January 2006, an alert team of North Carolina physicians published online a startling discovery that has put a cloud over Ketek, Sanofi-Aventis and the FDA. Three patients treated at the same hospital had come down with serious liver problems within days of taking Ketek. A 51-year-old woman required a liver transplant, a 26-year-old man died, and a 46-year-old man became ill with jaundice until he stopped taking the pills.

"When you see three cases, same county, same hospital, you start to connect the dots," said Charlotte, N.C., liver specialist Dr. John Hanson, who co-authored an account of the three cases in the journal *Annals of Internal Medicine*.

By the end of that year, 53 cases of liver toxicity nationwide had been traced to Ketek, including 12 cases of acute liver

To Learn More

Two bills designed to address antibiotic resistance are currently making their way through Congress. They are:

Preservation of Antibiotics for Medical Treatment Act: A bill introduced by Sens. Edward Kennedy, D-Mass., and Olympia Snow, R-Maine, would over a two-year period phase out of animal feeds antibiotics that are deemed important to human medicine.

- To learn more: Go to links.sfgate.com/ZCGK.

The STAAR Act: A bill sponsored by Sens. Sherrod Brown, D-Ohio, and Orrin Hatch, R-Utah, was introduced last fall to address the problem of antibiotic-resistant bacteria through research and enhanced federal surveillance, prevention and control.

- To learn more: Go to links.sfgate.com/ZCEF.

How to Get Involved

Want to tell your representatives in the U.S. Senate or the House of Representatives how you stand on the Preservation of Antibiotics for Medical Treatment Act or the STAAR Act?

- Call Sen. Barbara Boxer at her Washington office at (202) 224-3553 or her San Francisco office at (415) 403-0100, or e-mail her by going to links.sfgate.com/ZCEK.
- Call Sen. Dianne Feinstein at her Washington office at (202) 224-3841 or her San Francisco office at (415) 393-0707, or find her e-mail address at links.sfgate.com/ZCEL.
- Contact information for your House representative can be found at links.sfgate.com/ZCEJ.

failure, resulting in four deaths. A congressional investigation revealed that the drug was approved by the FDA despite the discovery that one doctor had fabricated data in a clinical trial. In February, under pressure from Congress, the FDA imposed a "black box" warning on the label of the drug, and dropped its approval for the two conditions for which it was most often prescribed: bronchitis and sinus infection.

Dr. Sidney Wolfe, director of Public Citizen's Health Research Group, acknowledged that antibiotic resistance is a serious problem, but warned that boosting financial incentives for drug companies to make new drugs could yield nothing more than other me-too products with problematic side effects. "I'm tired of all these lures to an industry making so much money today that they can't even see straight," he said.

Although doctors are frustrated by the slow pace of drug development, the biggest challenge may not be lawmakers or drugmakers, but the bugs themselves.

Among the most troubling forms of antibiotic resistance was one of the first encountered in the early days of penicillin. The penicillin molecule features a ring of carbon, called beta-lactam, which hobbles the ability of new bacteria to form cell membranes. But penicillin-resistant strains of bacteria quickly emerged equipped with enzymes called beta-lactamases, which can knock out those carbon rings.

UC Berkeley biochemistry Professor Hiroshi Nikaido, who has studied resistance mechanisms for 40 years, noted that some of the most difficult-to-treat classes of bacteria carry genes to make some form of beta-lactamase. "It must have been there for a long, long time, performing functions that nobody really understands," he said.

Bacteria have shown an extraordinary ability to develop new forms of the enzyme. At least 532 different types of beta-lactamases have been identified, according to Johns Hopkins University epidemiologist Dr. John Bartlett, who reported his findings last year in the journal *Clinical Infectious Diseases*.

Bacteria containing gangs of these enzymes—known as extended-spectrum beta-lactamases—are a growing threat, particularly in pneumonia-causing bugs such as Klebsiella and Pseudomonas. Bartlett said Pseudomonas aeruginosa seems to have a greater ability than most bacteria to develop resistance to "virtually any antibiotic." Some lab tests from hospitalized patients, he said, have already turned up resistance "to all available FDA-approved antibiotics."

It's a step toward what some researchers have labeled "the post-antibiotic era."

Martin Mackay, president for Global Research and Development at Pfizer Inc., the world's largest pharmaceutical firm, remembers as a young bacteriologist in the 1970s the optimism surrounding new antibiotics. "A lot of companies thought we'd cured infectious diseases," he said.

The years since have been a sobering reminder of the power of bacterial evolution. Mackay said neither he nor his company have given up, but instead have a new respect for the challenges ahead.

"I doubt there will ever be a time we really crack this resistance problem," he said. "I think this is going to be a constant war."

Super Bugged

**How DNA pollution may spawn deadly antibiotic resistance:
the story behind the headlines.**

JESSICA SNYDER SACHS

On a bright winter morning high in the Colorado Rockies, a slight young woman in oversize hip boots sidles up to a gap of open water in the icy Cache la Poudre River. Heather Storteboom, a 25-year-old graduate student at nearby Colorado State University, is prospecting for clues to an invisible killer.

Storteboom snaps on a pair of latex gloves and stretches over the frozen ledge to fill a sterile plastic jug with water. Then, setting the container aside, she swings her rubber-clad legs into the stream. "Ahh, no leaks," she says, standing upright. She pulls out a clean trowel and attempts to collect some bottom sediment; in the rapid current, it takes a half dozen tries to fill the small vial she will take back to the DNA laboratory of her adviser, environmental engineer Amy Pruden. As Storteboom packs to leave, a curious hiker approaches. "What were you collecting?" he asks. "Antibiotic resistance genes," she answers.

Storteboom and Pruden are at the leading edge of an international forensic investigation into a potentially colossal new health threat: DNA pollution. Specifically, the researchers are seeking out snippets of rogue genetic material that transforms annoying bacteria into unstoppable supergerms, immune to many or all modern antibiotics. Over the past 60 years, genes for antibiotic resistance have gone from rare to commonplace in the microbes that routinely infect our bodies. The newly resistant strains have been implicated in some 90,000 potentially fatal infections a year in the United States, higher than the number of automobile and homicide deaths combined.

Among the most frightening of the emerging pathogens is invasive MRSA, or methicillin-resistant *Staphylococcus aureus*. Outbreaks of MRSA in public schools recently made headlines, but that is just the tip of the iceberg. Researchers estimate that invasive MRSA kills more than 18,000 Americans a year, more than AIDS, and the problem is growing rapidly. MRSA caused just 2 percent of staph infections in 1974; in the last few years, that figure has reached nearly 65 percent. Most reported staph infections stem from MRSA born and bred in our antibiotic-drenched hospitals and nursing homes. But about 15 percent now involve strains that arose in the general community.

It is not just MRSA that is causing concern; antibiotic resistance in general is spreading alarmingly. A 2003 study of the mouths of healthy kindergartners found that 97 percent harbored bacteria with genes for resistance to four out of six tested antibiotics. In all, resistant microbes made up around 15 percent of the children's oral bacteria, even though none of the children had taken antibiotics in the previous three months. Such resistance genes are rare to nonexistent in specimens of human tissue and body fluid taken 60 years ago, before the use of antibiotics became widespread.

In part, modern medicine is paying the price for its own success. "Antibiotics may be the most powerful evolutionary force seen on this planet in billions of years," says Tufts University microbiologist Stuart Levy, author of *The Antibiotic Paradox: How the Misuse of Antibiotics Destroys Their Curative Powers.* By their nature, antibiotics support the rise of any bug that can shrug off their effects, by conveniently eliminating the susceptible competition.

But the rapid rise of bacterial genes for drug resistance stems from more than lucky mutation, Levy adds. The vast majority of these genes show a complexity that could have been achieved only over millions of years. Rather than rising anew in each species, the genes spread via the microbial equivalent of sexual promiscuity. Bacteria swap genes, not only among their own kind but also between widely divergent species, Levy explains. Bacteria can even scavenge the naked DNA that spills from their dead compatriots out into the environment.

The result is a microbial arms-smuggling network with a global reach. Over the past 50 years, virtually every known kind of disease-causing bacterium has acquired genes to survive some or all of the drugs that once proved effective against it. Analysis of a strain of vancomycin-resistant enterococcus, a potentially lethal bug that has invaded many hospitals, reveals that more than one-quarter of its genome—including virtually all its antibiotic-thwarting genes—is made up of foreign DNA. One of the newest banes of U.S. medical centers, a supervirulent and multi-drug-resistant strain of *Acinetobacter baumannii,* likewise appears to have picked up most of its resistance in gene swaps with other species.

So where in Hades did this devilishly clever DNA come from? The ultimate source may lie in the dirt beneath our feet.

For the past decade, Gerry Wright has been trying to understand the rise of drug resistance by combing through the world's richest natural source of resistance-enabling DNA: a clod of dirt. As the head of McMaster University's antibiotic research center in Hamilton, Ontario, Wright has the most tricked-out laboratory a drug designer could want, complete with a $15 million high-speed screening facility for simultaneously testing potential drugs against hundreds of bacterial targets. Yet he says his technology pales in comparison with the elegant antibiotic-making abilities he finds encoded in soil bacteria. The vast majority of the antibiotics stocking our pharmacy shelves—from old standards like tetracycline to antibiotics of last resort like vancomycin and, most recently, daptomycin—are derived from soil organisms.

Biologists assume that soil organisms make antibiotics to beat back the microbial competition and to establish their territory, Wright says, although the chemicals may also serve other, less-understood functions. Whatever the case, Wright and his students began combing through the DNA of soil microbes like streptomyces to better understand their impressive antibiotic-making powers. In doing so the researchers stumbled upon three resistance genes embedded in the DNA that *Streptomyces toyocaensis* uses to produce the antibiotic teicoplanin. While Wright was not surprised that the bug would carry such genes as antidotes to its own weaponry, he was startled to see that the antidote genes were nearly identical to the resistance genes in vancomycin-resistant enterococcus (VRE), the scourge of American and European hospitals.

"Yet here they were in a soil organism, in the exact same orientation as you find in the genome of VRE," Wright says. "That sure gave us a head-slap moment. If only we had done this experiment 15 years ago, when vancomycin came into widespread use, we might have understood exactly what kind of resistance mechanisms would follow the drug into our clinics and hospitals." If nothing else, that foreknowledge might have prepared doctors for the inevitable resistance they would encounter soon after vancomycin was broadly prescribed.

Wright wondered what else he might find in a shovelful of dirt. So he handed out plastic bags to students departing on break, telling them to bring back soil samples. Over two years his lab amassed a collection that spanned the continent. It even included a thawed slice of tundra mailed by Wright's brother, a provincial policeman stationed on the northern Ontario-Manitoba border.

Every tested strain in a dirt sample proved resistant to multiple antibiotics.

By 2005 Wright's team had combed through the genes of nearly 500 streptomyces strains and species, many never before identified. Every one proved resistant to multiple antibiotics, not just their own signature chemicals. On average, each could neutralize seven or eight drugs, and many could shrug off 14 or 15. In all, the researchers found resistance to every one of the 21 antibiotics they tested, including Ketek and Zyvox, two synthetic new drugs.

"These genes clearly didn't jump directly from streptomyces into disease-causing bacteria," Wright says. He had noted subtle variations between the resistance genes he pulled out of soil organisms and their doppelgängers in disease-causing bacteria. As in a game of telephone, each time a gene gets passed from one microbe to another, slight differences develop that reflect the DNA dialect of its new host. The resistance genes bedeviling doctors had evidently passed through many intermediaries on their way from soil to critically ill patients.

Wright suspects that the antibiotic-drenched environment of commercial livestock operations is prime ground for such transfer. "You've got the genes encoding for resistance in the soil beneath these operations," he says, "and we know that the majority of the antibiotics animals consume get excreted intact." In other words, the antibiotics fuel the rise of resistant bacteria both in the animals' guts and in the dirt beneath their hooves, with ample opportunity for cross-contamination.

A 2001 study by University of Illinois microbiologist Roderick Mackie documented this flow. When he looked for tetracycline resistance genes in groundwater downstream from pig farms, he also found the genes in local soil organisms like *Microbacterium* and *Pseudomonas,* which normally do not contain them. Since then, Mackie has found that soil bacteria around conventional pig farms, which use antibiotics, carry 100 to 1,000 times more resistance genes than do the same bacteria around organic farms.

"These animal operations are real hot spots," he says. "They're glowing red in the concentrations and intensity of these genes." More worrisome, perhaps, is that Mackie pulled more resistance genes from his deepest test wells, suggesting that the genes percolated down toward the drinking water supplies used by surrounding communities.

An even more direct conduit into the environment may be the common practice of irrigating fields with wastewater from livestock lagoons. About three years ago, David Graham, a University of Kansas environmental engineer, was puzzled in the fall by a dramatic spike in resistance genes in a pond on a Kansas feedlot he was studying. "We didn't know what was going on until I talked with a large-animal researcher," he recalls. At the end of the summer, feedlots receive newly weaned calves from outlying ranches. To prevent the young animals from importing infections, the feedlot operators were giving them five-day "shock doses" of antibiotics. "Their attitude had been, cows are big animals, they're pretty tough, so you give them 10 times what they need," Graham says.

The operators cut back on the drugs when Graham showed them that they were coating the next season's alfalfa crop with highly drug-resistant bacteria. "Essentially, they were feeding resistance genes back to their animals," Graham says. "Once they realized that, they started being much more conscious. They still used antibiotics, but more discriminately."

While livestock operations are an obvious source of antibiotic resistance, humans also take a lot of antibiotics—and their waste is another contamination stream. Bacteria make up about one-third of the solid matter in human stool, and Scott Weber, of the State University of New York at Buffalo, studies what happens to the antibiotic resistance genes our nation flushes down its toilets.

Conventional sewage treatment skims off solids for landfill disposal, then feeds the liquid waste to sewage-degrading bacteria. The end result is around 5 billion pounds of bacteria-rich slurry, or waste sludge, each year. Around 35 percent of this is incinerated or put in a landfill. Close to 65 percent is recycled as fertilizer, much of it ending up on croplands.

Weber is now investigating how fertilizer derived from human sewage may contribute to the spread of antibiotic-resistant genes. "We've done a good job designing our treatment plants to reduce conventional contaminants," he says. "Unfortunately, no one has been thinking of DNA as a contaminant." In fact, sewage treatment methods used at the country's 18,000-odd wastewater plants could actually affect the resistance genes that enter their systems.

Most treatment plants, Weber explains, gorge a relatively small number of sludge bacteria with all the liquid waste they can eat. The result, he found, is a spike in antibiotic-resistant organisms. "We don't know exactly why," he says, "but our findings have raised an even more important question." Is the jump in resistance genes coming from a population explosion in the resistant enteric, or intestinal, bacteria coming into the sewage plant? Or is it coming from sewage-digesting sludge bacteria that are taking up the genes from incoming bacteria? The answer is important because sludge bacteria are much more likely to thrive and spread their resistance genes once the sludge is discharged into rivers (in treated wastewater) and onto crop fields (as slurried fertilizer).

Weber predicts that follow-up studies will show the resistance genes have indeed made the jump to sludge bacteria. On a hopeful note, he has shown that an alternative method of sewage processing seems to decrease the prevalence of bacterial drug resistance. In this process, the sludge remains inside the treatment plant longer, allowing dramatically higher concentrations of bacteria to develop. For reasons that are not yet clear, this method slows the increase of drug-resistant bacteria. It also produces less sludge for disposal. Unfortunately, the process is expensive.

Drying sewage sludge into pellets—which kills the sludge bacteria—is another way to contain resistance genes, though it may still leave DNA intact. But few municipal sewage plants want the extra expense of drying the sludge, and so it is instead exported "live" in tanker trucks that spray the wet slurry onto crop fields, along roadsides, and into forests.

Trolling the waters and sediments of the Cache la Poudre, Storteboom and Pruden are collecting solid evidence to support suspicions that both livestock operations and human sewage are major players in the dramatic rise of resistance genes in our environment and our bodies. Specifically, they have found unnaturally high levels of antibiotic resistance genes in sediments where the river comes into contact with treated municipal wastewater effluent and farm irrigation runoff as it flows 126 miles from Rocky Mountain National Park through Fort Collins and across Colorado's eastern plain, home to some of the country's most densely packed livestock operations.

"Over the course of the river, we saw the concentration of resistance genes increase by several orders of magnitude," Pruden says, "far more than could ever be accounted for by chance alone." Pruden's team likewise found dangerous genes in the water headed from local treatment plants toward household taps.

Presumably, most of these genes reside inside live bacteria, but a microbe doesn't have to be alive to share its dangerous DNA. As microbiologists have pointed out, bacteria are known to scavenge genes from the spilled DNA of their dead.

Nobody knows how long free-floating DNA might persist in the water.

"There's a lot of interest in whether there's naked DNA in there," Pruden says of the Poudre's waters. "Current treatment of drinking water is aimed at killing bacteria, not eliminating their DNA." Nobody even knows exactly how long such free-floating DNA might persist.

All this makes resistance genes a uniquely troubling sort of pollution. "At least when you pollute a site with something like atrazine," a pesticide, "you can be assured that it will eventually decay," says Graham, the Kansas environmental engineer, who began his research career tracking chemical pollutants like toxic herbicides. "When you contaminate a site with resistance genes, those genes can be transferred into environmental organisms and actually increase the concentration of contamination."

Taken together, these findings drive home the urgency of efforts to reduce flagrant antibiotic overuse that fuels the spread of resistance, whether on the farm, in the home, or in the hospital.

For years the livestock pharmaceutical industry has played down its role in the rise of antibiotic resistance. "We approached this problem many years ago and have seen all kinds of studies, and there isn't anything definitive to say that antibiotics in livestock cause harm to people," says Richard Carnevale, vice president of regulatory and scientific affairs at the Animal Health Institute, which represents the manufacturers of animal drugs, including those for livestock. "Antimicrobial resistance has all kinds of sources, people to animals as well as animals to people."

The institute's own data testify to the magnitude of antibiotic use in livestock operations, however. Its members sell an estimated 20 million to 25 million pounds of antibiotics for use in animals each year, much of it to promote growth. (For little-understood reasons, antibiotics speed the growth of young animals, making it cheaper to bring them to slaughter.) The Union of Concerned Scientists and other groups have long urged the United States to follow the European Union, which in 2006 completed its ban on the use of antibiotics for promoting livestock growth. Such a ban remains far more contentious in North America, where the profitability of factory-farm operations depends on getting animals to market in the shortest possible time.

On the other hand, the success of the E.U.'s ban is less than clear-cut. "The studies show that the E.U.'s curtailing of these compounds in feed has resulted in more sick animals needing higher therapeutic doses," Carnevale says.

"There are cases of that," admits Scott McEwen, a University of Guelph veterinary epidemiologist who advises the Canadian government on the public-health implications of livestock antibiotics. At certain stressful times in a young animal's life, as when it is weaned from its mother, it becomes particularly susceptible to disease. "The lesson," he says, "may be that we would do well by being more selective than a complete ban."

McEwen and many of his colleagues see no harm in using growth-promoting livestock antibiotics known as ionophores. "They have no known use in people, and we see no evidence that they select for resistance to important medical antibiotics," he says. "So why not use them? But if anyone tries to say that we should use such critically important drugs as cephalosporins or fluoroquinolones as growth promoters, that's a no-brainer. Resistance develops quickly, and we've seen the deleterious effects in human health."

A thornier issue is the use of antibiotics to treat sick livestock and prevent the spread of infections through crowded herds and flocks. "Few people would say we should deny antibiotics to sick animals," McEwen says, "and often the only practical way to administer an antibiotic is to give it to the whole group." Some critics have called for restricting certain classes of critically important antibiotics from livestock use, even for treating sick animals. For instance, the FDA is considering approval of cefquinome for respiratory infections in cattle. Cefquinome belongs to a powerful class of antibiotic known as fourth-generation cephalosporins, introduced in the 1990s to combat hospital infections that had grown resistant to older drugs. In the fall of 2006, the FDA's veterinary advisory committee voted against approving cefquinome, citing concerns that resistance to this vital class of drug could spread from bacteria in beef to hospital superbugs that respond to little else. But the agency's recently adopted guidelines make it difficult to deny approval to a new veterinary drug unless it clearly threatens the treatment of a specific food-borne infection in humans. As of press time, the FDA had yet to reach a decision.

Consumers may contribute to the problem of DNA pollution whenever they use soaps and cleaning products containing antibiotic-like compounds.

Consumers may contribute to the problem of DNA pollution whenever they use antibacterial soaps and cleaning products. These products contain the antibiotic-like chemicals triclosan and triclocarban and send some 2 million to 20 million pounds of the compounds into the sewage stream each year. Triclosan and triclocarban have been shown in the lab to promote resistance to medically important antibiotics. Worse, the compounds do not break down as readily as do traditional antibiotics. Rolf Halden, cofounder of the Center for Water and Health at Johns Hopkins University, has shown that triclosan and triclocarban show up in many waterways that receive treated wastewater—more than half of the nation's rivers and streams. He has found even greater levels of these two chemicals in sewage sludge destined for reuse as crop fertilizer. According to his figures, a typical sewage treatment plant sends more than a ton of triclocarban and a slightly lesser amount of triclosan back into the environment each year.

For consumer antibacterial soaps the solution is simple, Halden says: "Eliminate them. There's no reason to have these chemicals in consumer products." Studies show that household products containing such antibacterials don't prevent the spread of sickness any better than ordinary soap and water. "If there's no benefit, then all we're left with is the risk," Halden says. He notes that many European retailers have already pulled these products from their shelves. "I think it's only a matter of time before they are removed from U.S. shelves as well."

Finally, there is the complicated matter of the vast quantity of antibiotics that U.S. doctors prescribe each year: some 3 million pounds, according to the Union of Concerned Scientists. No doctor wants to ignore an opportunity to save a patient from infectious disease, yet much of what is prescribed is probably unnecessary—and all of it feeds the spread of resistance genes in hospitals and apparently throughout the environment.

"Patients come in asking for a particular antibiotic because it made them feel better in the past or they saw it promoted on TV," says Jim King, president of the American Academy of Family Physicians. The right thing to do is to educate the patient, he says, "but that takes time, and sometimes it's easier, though not appropriate, to write the prescription the patient wants."

Curtis Donskey, chief of infection control at Louis Stokes Cleveland VA Medical Center, adds that "a lot of antibiotic overuse comes from the mistaken idea that more is better. Infections are often treated longer than necessary, and multiple antibiotics are given when one would work as well." In truth, his studies show, the longer hospital patients remain on antibiotics, the more likely they are to pick up a multidrug-resistant superbug. The problem appears to lie in the drugs' disruption of a person's protective microflora—the resident bacteria that normally help keep invader microbes at bay. "I think the message is slowly getting through," Donskey says. "I'm seeing the change in attitude."

Meanwhile, Pruden's students at Colorado State keep amassing evidence that will make it difficult for any player—medical, consumer, or agricultural—to shirk accountability for DNA pollution.

Late in the afternoon, Storteboom drives past dairy farms and feedlots, meatpacking plants, and fallow fields, 50 miles downstream from her first DNA sampling site of the day. Leaving her Jeep at the side of the road, she strides past cow patties and fast-food wrappers and scrambles down an eroded embankment of the Cache la Poudre River. She cringes at the sight of two small animal carcasses on the opposite bank, then wades in, steering clear of an eddy of gray scum. "Just gross," she mutters, grateful for her watertight hip boots.

Of course, the invisible genetic pollution is of greater concern. It lends an ironic twist to the river's name. According to local legend, the appellation comes from the hidden stashes (*cache*) of gunpowder (*poudre*) that French fur trappers once buried along the banks. Nearly two centuries later, the river's hidden DNA may pose the real threat.

UNIT 4

New Antimicrobial Drug Development

Unit Selections

Learning Objectives

- Describe three alternative strategies for developing antimicrobial drugs that are least likely to result in the development of drug-resistant microbes.

- What is selective toxicity? How is this principle applied when formulating new therapeutics?

- Why is it important to use novel antibiotics with caution? Under what circumstances would you prescribe the latest drug to a patient?

Student Website
www.mhcls.com

Internet References

U.S. Food and Drug Administration
  http://www.fda.gov
Agricultural Research Service: U.S. Department of Agriculture
  http://www.ars.usda.gov
Institute for One World Health
  http://www.oneworldhealth.org/story
Office of AIDS Research
  http://www.oar.nih.gov

As discussed in the previous unit, microbes have become increasingly resistant to the weaponry of modern medicine as they acquire mutations that give them the ability to inactivate antimicrobial agents. What new strategies can we employ to overcome drug-resistant strains of bacteria? Developing any type of antimicrobial compound requires detailed knowledge of the pathogen's physiology and structure. Since the earliest days of drug development, the principle of selective toxicity has been of utmost importance. That is, a drug should be aimed at killing the pathogen without adversely affecting host cells. Our knowledge of microbial genomes (as described in Unit 2) may aid in efforts to find microbial-specific drug targets. However, to date this approach has yielded limited results. Since the advent of penicillin, many other natural products have proven to be a good source of antibiotics. Researchers hope to find the next new "wonder drug" in the soil beneath our feet or even in the depths of the ocean, as marine microbes are emerging as new sources of pharmaceuticals. Another method being used to combat drug-resistant microbes is the reformulation of older drugs, since current pathogens have had little exposure to drugs used decades ago and would therefore be less likely to exhibit resistance. In addition to these strategies, researchers are asking the question, "What does nature use to kill bacteria?", hoping to find answers from the microbial world itself. The articles in this unit have been selected to give the reader an idea of the diversity of approaches being taken in the field of antimicrobial drug development. The evidence is clear—we need new antimicrobial drugs and we need them now to surmount the growing problem of drug-resistant microbes. It should be evident from this unit that although developing new drugs is a challenging venture, we have promising leads for novel therapeutics.

The review article by Aksoy and Unal provides a summary about several new drugs available for treatment of infections caused by Gram-positive bacteria. The merits and disadvantages of each are discussed, with emphasis given to possible side effects, since many of these drugs are currently being used in clinical trials. One of the drugs, Daptomycin, is derived from the fermentation of *Streptomyces roseosporus,* and it works by causing membrane depolarization and inhibition of protein synthesis, but without concomitant cell lysis. This is a unique mechanism of action, because most bactericidal antibiotics cause the lysis of target bacteria resulting in endotoxin release, which can be quite harmful to the host. The FDA has approved Daptomycin for use in the treatment of skin infections caused by MRSA. Development of new glycopeptides that inhibit bacterial cell wall synthesis are also described. The authors caution that these new drugs should be used with caution only if standard treatments fail. The judicious use of existing drugs is of critical importance in our battle against "superbugs." This theme is highlighted again in the article "HIV Integrase Inhibitors—Out of the Pipeline and into the Clinic." A new class of antiretroviral agents that inhibit integration of the HIV provirus into the host cell's genome has been approved for clinical trials. However, the author emphasizes that this new class of drugs should only be used in patients who have failed to respond to conventional antiretroviral therapy.

© Keith Brofsky/Getty Images

One example of exploiting nature's tools for drug discovery is the use of bacteriophage lysins. Bacteriophage are viruses that specifically attack bacteria and may be used as nontraditional antimicrobial agents. Chemicals from bacteriophage, called lysins, may be used to treat bacterial infections without resulting in drug resistance or causing harm to normal microbiota. Lysins form holes in the bacterial cell wall by digesting peptidoglycan. Phage lysins are highly specific and only kill the species of bacteria from which they were isolated. This specificity ensures the selective toxicity of this new type of antimicrobial drug and prevents damage to surrounding normal flora. "Bacteriophage Lysins as Effective Antibacterials" elaborates on the background, structure, mechanism, and practical clinical applications of this alternative infection treatment strategy.

Our understanding of microbial mechanisms of pathogenicity has greatly increased in recent years, and another possible strategy for attacking microbes is to undermine their virulence without actually killing them. An article in the *New England Journal of Medicine* examines the concept of deterring microbial pathogenicity by disrupting the synthesis of a bacterial virulence factor. "Removing the Golden Coat of *Staphylococcus aureus*" describes the inhibition of staphyloxanthin biosynthesis. Staphyloxanthin is the pigment that gives *S. aureus* its golden color, and it may function as a virulence factor because it can detoxify antibacterial chemicals produced by the host's immune cells. This research has not yet been translated into clinical results, but the author is cautiously optimistic that this new approach may yield clinically relevant compounds.

Another alternative strategy for antimicrobial drug development is to use antibodies as antimicrobial agents. This is a new avenue of research, and the authors discuss both pros and cons of using antibodies in the treatment of bacterial infections. Antibodies may be used either directly to inactivate pathogens, or indirectly to neutralize virulence factors such as toxins. Several types of antibodies are under investigation; however, none have been approved by the FDA at this time. An excellent summary of antibodies currently being studied in clinical trials is provided. Finally, a report from the AAAS meeting in Chicago describes new research on a novel compound isolated from ocean-dwelling sponges. This new compound, ageliferin, resensitizes a variety of drug-resistant microbes to become susceptible to antibiotics. This phenomenon has never been observed before and the mechanism by which this new compound "reprograms" microbes to become sensitive to antibiotics is under investigation. Although it may be many years before sea sponges are turned into useful pharmaceutical agents, this finding symbolizes the wealth of new antimicrobial drugs awaiting discovery in the natural world.

New Antimicrobial Agents for the Treatment of Gram-Positive Bacterial Infections

D. Y. AKSOY AND S. UNAL

Introduction

The introduction of benzylpenicillin in the 1940s ushered in an era of effective treatment of bacterial infections, but it soon became apparent that some strains of *Staphylococcus aureus* were drug-resistant because of the production of β-lactamase. Since the mid-1970s, resistance to antimicrobial agents has become an escalating problem.[1] A striking change during the past quarter-century has been the increasing importance of infections caused by Gram-positive bacteria, which might have resulted from the previous emphasis placed on controlling Gram-negative bacterial infections. Today, it is necessary to deal with infections caused by multidrug-resistant organisms, particularly methicillin-resistant staphylococci, penicillin- and erythromycin-resistant pneumococci, and vancomycin-resistant enterococci (VRE). The emergence and rapid spread of strains of methicillin-resistant *Staph. aureus* (MRSA), which are resistant not only to all β-lactams, but also to the other main antibiotic classes, has resulted in an increased use of glycopeptide antibiotics, i.e., vancomycin and teicoplanin. Unfortunately, VRE were detected in 1986, both in France and the UK, and multidrug-resistant strains of enterococci are now encountered throughout Europe and the USA.[2–4] Within 1 year of their initial detection, resistant strains were associated with infections in the USA, and they accounted for 26% of isolates of enterococci from blood in the USA by the year 2000. Since 1996, vancomycin-intermediate *Staph. aureus* (VISA) isolates with a vancomycin MIC of 8–16 mg/L have been reported, and this has been followed since 2002 by reports of vancomycin-resistant *Staph. aureus* (MIC ≥ 32 mg/L).[5–8]

As a result of these developments, there is an urgent need for effective new antimicrobial agents, as well as for prudent use of existing agents. Linezolid, quinupristin–dalfopristin, daptomycin, tigecyline, new glycopeptides and ceftobiprole are the main new agents that have recently become available for use or are under clinical development. This review summarises their properties, the results of recent studies with these agents, and future treatment possibilities.

Linezolid

Linezolid (Zyvox®) is the first of a new class of antimicrobial agents, the oxazolidinones. Linezolid is a synthetic antibiotic that, depending on the organism and the linezolid concentration, has a bacteriostatic or bactericidal effect by inhibiting protein synthesis at the ribosomal level and by preventing the formation of the protein initiation complex.[9] Linezolid is active against most Gram-positive bacteria, including methicillin-sensitive *Staph. aureus* (MSSA) and MRSA, *Streptococcus pneumoniae* (including multidrug-resistant strains), *Streptococcus pyogenes* and *Streptococcus agalactiae*. Rare occurrences of linezolid resistance among VRE in North America (0.8–1.8%), caused by a G2576U ribosomal mutation, have been reported, and an intrinsic resistance gene rendering a clinical strain of MRSA resistant to linezolid has been described.[10–13] Although resistant strains of *Staph. aureus* have been reported, linezolid remains highly active against MRSA (MIC$_{90}$ 2mg/L).[12–15]

Linezolid was approved by the US Food and Drug Administration (FDA) in 2000 for the treatment of uncomplicated and complicated skin and soft-tissue infections, including diabetic foot infections without concomitant osteomyelitis, community-acquired and nosocomial pneumonia, and vancomycin-resistant *Enterococcus faecium* infections, including cases with concurrent bacteraemia.[9] Activity against *Nocardia* spp. and *Mycobacterium* spp. has also been demonstrated, including cases of central nervous system infection and infective endocarditis.[14–18]

The high bioavailability of linezolid, whether administered orally or intravenously, with a standard twice-daily dose of 600 mg, makes it a suitable alternative for the treatment of infections that require the prolonged use of antimicrobial agents.[19,20] Tissue penetration is good, with marked accumulation in sweat, saliva and epithelial lining. Linezolid is reported to be equally effective or even superior to vancomycin for the treatment of pneumonia and soft-tissue infections.[21–23] However, recent meta-analyses have suggested that linezolid's apparent superiority for the treatment of pneumonia and bone and joint infections has limitations. Cases of MRSA endocarditis that failed to respond to linezolid treatment have also been reported.[15,24,25]

Linezolid is generally well-tolerated, with the most common adverse effects being diarrhoea, headache and nausea. Like many other antibiotics, it may predispose to pseudo-membranous colitis following the overgrowth of *Clostridium difficile*.[26] Myelosuppression (including anaemia, leukopenia, pancytopenia, and thrombocytopenia) has been reported in patients taking linezolid. In cases with a known treatment outcome, the affected haematological parameters rose to pre-treatment levels when linezolid was discontinued. Complete blood counts should be monitored weekly in patients taking linezolid, and particularly in patients who are prescribed linezolid for >2 weeks, patients with pre-existing myelosuppression, patients receiving concomitant drugs that produce bone marrow suppression, and patients with a chronic infection who have received previous or concomitant antibiotic therapy. Discontinuation of linezolid therapy should be considered for patients who develop, or have worsening, myelosuppression. Spontaneous reports of serotonin syndrome associated with the co-administration of linezolid and serotonergic agents, including such anti-depressants as selective serotonin re-uptake inhibitors, have been reported. Linezolid has not been studied in patients with uncontrolled hypertension, pheochromocytoma, carcinoid syndrome or untreated hyper-thyroidism.[27]

Lactic acidosis, peripheral neuropathy and optic neuropathy are other adverse effects of linezolid treatment. A recent study that compared linezolid to vancomycin, oxacillin and dicloxacillin (comparator antibiotics) for the treatment of seriously-ill patients with intravascular catheter-related bloodstream infections, including patients with catheter-site infections, reported a higher mortality rate in the linezolid arm. There was no difference in the mortality rate of patients with Gram-positive infections who were treated with other antibiotics. Furthermore, the mortality rate was higher only for patients who were infected only with Gram-negative organisms, or who were infected with both Gram-positive and Gram-negative organisms, or who had no infection when they entered the study and were treated with linezolid. Therefore, if infection with Gram-negative bacteria is known or suspected, appropriate therapy should be started immediately (http://www.fda.gov/cder/drug/InfoSheets/HCP/linezolidHCP.htm).

Linezolid is an effective alternative antimicrobial agent for the treatment of Gram-positive infections, but because of the absence of placebo-controlled double-blind studies, linezolid should be used with extreme caution in patients with vancomycin-induced nephrotoxicity or a documented absence of response to vancomycin.

Quinupristin–Dalfopristin

Quinupristin–dalfopristin (Synercid®) is a fixed mixture (30:70) of semi-synthetic streptogramin derivatives. These compounds enter bacterial cells by diffusion and bind to different sites on the 50S ribosomal subunit, resulting in an irreversible inhibition of bacterial protein synthesis.[28] Dalfopristin blocks the reaction catalysed by the peptidyl transferase catalytic centre of the 50S ribosome by inhibiting substrate attachment to the P-site and A-site of the ribosome. Quinupristin inhibits peptide chain elongation. The synergic effect of this drug combination appears to result from the fact that it targets both early and late stages of protein synthesis.[29] Thus, quinupristin–dalfopristin is bacteriostatic against *E. faecium* and bactericidal against MSSA, MRSA and *Strep. pyogenes*. It is ineffective against *Enterococcus faecalis*. In the USA, the only FDA-approved use of quinupristin–dalfopristin as an anti-staphylococcal agent is for the treatment of adults with complicated skin and skin-structure infections caused by MSSA. Otherwise, the drug is approved for complicated skin and skin-structure infections caused by *Strep. pyogenes,* and for serious VRE infections associated with bacteraemia.[30] Resistance in *E. faecium* was reported by the SENTRY antimicrobial surveillance programme in Europe and North America to be 10.0% and 0.6%, respectively.[10]

The results of studies in which the effectiveness of quinupristin–dalfopristin for the treatment of skin and soft-tissue infections was compared to that of other agents have generally been satisfactory, despite the fact that quinupristin–dalfopristin is inferior to other agents for the treatment of pneumonia and infective endocarditis.[24,31]

Quinupristin–dalfopristin can only be administered by the intravenous route in a dextrose 5% w/v solution. Drug elimination is through bile into faeces, but clearance may be slightly impaired in patients with renal insufficiency. The drug has significant toxicity problems, including arthralgia–myalgia syndrome and venous intolerance. Pain and inflammation at the infusion site is experienced by up to 74% of patients.[31–33] Hyperbilirubinaemia and liver toxicity can also occur. Quinupristin–dalfopristin has been shown to be a major inhibitor of the activity of the cytochrome P450 3A4 isoenzyme, and drug interactions (especially with cyclosporine) should be monitored carefully during therapy. Although the drug itself does not induce QTc prolongation, it can interfere with the metabolism of other drug products that are associated with QTc prolongation.[33,34]

Quinupristin–dalfopristin is the first parenteral streptogramin and offers a unique alternative treatment for infections caused by multidrug-resistant Gram-positive bacteria, but the broad spectrum of adverse effects makes it an inferior choice to other agents.

Daptomycin

Daptomycin (Cubicin®) is the first of a new class of cyclic lipopeptides. This agent is derived from the fermentation of a strain of *Streptomyces roseosporus*. Originally developed during the early 1980s, daptomycin was initially shelved because of concerns about skeletal-muscle toxicity. This effect was not seen with lower doses, and daptomycin received approval by the FDA in September 2003 for the treatment of complicated skin and soft-tissue infections.

Daptomycin attaches to the cytoplasmic membrane of Gram-positive bacteria, forming a channel that allows the efflux of potassium, causing depolarisation of the membrane, along with dysfunction of macromolecular synthesis and collapse of the organism without lysis. This mechanism of action is concentration-dependent, free calcium ion-dependent, rapid and unique.[35] Some other bactericidal antibiotics, most notably β-lactams, cause bacterial cells to lyse. This is potentially harmful, as it may release bacterial endotoxins and other inflammatory cell components into the circulation, triggering cytokine cascades, and potentially leading to septic shock and multiple organ failure.[36] The antimicrobial effect of daptomycin is concentration-dependent, with the optimal dose for serious bloodstream and endovascular infections reported to be 6 mg/kg once-daily. The recommended dose for skin and skin-structure infections is 4 mg/kg once-daily. Daptomycin also has a post-antibiotic effect lasting 1.5–6 h.[37]

Although daptomycin is approved by the FDA for the treatment of complicated skin and skin-structure infections that involve MRSA and other Gram-positive bacteria, its rapid bactericidal effect makes it appropriate for the treatment of other kinds of infection. The bactericidal activity of daptomycin, vancomycin, linezolid and quinupristin–dalfopristin against MRSA and VISA has been compared using in-vitro time-kill studies.[38] Against all organisms tested, daptomycin had bactericidal activity equal to or greater than that of the other agents.[39]

Daptomycin has efficacy comparable to standard therapy for the treatment of skin and skin-structure infections. It has been used successfully to treat bone and joint infections.[40,41] Successful results have also been reported for the treatment of bacteraemia and right-sided infective endocarditis, but not for the treatment of community-acquired pneumonia. When daptomycin was compared with standard therapy for bacteraemia and endocarditis caused by *Staph. aureus*, a successful outcome was documented in 53 of 120 of patients who received daptomycin, compared with 48 of 115 patients who received standard therapy (44.2% vs. 41.7%, respectively; absolute difference, 2.4%; 95% CI 10.2–15.1).[42–44] The lack of efficacy of daptomycin in treating community-acquired pneumonia is thought to be a consequence of a reduction of daptomycin activity in the presence of lung surfactant.[45]

Resistance to daptomycin is rare. Spontaneous resistance is uncommon, emerging *in vitro* at a rate of $<1 \times 10^{-10}$, but resistance can be induced by serial passage in increasing concentrations of the antibiotic.[46] There have been several reports of daptomycin resistance emerging in clinical isolates of MRSA from patients who received prolonged treatment.[47,48] The once-daily dosing schedule and the favourable safety profile (except for some concerns regarding rhabdomyolysis and neuropathy) make daptomycin an attractive option for the treatment of Gram-positive infections.[44]

Tigecycline

Tigecycline (Tygacil®) is a new, semi-synthetic glycylcycline, which is a new class of antimicrobial agent. Glycocides are derivatives of tetracyclines, and have broad-spectrum activity against susceptible and multidrug-resistant strains of bacteria. Tigecycline, the first glycylcycline, demonstrates potent in-vitro activity against a wide range of Gram-positive and Gram-negative bacteria, including MRSA and tetracycline-resistant *Staph. aureus,* as well as many multidrug-resistant

Gram-negative pathogens and anaerobes.[49] The potent in-vitro activity of tigecycline resulted in its approval by the FDA for the treatment of skin, soft-tissue and intra-abdominal infections.[50]

Like tetracycline, tigecycline exerts its mechanism of action by reversibly binding to the 30S ribosomal subunit and inhibiting protein translation in bacteria. This blockade prevents the entry of aminoacyl tRNA molecules into the A-site of the ribosome, resulting in the loss of peptide formation.[51,52] Tigecycline is highly effective *in vivo* against most Gram-positive organisms, including *Staph. aureus,* coagulase-negative staphylococci, *Enterococcus* spp., *Strep. pneumoniae,* and group A, group B and viridans streptococci. It also has good in-vitro activity against most Gram-negative organisms and anaerobes. Interestingly, some atypical microorganisms, e.g., *Mycobacterium abscessus, Mycobacterium chelonae,* the *Myco-bacterium fortuitum* group, *Mycobacterium marinum, Chlamydia pneumoniae, Mycoplasma hominis, Mycoplasma pneumonia* and *Ureaplasma urealyticum,* are also susceptible to tigecycline.[53] Tigecycline has demonstrated excellent overall tissue penetration in animal studies, including bone, bone marrow, salivary gland, thyroid, spleen, kidney and cerebrospinal fluid.[53]

Phase 3 randomised double-blind studies have confirmed the efficacy of tigecycline for the treatment of skin and skin-structure infections, and for intra-abdominal infections. Sacchidanand et al. [54] reported that tigecycline was not inferior to a combination of aztreonam and vancomycin for the treatment of complicated skin and skin-structure infections. Cure rates and microbiological eradication rates were similar for tigecycline and a comparator group of antibiotics, at 82.9% and 82.3%, respectively.[54] In two other phase 3 double-blind studies involving 833 clinically evaluable patients (422 treated with tigecycline and 411 treated with vancomycin and aztreonam), clinical response rates were similar for tigecycline (86.5%) (95% CI 82.9–89.6%) and vancomycin–aztreonam (88.6%) (95% CI 85.1–91.5%).[55]

In a pooled analysis of two phase 3 double-blind trials designed to evaluate the safety and efficacy of tigecycline in comparison with that of imipenem–cilastatin, clinical cure rates for the microbiologically evaluable group among 1642 adults with complicated intra-abdominal infections were 86.1% for tigecycline and 86.2% for imipenem–cilastatin (95% CI for the difference–4.5 to 4.4%; p <0.0001 for non-inferiority).[56] Similar results were obtained in another study which demonstrated that tigecycline was not inferior for treating complicated intra-abdominal infections.[57]

Clinically significant organ toxicity has not been observed in association with tigecycline use during clinical trials. Nausea and vomiting are the most common side-effects, and these are dose-limited and are not diminished by reducing the rate of drug infusion.[58] The low potential of the drug for the development of resistance, which is almost a rule for other tetracyclines, combined with its broad-spectrum activity against multidrug-resistant pathogens, provides key advantages. Therefore, tigecycline presents a genuinely new therapeutic option for the treatment of infections caused by multidrug-resistant bacteria.

New Glycopeptides

Oritavancin, telavancin and dalbavancin are new glycopeptides currently in clinical development, and appear to be potent molecules with favourable pharmacokinetic and pharmacodynamic properties.

Oritavancin

This agent was obtained by reductive alkylation with 4'-chloro-biphenylcarboxaldehyde of chloroeremomycin, which differs from vancomycin by the addition of a 4-epi-vancosamine sugar, and the replacement of vancosamine of the disaccharide moiety by an epivancosamine.[59] Although oritavancin has a general spectrum of activity comparable to that of vancomycin, it offers considerable advantages in terms of intrinsic activity (especially against streptococci). Unlike vancomycin, oritavancin can dimerise, leading to a cooperative interaction with two stems of the growing peptidoglycan chain. The lipophilic side-chain assists in membrane-anchoring by hydrophobic interactions, stabilising the dimer in the most favourable position. Evidence also exists for another mechanism of action of oritavancin, namely inhibition of the transglycosylation step of cell-wall synthesis.[60,61]

Oritavancin is active against most isolates of streptococci and staphylococci, as well as *Pepto-streptococcus* spp., *Propionibacterium acnes*, *Clostridium perfringens* and *Corynebacterium jeikeium*. Its activity against enterococci is not affected by the presence of *vanA*-, *vanB*- and *vanC*-encoded vancomycin resistance, or by aminoglycoside resistance. Its activity against pneumococci and viridans streptococci is not affected by the presence of intermediate- or high-level penicillin resistance. In addition, the presence of methicillin resistance does not affect the activity of oritavancin against *Staph. aureus* or coagulase-negative staphylococci. Oritavancin is inactive against Gram-negative aerobes and anaerobes.[62]

Oritavancin is active in a number of animal models of infection, including a central venous catheter-associated infection model in rats (vancomycin-resistant *E. faecalis*), a rabbit endocarditis model (MRSA, vancomycin-sensitive and vancomycin-resistant *E. faecalis*), a neutropenic mouse thigh model (*Staph. aureus* ATCC 13709), and a rabbit meningitis model (penicillin-sensitive *Strep. pneumoniae*).[63–68]

In a phase 2 clinical trial involving complicated skin and skin-structure infections, oritavancin was not inferior to vancomycin and cephalexin, but had a shorter mean duration of activity.[69] In a phase 2 open-label randomised trial comparing oritavancin (5–10 mg/kg once-daily for 10–14 days) with vancomycin (15 mg/kg twice-daily) and a β-lactam for 10–14 days in patients with *Staph. aureus*-associated bacteraemia, oritavancin was as effective as the comparators, with higher clinical and bacteriological success in the cohort receiving 10 mg/kg, and with no evidence of increased side-effects.[70] The exceptionally long terminal half-life suggests the existence of storage sites within the organism. Studies using cultured macrophages indicate that the drug accumulates slowly (by an endocytic process), but importantly, in the lysosomes, from which efflux is extremely slow.[71] The clinical importance of this property is currently unknown.

Telavancin

Telavancin (TD-6424) is a semi-synthetic derivative of vancomycin, possessing a hydrophobic side-chain on the vancosamine sugar (decylaminoethyl) and a (phosphonomethyl) aminoethyl substituent on the cyclic peptidic core.[72] Pharmacological studies suggest that the enhanced activity of telavancin against *Strep. pneumoniae*, *Staph. aureus* (to a lesser extent) and staphylococci and enterococci harbouring the *vanA* gene cluster results from a complex mechanism of action, which, on the basis of data obtained with close analogues, involves a perturbation of lipid synthesis and possible membrane disruption.[73]

Telavancin is active *in vitro* against all Gram-positive pathogens, including *vanA*-positive enterococci. As with oritavancin, extreme potency is observed for streptococci, particularly *Strep. pneumoniae*.[74] Telavancin is active in an in-vitro biofilm model, whereas vancomycin and a number of other antibiotics were much less effective.[75] Telavancin is highly effective in animal models of relevant infections,[76–78] and tissue penetration of telavancin is high after intravenous administration in healthy subjects.[79] Telavancin has proven to be effective and safe for patients with skin and soft-tissue infections, with cure rates of up to 96%, compared with 90% for standard therapy. Concerns related to QTc elevation will be clarified following further studies.[80,81]

Dalbavancin

This agent is a semi-synthetic derivative of the teicoplanin-related glycopeptide A40925, modified with an amide appendage at the C-terminus and an alteration of the hydrophobic acylglucosamine substituent. Like teicoplanin, dalbavancin is active against *vanB*-positive enterococci, as well as staphylococci and other important species.[73] The pharmacokinetic profile of dalbavancin is characterised by a long half-life of c. 7 days. This allows a once-weekly regimen for the treatment of infections caused by Gram-positive bacteria. This weekly regimen may offer certain advantages, e.g., improved patient compliance and reduced use of resources compared with antimicrobial agents administered more frequently.[82] Dalbavancin has been shown to be as effective as linezolid twice-daily for the treatment of complicated skin and skin-structure infections,[83] and is more effective than most anti-Gram-positive agents, both *in vitro* and *in vivo*.[84,85]

In addition to complicated skin and skin-structure infections, dalbavancin has been used to treat catheter-related bloodstream infection. In a randomised controlled open-label multicentre phase 2 trial that involved 75 adult patients with catheter-related bloodstream infection, the results were overwhelmingly in favour of dalbavancin, albeit with limitations.[86] To date, adverse events have been mild and limited, with the most common being pyrexia, headache, nausea, oral candidiasis, diarrhoea and constipation. Consequently, dalbavancin appears to be a promising antimicrobial agent for the treatment of infections caused by Gram-positive bacteria.

Ceftobiprole

Ceftobiprole is a novel broad-spectrum β-lactamase-stable parenteral cephalosporin with strong affinity for the penicillin-binding proteins PBP2a and PBP2x, responsible for resistance in staphylococci and pneumococci, respectively. Ceftobiprole can also bind to relevant penicillin-binding proteins of resistant Gram-positive and Gram-negative bacteria, and appears to have a low ability to select for resistance.[87,88] In-vivo screening models suggest good activity of ceftobiprole against Gram-positive and Gram-negative bacteria. Several animal models have been used in the evaluation of ceftobiprole, including mouse sepsis, abscess and pneumonia models, rat endocarditis and tissue cage models, and a rabbit endocarditis model.[89,90] These models suggest that ceftobiprole should have clinical efficacy as an empirical treatment for severe clinical infections.

Several phase 3 studies of ceftobiprole for the treatment of complicated skin and skin-structure infections have been performed. One study compared intravenous ceftobiprole (500 mg every 12 h) with intravenous vancomycin (1 g every 12 h) in patients with complicated skin and skin-structure infections caused by Gram-positive bacteria. Staphylococci were the predominant pathogens, and >25% of the microbiologically evaluable patients had infections caused by MRSA. In the clinically evaluable population, efficacy and adverse events were comparable between treatment arms. Additional clinical trials involving complicated skin and skin-structure infections and pneumonia patients are underway to evaluate ceftobiprole as a treatment for infections caused by both Gram-positive and Gram-negative bacteria. The broad-spectrum activity of ceftobiprole may allow it to be used as monotherapy for serious nosocomial infections for which combination therapy would otherwise be required.[91,92]

Conclusions

Linezolid, quinupristin–dalfopristin, daptomycin, tigecycline, new glycopeptides and ceftobiprole are the most important novel agents that are currently being considered as alternatives for the treatment of infections caused by Gram-positive bacteria. Doripenem, iclaprim (a recent example of a novel diamonipyrimidine), ranbezolide (a novel oxazolidinone) and ceftaroline (a novel cephalosporin with impressive anti-MRSA and anti-pneumococcal activity) are other agents that are currently under investigation.[93–96]

Changing patterns of resistance have compounded and exacerbated the need for new antimicrobial agents. Each of the above compounds has its own unique advantages and disadvantages, so each agent should be used cautiously when conventional treatment fails. The appropriate indications and cost-effectiveness of these molecules will determine future treatment options. Until then, prudent use of existing antibiotics, with strict reinforcement of infection control precautions, should continue to be the rule in the treatment of Gram-positive infections.

References

1. Johnson AP, Livermore DM, Tillotson GS. Antimicrobial susceptibility of Gram-positive bacteria: what's current, what's anticipated? *J Hosp Infect* 2001; **49** (suppl A): S3–S11.

2. Leclercq R, Derlot E, Duval J, Courvalin P. Plasmid-mediated resistance to vancomycin and teicoplanin in *Enterococcus faecium. N Engl J Med* 1988; **319:** 157–160.

3. Uttley AH, Collins CH, Naidoo J, George RC. Vancomycin-resistant enterococci. *Lancet* 1988; **i:** 57–58.

4. Aksoy DY, Tanriover MD, Unal S. Antimicrobial resistance: preventable or inevitable. Problem of the era from two perspectives. In: Gould I, van der Meer J, eds, *Antibiotic policies: fighting resistance,* 1st edn. New York: Springer, 2007: 113–133.

5. Anonymous. CDC National Nosocomial Infections Surveillance (NNIS) system report, data summary from January 1990 to May 1999. *Am J Infect Control* 1999; **27:** 520–532.

6. Kobayashi K, Rao M, Keis S, Reiney FA, Smith JM, Cook GM. Enterococci with reduced susceptibility to vancomycin in New Zealand. *J Antimicrob Chemother* 2000; **46:** 405–410.

7. Appelbaum PC. MRSA—the tip of the iceberg. *Clin Microbiol Infect* 2006; **12** (suppl 2): 3–10.

8. Chen CY, Tang JL, Hsueh PR *et al.* Trends and antimicrobial resistance of pathogens causing bloodstream infections among febrile neutropenic adults with hemato-logical malignancy. *J Formos Med Assoc* 2004; **103:** 526–532.

9. The European Committee on Antimicrobial Susceptibility Testing (EUCAST) Steering Committee. EUCAST technical note on linezolid. *Clin Microbiol Infect* 2006; **12:** 1243–1245.

10. Deshpande LM, Fritsche TR, Moet GJ, Biedenbach DJ, Jones RN. Antimicrobial resistance and molecular epidemiology of vancomycin-resistant enterococci from North America and Europe: a report from the SENTRY antimicrobial surveillance program. *Diagn Microbiol Infect Dis* 2007; **58:** 163–170.

11. Gonzales RD, Schreckenberger PC, Graham MB, Kelkar S, DenBesten K, Quinn JP. Infections due to vancomycin-resistant *Enterococcus faecium* resistant to linezolid. *Lancet* 2001; **357:** 1179.

12. Toh SM, Xiong L, Arias CA *et al.* Acquisition of a natural resistance gene renders a clinical strain of methicillin-resistant *Staphylococcus aureus* resistant to the synthetic antibiotic linezolid. *Mol Microbiol* 2007; **64:** 1506–1514.

13. Ross JE, Fritsche TR, Sader HS, Jones RN. Oxazolidinone susceptibility patterns for 2005: international report from Zyvox Annual Appraisal of Potency and Spectrum study. *Int J Antimicrob Agents* 2007; **29:** 295–301.

14. Mylona E, Fanourgiakis P, Vryonis E *et al.* Linezolid-based therapy in *Staphylococcus epidermidis* endocarditis. *Int J Antimicrob Agents* 2007; **29:** 597–598.

15. Munoz P, Rodrigues-Creixems M, Moreno M *et al.* Linezolid therapy for infective endocarditis. *Clin Microbiol Infect* 2007; **13:** 211–215.

16. Moitra RK, Auckland C, Cawley MI, Jones G, Cooper C. Systemic nocardiosis in a splenectomized patient with systemic lupus erythematosus. Successful treatment using linezolid. *J Clin Rheumatol* 2003; **9:** 47–50.

17. Sood R, Bhadauriya T, Rao M *et al.* Antimycobacterial acitivities of oxazolidinones: a review. *Infect Disord Drug Targets* 2006; **6:** 343–354.

18. Ntziora F, Falagas ME. Linezolid for the treatment of patients with central nervous system infection. *Ann Pharmacother* 2007; **41:** 296–308.

19. Falagas ME, Siempos II, Papagelopoulos PJ, Vardakas KZ. Linezolid for the treatment of adults with bone and joint infections. *Int J Antimicrob Agents* 2007; **29:** 233–239.

20. Kutscha-Lissberg F, Hebler U, Muhr G, Koller M. Linezolid penetration into bone and joint tissues infected with methicillin-resistant staphylococci. *Antimicrob Agents Chemother* 2003; **47:** 3964–3966.

21. Wunderink RG, Rello J, Cammarata SK, Croos-Dabrera RV, Kollef MH. Linezolid vs vancomycin, analysis of two double blind studies of patients with methicillin-resistant *Staphylococcus aureus* nosocomial pneumonia. *Chest* 2003; **124:** 1789–1797.

22. Wilson SE. Clinical trial results with linezolid, an oxazolidinone, in the treatment of soft tissue and post-operative Gram positive infections. *Surg Infec (Larchmt)* 2001; **2:** 25–35.

23. Hau T. Efficacy and safety of linezolid in the treatment of skin and soft tissue infections. *Eur J Clin Microbiol Infect Dis* 2002; **21:** 491–498.

24. Maclayton DO, Hall RG. Pharmacologic treatment options for nosocomial pneumonia involving methicillin-resistant *Staphylococcus aureus*. *Ann Pharmacother* 2007; **41:** 235–244.

25. Ruiz ME, Guerrero IC, Tuazon CU. Endocarditis caused by methicillin-resistant *Staphylococcus aureus:* treatment failure with linezolid. *Clin Infect Dis* 2002; **35:** 1018–1020.

26. Schweiger ES, Weinberg JM. Novel antibacterial agents for skin and skin structrure infections. *J Am Acad Dermatol* 2004; **50:** 331–340.

27. Bishop E, Melvani S, Howden BP, Charles PG, Grayson ML. Good clinical outcomes but high rates of adverse reactions during linezolid therapy for serious infections: a proposed protocol for monitoring therapy in complex patients. *Antimicrob Agents Chemother* 2006; **50:** 1599–1602.

28. Bryson HM, Spencer CM. Quinupristin–dalfopristin: antibacterial activity, pharmacokinetic profile, therapeutic trials and tolerability and current status. *Drugs* 1996; **52:** 406–415.

29. Cocito C, Di Giambattista M, Nyssen E, Vannuffel P. Inhibition of protein synthesis by streptogramins and related antibiotics. *J Antimicrob Chemother* 1997; **39** (suppl A): 7–13.

30. Johnson AP, Livermore DM. Quinupritin/dalfopristin: a new addition to antimicrobial arsenal. *Lancet* 1999; **354:** 2012–2013.

31. Nichols RL, Graham DR, Barriere SL *et al.* Treatment of hospitalized patients with complicated gram-positive skin and skin structure infections: two randomized, multicentre studies of quinupristin–dalfopristin versus cefazolin, oxacillin or vancomycin. Synercid Skin and Skin Structure Infection Group. *J Antimicrob Chemother* 1999; **44:** 263–273.

32. Eliopoulos G. Antimicrobial agents for treatment of serious infections caused by resistant *Staphylococcus aureus* and enterococci. *Eur J Clin Microbiol Infect Dis* 2005; **24:** 826–831.

33. Moellering RC, Linden PK, Reinhardt J, Blumberg EA, Bompart F, Talbot GH. The efficacy and safety of quinupristin/dalfopristin for the treatment of infections caused by vancomycin-resistant *Enterococcus faecium.* Synercid Emergency-Use Study Group. *J Antimicrob Chemother* 1999; **44:** 251–261.

34. Stamatakis MK, Richards JG. Interaction between quinupristin/dalfopristin and cyclosporine. *Ann Pharmacother* 1997; **31:** 576–578.

35. Torres-Viera C, Dembry LM. Approaches to vancomycin-resistant enterococci. *Curr Opin Infect Dis* 2004; **17:** 541–547.

36. Ginsburg I. The role of bacteriolysis in the pathophysiology of inflammation, infection and post infectious sequelas. *APMIS* 2002; **110:** 753–770.

37. Akins RL, Rybak MJ. Bactericidal activities of two daptomycin regimens against clinical strains of glycopeptide intermediate-resistant *Staphylococcus aureus,* vancomycin-resistant *Enterococcus faecium,* and methicillin-resistant *Staphylococcus aureus* isolates in an *in vitro* pharmacodynamic model with simulated endocardial vegetations. *Antimicrob Agents Chemother* 2001; **45:** 454–459.

38. Rybak MJ, Hershberger E, Moldovan T, Grucz RG. *In vitro* activities of daptomycin, vancomycin, linezolid and quinupristin/dalfopristin against staphylococci and enterococci including vancomycin-intermediate and resistant strains. *Antimicrob Agents Chemother* 2000; **44:** 1062–1066.

39. Cha R, Grucz RG, Rybak MJ. Daptomycin dose–effect relationship against resistant gram-positive organisms. *Antimicrob Agents Chemother* 2003; **47:** 1598–1603.

40. Arbeit RD, Maki D, Tally FP *et al.* The safety and efficacy of daptomycin for the treatment of complicated skin and skin-structure infections. *Clin Infect Dis* 2004; **38:** 1673–1661.

41. Martone WJ, Lamp KC. Efficacy of daptomycin in complicated skin and skin-structure infections due to methicillin-sensitive and -resistant *Staphylococcus aureus:* results from the CORE registry. *Curr Med Res Opin* 2006; **22:** 2337–2343.

42. Fowler VG, Boucher HW, Corey GR *et al. S. aureus* Endocarditis and Bacteremia Study Group. Daptomycin versus standard therapy for bacteremia and endocarditis caused by *Staphylococcus aureus. N Engl J Med* 2006; **355:** 653–665.

43. Segreti JA, Crank CW, Finney MS. Daptomycin for the treatment of gram-positive bacteremia and infective endocarditis. A retrospective case series of 31 patients. *Pharmacotherapy* 2006; **26:** 347–352.

44. Vouillamoz J, Moreillon P, Giddey M, Entenza JM. Efficacy of daptomycin in the treatment of experimental endocarditis due to susceptible and multi-drug resistant enterococci. *J Antimicrob Chemother* 2006; **58:** 1208–1214.

45. Silverman JA, Mortin LI, Vanpraagh AD, Li T, Alder J. Inhibition of daptomycin by pulmonary surfactant: *in vitro* modeling and clinical impact. *J Infect Dis* 2005; **191:** 2149–2152.

46. Falagas ME, Giannopoulou KP, Ntziora F, Vardakas KZ. Daptomycin for endocarditis and/or bacteraemia: a systematic review of the experimental and clinical evidence. *J Antimicrob Chemother* 2007; **60:** 7–19.

47. Mangili A, Bica I, Snydman DR, Hamer DH. Daptomycin-resistant methicillin-resistant *Staphylococcus aureus* bacteremia. *Clin Infect Dis* 2005; **40:** 1058–1060.

48. Skiest DJ. Treatment failure resulting from resistance of *Staphylococcus aureus* to daptomycin. *J Clin Microbiol* 2006; **44**: 655–656.

49. Hoban DJ, Bouchillon SK, Johnson BM *et al. In vitro* activities of tigecycline against 6792 gram-negative and gram-positive clinical isolates from the global Tigecyline Evaluation and Surveillance Trial (TEST Program, 2004). *Diagn Microbiol Infect Dis* 2005; **52**: 215–227.

50. Pankey GA. Tigecycline. *J Antimicrob Chemother* 2005; **56**: 470–480.

51. Projan SJ. Preclinical pharmacology of GAR-936, a novel glycylcyline antibacterial agent. *Pharmacotherapy* 2000; **20**: 219S–223S.

52. Sum PE, Petersen P. Synthesis and structure–activity relationship of novel glycylcyline derivatives leading to the discovery of GAR-936. *Bioorg Med Chem Lett* 1999; **9**: 1459–1462.

53. Al-Tatari H, Abdel-Haq N, Chearskul P, Asmar B. Antibiotics for treatment of resistant Gram-positive coccal infections. *Ind J Pediatr* 2006; **73**: 323–334.

54. Sacchidanand S, Penn RL, Embil JM *et al.* Efficacy and safety of tigecycline monotherapy compared with vancomycin plus aztreonam in patients with complicated skin and skin structure infections: results from a phase 3, randomized, double-blind trial. *Int J Infect Dis* 2005; **9**: 251–261.

55. Ellis-Grosse EJ, Babichak T, Dartois N *et al.* The efficacy and safety of tigecycline in the treatment of skin and skin structure infections: results of 2 double-blind phase 3 comparison studies with vancomycin–aztreonam. *Clin Infect Dis* 2005; **41** (suppl 5): S341–S353.

56. Babinchak T, Ellis-Grosse E, Dartois N *et al.* The efficacy and safety of tigecycline for the treatment of complicated intra-abdominal infections: analysis of pooled clinical trial data. *Clin Infect Dis* 2005; **41** (suppl 5): S354–S367.

57. Oliva ME, Rekha A, Yellin A *et al.* A multicenter trial of the efficacy and safety of tigecycline versus imipenem/cilastatin in patients with complicated intra-abdominal infections. *BMC Infect Dis* 2005; **5**: 88.

58. Stein GE, Craig WA. Tigecycline: a critical analysis. *Clin Infect Dis* 2006; **43**: 518–524.

59. Cooper RD, Snyder NJ, Zweifel MJ *et al.* Reductive alkylation of glycopeptide antibiotics: synthesis and antibacterial activity. *J Antibiot* 1996; **49**: 575–581.

60. Malabarba A, Ciabatti R. Glycopeptide derivatives. *Curr Med Chem* 2001; **8**: 1759–1773.

61. Allen NE, Nicas TI. Mechanism of action of oritavancin and related glycopeptide antibiotics. *FEMS Microbiol Rev* 2003; **26**: 511–532.

62. Guay DR. Oritavancin and tigecyline: investigational antimicrobials for multi-drug resistant bacteria. *Pharmacotherapy* 2004; **24**: 58–68.

63. Saleh-Mghir A, Lefort A, Petegnief Y *et al.* Activity and diffusion of LY333328 in experimental endocarditis due to vancomycin-resistant *Enterococcus faecalis. Antimicrob Agents Chemother* 1999; **43**: 115–120.

64. Lefort A, Saleh-Mghir A, Garry L, Carbon C, Fantin B. Activity of LY333328 combined with gentamicin *in vitro* and in rabbit experimental endocarditis due to vancomycin-susceptible or -resistant *Enterococcus faecalis. Antimicrob Agents Chemother* 2000; **44**: 3017–3021.

65. Rupp ME, Fey PD, Longo GM. Effect of LY333328 against vancomycin-resistant *Enterococcus faecium* in a rat central venous catheter-associated infection model. *J Antimicrob Chemother* 2001; **47**: 705–707.

66. Kaatz GW, Seo SM, Aeschlimann JR, Houlihan HH, Mercier RC, Rybak MJ. Efficacy of LY3333328 against experimental methicillin-resistant *Staphylococcus aureus* endocarditis. *Antimicrob Agents Chemother* 1998; **42**: 981–983.

67. Gerber J, Smirnov A, Wellmer A *et al.* Activity of LY333328 in experimental meningitis caused by a *Streptococcus pneumoniae* strain susceptible to penicillin. *Antimicrob Agents Chemother* 2001; **45**: 2169–2172.

68. Boylan CJ, Campanale K, Iverson PW, Phillips DL, Zeckel ML, Parr TR. Pharmacodynamics of oritavancin (LY333328) in a neutropenic-mouse thigh model of *Staphylococcus aureus* infection. *Antimicrob Agents Chemother* 2003; **47**: 1700–1706.

69. Mercier RC, Hrebickova L. Oritavancin: a new avenue for resistant Gram-positive bacteria. *Expert Rev Anti-Infect Ther* 2005; **3**: 325–332.

70. Loutit JS, O'Riordan W, San Juan J *et al.* Phase 2 trial comparing four regimens of oritavancin vs. comparator in the treatment of patients with *S. aureus* bacteremia. *Clin Microbiol Infect* 2004; **10** (suppl 3): 122.

71. Van Bambeke F, Carryn S, Seral C *et al.* Cellular pharmacokinetics and pharmacodynamics of oritavancin (LY333328) glycopeptide in a model of J774 mouse macrophages. *Antimicrob Agents Chemother* 2004; **48**: 2853–2860.

72. Judice JK, Pace JL. Semi-synthetic glycopeptide antibacterials. *Bioorg Med Chem Lett* 2003; **13**: 4165–4168.

73. Van Bambeke F. Glycopeptides in clinical development: pharmacological profile and clinical perspectives. *Curr Opin Pharmacol* 2004; **4**: 471–478.

74. Pace JL, Yang G. Glycopeptides: update on an old successful antibiotic class. *Biochem Pharmacol* 2006; **71**: 968–980.

75. Gander S, Kinnaird A, Finch R. Telavancin: *in vitro* activity against staphylococci in a biofilm model. *J Antimicrob Chemother* 2005; **56**: 337–343.

76. Reyes N, Skinner R, Benton BM *et al.* Efficacy of telavancin in a murine model of bacteraemia induced by methicillin-resistant *Staphylococcus aureus. J Antimicrob Chemother* 2006; **58**: 462–465.

77. Stucki A, Gerber P, Acosta F, Cottagnoud M, Cottagnaud P. Efficacy of telavancin against penicillin-resistant pneumococci and *Staphylococcus aureus* in a rabbit meningitis model and determination of kinetic parameters. *Antimicrob Agents Chemother* 2006; **50**: 770–773.

78. Madrigal AG, Basuino L, Chambers HF. Efficacy of telavancin in a rabbit model of aortic valve endocarditis due to methicillin-resistant *Staphylococcus aureus* or vancomycin-intermediate *Staphylococcus aureus. Antimicrob Agents Chemother* 2005; **49**: 3163–3165.

79. Sun HK, Duchin K, Nightingale CH, Shaw JP, Seroogy J, Nicolau DP. Tissue penetration of telavancin after intravenous

administration in healthy subjects. *Antimicrob Agents Chemother* 2006; **50:** 788–790.

80. Stryjewski ME, Chu VH, O'Riordan WD *et al.* Telavancin versus standard therapy for treatment of complicated skin and skin structure infections caused by gram positive bacteria; FAST 2 study. *Antimicrob Agents Chemother* 2006; **50:** 862–867.

81. Stryjewski ME, O'Riordan WD, Lau WK *et al.* Telavancin versus standard therapy for treatment of complicated skin and skin structure infections due to gram positive bacteria. *Clin Infect Dis* 2005; **40:** 1601–1607.

82. Dorr MB, Jabes D, Cavaleri M *et al.* Human pharmacokinetics and rationale for once-weekly dosing of dalbavancin, a semisynthetic glycopeptide. *J Antimicrob Chemother* 2005; **55** (suppl 2): 25–30.

83. Jauregui LE, Babazadeh S, Seltzer E *et al.* Randomized, double-blind comparison of once weekly dalbavancin versus twice-daily linezolid therapy for the treatment of complicated skin and skin structure infections. *Clin Infect Dis* 2005; **41:** 1407–1415.

84. Lefort A, Pavie J, Garry L, Chau F, Fantin B. Activities of dalbavancin *in vitro* and in a rabbit model of experimental endocarditis due to *Staphylococcus aureus* with or without reduced susceptibility to vancomycin and teicoplanin. *Antimicrob Agents Chemother* 2004; **48:** 1061–1064.

85. Malabarba A, Goldstein BP. Origin, structure and activity *in vitro* and *in vivo* of dalbavancin. *J Antimicrob Chemother* 2005; **55** (suppl 2): 15–20.

86. Raad I, Darouiche R, Vazquez J *et al.* Efficacy and safety of weekly dalbavancin therapy for catheter-related bloodstream infections caused by Gram-positive pathogens. *Clin Infect Dis* 2005; **40:** 374–380.

87. Hebeisen P, Heinze-Krauss I, Angehrn P, Hohl P, Page MG, Then RL. *In vitro* and *in vivo* properties of Ro-63-9141, a novel broad-spectrum cephalosporin with activity against methicillin-resistant staphylococci. *Antimicrob Agents Chemother* 2001; **45:** 825–836.

88. Bogdanovich T, Ednie LM, Shapiro S, Appelbaum PC. Antistaphylococcal activity of ceftobipirole: a new broad spectrum cephalosporin. *Antimicrob Agents Chemother* 2005; **49:** 4210–4219.

89. Chambers HF. Evaluation of ceftobiprole in a rabbit model of aortic valve endocarditis due to methicillin-resistant and vancomycin intermediate *Staphylococcus aureus. Anti-microb Agents Chemother* 2005; **49:** 884–888.

90. Vaudaux P, Gjinovci A, Bento M, Li D, Schrenzel J, Lew DP. Intensive therapy with ceftobiprole medocaril of experimental foreign-body infection by methicillin-resistant *Staphylococcus aureus. Antimicrob Agents Chemother* 2005; **49:** 3789–3793.

91. Noel GJ. Clinical profile of ceftobiprole, a novel beta-lactam antibiotic. *Clin Microbiol Infect* 2007; **13** (suppl 2): 25–29.

92. Chambers HF. Ceftobiprole: *in vivo* profile of a bactericidal cephalosporin. *Clin Microbiol Infect* 2006; **12** (suppl 2): 17–22.

93. Mushtaq S, Warner M, Ge Y, Kaniga K, Livermore DM. *In vitro* activity of ceftaroline (PPI-0903M, T-91825) against bacteria with defined resistance mechanisms and phenotypes. *J Antimicrob Chemother* 2007; **60:** 300–311.

94. Poulakou G, Giamarellou H. Investigational treatments for postoperative surgical site infections. *Expert Opin Investig Drugs* 2007; **16:** 137–155.

95. Hawser S, Lociuro S, Islam K. Dihydrofolate reductase inhibitors as antibacterial agents. *Biochem Pharmacol* 2006; **71:** 941–948.

96. Mathur T, Bhateja P, Pandya M, Fatma T, Rattan A. *In vitro* activity of RBx 7644 (ranbezolid) on biofilm producing bacteria. *Int J Antimicrob Agents* 2004; **24:** 369–373.

Acknowledgements—S.Unal is the Turkish coordinator of the linezolid nosocomial pneumoniae study and the TEST tigecycline study group. He has also received travel grants to attend conferences from Pfizer. The authors declare that they have no other conflicts of interest in relation to this manuscript.

Bacteriophage Lysins as Effective Antibacterials

Vincent A. Fischetti

Lysins are highly evolved enzymes produced by bacterio-phage (phage for short) to digest the bacterial cell wall for phage progeny release. In Gram-positive bacteria, small quantities of purified recombinant lysin added externally results in immediate lysis causing log-fold death of the target bacterium. Lysins have been used successfully in a variety of animal models to control pathogenic antibiotic resistant bacteria found on mucosal surfaces and infected tissues. The advantages over antibiotics are their specificity for the pathogen without disturbing the normal flora, the low chance of bacterial resistance to lysins, and their ability to kill colonizing pathogens on mucosal surfaces, a capacity previously unavailable. Thus, lysins may be a much needed anti-infective in an age of mounting antibiotic resistance.

Background

Bacteriophages or phages are viruses that specifically infect bacteria. After replication inside its bacterial host the phage is faced with a problem, it needs to exit the bacterium to disseminate its progeny phage. To solve this, double-stranded DNA phages have evolved a lytic system to weaken the bacterial cell wall resulting in bacterial lysis. Phage lytic enzymes or lysins are highly efficient molecules that have been refined over millions of years of evolution for this very purpose. These enzymes target the integrity of the cell wall, and are designed to attack one of the five major bonds in the peptidoglycan. With few exceptions,[1] lysins do not have signal sequences, so they are not translocated through the cytoplasmic membrane to attack their substrate in the peptidoglycan, this movement is controlled by a second phage gene product in the lytic system, the holin.[2] During phage development in the infected bacterium, lysin accumulates in the cytoplasm in anticipation of phage maturation. At a genetically specified time, holin molecules are inserted in the cytoplasmic membrane forming patches, ultimately resulting in generalized membrane disruption,[3] allowing the cytoplasmic lysin to access the peptidoglycan, thereby causing cell lysis and the release of progeny phage.[2] In contrast to large DNA phage, small RNA and DNA phages use a different release strategy. They call upon phage-encoded proteins to interfere with bacterial host enzymes

responsible for peptidoglycan biosynthesis,[4,5] resulting in mis-assembled cell walls and ultimate lysis. Scientists have been aware of the lytic activity of phage for nearly a century, and while whole phage has been used to control infection,[6] not until recently have lytic enzymes been exploited for bacterial control *in vivo*.[7••,8•,9] One of the main reasons that such an approach is now even being considered is the sharp increase in antibiotic resistance among pathogenic bacteria. Current data indicate that lysins work only with Gram-positive bacteria, since they are able to make direct contact with the cell wall carbohydrates and peptidoglycan when added externally, whereas the outer membrane of Gram-negative bacteria prevents this interaction. This review will outline the remarkable potency these enzymes have in killing bacteria both *in vitro* and *in vivo*.

The great majority of human infections (viral or bacterial) begin at a mucous membrane site (upper and lower respiratory, intestinal, urogenital, and ocular). In addition, the human mucous membranes are the reservoir (and sometimes the only reservoir) for many pathogenic bacteria found in the environment (i.e., pneumococci, staphylococci, and streptococci) some of which are resistant to antibiotics. In most instances, it is this mucosal reservoir that is the focus of infection in the population.[10•,11,12] To date, except for polysporin and mupirocin ointments, which are the most widely used topically, there are no anti-infectives that are designed to control colonizing pathogenic bacteria on mucous membranes.[13] Current practice is to wait for infection to occur before treating. Because of the fear of increasing the resistance problem, antibiotics are not indicated to control the carrier state of disease bacteria. It is acknowledged, however, that by reducing or eliminating this human reservoir of pathogens in the community and controlled environments (i.e., hospitals and nursing homes), the incidence of disease will be markedly reduced.[10•,13] Toward this goal, lysins have been developed to prevent infection by safely and specifically destroying disease bacteria on mucous membranes. For example, on the basis of extensive animal results, enzymes specific for *S. pneumoniae* and *S. pyogenes* may be used nasally and orally to control these organisms in the community as well as in nursing homes and hospitals to prevent or markedly reduce serious infections caused by these bacteria. This has been accomplished by capitalizing on the efficiency by which phage lysins kill bacteria.[14] Like antibiotics, which are used by bacteria to control the organisms around them in the environment, phage lysins are the culmination of millions of years of development by the bacteriophage in their association with bacteria. Specific lysins have now been identified and purified that are able to kill specific Gram-positive bacteria seconds after contact.[7••,15] For example, nanogram quantities of lysin could reduce 10^7 S. pyogenes by >6 logs seconds after enzyme addition. No known biological compounds, except chemical agents, kill bacteria this quickly.

Structure

Lysins from DNA-phage that infect Gram-positive bacteria are generally between 25 and 40 kDa in size except the PlyC for streptococci that is 114 kDa. This enzyme is unique because it is composed of two separate gene products, PlyCA and PlyCB. On the basis of biochemical and biophysical studies, the catalytically active PlyC holoenzyme is composed of eight PlyCB subunits for each PlyCA.[16] A feature of all other Gram-positive phage lysins is their two-domain structure.[17,18••] With some exceptions, the N-terminal domain contains the catalytic activity of the enzyme. This activity may be either an endo-β-N-acetylglucosaminidase or N-acetylmuramidase (lysozymes), both of which act on the sugar moiety of the bacterial wall, an endopeptidase that acts on the peptide moiety, or an N-acetylmuramoyl-L-alanine amidase (or amidase), which hydrolyzes the amide bond connecting the glycan strand and peptide moieties.[14,19] Recently an enzyme with γ-D-glutaminyl-L-lysine endopeptidase activity has also been reported.[20] In some cases, particularly staphylococcal lysins, two and perhaps even three different catalytic domains may be linked to a single binding domain.[21] The C-terminal cell binding domain (termed the CBD), however, binds to a specific substrate (usually carbohydrate) found in the cell wall of the host bacterium.[22,23•,24] Efficient cleavage requires that the binding domain binds to its cell wall substrate, offering some degree of specificity to the enzyme since these substrates are only found in enzyme-sensitive bacteria. The first complete crystal structure for the free and choline bound states of the Cpl-1 lytic enzyme has recently been published.[25••] As suspected, the data suggest that choline recognition by the choline binding domain of Cpl-1 may allow the catalytic domain to be properly oriented for efficient cleavage. An interesting feature of this lysin is its hairpin conformation, suggesting that the two domains interact with each other before the interaction of the binding domain with its substrate in the bacterial cell wall. Other lytic enzymes need to be crystallized to determine if this is a common feature of all lysins.

When the sequences between lytic enzymes of the same enzyme class were compared, it showed high sequence homology within the N-terminal catalytic region and very little homology within in the C-terminal cell binding region. It seemed counterintuitive that the phage would design a lysin that was uniquely lethal for its host organism; however, as more is learned about how these enzymes function, a possible reason for this specificity became apparent (see below, Resistance). However, because of the specificity, enzymes that spilled out after cell lysis had a good chance of killing potential bacterial hosts in the vicinity of the released phage progeny. Because of this, the enzymes have evolved to bind to their cell wall binding domains at a high affinity[26••] perhaps to limit the release of free enzyme.

Because of their domain structure, it seemed plausible that different enzyme domains could be swapped resulting in lysins with different bacterial and catalytic specificities. This was actually accomplished by excellent detailed studies of Garcia et al.,[18••,27] in which the catalytic domains of lytic enzymes for S. pneumoniae phage could be swapped, resulting in a new enzyme having the same binding domain for pneumococci, but able to cleave a different bond in the peptidoglycan. This capacity allows for enormous potential in creating designer enzymes with high specificity and equally high cleavage potential.

Though uncommon, introns have been associated with certain lysins. For example, 50% of S. thermophilus phages have been reported to have their lysin gene interrupted by a self-splicing group I intron.[28] This also appears to be the case for a S. aureus lytic enzyme[29] and perhaps the C1 lysin for group C streptococci.[30] While introns have been previously reported in phage genes, they have rarely been identified in the host genome.[31,32]

Mode of Action

Thin section electron microscopy of lysin-treated bacteria reveals that lysins exert their lethal effects by forming holes in the cell wall through peptidoglycan digestion. The high internal pressure of bacterial cells (roughly 3–5 atmospheres) is controlled by the highly cross-linked cell wall. Any disruption in the wall's integrity will result in the extrusion of the cytoplasmic membrane and ultimate hypotonic lysis. Catalytically, a single enzyme molecule should be sufficient to cleave an adequate number of bonds to kill an organism; however, it is uncertain at this time whether this theoretical limit is possible. The reason comes from the work of Loessner et al.,[26••] showing that a listeria phage enzyme had a binding affinity approaching that of an IgG molecule for its substrate, suggesting that phage enzymes, like cellulases[33] are one-use enzymes, probably requiring several molecules attacking a local region to sufficiently weaken the cell wall.

Lysin Efficacy

In general, lysins only kill the species (or subspecies) of bacteria from which they were produced. For instance, enzymes produced from streptococcal phage kill certain streptococci, and enzymes produced by pneumococcal phage kill pneumococci.[15,7••] Specifically, a lysin from a group C streptococcal phage (PlyC) will kill group C streptococci as well as groups A and E streptococci, the bovine pathogen S. uberis and the horse pathogen, S. equi, but essentially no effect on streptococci normally found in the oral cavity of humans and other Gram-positive bacteria. Similar results are seen with a pneumococcal-specific lysin; however in this case, the enzyme was also tested against strains of penicillin-resistant pneumococci and the killing efficiency was the same. Unlike antibiotics, which are usually broad spectrum and kill many different bacteria found in the human body, some of which are beneficial, lysins that kill only the disease organism with little to no effect on the normal human bacterial flora may be identified. In some cases however, phage enzymes may be isolated with broad lytic activity. For example, an enterococcal phage lysin has recently been reported to not only kill enterococci but also a number of other Gram-positive pathogens such as *S. pyogenes*, group B streptococci, and *Staphylococcus aureus,* making it one of the broadest acting lysins identified.[34] However, its activity for these other pathogens was lower than for enterococci.

An important lysin with respect to infection control is a lysin directed to *Staphylococcus aureus*[35–39]. However, in most cases these enzymes show low activity or are difficult to produce large quantities. In one recent publication,[35] a staphylococcal enzyme was described that could be easily produced recombinantly and had a significant lethal effect on methicillin resistant *Staphylococcus aureus* (MRSA) both *in vitro* and in a mouse model. In the animal experiments the authors show that the enzymes may be used to decolonize staphylococci from the nose of the mice as well as protect the animals from an intraperitoneal challenge with MRSA. However, in the latter experiments, the best protection was observed if the lysine was added up to 30 min after the MRSA.

Antibiotic and Lysin Synergy

Several lysins have been identified from pneumococcal bacteriophage that are classified into two groups: amidases and lysozymes. Exposure of pneumococci to either of these enzymes leads to efficient lysis. Both enzymes have very different N-terminal catalytic domains but share a similar C-terminal choline cell binding domain. These enzymes were tested to determine whether their simultaneous use is competitive or synergistic.[40]

To accomplish this, three different methods of analyses were used to determine synergy, time kill in liquid, disk diffusion, and checkerboard broth microdilution analysis. All three are standard methods used in the antibiotic industry to determine synergy.[41] In all three assays, the results revealed a clear synergistic effect in the efficiency of killing when both enzymes were used.[40] *In vivo,* the combination of two lysins with different peptidoglycan specificities was found to be more effective in protecting against disease than each of the single enzymes.[40,42] Thus, in addition to more effective killing, the application of two different lysins may significantly retard the emergence of enzyme-resistant mutants.

When the pneumococcal lysin Cpl-1 was used in combination with certain antibiotics, a similar synergistic effect was seen. Cpl-1 and gentamicin were found to be increasingly synergistic in killing pneumococci with a decreasing penicillin MIC, while Cpl-1 and penicillin showed synergy against an extremely penicillin-resistant strain.[43] Synergy was also observed with a staphylococcal-specific enzyme and glycopeptide antibiotics.[35] Thus, the right combination of enzyme and antibiotic could help in the control of antibiotic resistant bacteria as well as reinstate the use of certain antibiotics for which resistance has been established.

Effects of Antibodies

A concern in the use of lysins is the development of neutralizing antibodies that could reduce the *in vivo* activity of enzyme during treatment. Unlike antibiotics, which are small molecules that are not generally immunogenic, enzymes are proteins that stimulate an immune response, when delivered mucosally or systemically, which could interfere with the lysin's activity. To address this, rabbit hyperimmune serum raised against the pneumococcal-specific enzyme Cpl-1 was assayed for its effect on lytic activity.[8] It was found that highly immune serum slows, but does not block the lytic activity of Cpl-1. When similar *in vitro* experiments were performed with antibodies directed to an anthrax-specific and an *S. pyogenes*-specific enzyme, similar results were obtained (unpublished data). These results were also verified with a staphylococcal-specific lysin.[35]

To test the relevance of this *in vivo,* mice that received three intravenous doses of the Cpl-1 enzyme had tested positive for IgG against Cpl-1 in 5 of 6 cases with low but measurable titers of about 1:10. These vaccinated and naïve control mice were then challenged intravenously with pneumococci and treated by the same route with 200 μg Cpl-1 after 10 h. Within a minute, the treatment reduced the bacteremic titer of Cpl-1-immunized mice to the same degree as the naive mice, supporting the *in vitro* data that antibody to lysins have little to no neutralizing effect. A similar experiment by Rashel *et al.* with a staphylococcal enzyme,[35] showed the same result and that the animals injected with lysin multiple times exhibited no adverse events.

This unexpected effect may be partially explained if the binding affinity of the enzyme for its substrate in the bacterial cell wall is higher than the antibody's affinity for the enzyme. This is supported by the results of Loessner *et al.*,[26] showing that the cell wall binding domain of a listeria-specific phage enzyme binds to its wall substrate at the affinity of an IgG molecule (nanomolar affinities). However, while this may explain the inability of the antibody to neutralize the binding domain, it does not explain why antibodies to the catalytic domain do not neutralize. Nevertheless, these results are encouraging since they suggest that such enzymes may be used repeatedly in certain situations to control colonizing bacteria on mucosal surfaces in susceptible populations such as hospitals, day care centers, and nursing homes, or in blood to eliminate antibiotic resistant bacteria in cases of septicemia and bacteremia.

Animal Models of Infection

Animal models of mucosal colonization were used to test the capacity of lysins to kill organisms on these surfaces; perhaps the most important use for these enzymes. An oral colonization model was developed for *S. pyogenes*,[7] a nasal model for pneumococci,[15] and a vaginal model for group B streptococci.[44] In all three cases, when the animals were colonized with their respective bacteria and treated with a single dose of lysin, specific for the colonizing organism, these organisms were reduced by several logs (and in some cases below the detection limit of the assay) when tested again two to four hours after lysin treatment. These results lend support to the idea that such enzymes may be used in specific high-risk populations to control the reservoir of pathogenic bacteria and thus control disease.

Sepsis, Pneumonia, Endocarditis, and Meningitis

Similar to other proteins delivered intravenously to animals and humans, lysins have a short half-life (c.a. 15–20 min).[8] However, the action of lysins for bacteria is so rapid that this may be sufficient time to observe a therapeutic effect.[8,42] Mice intravenously infected with type 14 *S. pneumoniae* and treated 1 h later with a single bolus of 2.0 mg of Cpl-1 survived through the 48 h endpoint, whereas the median survival time of buffer-treated mice was only 25 h, and only 20% survival at 48 h. Blood and organ cultures of the euthanized surviving mice showed that only one Cpl-1-treated animal was totally free of infection at 48 h, suggesting that multiple enzyme doses or a constant infusion of enzyme would be required to eliminate the organisms completely in this application. Similar results were obtained when animals were infected and treated intraperitoneally with lysin.[42,35] Because of lysin's short half-life, it may be necessary to modify the lysins with polyethylene glycol or the Fc region of IgG, to extend the residence time *in vivo* to several hours.[45] In recent studies, phage lysins have also been shown to be successful in the treatment of meningitis by adding the lysin directly to the brain, intrathecally,[46] and endocarditis by delivering the lysin intravenously by constant IV infusion.[47] Both these applications would also benefit from modified long-acting lysins.

The ultimate challenge for lysins would be to determine whether they are able to cure an established infection. To approach this, a mouse pneumonia model was developed in which mice were transnasally infected with pneumococci and treated with Cpl-1 by repeated intraperitoneal injections after infection was established (Witzenrath *et al.*, in press). From a variety of clinical measurements, as well as morphologic changes in the lungs, it was shown that at 24 h mice suffered from severe pneumonia. When treatment was initiated at 24 h and every 12 h thereafter, 100% of the mice survived otherwise fatal pneumonia and showed rapid recovery. Cpl-1 dramatically reduced pulmonary bacterial counts and prevented bacteremia.

Using lysins systemically to kill bacteria could result in an increase in cytokine production as a result of bacterial debris being release. In one study to address this issue,[47] untreated pneumococcal endocarditis induced the release of interleukin-1α (IL-1α), IL-1β, IL-6, IL-10, gamma interferon, and tumor necrosis factor, but not IL-2, IL-4, or granulocyte-macrophage colony- stimulating factor. However, in a mouse model of pneumonia (Witzenrath *et al.*, in press) Witzenrath showed that transnasal infection with pneumococci caused an increase of proinflammatory cytokines and chemokines within 36 h. However, treatment with Cpl-1 was associated with reduced synthesis of IL-1β, IL-6, and the chemokines KC, Mip-1α, and MCP-1, as well as G-CSF in the lung. In addition, decreases were observed in plasma concentrations of IL-6, KC, and Mip-1α, and MCP-1, G-CSF, and IFNγ concentrations. Pulmonary and systemic IL-10 synthesis was only found in septic animals 60 h after infection, and was completely prevented by Cpl-1 treatment. The reason for this difference has not been determined as yet; however, it may depend on the amount of lysin used for the treatments. It is possible that in the former study,

in which the animals were treated with a constant IV infusion of lysin that the high concentration of the enzyme resulted in the fragmentation of the bacterial cell wall while in the second study, enzyme was delivered in 12 h intervals resulting in producing holes in the bacterial cell wall without forming wall pieces that are clearly more inflammatory.[48,49]

Since lysins are proteins it is unlikely that they are able to enter cells. However, there are no publications to date that address the issue of the effects of lysins on intracellular bacteria.

Anthrax

Because lysins are able to kill pathogenic bacteria rapidly; they may be a valuable tool in controlling biowarfare bacteria. To determine the feasibility of this approach a lytic enzyme was identified from the gamma phage, a lytic phage that is highly specific for *Bacillus anthracis*.[50] The gamma lysin, termed PlyG, was purified to homogeneity by a two-step chromatography procedure and tested for its lethal action on gamma phage-sensitive bacilli.[9] Three seconds after contact, as little as 100 units (about 100 μg/ml) of PlyG mediated a 5000-fold decrease in viable counts of a suspension of $\sim 10^7$ bacilli. When the enzyme was tested against ten *B. anthracis* strains from different clonal types isolated worldwide, all could be killed. In addition, the PlyG lysin was also lethal for five mutant *B. anthracis* strains lacking either capsule or toxin-associated plasmids.

On the basis of physiological characteristics, sensitivity to gamma phage and mouse lethality, a strain of *B. cereus* was identified that is closely related to *B. anthracis*. *In vivo* experiments revealed that when 10^7 of these organisms were administered intraperitoneally (IP) to 10 mice, all except one died of a rapidly fatal septicemia within four hours (10% survival). When a second set of 18 mice were also challenged IP with these bacilli, but given a single 100 μg dose of PlyG 15 min later by the same route, only five animals died (72% survival) and in three of these animals death was delayed >24 h. It is anticipated that based on the half-life of lysins (see above) higher doses, multiple doses, or constant IV infusion of lysin will result in increased survival.

Because the treatment window for individuals exposed to anthrax is about 48 h, PlyG may be used intravenously in post exposure individuals to increase the treatment window beyond the 48 h period. This would allow more time to test the infecting bacillus for its antibiotic resistance spectrum before treatment.

Bacterial Resistance to Lysins

Exposure of bacteria grown on agar plates to low concentrations of lysin did not lead to the recovery of resistant strains even after over 40 cycles. Organisms in colonies isolated at the periphery of a clear lytic zone created by a 10 μl drop of dilute lysin on a lawn of bacteria that always resulted in enzyme-sensitive bacteria. Enzyme resistant bacteria could also not be identified after >10 cycles of bacterial exposure to low concentrations of lysin (5–20 units) in liquid culture.[9,15] These results may be explained for example by the fact that the cell wall receptor for the pneumococcal lysin is choline,[51] a molecule that is essential for pneumococcal viability. While not yet proven, it is possible that during a phage's association with bacteria over the millennia, to avoid becoming trapped inside the host, the binding domain of their lytic enzymes has evolved to target a unique and essential molecule in the cell wall, making resistance to these enzymes a rare event. Since through evolution the phage has performed the 'high throughput' analysis to identify the 'Achilles heel' of these bacteria, we may take advantage of this by identifying the pathway for the synthesis of the lytic enzyme's cell wall receptor and identify inhibitors for this pathway. This would theoretically result in lead compounds that may be used to identify new anti-microbials that would be difficult to become resistant against.

Secondary Bacterial Infections

Secondary bacterial infections following upper respiratory viral infections such as influenza, are a major cause of morbidity and mortality.[52,53] The organisms responsible for most of these complications *are S. aureus and S. pneumonia*. Furthermore, otitis media due to *S. pneumonia* is a leading cause of morbidity and health care expenditures worldwide and also increases after an upper respiratory viral infection.[54] Eliminating or reducing the bacterial burden by these organisms will significantly reduce or eliminate these secondary infections. However, except for mupirocin and polysporin ointments, for which resistance is being developed, there is no effective way to eliminate these organisms from the upper respiratory mucosa.

In a mouse model of otitis media, using a non-invasive bioluminescent imaging technique, 80% of mice colonized with *Streptococcus pneumoniae* naturally develop otitis media upon infection with influenza virus. Treatment of these mice with Cpl-1 lysin before influenza challenge was 100% effective at preventing the development of otitis media.[55] Thus, treatment of high risk individuals with lysin during influenza season to decolonize them from pneumococci and staphylococci could prove effective in reducing secondary infections by these bacteria.

Conclusion

Lysins are a new reagent to control bacterial pathogens, particularly those found on the human mucosal surface. For the first time we may be able to specifically kill pathogens on mucous membranes without affecting the surrounding normal flora thus reducing a significant pathogen reservoir in the population. Since this capability has not been previously available, its acceptance may not be immediate. Nevertheless, like vaccines, we should be striving to develop methods to prevent rather than treat infection. Whenever there is a need to kill bacteria, and contact can be made with the organism, lysins may be freely utilized. Such enzymes will be of direct benefit in environments where antibiotic resistant Gram-positive pathogens are a serious problem, such as hospitals, day care centers, and nursing homes. The lysins isolated thus far are remarkably heat stable (up to 60°C) and are relatively easy to produce in a purified state and in large quantities, making them amenable to these applications. The challenge for the future is to use this basic strategy and improve upon it, as was the case for second and third generation antibiotics. Protein engineering, domain swapping, and gene shuffling all could lead to better lytic enzymes to control bacterial pathogens in a variety of environments. Since it is estimated that there are 10^{31} phage on earth, the potential to identify new lytic enzymes as well as those that kill Gram-negative bacteria is enormous. Perhaps some day phage lytic enzymes will be an essential component in our armamentarium against pathogenic bacteria.

References and Recommended Reading

Papers of particular interest, published within the period of review, have been highlighted as:
• of special interest
•• of outstanding interest

1. Loessner MJ, Maier SK, Daubek-Puza H, Wendlinger G, Scherer S: Three Bacillus cereus bacteriophage endolysins are unrelated but reveal high homology to cell wall hydrolases from different bacilli. *J Bacteriol* 1997, **179**:2845–2851.

2. Wang I-N, Smith DL, Young R: Holins: the protein clocks of bacteriophage infections. *Annu Rev Microbiol* 2000, **54**:799–825.

3. Wang I-N, Deaton J, Young R: Sizing the Holin Lesion with an Endolysin-beta-Galactosidase fusion. *J Bacteriol* 2003, **185**:779–787.

4. Young R, Wang I-N, Roof WD: Phages will out: strategies of host cell lysis. *Trends Microbiol* 2000, **8**:120–128.

5. Bernhardt TG, Wang IN, Struck DK, Young R: A protein antibiotic in the phage Q-beta virion: diversity in lysis targets. *Science* 2001, **292**:2326–2329.

6. Matsuzaki S, Rashel M, Uchiyama J, Sakurai S, Ujihara T, Kuroda M, Ikeuchi M, Tani T, Fujieda M, Wakiguchi H, Imai S: Bacteriophage therapy: a revitalized therapy against bacterial infectious diseases. *J Infect Chemother* 2005, **11**:211–219.

7. Nelson D, Loomis L, Fischetti VA: Prevention and elimination
•• of upper respiratory colonization of mice by group A streptococci by using a bacteriophage lytic enzyme. *Proc Natl Acad Sci U S A* 2001, **98**:4107–4112.

This is the first publication showing that phage lysins may be used in vivo for bacterial decolonization of mucous membranes.

8. Loeffler JM, Djurkovic S, Fischetti VA: Phage lytic enzyme
• Cpl-1 as a novel antimicrobial for pneumococcal bacteremia. *Infect Immun* 2003, **71**:6199–6204.

This paper shows that lysins function in blood to kill bacteria.

9. Schuch R, Nelson D, Fischetti VA: A bacteriolytic agent that detects and kills *Bacillus anthracis*. *Science* 2002, **418**:884–889.

10. Eiff CV, Becker K, Machka K, Stammer H, Peters G: Nasal
• carriage as a source of *Staphlococcus aureus* bacteremia. *N Engl J Med* 2001, **344**:11–16.

This paper shows that the staphylococci you carry are the ones that cause infection.

11. Coello R, Jimenez J, Garcia M, Arroyo P, Minguez D, Fernandez C, Cruzet F, Gaspar C: Prospective study of infection, colonization and carriage of methicillin-resistant *Staphylococcus aureus* in an outbreak affecting 990 patients. *Eur J Clin Microbiol Infect Dis* 1994, **13**:74–81.

12. de Lencastre H, Kristinsson KG, Brito-Avo A, Sanches IS, Sa-Leao R, Saldanha J, Sigvaldadottir E, Karlsson S, Oliveira D, Mato R et al.: Carriage of respiratory tract pathogens and molecular epidemiology of Streptococcus pneumoniae colonization in healthy children attending day acre centers in lisbon, portugal. Microb Drug Resist 1999, 5:19–29.

13. Hudson I: The efficacy of intranasal mupirocin in the prevention of staphylococcal infections: a review of recent experience. J Hosp Infect 1994, 28:235.

14. Young R: Bacteriophage lysis: mechanism and regulation. Microbiol Rev 1992, 56:430–481.

15. Loeffler JM, Nelson D, Fischetti VA: Rapid killing of Streptococcus pneumoniae with a bacteriophage cell wall hydrolase. Science 2001, 294:2170–2172.

16. Nelson D, Schuch R, Chahales P, Zhu S, Fischetti VA: PlyC: a multimeric bacteriophage lysin. Proc Natl Acad Sci U S A 2006, 103:10765–10770.

17. Diaz E, Lopez R, Garcia JL: Chimeric phage-bacterial enzymes: a clue to the modular evolution of genes. Proc Natl Acad Sci U S A 1990, 87:8125–8129.

18. Garcia P, Garcia JL, Garcia E, Sanchez-Puelles JM, Lopez R:
•• Modular organization of the lytic enzymes of Streptococcus pneumoniae and its bacteriophages. Gene 1990, 86:81–88.
One of the earliest studies showing that phage lysins have a modular organisation.

19. Loessner MJ: Bacteriophage endolysins—current state of research and applications. Curr Opi Microbiol 2005, 8:480–487.

20. Pritchard DG, Dong S, Kirk MC, Cartee RT, Baker JR: LambdaSa1 and LambdaSa2 prophage lysins of Streptococcus agalactiae. Appl Environ Microbiol 2007, 73:7150–7154.

21. Navarre WW, Ton-That H, Faull KF, Schneewind O: Multiple enzymatic activities of the murein hydrolase from staphylococcal phage phi11. Identification of a D-alanyl-glycine endopeptidase activity. J Biol Chem 1999, 274:15847–15856.

22. Lopez R, Garcia E, Garcia P, Garcia JL: The pneumococcal cell wall degrading enzymes: a modular design to create new lysins? Microb Drug Resist 1997, 3:199–211.

23. Lopez R, Garcia JL, Garcia E, Ronda C, Garcia P: Structural
• analysis and biological significance of the cell wall lytic enzymes of Streptococcus pneumoniae and its bacteriophage. FEMS Microbiol Lett 1992, 79:439–447.
This paper along with ref [18••] show that the lysin domains may be swapped to make new enzymes.

24. Garcia E, Garcia JL, Arraras A, Sanchez-Puelles JM, Lopez R: Molecular evolution of lytic enzymes of Streptococcus pneumoniae and its bacteriophages. Proc Natl Acad Sci U S A 1988, 85:914–918.

25. Hermoso JA, Monterroso B, Albert A, Galan B, Ahrazem O,
•• Garcia P, Martinez-Ripoli M, Garcia JL, Menendez M: Structural basis for selective recognition of pneumococcal cell wall by modular endolysin from phage Cp-1. Structure 2003, 11:1239–1249.
This is the first complete structure of a lysine.

26. Loessner MJ, Kramer K, Ebel F, Scherer S: C-terminal domains
•• of Listeria monocytogenes bacteriophage murein hydrolases determine specific recognition and high-affinity binding to bacterial cell wall carbohydrates. Mol Microbiol 2002, 44:335–349.
This is the first paper to show that lysins bind to their cell wall substrates at high affinity.

27. Weiss K, Laverdiere M, Lovgren M, Delorme J, Poirier L, Beliveau C: Group A Streptococcus carriage among close contacts of patients with invasive infections. Am J Epidemiol 1999, 149:863–868.

28. Foley S, Bruttin A, Brussow H: Widespread distribution of a group I intron and its three deletion derivatives in the lysin gene of Streptococcus thermophilus bacteriophages. J Virol 2000, 74:611–618.

29. Flaherty SO, Coffey A, Meaney W, Fitzgerald GF, Ross RP: Genome of staphylococcal phage K: a new lineage of Myoviridae infecting Gram- positive bacteria with a low G + C content. J Bacteriol 2004, 186(9):2862–2871 Ref Type: Generic.

30. Nelson D, Schuch R, Zhu S, Tscherne DM, Fischetti VA: Genomic sequence of C1, the first streptococcal phage. J Bacteriol 2003, 185(11):3325–3332.

31. Fernandez-Lopez M, Munoz-Adelantado E, Gillis M, Williams A, Toro N: Dispersal and evolution of the Sinorhizobium meliloti group II RmInt1 Intron in bacteria that interact with plants. Mol Biol Evol 2005, 22:1518–1528.

32. Tan K, Ong G, Song K: Introns in the cytolethal distending toxin gene of Actinobacillus Actinomycetemcomitans. J Bacteriol 2005, 187:567–575.

33. Jervis EJ, Haynes CA, Kilburn DG: Surface diffusion of cellulases and their isolated binding domains on cellulose. J Biol Chem 1997, 272:24016–24023.

34. Yoong P, Schuch R, Nelson D, Fischetti VA: Identification of a broadly active phage lytic enzyme with lethal activity against antibiotic-resistant Enterococcus faecalis and Enterococcus faecium. J Bacteriol 2004, 186:4808–4812.

35. Rashel M, Uchiyama J, Ujihara T, Uehara Y, Kuramoto S, Sugihara S, Yagyu K, Muraoka A, Sugai M, Hiramatsu K et al.: Efficient elimination of multidrug-resistant Staphylococcus aureus by cloned lysin derived from bacteriophage phi MR11. J Infect Dis 2007, 196:1237–1247.

36. Sass P, Bierbaum G: Lytic activity of recombinant bacteriophage phi11 and phi12 endolysins on whole cells and biofilms of Staphylococcus aureus. Appl Environ Microbiol 2007, 73:347–352.

37. O'Flaherty S, Coffey A, Meaney W, Fitzgerald GF, Ross RP: The recombinant phage lysin LysK has a broad spectrum of lytic activity against clinically relevant staphylococci, including methicillin-resistant Staphylococcus aureus. J Bacteriol 2005, 187:7161–7164.

38. Clyne M, Birkbeck TH, Arbuthnott JP: Characterization of staphylococcal Y-lysin. J Gen Microbiol 1992, 138: 923–930.

39. Sonstein SA, Hammel JM, Bondi A: Staphylococcal bacteriophage-associated lysin: a lytic agent active against Staphylococcus aureus. J Bacteriol 1971, 107:499–504.

40. Loeffler JM, Fischetti VA: Synergistic lethal effect of a
• combination of phage lytic enzymes with different activities on penicillin-sensitive and -resistant Streptococcus pneumoniae strains. Antimicrob Agents Chemother 2003, 47:375–377.
First paper to show synergy between lysins and along with reference [43], synergy with antibiotics.

41. Eliopoulos G, Moellering R: Antimicrobial combinations. In Antibiotics in Laboratory Medicine. Edited by Lorian V. Baltimore: Williams and Wilkins; 1991.

42. Jado I, Lopez R, Garcia E, Fenoll A, Casal J, Garcia P: Phage lytic enzymes as therapy for antibiotic-resistant Streptococcus pneumoniae infection in a murine sepsis model. J Antimicrob Chemother 2003, 52:967–973.

43. Djurkovic S, Loeffler JM, Fischetti VA: Synergistic killing of Streptococcus pneumoniae with the bacteriophage lytic enzyme Cpl-1 and penicillin or gentamicin depends on the level of penicillin resistance. Antimicrob Agents Chemother 2005, 49:1225–1228.

44. Cheng Q, Nelson D, Zhu S, Fischetti VA: Removal of group B streptococci colonizing the vagina and oropharynx of mice with a bacteriophage lytic enzyme. Antimicrob Agents Chemother 2005, 49:111–117.

45. Walsh S, Shah A, Mond J: Improved Pharmacokinetics and reduced antibody reactivity of lysostaphin conjugated to polyethylene glycol. *Antimicrob Agents Chemother* 2003, **47:**554–558.

46. Grandgirard D, Loeffler JM, Fischetti VA, Leib SL: Phage lytic enzyme cpl-1 for antibacterial therapy in experimental pneumococcal meningitis. *J Infect Dis* 2008, **197:**1519–1522.

47. Entenza JM, Loeffler JM, Grandgirard D, Fischetti VA, Moreillon P: Therapeutic effects of bacteriophage Cpl-1 lysin against *Streptococcus pneumoniae* endocarditis in rats. *Antimicrob Agents Chemother* 2005, **49:**4789–4792.

48. Kengatharan KM, De KS, Robson C, Foster SJ, Thiemermann C: Mechanism of Gram-positive shock: identification of peptidoglycan and lipoteichoic acid moieties essential in the induction of nitric oxide synthase, shock, and multiple organ failure. *J Exp Med* 1998, **188:**305–315.

49. Tuomanen E, Liu H, Hengstler B, Zak O, Tomasz A: The induction of meningeal inflammation by components of the pneumococcal cell wall. *J Infect Dis* 1985, **151:**859–868.

50. Watanabe T, Morimoto A, Shiomi T: The fine structure and the protein composition of gamma phage of Bacillus anthracis. *Can J Microbiol* 1975, **21:**1889–1892.

51. Garcia P, Garcia E, Ronda C, Tomasz A, Lopez R: Inhibition of lysis by antibody against phage-associated lysin and requirement of choline residues in the cell wall for progeny phage release in *Streptococcus pneumoniae*. *Curr Microbiol* 1983, **8:**137–140.

52. Brundage JF, Shanks GD: What really happened during the 1918 influenza pandemic? The importance of bacterial secondary infections. *J Infect Dis* 2007, **196:**1717–1718.

53. Brundage JF, Shanks GD: Deaths from bacterial pneumonia during 1918–19 influenza pandemic. *Emerg Infect Dis* 2008, **14:**1193–1199.

54. McCullers JA: Insights into the interaction between influenza virus and pneumococcus. *Clin Microbiol Rev* 2006, **19:**571–582.

55. McCullers JA *et al.*: Novel strategy to prevent otitis media caused by colonizing Streptococcus pneumoniae. *PLoS Pathog* 2007, **3:**p. e28.

Acknowledgments—I wish to acknowledge the members of my laboratory, Qi Chang, Mattias Collin, Anu Daniel, Sherry Kan, Jutta Loeffler, Daniel Nelson, Jonathan Schmitz, Raymond Schuch, and Pauline Yoong, who are responsible for much of the phage lysin work and Peter Chahales, Adam Pelzek, Rachel Shively, Mary Windels, and Shiwei Zhu for the excellent technical assistance. I am indebted to my collaborators Philippe Moreillon, John McCullars, Stephen Leib, and Martin Witzenrath for their excellent work with the lysins in their model systems. I also wish to thank Abraham Turetsky at the Aberdeen Proving Grounds and Leonard Mayer of the CDC for testing the gamma lysin against authentic *B. anthracis* strains and Richard Lyons and Julie Lovchik for the animal protection studies with anthrax. The work is supported by DARPA and USPHS Grants AI057472 and AI11822.

HIV Integrase Inhibitors— Out of the Pipeline and into the Clinic

DIANE V. HAVLIR, MD

Human immunodeficiency virus (HIV) infection has been transformed over the past two decades from a fatal to a chronic disease, because of combination antiretroviral therapy—a medical triumph.[1] However, HIV has proven to be a masterful escape artist with regard to the pharmacologic agents strategically deployed to block its replication, and the counterpoint to the antiretroviral success story is one of drug resistance and toxicity. For a sizable number of patients who have developed or acquired highly drug-resistant HIV, suffering the ill effects of HIV disease is either a reality or a looming threat.

In most patients with highly drug-resistant HIV, the resistance develops because of sequential exposure to HIV drugs in the context of incomplete virologic suppression. The HIV regimens currently in use are designed to be sufficiently potent to suppress the virus and thus the emergence of drug-resistant mutants. When the adherence to medication is suboptimal or drug absorption or metabolism is altered, viral replication crosses the threshold necessary for the selection and outgrowth of drug-resistant mutants. Unfortunately, the genetic barrier to drug resistance for several of the most important HIV agents is low, requiring only a single point mutation to confer loss of activity. Drug-resistant virus can also be transmitted from person to person and from mother to child, although the transmission of strains resistant to multiple classes of drugs is rare, to date.[2]

For over a decade, we have approached the treatment of drug-resistant HIV as a Sisyphean task, using a strategy that we knew from the start was doomed to fail—the addition of a single new agent to an ailing regimen. This approach was based not on ignorance but on the lack of new classes of antiretroviral agents. The situation is complicated further by our incomplete understanding of cross-resistance and our incomplete ability to predict residual antiretroviral activity associated with the individual agents in the drug classes targeting HIV nucleoside reverse-transcriptase inhibitors and protease inhibitors.

For patients infected with drug-resistant HIV, adding a single new drug to a failing regimen provides temporary benefit yet inevitably leads to the selection of a virus that is even more drug resistant. Some strains that are highly drug resistant replicate less efficiently in vitro than others, but ongoing replication of these strains in vivo is still associated with a decline in immune function. Thus, delaying a switch in regimen places the patient at jeopardy for serious infections and cancers as well as for a circulating strain of HIV more resistant to a future antiretroviral "backbone" regimen. These circumstances back providers and patients into a corner, forcing them to gamble on either a wait-and-see strategy or a switch strategy.

Until recently, the pace of HIV-drug development was not conducive to an aggressive approach to treating patients with a highly drug-resistant virus. Since 2003, nine new drugs and three new drug classes, including HIV integrase inhibitors, were approved for HIV treatment. HIV integrase was a natural target for HIV chemotherapy because of both its central role in the HIV life cycle and the absence of a human homologue. The development of this class of drugs exploited and shed light on the complex multistep process of integration of the HIV provirus into the host genome.[3] Raltegravir, the first compound of this class to be approved for clinical use, inhibits strand transfer, the third and final step of the provirus integration.

In this issue of the *Journal,* Steigbigel et al.[4] report the findings of the BENCHMRK-1 and BENCHMRK-2 studies (the Blocking Integrase in Treatment Experienced Patients with a Novel Compound against HIV, Merck studies; ClinicalTrials.gov numbers, NCT00293267 and NCT00293254), which evaluated the activity of raltegravir among 699 patients infected with HIV. Only patients infected with HIV that had documented resistance to three classes of HIV drugs were eligible for these studies. The patients were randomly assigned to receive an optimized antiretroviral regimen either alone or in conjunction with raltegravir. The antiretroviral regimens were individually designed on the basis of previous antiretroviral history and results of drug-resistance testing. The regimens were permitted to include darunavir, an HIV protease inhibitor that was not yet licensed at the time of the study.

Patients who received raltegravir had higher rates of virologic suppression than those who received placebo, and the overall rates of viral suppression are among the highest reported for patients infected with HIV with triple-class resistance. Increases in the CD4 cell count were more pronounced in the raltegravir group than in the placebo group; the overall adverse-event profile did not differ between the two groups. Rates of cancer were higher in the raltegravir group. However, the rates of adverse events were low, and differences in the rates from the combined BENCHMRK studies and from a larger data set including other studies of raltegravir were not significant. One could speculate that the earlier occurrence of clinical events in the raltegravir group reflects a more robust immunologic response and unmasking of underlying conditions. Continued monitoring for these and other adverse events in patients receiving raltegravir will be important during its expanded use.

The companion article by Cooper et al.[5] in this issue of the *Journal* underscores the importance of fully active multidrug therapy in sustaining virologic suppression in patients infected with HIV that has multiclass drug resistance. Virologic response rates were 51%, 61%, and 71% when raltegravir was provided in combination with no, one, or two other active drugs, respectively. Data on the high end of this dose–response curve could potentially be even greater when patients are candidates for the CCR5 inhibitor maraviroc, an agent newly approved by the Food and Drug Administration in a new class of HIV drugs expected to have activity against the HIV infecting this population. The 51% response rates on the low end should not encourage the sole addition of raltegravir to the regimen of a patient with multiclass-resistant HIV, which could lead to rapid development of drug resistance. Resistance to raltegravir requires only a single point mutation, and among 94 subjects in the raltegravir group with virologic failure who underwent genotyping, approximately two thirds showed at least one of the known resistance mutations to raltegravir by week 48. As expected, the prevalence of resistance mutations was lower among isolates from patients who were given (in addition to raltegravir) two or more active drugs (33%), as compared with none (78%).

What do these results mean for clinical practice? The results of the BENCHMRK studies usher in a new era for HIV therapy—the expectation that combination regimens involving new agents can suppress even the most drug-resistant HIV. Current HIV guidelines endorse this approach, and clinicians are eager to put it into practice.[6]

At present, raltegravir and other new classes of HIV drugs will be used principally for patients in whom available antiretroviral agents have failed to achieve HIV suppression. It is crucial that new HIV agents are used wisely to maximize benefit and minimize resistance. It is also important to rethink our current HIV treatment strategies and to extend research on these new agents in other populations, such as patients infected with HIV who have never been treated. Given the unmet treatment needs in the global HIV epidemic, it is further incumbent on researchers, policymakers, and advocacy groups to evaluate the optimal use of new agents in low- and middle-income populations and to work toward making the drugs available for both children and adults.

Raltegravir is a powerful weapon for patients combating drug-resistant HIV. However, the low genetic barrier to drug resistance against raltegravir represents a major point of vulnerability.[7] With attention to adherence and with the use of accompanying active antiretroviral agents, the risk can be minimized. Nonetheless, we must acknowledge that even though regimens are sufficiently potent to treat drug-resistant HIV, they may require a substantial pill burden and use of the injectable drug enfuvirtide. These regimens can be overwhelming even for the most motivated patient. The fact that integrase inhibitors are emerging out of the pipeline and into the clinic shows that persistence and investment in new agents can shed light on basic science, open new doors for research, and most importantly, transform approaches to HIV treatment.

References

1. Palella FJ Jr, Delaney KM, Moorman AC, et al. Declining morbidity and mortality among patients with advanced human immunodeficiency virus infection. N Engl J Med 1998;338:853–60.
2. Transmission of drug-resistant HIV-1 in Europe remains limited to single classes. AIDS 2008;22:625–35.
3. Hazuda DJ, Felock P, Witmer M, et al. Inhibitors of strand transfer that prevent integration and inhibit HIV-1 replication in cells. Science 2000;287:646–50.
4. Steigbigel RT, Cooper DA, Kumar PN, et al. Raltegravir with optimized background therapy for resistant HIV-1 infection. N Engl J Med 2008;359:339–54.
5. Cooper DA, Steigbigel RT, Gatell JM, et al. Subgroup and resistance analyses of raltegravir for resistant HIV-1 infection. N Engl J Med 2008;359:355–65.
6. Hammer SM, Saag MS, Schechter M, et al. Treatment for adult HIV infection: 2006 recommendations of the International AIDS Society–USA panel. JAMA 2006;296:827–43.
7. Malet I, Delelis O, Valantin MA, et al. Mutations associated with failure of raltegravir treatment affect integrase sensitivity to the inhibitor in vitro. Antimicrob Agents Chemother 2008;52:1351–8.

Diane V. Havlir, MD from the Department of Medicine, San Francisco General Hospital, University of California at San Francisco, San Francisco.

No potential conflict of interest relevant to this article was reported.

Removing the Golden Coat of *Staphylococcus aureus*

ROBERT S. DAUM, MD

Infections caused by *Staphylococcus aureus,* particularly those that are resistant to methicillin and all available β-lactam antibiotics—the so-called methicillin-resistant *S. aureus* (MRSA) infections—have been declared a public health imperative. A recent report from the Centers for Disease Control and Prevention[1] estimated that 18,650 persons in the United States died from invasive MRSA infections in 2005. A new therapeutic approach to the management of these infections would therefore be welcome.

The development of new antibiotics to treat MRSA infections has slowed for complex reasons.[2] Some researchers have pointed to a lack of "new" microbial targets. Indeed, most antimicrobial agents currently in use are directed against bacterial cell-wall metabolism, the machinery of bacterial protein synthesis, or a biosynthetic pathway unique to bacteria. Targeting a pathway that is common to bacteria and humans may kill the bacteria but may also incur unacceptable adverse events.

Liu and colleagues[3] recently observed that the synthesis of staphyloxanthin, the carotenoid pigment that bestows the golden yellow color of clinical isolates of *S. aureus,* can be greatly diminished by a candidate inhibitor of human squalene synthase; this agent was previously developed to inhibit cholesterol synthesis. The inhibitor is likely to have few toxic effects, since only cholesterol synthesis is targeted and cholesterol is likely to be present in the serum or the diet. Liu et al. also observed that the human squalene synthase inhibitor greatly diminished staphyloxanthin biosynthesis in vitro. They inoculated mice intraperitoneally with a wild-type, pigmented *S. aureus* isolate called ATCC 27659 or an isogenic mutant that lacked the dehydrosqualene synthase gene. Only the mutant strain was cleared by the host. A comparison of experimental intranasal inoculation showed that the rate and magnitude of colonization for the two strains did not differ. Liu et al. also showed that treatment of mice with the dehydrosqualene synthase inhibitor after intraperitoneal inoculation resulted in enhanced bacterial killing, as evidenced by lower bacterial counts in the mouse kidney.

The success of this approach is largely dependent on a pathway used in the early steps of synthesis of both staphyloxanthin and squalene, which is a cholesterol precursor. Liu et al. posit that staphyloxanthin may be a critical virulence factor in *S. aureus* infections because of its ability to detoxify antibacterial molecular species such as the superoxide anion, hydrogen peroxide, and hypochlorous acid, which are generated mainly by neutrophils of the host's immune system.

The concept of countering pathogenicity by inhibiting the synthesis of a virulence factor is new. Whether it will be clinically applicable requires further testing, possibly including the targeting of other components of staphyloxanthin synthesis or other virulence factors, toxins, or adhesins. The relative importance of these factors remains to be seen.

Many steps would be required to translate this observation for use in the clinical arena. Liu et al. studied just one strain of *S. aureus;* its relationship to other currently circulating strains—epidemic MRSA strains in particular—is uncertain. Some clinical isolates do not produce staphyloxanthin, which is inferred from their lack of yellow pigment on solid mediums. A more thorough characterization of the animal model used by Liu et al. would be informative. The decrease in bacteria surviving in the kidney after parenteral inoculation observed by Liu et al. is consistent with a therapeutic effect, but other organs and other routes of inoculation should be tested. Moreover, the effect of renal bacteria on survival varied among the animals studied; identifying the causes of this variation may help to guide further research.

S. aureus is a commensal bacterium; it colonizes the skin and mucosal surfaces such as the nares, pharynx, and vagina. At any given time, approximately 25 to 40% of people are colonized with *S. aureus.* The mere presence of the organism in the nose or on the skin does not seem to provoke a host response; thus, the staphyloxanthin inhibitor used by Liu et al. had no effect on *S. aureus* in this setting.

Disease is produced by one of several mechanisms. A breach in the mucosal barrier or skin may allow opportunistic multiplication of bacteria in the subcutaneous tissues and thereby result in an infection of the skin or soft tissue; invasive disease infrequently follows. Many neutrophils are recruited to the site of this event. An inhibitor of staphyloxanthin synthesis may halt the progression of a recognized skin and soft-tissue infection. Another route of infection is through inhalation; infection may establish a primary pneumonia that may be necrotizing. A preceding infection with a virus such as influenza may predispose persons to this form of infection. It is uncertain whether the production of staphyloxanthin would influence this scenario. Occasionally, severe disease (e.g., a systemic toxinosis such as toxic shock syndrome) may occur without the migration of *S. aureus* from its superficial mucosal perch; one of a variety of toxins or superantigens that can initiate a cascade of host responses may be secreted. It is unlikely that the enhanced killing of neutrophils would critically affect the course of such a toxinosis.

Although most *S. aureus* infections, on reaching the bloodstream, are probably self-limited by host defense mechanisms, an intravascular infection or metastatic infection may occur. The intraperitoneal inoculation route used by Liu et al. most closely resembles invasive disease, and it is in this context that their approach, perhaps in combination with another antiinfective agent such as a conventional antibacterial compound, shows the most promise.

S. aureus is a dynamic species, endowed with an array of adhesins and virulence factors, and thus it facilely adapts to a variety of environments. Therefore, it seems unlikely that the approach used by Liu et al. will singly solve the therapeutic dilemma created by antibacterial-resistant isolates. It does, however, open the door to a new line of clinically relevant research.

References

1. Klevens RM, Morrison MA, Nadle J, et al. Invasive methicillin-resistant Staphylococcus aureus infections in the United States. JAMA 2007;298:1763–71.
2. Spellberg B, Guidos R, Gilbert D, et al. The epidemic of antibiotic-resistant infections: a call to action for the medical community from the Infectious Diseases Society of America. Clin Infect Dis 2008;46:155–64.
3. Liu CI, Liu GY, Song Y, et al. A cholesterol biosynthesis inhibitor blocks Staphylococcus aureus virulence. Science 2008;319:1391–4.

ROBERT S. DAUM, MD, from the Department of Pediatrics, Section of Infectious Diseases, University of Chicago, Chicago.

Acknowledgments—Dr. Daum reports receiving fees for serving on advisory boards of Pfizer and Astellas Pharma and receiving grant support from Pfizer. No other potential conflict of interest relevant to this article was reported.

I thank Henry Chambers and Richard Proctor for helpful comments on an earlier version of the manuscript.

Antibodies for the Treatment of Bacterial Infections: Current Experience and Future Prospects

CHRISTOPHER BEBBINGTON AND GEOFFREY YARRANTON

Antibodies can be used for the prevention and treatment of bacterial infections in animal models of disease. Current antibody technology allows the generation of high affinity human/humanized antibodies that can be optimized for antibacterial activity and *in vivo* biodistribution and pharmacokinetics. Such antibodies have exquisite selectivity for their bacterial target antigen and promise efficacy and safety. Why are there no monoclonal antibody products approved for the treatment or prevention of bacterial infections? Can antibodies succeed where antibiotics are failing? Some antibody therapies are currently being evaluated in clinical trials but several have failed despite positive data in animal disease models. This review will discuss the pros and cons of antibody therapeutics targeted at bacterial infections.

Introduction

An increasing number of serious bacterial infections are caused by organisms resistant to antibiotics, including vancomycin-resistant *Enterococcus* (a significant problem in immunocompromised patients), multiply resistant *Staphylococcus aureus, Acinetobacter baumannii,* and *Pseudomonas aeruginosa,* common causative agents in hospital-acquired pneumonia and other infections. Chronic *P. aeruginosa* infection is also a major factor in the progressive decline in lung function in cystic fibrosis (CF) patients. Antibodies that kill drug-resistant bacteria typically exploit mechanisms distinct from those of antibiotics and are therefore unlikely to select for cross-resistance. Such antibodies thus provide attractive new therapeutic options for some of the most intractable bacterial infections. Two alternative approaches to antibacterial antibody development are discussed here: exploiting antibodies directly targeting the bacterial surface; or antibodies that act indirectly, protecting normal host immunity by neutralizing bacterial toxins or other virulence factors essential for pathogenicity and maintenance of infection.

Direct Targeting to Bacterial Cell Surfaces

Antigens on the bacterial cell surface are appealing targets for antibody-based intervention because they provide the potential for blocking antibodies to interfere with bacterial colonization as well as marking the bacteria for destruction by immune effector responses (opsonization). In some cases, antibodies with direct bactericidal activity have been identified, such as a single-chain antibody to the spirochete *Borellia*.[1] Similarly, anti-idiotype antibodies which mimic a yeast killer toxin have been reported to have direct cell-killing activity against a broad spectrum of Gram-positive bacteria in addition to yeasts, fungi, and mycobacteria.[2] These antibodies are yet to be tested in the clinic but they may overcome the need for a competent host immune response for efficacy. In most cases, however, antibodies to bacterial surface components rely on the recruitment of an array of antibody-directed immune effector functions. Antibodies with appropriate constant region isotypes (chiefly human IgG1 and IgG3) can recruit cells of the immune system with multiple activities: stimulating phagocytosis of bacteria by macrophages; inducing antibody-directed cellular cytotoxicity (ADCC) by macrophages or NK cells; activating the complement cascade; and generating the oxidative burst from neutrophils. (IgM antibodies also potently induce complement and ADCC activities). Furthermore, all antibody variable regions are capable of catalyzing redox reactions from singlet oxygen provided by activated neutrophils, leading to the generation of several highly potent oxidizing agents directly harmful to bacteria,[3] including ozone, a potent antibacterial agent which also stimulates inflammatory responses.[4] Indeed, the inflammation induced by complement activation and ozone generation has the potential to recruit additional elements of the immune system to further boost immunity.

Direct Cell-Targeting Antibodies in Development

Antibodies Targeting Staphylococcus

Two hyperimmune human polyclonal antibody preparations (Altastaph and Veronate) have recently been evaluated for the prevention of S. aureus infection in very low birth weight (VLBW) infants but both have failed to show sufficient efficacy.[5] Inadequate immune effector cell responses in premature infants may have contributed to the lack of efficacy in these trials. It is also possible that naturally occurring antibodies may have low activity because bacteria have evolved to be able to evade the human immune system. Monoclonal antibodies specifically targeting surface components with important functions in bacterial colonization or infection may be more effective. There are three monoclonal antibodies against S. aureus currently in clinical development, Pagibaximab, Aurexis, and Aurograb (see Table 1), but none have yet demonstrated significant efficacy.

Pagibaximab is a chimeric IgG1 antibody recognizing the surface component lipoteichoic acid of S. aureus and S. epidermidis. A Phase 2 clinical study in which VLBW infants were given up to three doses at 60 or 90 mg/kg showed evidence of prophylactic activity but only at the highest dose of antibody,[6] again perhaps reflecting the comparatively poor immune response in premature infants.

Aurexis, a humanized IgG1 antibody to S. aureus Clumping Factor A (ClfA), a protein implicated in mediating adherence to damaged endothelia and inhibiting phagocytosis, has activity in prophylactic and therapeutic models in rabbits.[7] Results of a Phase 2 clinical study in which a single 20 mg/kg dose of Aurexis was administered in adult patients with S. aureus bacteremia showed no statistically significant responses, although fewer deaths occurred and no worsening of sepsis occurred in the antibody-treated group.[8]

Aurograb is a single-chain antibody variable fragment (scFv) that binds to the S. aureus surface protein GrfA.[9] GrfA

Table 1 Antibodies to Bacterial Targets in Clinical Studies.

Pathogen	Antibody	Target	Antibody Form	Company	Ref	Clinical Studies
Staphylococcus	Veronate	S. aureus surface components	Human polyclonal	Inhibitex	[5]	Failed Phase 3
Staphylococcus	Altastaph	S. aureus vaccine	Polyclonal purified from vaccinees	Nabi	[5]	Failed Phase 2
Staphylococcus	Pagibaximab	Lipoteichoic acid (surface antigen)	Chimeric IgG1k	Medimmune	[6]	Phase 2 completed
Staphylococcus	Aurexis (tefibazumab)	Clumping Factor A (surface antigen)	Humanized IgG1k	Inhibitex	[7, 8]	Phase 2 completed
Staphylococcus	Aurograb	GrfA (surface efflux pump)	Recombinant human scFv	Novartis	[9]	Failed Phase 2
B. anthracis	ABthrax (PAmAb; raxibacumab)	Protective antigen (toxin)	Recombinant human IgG1λ	HGS	[36, 37]	Surrogate efficacy and Phase 1 completed [a]
B. anthracis	Anthim	Protective antigen (toxin)	Deimmunized Mab	Elusys	[38]	Surrogate efficacy and Phase 1 completed [s]
B. anthracis	Valortim (MDX-1303)	Protective antigen (toxin)	Human IgG1k	Medarex/ Pharmathene	[39]	Surrogate efficacy and Phase 1 completed [s]
C. difficile	MDX-066 MDX-1388	Toxins A and B	Human IgG1	Medarex and MBL	[41]	Phase 2 ongoing
E. coli (Shiga toxin producing)	Shigamabs	Shiga toxins Stx1 and Stx2	NA	Thallion	[40]	Phase 1 completed
P. aeruginosa	Anti-Pseudomonas IgY	P. aeruginosa	Chicken polyclonal	Immunsystem AB	[10]	Phase 1/2 ongoing
P. aeruginosa serotype 011	KBPA101	LPS	Human monoclonal IgM	Kenta Biotech	Kenta website	Phase 2 ongoing
P. aeruginosa	KB001	PcrV (toxin delivery by the TTSS)	PEGylated Humaneered™ Fab′	KaloBios	KaloBios website	Phase 2 ongoing

NA: not available.

[a]Therapies for inhalation anthrax are evaluated under the FDA 'animal rule'; safety and pharmacokinetics are tested in human volunteers.

is a drug-efflux pump and Aurograb increases the sensitivity of *S. aureus* to vancomycin by the inhibition of the pump. Because Aurograb does not have an Fc-region, it is expected to have a half-life of only a few minutes *in vivo,* requiring frequent dosing. A Phase 2 study of Aurograb (dosed twice-daily at 1 mg/kg) in combination with vancomycin for deep-seated *S. aureus* infections failed to show efficacy (Novartis new release URL: http://www.novartis.com/newsroom/media-releases/en/2008/1247203.shtml).

Antibodies to *P. aeruginosa* Surface Antigens in Development

A polyclonal antibody preparation (IgY; Table 1) from the eggs of hens immunized with *P. aeruginosa* has been evaluated in an open-label study in CF children whose first *P. aeruginosa* colonization was eradicated with antibiotics. Recently published data from patients treated for up to 12 years provide evidence of a trend to efficacy with only 2/17 patients who gargled daily with the chicken antibody showing recolonization with *P. aeruginosa* compared with 7/23 in a control group.[10]

For the development of monoclonal antibodies, there is a paucity of surface antigens conserved across different strains of *P. aeruginosa* to which useful antibodies can be generated. Several groups have raised antibodies against the *O*-linked polysaccharide side chains of lipopolysaccharide (LPS) of *P. aeruginosa* but the LPS side chains are highly variable and contribute at least 20 serotypes among clinical isolates. Multiple human antibodies raised against a total of 10 serotypes in mice transgenic for human immunoglobulin genes have shown activity in animals[11,12]), raising the prospect of an oligoclonal antibody preparation to treat at least the majority of serotypes. A human IgM antibody specific for serotype 011 (KBPA101) developed by Kenta Biotech, is currently in Phase 2 clinical trials, with other antibodies against other serotypes in preclinical development (Kenta Biotech website; URL: http://www.kentabiotech.com). IgM antibodies, however, have shorter *in vivo* half-lives than IgG antibodies, are sensitive to the many secreted bacterial proteases (see below), and may be costly to deliver. Furthermore, expression of *O*-linked LPS side chains is reduced in mucoid strains that are prevalent in the lungs of CF patients.[13]

The bacterial flagellum is a comparatively conserved structure and two monoclonal antibodies against two common flagellum variants have shown activity in a mouse burn model of *P. aeruginosa* infection.[14] The antibodies collectively recognize up to 98% of clinical isolates but the need to combine two antibodies and the possibility of selection of flagellum-negative variants may have been factors in the lack of further development.

Drawbacks of Antibodies to Bacterial Surface Components

Many bacterial infections are opportunistic and the existence of persistent infection indicates effective evasion of host immune response. In almost all CF patients having chronic lung infections with *P. aeruginosa,* high titers of opsonizing antibodies can be measured in the systemic circulation yet the infectious pathology still progresses. Indeed, antibodies from CF patients have been demonstrated to be ineffective at mediating phagocytic killing.[15]

One of the major concerns in targeting the bacterium directly is the selection of escape variants, which may arise through a wide variety of mechanisms in addition to the selection of variants that do not retain the antibody-binding site. Production of a bacterial capsule, for example, can provide effective protection against opsonization. Capsules are typically permeable to antibodies but mask antibodies bound to the bacterial surface, preventing contact with receptor-bearing effector cells.[16]

Bacteria typically have multiple mechanisms to evade immune responses and block antibody effector functions. *S. aureus* produces two proteins, Protein A and Sbi, that bind directly to antibody Fc regions and inhibit complement activation and Fc-receptor binding.[17] Protein A also binds variable regions of some antibodies and so can inhibit the activity of some Fab and scFv molecules as well.[17] Similar antibody-binding proteins are found in other bacteria, such as the Protein G and Protein H of *Streptococcus pyogenes.*

The importance of neutrophils in the antibody response makes these cells a major target for bacterial evasion. The chemotaxis inhibitory protein of *S. aureus* (CHIPS) binds both the formyl peptide receptor and complement C5a receptor to inhibit neutrophil recruitment to sites of infection.[18●] Similarly, complement activation is a frequent target for bacterial immune evasion mechanisms. *S. aureus,* for example, has at least five complement-inhibitory proteins.[18●] Both CHIPS and the staphylococcal complement inhibitor (SCIN) show strict specificity for human complement[19] which may be a significant factor in determining the differences in the responses to some antibodies in animal models compared with human disease.

Many bacteria secrete proteases that inactivate antibody effector functions by cleaving at the hinge region of IgG and IgM, separating the Fc-region from the antigen-binding site. The resulting Fab or F(ab')$_2$ fragments compete for binding to the infecting bacteria further preventing effective immune clearance.[20] *P. aeruginosa* secretes at least four distinct proteases that are pathogenic determinants abundant in the sputum of CF patients.[21,22,23] Interestingly, in the human airway a number of antiproteinases, for example α_1 protease inhibitor (α_1-PI) are secreted as a protective mechanism for the epithelial surface, but *P. aeruginosa* secretes a potent inhibitor of α_1-PI, the phenazine derivative, pyocyanin.[24] *S. aureus* also produces four major proteases that directly or indirectly degrade antibodies.[25,26] Thus, passive immunotherapy that relies on antibody effector functions is likely to be negatively impacted by secreted proteases capable of inactivating IgG. Antibody therapies that do not rely on Fc effector function may have a greater chance of success.

Lastly, the access of antibodies may be limited by biofilm formation. Many pathogenic bacteria switch morphology between a free-swimming planktonic form and a sessile mucoid form that forms the biofilm, creating a reservoir of organisms comparatively inaccessible to antibiotics. Biofilm formation is triggered by a variety of environmental conditions and can be induced by the immune response.[27] In CF, the lungs tend to be colonized by mucoid forms of *P. aeruginosa*. Although these are less pathogenic they can seed pathogenic planktonic bacteria throughout the lungs leading to progressive lung damage. The switch to biofilm growth involves a profound antigenic shift with the downregulation of *O*-linked LPS side chains and upregulation of alginate production making the identification of antibodies to *P. aeruginosa* surfaces that will detect all clinical strains particularly challenging.[13]

Antibodies to an exopolysaccharide of *S. epidermidis,* poly *N*-acetyl glucosamine (PNAG) have been shown to be ineffective in opsonic killing of bacteria in a biofilm because of high levels of non-cell-associated soluble antigen.[28] These authors showed by immunofluorescence analysis that the biofilm did not present a significant diffusion barrier and suggest that the PNAG shed from cells in the biofilm simply acts as a decoy to reduce the level of antibodies capable of reaching the bacterial cell surface. This has important implications because it suggests that appropriate choice of antibodies to epitopes not shed from the cell surface may permit better access to bacteria growing in biofilms. Thus, for example, Alopexx is in preclinical development with an antibody, F598, specific for a deacetylated form of *S. aureus* PNAG preferentially retained on the bacterial surface, in contrast to the acetylated form abundant in biofilms. The antibody has potent activity in animal models and deacetylated PNAG is also found in *S. epidermidis* and even *E. coli,* implying that such antibodies may have a broad spectrum of activity.[29] The conserved alginate molecule of *P. aeruginosa* is another target for which choice of the precise epitope may be crucial. One such antibody, F429, against a carboxylic-acid component of alginate mediates phagocyte-dependent killing and is protective against both mucoid and nonmucoid strains in a pulmonary infection model, because of rapid induction of alginate expression *in vivo.*[30] The antibody is currently in preclinical development at Aridis, designated Aerucin. However, it remains to be seen if improved access of antibodies to bacteria in biofilms will be sufficient for clinical efficacy; such antibodies are also subject to the other bacterial evasion strategies described above.

Indirect Neutralization of Pathogenic Bacteria

Bacteria produce a number of diffusible virulence factors such as pigments, proteases and toxins as well as 'signaling' molecules. Neutralizing virulence factors represent an 'indirect' targeting approach to bacterial infections, its success relying upon the ability of the host immune system to clear the infection. Several antibodies to toxins and signaling molecules are currently being evaluated.

It is now recognized that bacteria synthesize factors that regulate the metabolic activity of other bacteria within their vicinity. This 'communication' can be either intra-species or inter-species and is termed quorum sensing.

Gram-negative bacteria communicate through a number of diffusible autoinducers (AIs) including *N*-acyl homocysteine lactones and *N*-acyl homoserine lactones (C-12 inducers).[31] Expression of the AIs regulates the expression of virulence

factors that directly contribute to colonization and dissemination of the pathogen. The C-12 AIs have also been shown to interfere with host cell signaling in cells of the immune system and can induce apoptosis in macrophages.[32] In animal studies, mice with a high titer of antibodies to *N*-acyl homoserine lactone following immunization were shown to be more resistant to *P. aeruginosa* challenge[33] and an anti-C-12 antibody protects macrophages from C-12-induced cytotoxicity.[34] Gram-positive bacterial signaling is mainly through small peptides. A monoclonal antibody to *S. aureus* autoinducing peptide 4 (AIP-4) suppressed pathogenicity in murine disease models of infection.[35] The results in *in vitro* and *in vivo* models of bacterial infection suggest that neutralizing quorum sensing molecules may provide a new and effective treatment modality although the importance of quorum sensing in pathogenicity in humans is unknown. The concentrations of AIs in infected tissues are relatively high (μM) and it may be difficult to deliver sufficient antibody to provide effective neutralization during an ongoing infection. Success of the approach may depend on efficient antibody delivery and rapid clearance of immune complexes.

Many pathogenic bacteria secrete protein toxins that are key virulence factors. Neutralization of these toxins should render the bacteria less pathogenic and allow clearance of bacteria by the immune system. Currently, several antibody approaches targeting such toxins are in clinical trials (Table 1).

ABthrax is a recombinant human antibody with high affinity for *Bacillus anthracis* protective antigen, the receptor-binding component of anthrax toxin. The antibody provides complete protection against a lethal bacterial challenge in animal models.[36] A Phase 1 clinical trial has been completed with this antibody[37] and further development is in progress. Two other antibodies to protective antigen are also in clinical development having shown activity in animal models of inhalation anthrax; a deimmunized antibody[38] and a human antibody, Valortim.[39]

Hemolytic-uremic syndrome (HUS) is a serious complication of infection by Shiga toxin-producing *E. coli* (STEC). A chimeric antibody that neutralizes the toxin Stx2 is protective in animal models and has been tested in Phase 1 safety study in humans.[40] Further studies are planned with a pair of antibodies recognizing both Stx2 and Stx1 (Thallion website; URL: http://www.thallion.com).

The pathogenic role of *C. difficile* in intestinal disease is well established. This organism can cause mild self-limiting diarrhea or fulminating pseudomembranous colitis. The pathogenicity of this microbe is in part dependent upon two immunologically distinct toxins: A and B. Human antibodies that neutralize either toxin A or B have been evaluated in a hamster and mouse model of infection. Combination of an anti-A toxin antibody and anti-B antibody demonstrated the best *in vivo* efficacy.[41] Clinical trials with this pair of antibodies are ongoing (Medarex website; URL: http://www.medarex.com/Development/Pipeline.html).

Many Gram-negative bacteria are able to inject toxins directly into adjacent cells using a protein syringe-like structure called the type III secretion system (TTSS), providing a mechanism for blocking the activity of multiple toxins with a single therapeutic agent.[42•] Indeed, acquisition of an active TTSS can be causally related to pathogenicity, as in STEC described above.[40,42•] In the case of *P. aeruginosa,* an important pathogen in immunocompromised people and those on a ventilator, almost all environmental isolates possess the TTSS-encoding genes but different strains vary in the exotoxins secreted. Pathogenicity correlates with the expression of the TTSS and the exotoxins secreted.[43,44] A monoclonal antibody, Mab166, directed against the PcrV needle-tip protein of the *P. aeruginosa* TTSS, inhibits TTSS function and is therapeutically active in animal models of acute *P. aeruginosa* infection.[45,46] An engineered human Fab' fragment has also been generated against the PcrV protein, derived from Mab166 by a process of antibody humaneering™. This antibody fragment, conjugated to polyethylene-glycol (PEG), is currently in clinical development for the treatment of *P. aeruginosa* lung infections in CF and in mechanically ventilated patients (KB001; KaloBios website; URL: http://www.kalobios.com/kb_pipeline_001.php). This drug candidate has novel properties that address potential problems with antibody-based antibacterial therapeutics. The activity of the Fab fragment is important in reducing sensitivity to *Pseudomonas* proteases that degrade IgG antibodies and PEGylation both extends *in vivo* half-life and further reduces susceptibility to proteolytic inactivation. The ability of the Fab fragment to inactivate the TTSS results in prevention of toxin injection into cells and also prevents the direct killing of macrophages by the TTSS that occurs even in the absence of toxins.[47] KB001 is thus an exciting new approach to the treatment of *P. aeruginosa* infections that is not dependent on antibody effector function.

The neutralization and clearance of bacterial toxins provides a logical antibacterial therapeutic approach. A human immune serum that neutralizes botulinum toxin is approved for the treatment of infants with suspected botulism[48] and a horse serum is approved for the treatment of adults. The success of antitoxin approaches depends upon early diagnosis of infection and in some cases, the ability to neutralize multiple unrelated toxins. For the latter, a monoclonal antibody is likely to be less effective than a polyclonal approach. The direct comparison of single toxin neutralization and double toxin neutralization with *C. difficile* toxin monoclonal antibodies in the clinic will be instructive. The inhibition of the *P. aeruginosa* TTSS represents a monoclonal antibody approach that effectively neutralizes several protein toxins that correlate with pathogenicity in humans and should test the utility of antivirulence approaches in human bacterial infections.[49]

Conclusions and Future Prospects

There are currently no FDA-approved antibacterial monoclonal antibody products. Those evaluated to date may not have had optimal activity. Some, for example, have had short *in vivo* half-lives (scFv and IgM) requiring frequent dosing. Others may not have had appropriate specificity. Centoxin, the first monoclonal antibody to be developed against a bacterial cell-wall component, was ultimately shown to have inadequate specificity.[50•] In other cases, antibodies requiring immune effector functions have been tested in patients with weakened or immature immune systems. Species-specificity in bacterial immune evasion mechanisms may also have contributed to differences between the results in animal models and human clinical trials.

There is now a greater understanding of immune evasion mechanisms, and the need to match antibody target epitopes and mechanism of action, as well as optimizing antibody delivery to the site of infection.

One novel approach to addressing the problem of bacterial complement evasion is being pursued by Elusys. The major route of clearance of complement-opsonized immune complexes in humans is via the complement receptor 1 (CR1) expressed on erythrocytes. Erythrocytes coated with bacteria are rapidly cleared via the liver. Elusys has exploited this system to target immune complexes directly to CR1 using an anti-CR1 antibody. In mouse models, an anti-CR1 antibody cross-linked to an antibody to *S. aureus* Protein A to generate a heteropolymer (designated ETI-211) has shown rapid clearance of *S. aureus* from the blood stream (within 30 min) and demonstrated survival benefits in prophylactic and therapeutic models.[51] This approach may be attractive for enhancing bacterial clearance for the treatment of bacteremia although it may be less effective for tissue infections in which access to erythrocytes is limited.

Alternative methods for reducing virulence may provide the opportunity to interfere with bacterial defenses and allow more effective clearance by natural host mechanisms. Bacterial toxins in particular represent a class of highly conserved targets which play an essential role in pathogenicity. Neutralizing either the toxins or toxin delivery systems such as the TTSS of Gram-negative bacteria provides the potential to intervene in the pathology of the disease without selecting escape variants and may provide effective means of prevention or treatment of some of the most intractable bacterial diseases.

References and Recommended Reading

Papers of particular interest, published within the period of review, have been highlighted as:
• of special interest

1. LaRocca TJ, Katona LI, Thanassi, DG, Benach JL: Bactericidal action of a complement-independent antibody against relapsing fever *Borrelia* resides in its variable region. *J Immunol* 2008, **180**: 6222–6228.
2. Conti S, Magliani W, Arseni S, Dieci E, Frazzi R, Salati A, Varaldo PE, Polonelli L, In vitro activity of monoclonal and recombinant yeast killer toxin-like antibodies against antibiotic-resistant gram-positive cocci. *Mol Med* 2000, **6:**613–619.
3. Wentworth P Jr, Wentworth AD, Zhu X, Wilson IA, Janda KD, Eschenmoser A, Lerner RA: Evidence for the production of trioxygen species during antibody-catalyzed chemical modification of antigens *Proc Natl Acad Sci U S A* 2003, **100:**1490–1493.
4. Babior BM, Takeuchi C, Ruedi J, Gutierrez A, Wentworth P, Investigating antibody-catalyzed ozone generation by human neutrophils. *Proc Natl Acad Sci U S A* 2003, **100:**3031–3034.
5. Weisman LE, Antibody for the prevention of neonatal noscocomial staphylococcal infection: a review of the literature. *Arch Pediatr* 2007, **14 (Suppl 1):**S31–S34.
6. Thackray H, Lassiter H, Walsh B: Phase II randomized, double blind, placebo-controlled, safety, pharmacokinetics,

pharmacodynamics, and clinical activity study in very low birth weight (VLBW) neonates of BSYX-A110, an anti-staphylococcal monoclonal antibody for the prevention of staphylococcal infection. *Pediatr Res* 2006, **59**:3724 (Abstract).

7. Domanski PJ, Patel PR, Bayer AS, Zhang L, Hall AE, Syribeys PJ, Gorovits EL, Bryant D, Vernachio JH, Hutchins JT, Patti JM, Characterization of a humanized monoclonal antibody recognizing clumping factor A expressed by *Staphylococcus aureus*. *Infect Immun* 2005, **73**:5229–5232.

8. Weems JJ, Steinberg JP, Filler S, Baddley JW, Corey GR, Sampathkumar P, Winston L, John JF, Kubin CJ, Talwani R *et al.*: Phase II, randomized, double-blind, multicenter study comparing the safety and pharmacokinetics of tefibazumab to placebo for treatment of *Staphylococcus aureus* bacteremia. *Antimicrob Agents Chemother* 2006, **50**:2751–2755.

9. Burnie JP, Matthews RC, Carter T, Beaulieu E, Donohoe M, Chapman C, Williamson P, Hodgetts SJ, Identification of an immunodominant ABC transporter in methicillin-resistant *Staphylococcus aureus* infections. *Infect Immun* 2000, **68**:3200–3209.

10. Nilsson E, Larsson A, Olesen HV, Wejåker PE, Kollberg H: Good effect of IgY against *Pseudomonas aeruginosa* infections in cystic fibrosis patients. *Pediatr Pulmonol* 2008, **43**:892–899.

11. Hemachandra S, Kamboj K, Copfer J, Pier G, Green LL, Schreiber JR: Human monoclonal antibodies against *Pseudomonas aeruginosa* lipopolysaccharide derived from transgenic mice containing megabase human immunoglobulin loci are opsonic and protective against fatal pseudomonas sepsis. *Infect Immun* 2001, **69**:2223–2229.

12. Lai Z, Kimmel R, Petersen S, Thomas S, Pier G, Bezabeh B, Luo R, Schreiber JR, Multi-valent human monoclonal antibody preparation against *Pseudomonas aeruginosa* derived from transgenic mice containing human immunoglobulin loci is protective against fatal pseudomonas sepsis caused by multiple serotypes. *Vaccine* 2005, **23**:3264–3271.

13. Hatano K, Goldberg JB, Pier GB: Biologic activities of antibodies to the neutral-polysaccharide component of the *Pseudomonas aeruginosa* lipopolysaccharide are blocked by O side chains and mucoid exopolysaccharide (alginate). *Infect Immun* 1995, **63**:21–26.

14. Rosok MJ, Stebbins MR, Connelly K, Lostrom ME, Siadak AW: Generation and characterization of murine antiflagellum monoclonal antibodies that are protective against lethal challenge with *Pseudomonas aeruginosa*. *Infect Immun* 1990, **58**:3819–3828.

15. Meluleni GJ, Grout M, Evans DJ, Pier GB: Mucoid *Pseudomonas aeruginosa* growing in a biofilm in vitro are killed by opsonic antibodies to the mucoid exopolysaccharide capsule but not by antibodies produced during chronic lung infection in cystic fibrosis patients. *J Immunol* 1995, **155**:2029–2038.

16. Risley AL, Loughman A, Cywes-Bentley C, Foster TJ, Lee JC: Capsular polysaccharide masks clumping factor A-mediated adherence of *Staphylococcus aureus* to fibrinogen and platelets. *J Infect Dis* 2007 **196**:919–927.

17. Atkins KL, Burman JD, Chamberlain ES, Cooper JE, Poutrel B, Bagby S, Jenkins AT, Feil EJ, van den Elsen JM, *S. aureus* IgG-binding proteins SpA and Sbi: host specificity and mechanisms of immune complex formation. *Mol Immunol* 2008,**45**:1600–1611.

18. Lambris JD, Ricklin D, Geisbrecht BV, Complement evasion by
 • human pathogens. *Nat Rev Microbiol* 2008, **6**:132–142.
A review of the multiple mechanisms used by bacteria to interfere with complement activation.

19. Jongerius I, Köhl J, Pandey MK, Ruyken M, van Kessel KP, van Strijp JA, Rooijakkers SH: Staphylococcal complement evasion by various convertase-blocking molecules. *J Exp Med* 2007, **204**:2461–2471.

20. Fick RB Jr, Naegel GP, Squier SU, Wood RE, Gee JB, Reynolds HY: Proteins of the cystic fibrosis respiratory tract. Fragmented immunoglobulin G opsonic antibody causing defective opsonophagocytosis. *J Clin Invest* 1984,**74**:236–248.

21. Fick RB Jr, Baltimore RS, Squier SU, Reynolds HY: IgG proteolytic activity of *Pseudomonas aeruginosa* in cystic fibrosis. *J Infect Dis* 1985, **151**:589–598.

22. Engel LS, Hill JM, Caballero AR, Green LC, O'Callaghan RJ: Protease IV, a unique extracellular protease and virulence factor from *Pseudomonas aeruginosa*. *J Biol Chem* 1998, **273**:16792–16797.

23. Smith L, Rose B, Tingpej P, Zhu H, Conibear T, Manos J, Bye P, Elkins M, Willcox M, Bell S *et al.*: Protease IV production in *Pseudomonas aeruginosa* from the lungs of adults with cystic fibrosis. *J Med Microbiol* 2006, **55**:1641–1644.

24. Britigan BE, Railsback MA, Cox CD: The *Pseudomonas aeruginosa* secretory product pyocyanin inactivates alpha1 protease inhibitor: implications for the pathogenesis of cystic fibrosis lung disease. *Infect Immun* 1999, **67**:1207–1212.

25. Karlsson A, Arvidson S: Variation in extracellular protease production among clinical isolates of *Staphylococcus aureus* due to different levels of expression of the protease repressor sarA. *Infect Immun* 2002, **70**:4239–4246.

26. Rooijakkers SH, van Wamel WJ, Ruyken M, van Kessel KP, van Strijp JA: Anti-opsonic properties of staphylokinase. *Microbes Infect* 2005, **7**:476–484.

27. Mathee K, Ciofu O, Sternberg C, Lindum PW, Campbell JI, Jensen P, Johnsen AH, Givskov M, Ohman DE, Molin S *et al.*: Mucoid conversion of *Pseudomonas aeruginosa* by hydrogen peroxide: a mechanism for virulence activation in the cystic fibrosis lung. *Microbiology* 1999, **145**:1349–1357.

28. Cerca N, Jefferson KK, Oliveira R, Pier GB, Azeredo J: Comparative antibody-mediated phagocytosis of *Staphylococcus epidermidis* cells grown in a biofilm or in the planktonic state. *Infect Immun* 2006, **74**:4849–4855.

29. Cerca N, Jefferson KK, Maira-Litrán T, Pier DB, Kelly-Quintos C, Goldmann DA, Azeredo J, Pier GB: Molecular basis for preferential protective efficacy of antibodies directed to the poorly acetylated form of staphylococcal poly-*N*-acetyl-beta-(1-6)-glucosamine. *Infect Immun* 2007, **75**:3406–3413.

30. Pier GB, Boyer D, Preston M, Coleman FT, Llosa N, Mueschenborn-Koglin S, Theilacker C, Goldenberg H, Uchin J, Priebe GP *et al.*: Human monoclonal antibodies to *Pseudomonas aeruginosa* alginate that protect against infection by both mucoid and nonmucoid strains. *J Immunol* 2004, **173**:5671–5678.

31. March JC, Bentley WE: Quorum sensing and bacterial cross-talk in biotechnology. *Curr Opin Biotechnol* 2004, **15**:495–502.

32. Kravchenko VV, Kaufmann GF, Mathison JC, Scott DA, Katz AZ, Grauer DC, Lehmann M, Meijler MM, Janda KD, Ulevitch RJ, Modulation of gene expression via disruption of NF-kappaB signaling by a bacterial small molecule. *Science* 2008, **321**: 259–263.

33. Miyairi S, Tateda K, Fuse ET, Ueda C, Saito H, Takabatake T, Ishii Y, Horikawa M, Ishiguro M, Standiford TJ, Yamaguchi K, Immunization with 3-oxododecanoyl-L-homoserine lactone–protein conjugate protects mice from lethal *Pseudomonas aeruginosa* lung infection. *J Med Microbiol* 2006, **55**:1381–1387.

34. Kaufmann GF, Park J, Mee JM, Ulevitch RJ, Janda KD: The quorum quenching antibody RS2-1G9 protects macrophages from the cytotoxic effects of the *Pseudomonas aeruginosa* quorum sensing signalling molecule *N*-3-oxo-dodecanoyl-homoserine lactone. *Mol Immunol* 2008, **45**:2710–2714.

35. Park J, Jagasia R, Kaufmann GF, Mathison JC, Ruiz DI, Moss JA, Meijler MM, Ulevitch RJ, Janda KD: Infection control by antibody disruption of bacterial quorum sensing signaling. *Chem Biol* 2007, **14**:1119–1127.

36. Babin M, Ou Y, Roschke V, Yu A, Subramanian M, Choi G: Protection against inhalation anthrax-induced lethality by a human monoclonal antibody to protective antigen in rabbits and cynomolgus monkeys. In *43rd Interscience Conference on Antimicrobial Agents and Chemotherapy* 2003, (Abstract # 3836).

37. Subramanian GM, Cronin PW, Poley G, Weinstein A, Stoughton SM, Zhong J, Ou Y, Zmuda JF, Osborn BL, Freimuth WW: A phase 1 study of PAmAb, a fully human monoclonal antibody against *Bacillus anthracis* protective antigen, in healthy volunteers. *Clin Infect Dis* 2005, **41:**12–20.

38. Mohamed N, Clagett M, Li J, Jones S, Pincus S, D'Alia G, Nardone L, Babin M, Spitalny G, Casey L: A high-affinity monoclonal antibody to anthrax protective antigen passively protects rabbits before and after aerosolized *Bacillus anthracis* spore challenge. *Infect Immun* 2005, **73:**795–802.

39. Vitale L, Blanset D, Lowy I, O'Neill T, Goldstein J, Little SF, Andrews GP, Dorough G, Taylor RK, Keler T: Prophylaxis and therapy of inhalational anthrax by a novel monoclonal antibody to protective antigen that mimics vaccine-induced immunity. *Infect Immun* 2006, **74:**5840–5847.

40. Dowling TC, Chavaillaz PA, Young DG, Melton-Celsa A, O'Brien A, Thuning-Roberson C, R Edelman, Tacket CO: Phase 1 safety and pharmacokinetic study of chimeric murine-human monoclonal antibody c alpha Stx2 administered intravenously to healthy adult volunteers. *Antimicrob Agents Chemother* 2005, **49:**1808–1812.

41. Babcock GJ, Broering TJ, Hernandez HJ, Mandell RB, Donahue K, Boatright N, Stack AM, Lowy I, Graziano R, Molrine D, Ambrosino DM, Thomas WD: Human monoclonal antibodies directed against toxins A and B prevent *Clostridium difficile*-induced mortality in hamsters. *Infect Immun* 2006, **74:**6339–6347.

42. Müller S, Feldman MF, Cornelis GR, The Type III secretion
   • system of Gram-negative bacteria: a potential therapeutic target? *Expert Opin Ther Targets* 2001, **5:**327–339.
   This review provides a detailed description of the needle-like type III secretion systems used for the injection of exotoxins by many Gram-negative bacteria.

43. Feltman H, Schulert G, Khan S, Jain M, Peterson L, Hauser AR: Prevalence of type III secretion genes in clinical and environmental isolates of *Pseudomonas aeruginosa*. *Microbiology* 2001, **147:**2659–2669.

44. Hauser AR, Cobb E, Bodi M, Mariscal D, Vallés J, Engel JN, Rello J: Type III protein secretion is associated with poor clinical outcomes in patients with ventilator-associated pneumonia caused by *Pseudomonas aeruginosa*. *Crit Care Med* 2002, **30:**521–528.

45. Frank DW, Vallis A, Wiener-Kronish JP, Roy-Burman A, Spack EG, Mullaney BP, Megdoud M, Marks JD, Fritz R, Sawa T: Generation and characterization of a protective monoclonal antibody to *Pseudomonas aeruginosa* PcrV. *J Infect Dis* 2002, **186:**64–73.

46. Faure K, Fujimoto J, Shimabukuro DW, Ajayi T, Shime N, Moriyama K, Spack EG, Wiener-Kronish JP, Sawa T: Effects of monoclonal anti-PcrV antibody on *Pseudomonas aeruginosa*-induced acute lung injury in a rat model. *J Immune Ther Vaccines* 2003, **1:**2–10.

47. Dacheux D, Toussaint B, Richard M, Brochier G, Croize J, Attree I: *Pseudomonas aeruginosa* cystic fibrosis isolates induce rapid, type III secretion-dependent, but ExoU-independent, oncosis of macrophages and polymorphonuclear neutrophils. *Infect Immun* 2000, **68:**2916–2924.

48. Arnon SS, Schechter R, Maslanka SE, Jewell NP, Hatheway CL: Human botulism immune globulin for the treatment of infant botulism. *N Engl J Med* 2006, **354:**462–471.

49. Cegelski L, Marshall GR, Eldridge GR, Hultgren SJ: The biology and future prospects of antivirulence therapies. *Nat Rev Microbiol* 2008, **6:**17–27.

50. Edgington S, What went wrong with Centoxin? *Biotechnology*
   • 1992, **10:**617.
   This brief review describes some of the potential pitfalls of antibody development in the field of infectious disease.

51. Ferreira CS, Li J, Pincus SE, Casey LS, Mohamed N, Kreiswirth BN, Jones SM: A heteropolymer against Staphylococcal protein A prevents spread of blood-borne *Staphylococcus aureus* to target organs. *Presented at 45th Annual Interscience Conference on Antimicrobial Agents and Chemotherapy (ICAAC)*; December: 2005.

Sponge's Secret Weapon Revealed

Chemical restores bacteria's vulnerability to antibiotics.

LAURA SANDERS

A chemical from an ocean-dwelling sponge can reprogram antibiotic-resistant bacteria to make them vulnerable to medicines again, new evidence suggests.

Once-ineffective antibiotics proved lethal for bacteria treated with the compound, chemist Peter Moeller reported February 13.

"The potential is outstanding. This could revolutionize our approach to thinking about how infections are treated," comments Carolyn Sotka of the National Oceanic and Atmospheric Administration's Oceans and Human Health Initiative in Charleston, S.C.

Everything living in the ocean survives in a microbial soup, under constant bombardment from bacterial assaults. Researchers led by Moeller, of Hollings Marine Laboratory in Charleston, found a sea sponge thriving in the midst of dead marine creatures. This anomalous life amidst death raised an obvious question, said Moeller: "How is this thing surviving when everything else is dead?"

Analyses of the sponge's chemical defenses pointed to a compound called ageliferin. Biofilms, communities of bacteria notoriously resistant to antibiotics, dissolved when treated with fragments of the ageliferin molecule. And new biofilms did not form.

So far, the ageliferin offshoot has, in the lab, successfully resensitized bacteria that cause whooping cough, ear infections, septicemia and food poisoning. The compound also works on Pseudomonas aeruginosa, which causes horrible infections in wounded soldiers, and on MRSA, bacteria resistant to multiple drugs and known to wreak havoc in hospitals.

The compound is also able to reprogram antibiotic-resistant bacteria that don't form biofilms. After bacteria are treated with the compound, antibiotics that have had no effect are once again lethal. This substance may be the first that can restore bacterial vulnerability, Moeller said.

The problem of perpetuating a bacterial-resistance arms race, in which bacteria rapidly develop countermeasures against new antibiotics, may be avoided entirely with the new compound.

"Since the substance is nontoxic to the bacterium, it's not throwing up any red flags," Moeller said.

Other than "doing something really funky that we're excited about," the way this compound interferes with bacteria's resistance

Agelas conifera and other sponges make a compound that resensitizes bacteria to antibiotics.

to antibiotics is still unknown, Moeller said. The compound may sneak by bacteria's sensors that trigger new ways to combat antibiotics. The research is still in very a early phase and treatments for human infections are a long way off, Moeller said.

UNIT 5

The Immune Response: Natural and Acquired

Unit Selections

Learning Objectives

- Compare and contrast the innate and adaptive immune system.

- How is the process of autophagy connected with the immune system?

- Why is it important to vaccinate all school-age children against common childhood diseases? What arguments would you present to someone who did not want to immunize her children? What is herd immunity? How can someone else's decision to forgo vaccination affect other people in the community?

- What is some evidence for the "hygiene hypothesis"? Do you believe it? Why or why not?

- How can knowledge of the immune system be used in drug development? Why is this an important avenue of research given what you learned about in Unit 3?

Student Website
www.mhcls.com

Internet References

Immunization Action Coalition
http://www.immunize.org
Laboratory of Immunology (NIAID)
http://www3.niaid.nih.gov/labs/aboutlabs/li/
The Institute for Vaccine Safety
http://www.vaccinesafety.edu

For all the multitudes of microbes we encounter in our daily lives, most of them do not cause us harm when they enter our bodies. For this, we can thank our immune system. Ever on the alert for signs of foreign invaders, the innate immune response is the "first responder" system that recognizes generalized classes of molecules present on pathogens. This recognition triggers an inflammatory response to cordon off the offending organisms and prevent their spread to other parts of the body. In contrast, the adaptive immune response reacts to pathogens that it has encountered before by stimulating the renewed production of specific antibodies that "flag" the pathogens in order to facilitate their destruction by other cells of the immune system. Vaccines use inactivated or attenuated fragments of pathogens to stimulate the adaptive immune response so that it will respond quickly with antibody production upon later exposure to the same pathogen. Vaccination against infectious diseases remains one of the most effective tools of maintaining public health.

This unit opens with articles from *Scientific American* covering recent advances in immunology. "Immunity's Early Warning System" focuses on the role of Toll-like receptors (TLRs) in detecting foreign pathogens and signaling an inflammatory response. TLRs are crucial regulators of the balance between the adaptive immune system and inflammation. The manipulation of TLRs may be useful as drug targets, and the article outlines examples of candidate drug compounds that interact with TLRs. In "How Cells Clean House," the authors describe new insights into the process of autophagy, a mechanism by which cells get rid of invading pathogens as well as old proteins and cell debris. Autophagy also directly participates in the immune response by stimulating both the innate and adaptive branches of immunity. Research advances in this field promise to yield new treatments for various types of diseases.

There has been a disturbing resurgence in childhood diseases, such as measles, that were once common killers of young children, but that waned as all children received immunizations against these diseases. In recent years many parents have refused to immunize their children because of fear and misinformation about the safety of vaccines. While there is no scientific evidence supporting a link between vaccines and autism, many people believe that vaccines have led to a rise in the incidence of autism in the United States. An excellent article in *Time* magazine addresses the issue of vaccine safety. It also outlines the consequences for public health when some members of a population do not get vaccinated. For example, in Nigeria the number of polio cases has risen about thirty-fold after parents were dissuaded from vaccinating their children against this disease. Herd immunity protects the most vulnerable citizens (such as the elderly and newborns) in a population; when most of a community is immunized, pathogens are less likely to enter that group of people and find a susceptible host. Whenever a new vaccine appears, it takes some time for the public to accept its widespread use. A vaccine recommended for teenage girls, Gardisil offers protection against human papilloma virus (HPV) and decreases the risk

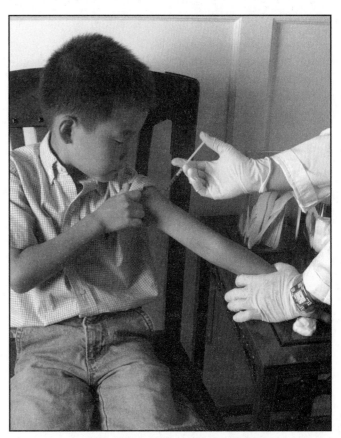

© The McGraw-Hill Companies, Inc./Jill Braaten, photographer

of getting cervical cancer. An article in *Nursing* offers advice on how to teach parents and adolescents about the benefits of this vaccine.

The next article in this unit, "Caution: Killing Germs May Be Hazardous to Your Health," describes our historical relationship with microbes and the consequences of our obsession with getting rid of them. It highlights the rise in drug-resistant strains of microbes (discussed in Unit 3), which have been growing every year, and also discusses the "hygiene hypothesis," which has been used to explain the rise in asthma and allergies seen in "supersanitized" industrialized environments. In order to develop a functional immune system, individuals need exposure to a variety of harmless microbes from an early age. This theme is also explored in "Why We're Sicker," which criticizes the modern lifestyle for not giving our developing immune systems enough practice in differentiating between true threats like pathogens and harmless "invaders" like pollen or certain foods. A variety of autoimmune disorders is examined.

Other articles in this unit examine how knowledge of the immune system can be used for developing therapeutics and vaccines. A review in *Current Opinion in Biotechnology* describes the rationale for targeting host innate immunity. Since the pathogen itself is not targeted, there is little chance that it would develop drug resistance. Furthermore, since innate immunity recognizes broad classes of pathogens, drugs aimed at components of this system could be effective against a variety of microbial groups. The clinical status of several new immunomodulatory drugs is summarized. The importance of this type of research can be appreciated given the alternative drug development strategies discussed in Unit 4. Finally, this unit ends with an article on the development of malaria vaccines and the challenges facing researchers in this field. Since the life cycle of the malaria parasite involves both mosquitoes and humans, success in eliminating malaria from a community requires the eradication of all parasites in various stages of their life cycle.

Immunity's Early-Warning System

The innate immune response constitutes the first line of defense against invading microbes and plays a role in inflammatory disease. Surprising insights into how this system operates could lead to new therapies for a host of infectious and immune-related disorders.

LUKE A. J. O'NEILL

A woman is riding an elevator when her fellow passengers start to sneeze. As she wonders what sort of sickness the other riders might be spreading, her immune system swings into action. If the bug being dispersed by the contagious sneezers is one the woman has met before, a battalion of trained immune cells—the foot soldiers of the so-called adaptive immune system—will remember the specific invader and clear it within hours. She might never realize she had been infected.

But if the virus or bacterium is one that our hapless rider has never wrestled, a different sort of immune response comes to the rescue. This "innate" immune system recognizes generic classes of molecules produced by a variety of disease-causing agents, or pathogens. When such foreign molecules are detected, the innate system triggers an inflammatory response, in which certain cells of the immune system attempt to wall off the invader and halt its spread. The activity of these cells—and of the chemicals they secrete—precipitates the redness and swelling at sites of injury and accounts for the fever, body aches and other flulike symptoms that accompany many infections.

The inflammatory assault, we now know, is initiated by Toll-like receptors (TLRs): an ancient family of proteins that mediate innate immunity in organisms from horseshoe crabs to humans. If TLRs fail, the entire immune system crashes, leaving the body wide open to infection. If they work too hard, however, they can induce disorders marked by chronic, harmful inflammation, such as arthritis, lupus and even cardiovascular disease.

Discovery of TLRs has generated an excitement among immunologists akin to that seen when Christopher Columbus returned from the New World. Scores of researchers are now setting sail to this new land, where they hope to find explanations for many still mysterious aspects of immunity, infection and disorders involving abnormal defensive activity. Study of these receptors, and of the molecular events that unfold after they encounter a pathogen, is already beginning to uncover

targets for pharmaceuticals that may enhance the body's protective activity, bolster vaccines, and treat a range of devastating and potentially deadly disorders.

Cinderella Immunity

Until about five years ago, when it came to the immune system, the adaptive division was the star of the show. Textbooks were filled with details about B cells making antibodies that latch onto specific proteins, or antigens, on the surface of an invading pathogen and about T cells that sport receptors able to recognize fragments of proteins from pathogens. The response is called adaptive because over the course of an infection, it adjusts to optimally handle the particular microorganism responsible for the disease.

Adaptive immunity also grabbed the spotlight because it endows the immune system with memory. Once an infection has been eliminated, the specially trained B and T cells stick around, priming the body to ward off subsequent attacks. This ability to remember past infections allows vaccines to protect us from diseases caused by viruses or bacteria. Vaccines expose the body to a disabled form of a pathogen (or harmless pieces of it), but the immune system reacts as it would to a true assault, generating protective memory cells in the process. Thanks to T and B cells, once an organism has encountered a microbe and survived, it becomes exempt from being overtaken by the same bug again.

The innate immune system seemed rather drab in comparison. Its components—including antibacterial enzymes in saliva and an interlocking set of proteins (known collectively as the complement) that kill bacteria in the bloodstream—were felt to be less sophisticated than targeted antibodies and killer T cells. What is more, the innate immune system does not tailor its response in the same way that the adaptive system does.

In dismissing the innate immune response as dull and uninteresting, however, immunologists were tiptoeing around a dirty

The Divisions of the Immune System

The mammalian immune system has two overarching divisions. The innate part acts near entry points into the body and is always at the ready. If it fails to contain a pathogen, the adaptive division kicks in, mounting a later but highly targeted attack against the specific invader.

Innate Immune System

This system includes, among other components, anti-microbial molecules and various phagocytes [cells that ingest and destroy pathogens]. These cells, such as dendritic cells and macrophages, also activate an inflammatory response, secreting proteins called cytokines that trigger an influx of defensive cells from the blood. Among the recruits are more phagocytes—notably monocytes [which can mature into macrophages] and neutrophils.

Adaptive Immune System

This system "stars" B cells and T cells. Activated B cells secrete antibody molecules that bind to antigens—specific components unique to a given invader—and destroy the invader directly or mark it for attack by others. T cells recognize antigens displayed on cells. Some T cells help to activate B cells and other T cells; other T cells directly attack infected cells. T and B cells spawn "memory" cells that promptly eliminate invaders encountered before.

Overview/Innate Immunity

- Innate immunity serves as a rapid response system for detecting and clearing infections by any infectious agent. The response is mediated by a family of molecules called Toll-like receptors (TLRs), made by many defensive cells.
- When TLRs detect an invader, they trigger the production of an array of signaling proteins that induce inflammation and direct the body to mount a full-fledged immune response.
- If TLRs are underactive, the immune system fails; if overactive, they can give rise to disorders such as rheumatoid arthritis and even cardiovascular disease. Learning how to manipulate TLRs or the proteins with which they interact could provide new options for treating infectious and inflammatory diseases.

little secret: the adaptive system does not work in the absence of the allegedly more crude innate response. The innate system produces certain signaling proteins called cytokines that not only induce inflammation but also activate the B and T cells that are needed for the adaptive response. The posh sister, it turns out, needs her less respected sibling to make her shine.

By the late 1990s immunologists knew a tremendous amount about how the adaptive immune system operates. But they had less of a handle on innate immunity. In particular, researchers did not understand how microbes activate the innate response—or exactly how this stimulation helps to drive the adaptive response of T and B cells. Soon after, though, they would learn that much of the answer lay with the TLRs, which are produced by various immune system cells. But the path scientists traveled to get to these proteins was a circuitous one, winding through studies of fruit fly development, the search for drugs to treat arthritis, and the dawn of the genomic era.

Weird Protein

The path actually had its beginnings in the early 1980s, when immunologists started to study the molecular activity of cytokines. These protein messengers are produced by various immune cells, including macrophages and dendritic cells. Macrophages patrol the body's tissues, searching for signs of infection. When they detect a foreign protein, they set off the inflammatory response. In particular, they engulf and destroy the invader bearing that protein and secrete a suite of cytokines, some of which raise an alarm that recruits other cells to the site of infection and puts the immune system in general on full alert. Dendritic cells ingest invading microbes and head off to the lymph nodes, where they present fragments of the pathogen's proteins to armies of T cells and release cytokines—activities that help to switch on the adaptive immune response.

To study the functions of various cytokines, researchers needed a way to induce the molecules' production. They found that the most effective way to get macrophages and dendritic cells to make cytokines in the laboratory was to expose them to bacteria—or more important, to selected components of bacteria. Notably, a molecule called lipopolysaccharide (LPS), made by a large class of bacteria, stimulates a powerful immune response. In humans, exposure to LPS causes fever and can lead to septic shock—a deadly vascular shutdown triggered by an overwhelming, destructive action of immune cells. LPS, it turns out, evokes this inflammatory response by prompting macrophages and dendritic cells to release the cytokines tumor necrosis factor-alpha (TNF-alpha) and interleukin-1 (IL-1).

Indeed, these two cytokines were shown to rule the inflammatory response, prodding immune cells into action. If left unchecked, they can precipitate disorders such as rheumatoid arthritis, an autoimmune condition characterized by excessive inflammation that leads to destruction of the joints. Investigators therefore surmised that limiting the effects of TNF alpha and IL-1 might slow the progress of the disease and alleviate the suffering of those with arthritis. To design such a therapy, though, they needed to know more about how these molecules work. And the first step was identifying the proteins with which they interact.

In 1988 John E. Sims and his colleagues at Immunex in Seattle discovered a receptor protein that recognizes IL-1. This receptor resides in the membranes of many different cells in the body, including macrophages and dendritic cells. The part of the receptor that juts out of the cell binds to IL-1, whereas the

Tolls in Charge

Toll-like receptors [TLRs], made by many cells of the innate system, have been found to both orchestrate the innate immune response and play a critical role in the adaptive response. TLR4, for example, elicits these defenses when gram-negative bacteria begin to invade. TLR4 detects the incursions by binding to lipopolysaccharide [LPS], a sugar unique to gram-negative bacteria. Having recognized LPS, pairs of TLR4s signal to four molecules inside the cell—MyD88, Mal, Tram and Trif—which, in turn, trigger molecular interactions that ultimately activate a master regulator of inflammation (NF-κB). This regulator then switches on genes that encode immune activators, including cytokines. These cytokines induce inflammation and also help to switch on the T and B cells of the adaptive immune division.

segment that lies inside the cell relays the message that IL-1 has been detected. Sims examined the inner part of the IL-1 receptor carefully, hoping it would yield some clue as to how the protein transmits its message—revealing, for example, which signaling molecules it activates within cells. But the inner domain of the human IL-1 receptor was unlike anything researchers had seen before, so he was stymied.

Then, in 1991, Nick J. Gay of the University of Cambridge—working on a completely unrelated problem—made a strange discovery. He was looking for proteins that were similar to a fruit-fly protein called Toll. Toll had been identified by Christiane Nusslein-Volhard in Tübingen, Germany, who gave the protein its name because flies that lack Toll look weird (*Toll* being the German word for "weird"). The protein helps the developing *Drosophila* embryo to differentiate its top from its bottom, and flies without Toll look jumbled, as if they have lost their sidedness.

Gay searched the database containing all the gene sequences then known. He was looking for genes whose sequences closely matched that of Toll and thus might encode Toll-like proteins. And he discovered that part of the Toll protein bears a striking resemblance to the inner part of the human IL-1 receptor, the segment that had mystified Sims.

At first the finding didn't make sense. Why would a protein involved in human inflammation look like a protein that tells fly embryos which end is up? The discovery remained puzzling until 1996, when Jules A. Hoffmann and his collaborators at CNRS in Strasbourg showed that flies use their Toll protein to defend themselves from fungal infection. In *Drosophila*, it seems, Toll multitasks and is involved in both embryonic development and adult immunity.

Worms, Water Fleas and You

The IL-1 receptor and the Toll protein are similar only in the segments that are tucked inside the cell; the bits that are exposed to the outside look quite different. This observation led researchers to search for human proteins that resemble Toll in its entirety. After all, evolution usually conserves designs that work well—and if Toll could mediate immunity in flies, perhaps similar proteins were doing the same in humans.

Acting on a tip from Hoffmann, in 1997 Ruslan Medzhitov and the late Charles A. Janeway, Jr., of Yale University discovered the first of these proteins, which they called human Toll. Within six months or so, Fernando Bazan and his colleagues at DNAX in Palo Alto, Calif., had identified five human Tolls, which they dubbed Toll-like receptors (TLRs). One, TLR4, was the same human Toll described by Medzhitov and Janeway.

At that point, researchers still did not know exactly how TLRs might contribute to human immunity. Janeway had found that stuffing the membranes of dendritic cells with TLR4 prompted the production of cytokines. But he could not say how TLR4 became activated during an infection.

The answer came in late 1998, when Bruce Beutler and his co-workers at the Scripps Institute in La Jolla, Calif., found that mutant mice unable to respond to LPS harbor a defective version of TLR4. Whereas normal mice die of sepsis within an hour of being injected with LPS, these mutant mice survive and behave as if they have not been exposed to the molecule at all; that is, the mutation in the TLR4 gene renders these mice insensitive to LPS.

This discovery made it clear that TLR4 becomes activated when it interacts with LPS. Indeed, its job is to sense LPS. That realization was a major breakthrough in the field of sepsis, because it revealed the molecular mechanism that underlies inflammation and provided a possible new target for treatment of a disorder that sorely needed effective therapies. Within two years, researchers determined that most TLRs—of which 10 are now known in humans—recognize molecules important to the survival of bacteria, viruses, fungi and parasites. TLR2 binds to lipoteichoic acid, a component of the bacterial cell wall. TLR3 recognizes the genetic material of viruses; TLR5 recognizes flagellin, a protein that forms the whiplike tails used by bacteria to swim; and TLR9 recognizes a signature genetic sequence called CpG, which occurs in bacteria and viruses in longer stretches and in a form that is chemically distinct from the CpG sequences in mammalian DNA.

TLRs, it is evident, evolved to recognize and respond to molecules that are fundamental components of pathogens. Eliminating or chemically altering any one of these elements could cripple an infectious agent, which means that the organisms cannot dodge TLRs by mutating until these components are unrecognizable. And because so many of these elements are shared by a variety of microbes, even as few as 10 TLRs can protect us from virtually every known pathogen.

Innate immunity is not unique to humans. In fact, the system is quite ancient. Flies have an innate immune response, as do starfish, water fleas and almost every organism that has been examined thus far. And many use TLRs as a trigger. The nematode worm has one that allows it to sense and swim away from infectious bacteria. And plants are rife with TLRs. Tobacco has one called N protein that is required for fighting tobacco mosaic virus. The weed *Arabidopsis* has more than 200. The first Toll-like protein most likely arose in a single-celled organism that

The Jobs of Toll-like Receptors

Each Toll-like receptor can detect some essential component of a broad class of disease-causing agents, and as a group, TLRs can apparently recognize almost every pathogen likely to cause infections. Different combinations appear in different kinds of cells, where the molecules act in pairs. Investigators have identified 10 human TLRs and know many of the molecules they recognize. The function of TLR10 and the partners of TLR3, 5, 7, 8 and 9 are unknown.

Mechnikov's Fleas

The discovery of Tolls and Toll-like receptors extends a line of research begun more than 100 years ago, when Russian biologist Ilya Mechnikov essentially discovered innate immunity. In the early 1880s, Mechnikov plucked some thorns from a tangerine tree and poked them into a starfish larva. The next morning he saw that the thorns were surrounded by mobile cells, which he surmised were in the process of engulfing bacteria introduced along with the foreign bodies. He then discovered that water fleas [Daphnia] exposed to fungal spores mount a similar response. This process of phagocytosis is a cornerstone of innate immunity, and its discovery earned Mechnikov a Nobel Prize in 1908.

Mechnikov was a character. Speaking of the era when he worked at the Pasteur Institute, his Nobel Prize biography notes, "It is said of him that at this time he usually wore overshoes in all weathers and carried an umbrella, his pockets being overfull with scientific papers, and that he always wore the same hat, and often, when he was excited, sat on it."

was a common ancestor of plants and animals. Perhaps these molecules even helped to facilitate our evolution. Without an efficient means of defense against infection, multicellular organisms might never have survived.

Storming the Castle

The innate system was once thought to be no more elaborate than the wall of a castle. The real action, researchers believed, occurred once the wall had been breached and the troops inside—the T and B cells—became engaged. We now know that the castle wall is studded with sentries—TLRs—that identify the invader and sound the alarm to mobilize the troops and prepare the array of defenses needed to fully combat the attack. TLRs, in other words, unleash both the innate and adaptive systems.

The emerging picture looks something like this. When a pathogen first enters the body, one or more TLRs, such as those on the surface of patrolling macrophages and dendritic cells, latch onto the foreign molecules—for example, the LPS of gram-negative bacteria. Once engaged, the TLRs prompt the cells to unleash particular suites of cytokines. These protein messengers then recruit additional macrophages, dendritic cells and other immune cells to wall off and nonspecifically attack the marauding microbe. At the same time, cytokines released by all these busy cells can produce the classic symptoms of infection, including fever and flulike feelings.

Macrophages and dendritic cells that have chopped up a pathogen display pieces of it on their surface, along with other molecules indicating that a disease causing agent is present. This display, combined with the cytokines released in response to TLRs, ultimately activates B and T cells that recognize those specific pieces, causing them—over the course of several days—to proliferate and launch a powerful, highly focused attack on the particular invader. Without the priming effect of TLRs, B and T cells would not become engaged and the body would not be able to mount a full immune response. Nor could the body retain any memory of previous infections.

Following the initial infection, enough memory T and B cells are left behind so that the body can deal more efficiently with the invader should it return. This army of memory cells can act so quickly that inflammation might not occur at all. Hence, the

victim does not feel as ill and might not even notice the infection when it recurs.

Innate and adaptive immunity are thus part of the same system for recognizing and eliminating microbes. The interplay between these two systems is what makes our overall immune system so strong.

Choose Your Weapon

To fully understand how TLRs control immune activity, immunologists need to identify the molecules that relay signals from activated TLRs on the cell surface to the nucleus, switching on genes that encode cytokines and other immune activators. Many investigators are now pursuing this search intensively, but already we have made some fascinating discoveries.

We now know that TLRs, like many receptors that reside on the cell surface, enlist the help of a long line of signaling proteins that carry their message to the nucleus, much as a bucket brigade shuttles water to a fire. All the TLRs, with the exception of TLR3, hand off their signal to an adapter protein called MyD88. Which other proteins participate in the relay varies with the TLR: my laboratory studies Mal, a protein we discovered that helps to carry signals generated by TLR4 and TLR2. TLR4 also requires two other proteins—Tram and Trif—to relay the signal, whereas TLR3 relies on Trif alone. Shizuo Akira of Osaka University in Japan has shown that mice engineered so that they do not produce some of these intermediary signaling proteins do not respond to microbial products, suggesting that TLR-associated proteins could provide novel targets for new anti-inflammatory or antimicrobial agents.

Interaction with different sets of signaling proteins allows TLRs to activate different sets of genes that hone the cell's response to better match the type of pathogen being encountered. For example, TLR3 and TLR7 sense the presence of viruses. They then trigger a string of molecular interactions that induce the production and release of interferon, the major antiviral cytokine. TLR2, which is activated by bacteria, stimulates the release of a blend of cytokines that does not include interferon but is more suited to activating an effective antibacterial response by the body.

The realization that TLRs can detect different microbial products and help to tailor the immune response to thwart the enemy is now overturning long-held assumptions that innate immunity is a static, undiscriminating barrier. It is, in fact, a dynamic system that governs almost every aspect of inflammation and immunity.

From *Legionella* to Lupus

On recognizing the central role that TLRs play in initiating immune responses, investigators quickly began to suspect that hobbled or overactive versions of these receptors could contribute to many infectious and immune-related disorders. That hunch proved correct. Defects in innate immunity lead to greater susceptibility to viruses and bacteria. People with an underactive form of TLR4 are five times as likely to have severe bacterial infections over a five-year period than those with a normal TLR4. And people who die from Legionnaire's disease often harbor a mutation in TLR5 that disables the protein, compromising their innate immune response and rendering them unable to fight off the *Legionella* bacterium. On the other hand, an overzealous immune response can be equally destructive. In the U.S. and Europe alone, more than 400,000 people die annually from sepsis, which stems from an overactive immune response led by TLR4.

Other studies are pointing to roles for TLRs in autoimmune diseases such as systemic lupus erythematosus and rheumatoid arthritis. Here TLRs might respond to products from damaged cells, propagating an inappropriate inflammatory response and promoting a misguided reaction by the adaptive immune system. In lupus, for example, TLR9 has been found to react to the body's own DNA.

Innate immunity and the TLRs could also play a part in heart disease. People with a mutation in TLR4 appear to be less prone to developing cardiovascular disease. Shutting down TLR4 could protect the heart because inflammation appears to contribute to the formation of the plaques that clog coronary arteries.

Manipulation of TLR4 might therefore be another approach to preventing or limiting this condition.

Volume Control

Many of the big pharmaceutical companies have an interest in using TLRs and their associated signaling proteins as targets for drugs that could treat infections and immune-related disorders. With the spread of antibiotic resistance, the emergence of new and more virulent viruses, and the rising threat of bioterrorism, the need to come up with fresh ways to help our bodies fight infection is becoming more pressing.

TLRs as Drug Targets

Agents that activate TLRs and thus enhance immune responses could increase the effectiveness of vaccines or protect against infection. They might even prod the immune system to destroy tumors. In contrast, drugs that block TLR activity might prove useful for dampening inflammatory disorders. Drugs of both types are under study (below).

Drug Type	Examples
TLR4 activator	MPL, an allergy treatment and vaccine adjuvant [immune system activator] from Corixa (Seattle), is in large-scale clinical trials
TLR7 activator	ANA245 [isatoribine], an antiviral agent from Anandys [San Diego], is in early human trials for hepatitis C
TLR7 and TLR8 activator	Imiquimod, a treatment for genital warts, basal cell skin cancer and actinic keratosis from 3M [St. Paul, Minn.], is on the market
TLR9 activator	ProMune, a vaccine adjuvant and treatment for melanoma skin cancer and non-Hodgkin's lymphoma from Coley [Wellesley, Mass.], is in large-scale clinical trials
TLR4 inhibitor	E5564, an antisepsis drug from Eisai [Teaneck, N.J.], is in early human trials
General TLR inhibitor	RDP58, a drug for ulcerative colitis and Crohn's disease from Genzyme [Cambridge, Mass.], is entering large-scale clinical trials
General TLR inhibitor	OPN201, a drug for autoimmune disorders from Opsona Therapeutics [Dublin, Ireland], is being tested in animal models of inflammation

Work on TLRs could, for example, guide the development of safer, more effective vaccines. Most vaccines depend on the inclusion of an adjuvant, a substance that kick-starts the inflammatory response, which in turn pumps up the ability of the adaptive system to generate the desired memory cells. The adjuvant used in most vaccines today does not provoke a full adaptive response; instead it favors B cells over T cells. To elicit a stronger response, several companies have set their sights on compounds that activate TLR9, a receptor that recognizes a broad range of bacteria and viruses and drives a robust immune response.

And TLRs are teaching us how to defend ourselves against biological weapons, such as poxviruses. A potential staple in the bioterrorist arsenal, these viruses can shut down TLRs and thereby avoid detection and elimination. In collaboration with Geoffrey L. Smith of Imperial College London, my lab found that by removing the viral protein that disables TLRs, we could generate a weakened virus that could serve as the basis of a vaccine unlikely to provoke an unintended fatal pox infection.

Armed with an understanding of TLRs and innate immunity, physicians might be able to predict which patients will fare poorly during infection and treat them more aggressively. If, for instance, patients came to a clinic with a bacterial infection and were found to have a mutant TLR4, the doctor might bombard them with antibiotics or with agents that could somehow bolster their immune response to prevent the infection from doing lasting damage.

Of course, a balance must be struck between stimulating an immune response that is sufficient to clear a microbe and precipitating an inflammatory response that will do more harm than good. Similarly, any medications that aim to relieve inflammation by quelling TLR activity and cytokine release must not, at the same time, undercut the body's defense against infection.

Anti-inflammatory drugs that interfere with TNF-alpha, one of the cytokines produced as a result of TLR4 activation, offer a cautionary tale. TNF-alpha produced during infection and inflammation can accumulate in the joints of patients with rheumatoid arthritis. The anti-inflammatory compounds alleviate the arthritis, but some people taking them wind up with tuberculosis. The infection is probably latent, but reining in the inflammatory response can also dampen the pathogen-specific responses and allow the bacterium to reemerge.

In short, TLRs are like the volume knob on a stereo, balancing adaptive immunity and inflammation. Researchers and pharmaceutical companies are now looking for ways to tweak these controls, so they can curtail inflammation without disabling immunity.

Given that TLRs were unheard of seven years ago, investigators have made enormous progress in understanding the central role these proteins play in the body's first line of defense. Innate immunity, long shrouded in oblivion, has suddenly become the belle of the ball.

More to Explore

Innate Immunity. Ruslan Medzhitov and Charles Janeway in *New England Journal of Medicine,* Vol. 343, No. 5, pages 338–344; August 3, 2000.

Inferences, Questions and Possibilities in Toll-like Receptor Signaling, Bruce Beutler in *Nature,* Vol. 430, pages 257–263; July 8, 2004.

Toll-like Receptor Control of the Adaptive Immune Responses. Akiko Iwasaki and Ruslan Medzhitov in *Nature Immunology,* Vol. 5, No. 10, pages 987–995; October 2004.

TLRs: Professor Mechnikov, Sit on Your Hat. L.A.J. O'Neill in *Trends in Immunology,* Vol. 25, No. 12, pages 687–693; December 2004.

LUKE A. J. O'NEILL received his PhD in pharmacology from the University of London in 1985 for work on the pro-inflammatory cytokine interleukin-l. O'Neill is Science Foundation Ireland Research Professor and head of the department of biochemistry at Trinity College in Dublin. He is founder of Opsona Therapeutics, a drug development company in Dublin.

How Cells Clean House

Worn-out proteins, malfunctioning organelles, invading microorganisms: All are swept up by tiny internal "vacuum cleaners" that keep a living cell healthy. If the process, called autophagy, can be kept in good working order, aging itself might be delayed.

VOJO DERETIC AND DANIEL J. KLIONSKY

E very once in a while biologists come to realize that what was at one time regarded as a minor and relatively obscure cellular process is, in fact, of central importance. Not only is the process ubiquitous, but by virtue of that ubiquity it also plays a role in a broad range of normal and disease states. So it was with the discovery of the role of nitric oxide in the circulatory system, a discovery that led to a Nobel Prize, as well as to many beneficial drugs. Now another formerly obscure process known as autophagy is suddenly claiming extraordinary scientific attention.

In basic outline, autophagy (from the Greek, meaning "self-eating") is simple enough. Within every cell but outside the nucleus lies the cytoplasm, a kind of formless jelly supported by a skeletal matrix, in which a vast and intricate population of large molecules, or macromolecules, and specialized functional subunits called organelles is suspended. The workings of the cytoplasm are so complex—rather like some of today's computer systems—that it is constantly becoming gummed up with the detritus of its ongoing operations. Autophagy is, in part, a cleanup process: the trash hauling that enables a cell whose cytoplasm is clotted with old bits of protein and other unwanted sludge to be cleaned out.

Refurbishing the cytoplasm can give new life to any cell, but it is particularly important to cells such as neurons that do not get replaced. A neuron that must live as long as the organism that hosts it has virtually no other way to renew and maintain its operations. Cell biologists have also determined that autophagy acts as a defense against harmful viruses and bacteria. Any foreign object or organism that evades the extracellular immune system and makes its way through the cell membrane into the cytoplasm becomes a potential target for the autophagy system.

By the same token, when autophagy runs too slow, runs too fast or otherwise malfunctions, the consequences can be dire indeed. Many of the millions of people who suffer from Crohn's disease, a form of inflammatory bowel disease, may have defective autophagy systems that cannot keep the microbial flora in the gut from growing uncontrollably. A breakdown in the autophagy system in

Key Concepts

- Inside the cytoplasm of a living cell, organelles called autophagosomes continually engulf bits of cytoplasm, along with damaged cell parts and invading bacteria and viruses. The "sweepings" are carried to digestive organelles for breakup and recycling. The process is called autophagy.
- Cell biologists are learning about autophagy in great detail by tracing the protein signals that drive and control the process.
- A fuller understanding of autophagy is opening up new options for treating cancer, infectious disease, immune disorders and dementia, and it may one day even help to slow down aging.

—The Editors

brain neurons has been linked to Alzheimer's disease, as well as to aging itself. Even a well-oiled autophagy system can be detrimental, enabling a cancer cell targeted by a blast of radiation or a toxic dose of chemotherapy to survive and repair itself, thereby perpetuating the cancer. Autophagy can sometimes act to eliminate a diseased cell for the greater good of the organism, but it can also become overzealous, consuming a cell even when the loss of that cell is not in the interest of the organism.

In the past decade investigators have been able to learn in great detail how the autophagy system works. Such insights are important not only because they enhance the basic understanding of how cells work, but also because they could lead to the design of drugs that might induce the system to ramp up or quiet down as needed. Controlling the rates of the process as well as the specific targets of its activities could have enormous therapeutic benefits and might even alleviate some of the decline in brain functioning people experience as they age.

Rescue Squad Turned Cleanup Crew

Biologists apply the term "autophagy" to several related processes, but here we mean the kind of cleanup technically known as macro-autophagy that has been most thoroughly studied so far. The process begins as various proteins and lipids, or fats, in the cytoplasm form sheets of double-layered membrane [*see box "Autophagy, Step by Step"*]. The sheets of membrane curl up on themselves into an open-ended globule that simply engulfs bits of cytoplasm along with whatever might be inside them. The globule, called a phagophore, then seals itself into a closed capsule known as an autophagosome. The autophagosome generally ferries its cargo to a lysosome, a kind of disposal plant, elsewhere within the cytoplasm. Typically the two organelles fuse into an "autolysosome," where the autophagosome gives up its cargo to the "digestive juices" of the lysosome. The useful molecular pieces that remain after digestion are recycled back into the cytoplasm.

In a general way, the process as an ongoing cellular activity has been recognized at least since the 1960s, when Christian de Duve of the Rockefeller University and others studied it under the electron microscope. Ten years ago one of us (Klionsky) and others (particularly Yoshinori Ohsumi of the National Institute for Basic Biology in Okazaki, Japan, and his co-workers) began to study its molecular biology in yeast, which is far simpler than studying the same function in higher animals. That strategy has exposed many of the otherwise elusive details of the autophagic machinery because many of the proteins that take part in autophagy or regulate it are virtually identical to their counterparts in people, having remained little changed throughout evolution.

Autophagy itself may have evolved as a response to cell starvation or as a primitive immune defense, or both. To appreciate the need for a starvation response, think about what happens when an entire organism is deprived of food. If a person restricts food intake, the body does not immediately cease functioning and die; instead it starts to break down its own nutritional reserves. Fat cells can go first, but ultimately even muscle cells are broken up and fed to the metabolic fires to keep essential processes running.

Similarly, when cells starve they, too, break down parts of themselves to maintain their essential activities. Autophagosomes are active continuously, whether a cell is starving or not, engulfing bits of cytoplasm and so repeatedly renewing much of the cytoplasmic content. But several kinds of stress—starvation, the absence of growth factors or lack of oxygen, to name a few—signal the cell to speed up its assembly of autophagosomes. Hence, when nutrients are scarce, autophagy intensifies; autophagosomes scavenge the cytoplasm for proteins and organelles (regardless, it seems, of their functional status) that can be digested into nutrients and energy the cell can use.

If autophagy evolved, in part, as a response to starvation, its housekeeping function—even when nutrients abound—has long since become just as vital to the cell. Autophagosomes help to rid the cell of various kinds of unwanted denizens of the cytoplasm. Proteins, for instance, which carry out all the work of the cell, are sometimes put together incorrectly, and they can "wear out" with time. As a result, they may not function or, worse, may malfunction. If so, they must be culled before they cause a problem. Continuous autophagy keeps their concentrations at a low level.

How It Works
Autophagy, Step by Step

Removal and degradation of intracellular material—often waste—are carried out by vesicles called autophagosomes that form in the cytoplasm, the jellylike material surrounding the cell nucleus. Here a damaged mitochondrion, an organelle that chemically packages energy to power metabolic processes, is swept up by an autophagosome and carried to another organelle called a lysosome that breaks down the cargo. By tracing proteins in the cell, the authors and their colleagues have been unraveling the details of the process.

1. **Induction:** A wide range of signals at the cell's outer membrane can speed up the baseline rate at which autophagy proceeds.
2. **Nucleation:** A double-layered membrane called a phagophore forms out of various large molecules in the cytoplasm.
3. **Expansion and Cargo Recognition:** The phagophore expands and closes in on itself, probably by adding new sheets of membrane. It then surrounds and engulfs a bit of cytoplasm, along with, perhaps, a damaged protein or organelle.
4. **Protein Recycling:** The double-layered membrane seals, and the resulting autophagosome sheds membrane proteins that took part in its formation. The proteins are cycled back into the cytoplasm.
5. **Fusion:** The outer layer of the autophagosome membrane fuses with a lysosome; enzymes in the lysosome cut through the inner membrane layer of the autophagosome.
6. **"Digestion":** Enzymes in the fused "autolysosome" gain access to the cargo of the autophagosome and break it down.
7. **Cargo Recycling:** The chemical building blocks of the cargo, particularly amino acids, are released back into the cytoplasm for reuse.

Autophagosomes not only remove damaged proteins, but they also seek out and sequester damaged organelles many times the size of a protein. Mitochondria, for instance, are the organelles primarily responsible for generating energy within a cell, and they can send signals to other parts of the cell that initiate apoptosis, or cellular suicide.

Cells induce apoptosis for a variety of reasons, all more or less for the greater good of the organism. For example, the body continually generates more cells than it needs, and they must be eliminated. An aging cell that has ceased functioning efficiently may kill itself to make room for younger, more robust cells. A cell that switches from normal growth to cancerous proliferation can also be induced to commit suicide, making apoptosis one of the most important built-in barriers against cancer. Apoptosis depends on a complex series of cellular events, rigorously orchestrated by numerous protein signals, and so the death of the cell by apoptosis is considered to be a programmed event.

But a faulty mitochondrion can wreak havoc if it sets off apoptosis at the wrong time [*see box "Making the Ultimate Decision"*]. Among the by-products of a functioning mitochondrion are reactive oxygen species (ROS)—oxygen ions and other oxygen-based molecular fragments. Working with such volatile chemicals often causes mitochondria to leak some of their contents, including the signaling proteins that initiate apoptosis. In other words, a minor flaw in a small part of the cell can lead, inadvertently, to the death of the entire cell. The accidental cellular demise of a few skin cells might not be a big deal, but such a loss of memory neurons in the brain would definitely spell trouble.

Autophagy is a fail-safe against such a destructive mistake. Autophagosomes can remove damaged mitochondria and other kinds of organelles from the cytoplasm and ensure that they are destroyed by lysosomal enzymes in an autolysosome before they can induce an unscheduled programmed cell death—or, worse, the disorganized cellular demise known as necrosis.

Mitochondria can also release ROS into the cytoplasm, which, as the name "reactive oxygen species" implies, tend to react with many other molecules. In a healthy cell ROS levels are kept under control by antioxidant molecules that scavenge ROS. According to Shengkan V. Jin of the University of Medicine and Dentistry of New Jersey, however, when mitochondria become damaged, they can flood the cell with 10 times the usual release of ROS, much more than normal cellular detoxification systems can handle. The escape of such large amounts of ROS poses a cancer threat, because ROS that reach the nucleus may induce malignant changes in genes. Once again, autophagy can come to the rescue, removing the dysfunctional mitochondria from the cell. Eileen White of Rutgers University believes that autophagy also mitigates genome damage in cancer cells, thereby helping to prevent new tumors from forming.

Double-Edged Sword

Soon after cell biologists unraveled the intricate molecular pathways of apoptosis, they recognized that cells can kill themselves by other means as well. Autophagy became a prime suspect. Current nomenclature reflects that history: apoptosis is also known as programmed cell death type I; autophagy is sometimes referred to as programmed cell death type II—although that designation remains controversial.

Autophagy could lead to cell death in two ways: the process might simply continue digesting the contents of the cytoplasm until the cell dies, or it may stimulate apoptosis. But why would a process that often prevents untimely cell death from accidental apoptosis sometimes be invoked to cause cell death itself? The puzzle may turn out to have a fascinating resolution. Apoptosis and autophagy may be closely interrelated and carefully balanced. For example, if organelle damage is too extensive for autophagy to bring under control, the cell must die for the sake of the entire organism. The cell may then rely on either of its suicide programs: it may allow autophagy to continue to the end, or it can signal for apoptosis, holding autophagy as a backup system if apoptosis is compromised. Two of the most intense

---

Surviving Starvation

Autophagosomes are constantly consuming parts of the cytoplasm, but nutrient scarcity boosts their baseline number. That increase speeds up the rate at which intracellular components, including intact proteins and other macromolecules, are digested by autolysosomes into basic biochemical building blocks that are delivered to the cytoplasm as nutrients. The nutrient scarcity also signals the cell to reduce its functioning volume. Without such literal "self-eating," the essential activities of the cell could not continue and the cell would die.

---

and somewhat controversial areas of current investigation are how autophagy and apoptosis interconnect and whether autophagy on its own should be considered a pathway for cell death.

Does autophagy contribute mainly to cell survival—or does it also act as an "angel of death"?

Work at the molecular level may help resolve whether autophagy is primarily a pathway for cell survival or whether it can, in addition, act as an "angel of death." Recent studies by Beth Levine of the University of Texas Southwestern Medical Center at Dallas and Guido Kroemer of the French National Scientific Research Center (CNRS) have shown how the two processes can be coordinated. One of the proteins that signals for autophagy to begin, known as Beclin 1, binds with a protein that prevents apoptosis from starting, Bcl-2. Life-and-death decisions are made as bonds between the two kinds of proteins are enhanced or broken. Levine's findings of that connection between autophagy and apoptosis have been further supported by the discovery that a fragment of a protein known as Atg5, which plays a leading role in the formation of autophagosomes, can make its way to mitochondria. Once there Atg5 can switch what was initially a purely autophagic response to an apoptotic one.

Every benefit seems to have its flaws, and autophagy is no exception. We noted earlier that cancer cells can sometimes invoke autophagy to save themselves. Anticancer treatments are often aimed at inducing malignant cells to commit suicide. Yet some cancer cells can defend against the treatments because autophagy jumps in to remove damaged mitochondria before they can trigger apoptosis. In fact, radiation and chemotherapy can actually induce higher-than-usual levels of autophagy.

Cancer cells can also take advantage of autophagy to avoid being starved. Few nutrients can reach the inside of a tumor, but as we mentioned earlier, a shortage of nutrients can trigger autophagy, prolonging the life of a cancer cell by enabling it to break down its own macromolecules for food. A straightforward treatment strategy might therefore be to suppress autophagy within a tumor or during radiation therapy or chemotherapy. Drugs for that purpose are in clinical trials. Unfortunately, as White points

Live or Let Die?
Making the Ultimate Decision

The last act of a badly damaged cell can be to trigger its own death for the greater good of the organism. One suicidal pathway called apoptosis begins when mitochondria in the cytoplasm release signaling proteins. Some investigators have proposed that autophagy can act to save the cell from unnecessary apoptosis. Paradoxically, autophagy may also act as a second suicidal pathway when cell death is needed but apoptosis fails. Moreover, apoptosis and autophagy share certain kinds of signaling proteins, suggesting that the two processes engage in cross talk and may best be regarded as parts of a more comprehensive system within the cell.

Autophagy as Safety Net

A damaged mitochondrion can send a spurious signal for the cell to begin apoptosis, even though the cellular damage is minimal. Autophagy can prevent the signal from causing unnecessary cell suicide.

1. Signaling proteins that initiate apoptosis leak out of a damaged mitochondrion in the cytoplasm of an otherwise healthy cell
2. An autophagosome engulfs the damaged mitochondrion and the leaking signaling proteins, squelching the spurious signal
3. The autophagosome fuses with a lysosome and the toxic cargo is destroyed, saving the healthy cell

Autophagy as "Decider"

In a badly damaged cell the system for triggering cell suicide responds dynamically to signals of stress. In the end, autophagy may throttle down, enabling the cell to survive; continue devouring the cell from the inside until it dies; signal for cell suicide by apoptosis or, if apoptosis fails, serve as a suicidal backup to prevent the disorganized cell demise known as necrosis.

out, suppressing autophagy could boost the number of genetic mutations in cancer cells and so increase the chances of a relapse. It may take some fine-tuning to get the treatments right.

Preventing Neuron Breakdown

Given the role of autophagy in keeping the cytoplasm clear of detritus and malfunctioning parts, it is hardly surprising that the process turns out to be particularly important to the well-being of long-lived cells such as neurons. Inefficient autophagy plays a pivotal role in neurodegenerative disorders such as Alzheimer's, Parkinson's and Huntington's diseases. All three cause slow but inexorable changes in the brain, but Alzheimer's, a form of dementia that afflicts 4.5 million people in the U.S. alone, is the most common.

One of the most frequent effects of normal aging is the accumulation of a brownish material called lipofuscin, a mix of lipids and proteins, in the bodies of brain cells. Superficially, the stuff can

be likened to liver spots on aging skin. The accumulation of such material, according to Ralph A. Nixon of the Nathan S. Kline Institute for Psychiatric Research, is a sign that aging brain cells can no longer remove abnormally modified or damaged proteins fast enough to keep pace with their buildup. In Alzheimer's patients, a yellowish or brownish pigment called ceroid also builds up inside neurites, or projections from nerve cell bodies. The neurites swell where ceroid collects, and amyloid, or senile, plaques characteristic of the disease form on the outside of the swollen neurites.

So far investigators have not fully deciphered the exact ways senile plaques or their precursors lead to neuron damage. But the latest research shows, tellingly, that enzymes that help to deposit the plaques in certain early-onset forms of Alzheimer's are present on the membranes of autophagosomes. According to Nixon, such plaques may stem in part from incomplete autophagy and the consequent failure of the neurons to digest substances that would normally be swept up from their cytoplasm, broken down and recycled for parts [see box "When the Cleaning Stops"]. Supporting Nixon's conclusion, electron micrographs of senile plaques in the brains of Alzheimer's patients show massive numbers of immature autophagosomes accumulating inside the parts of the neurons nearest the plaques. Precisely how the plaque material may collect on the outside of nerve cells has not been conclusively traced.

Given those results, it would seem that any means of promoting autophagy might slow the onset of the debilitating symptoms of Alzheimer's. Regretfully, however, no one yet knows whether activating autophagy in Alzheimer's patients would have any benefit, if the treatment cannot also ensure that autophagosomes fuse with lysosomes. But the good news is that such a treatment might be effective for Huntington's patients. A drug known as rapamycin, or sirolimus, which suppresses immunity and is used to block the rejection of organ transplants, particularly kidney transplants, turns out to induce autophagy as well. Rapamycin is now being tested for its effectiveness in stimulating autophagy to remove a kind of protein aggregate seen in Huntington's patients.

Getting Bugs out of the System

If an autophagosome can capture and destroy a leaky, cell-endangering mitochondrion, couldn't it do the same to unwanted parasites that invade the cellular interior—bacteria, protozoa and viruses that manage to get through the cell membrane? In fact, that hypothesis was recently verified experimentally. Taken together, studies by one of us (Deretic) and, nearly simultaneously, by two groups in Japan, one led by Tamotsu Yoshimori of Osaka University, the other by Chihiro Sasakawa of the University of Tokyo, have shown that autophagy can eliminate a diverse range of pathogens. The list includes *Mycobacterium tuberculosis,* the tuberculosis bacterium annually responsible for two million deaths worldwide; gut pathogens such as *Shigella* and *Salmonella;* group A streptococci; *Listeria,* which occurs in raw-milk cheeses; *Francisella tularensis,* which the Centers for Disease Control and Prevention has listed as a bioterrorism agent; and parasites such as *Toxoplasma gondii,* which is a major cause of illness in people with AIDS.

Yet just as cancer cells can exploit autophagy for their own survival, some microorganisms have evolved ways to subvert the process. For example, *Legionella pneumophila,* which causes Legionnaires'

Autophagy in Alzheimer's? When the Cleaning Stops

In an aging brain neuron, autophagosomes can fail to complete their development, leading to a buildup of damaged proteins and consequent swelling in a neurite, or projection from the cell body of the neuron. The immature autophagosomes collect at the same site. Enzymes that create protein fragments called amyloid beta seem to concentrate on the immature autophagosomes, and those fragments collect on the outer neurite surface. Aggregates of amyloid beta are the so-called senile plaques characteristic of neurons in the brains of Alzheimer's patients. Together those findings suggest that a breakdown in autophagy may contribute to Alzheimer's disease.

Cell Defense Repelling Invaders

Autophagy can mount several kinds of defenses against pathogens that enter the cytoplasm.

Pathogen Degradation
Vesicle that buds off the cell membrane with an invading microorganism inside can be "swallowed whole" by an autophagosome and digested into harmless fragments by a lysosome.

Innate Immune Response
1. Virus that evades the first line of autophagosome defenses releases its nucleic acid (RNA, for instance).
2. An autophagosome delivers some of the viral RNA to an endosome, or compartment in the cell.
3. Viral RNA in the endosome binds with a TLR, stimulating production of more autophagosomes and interferon (an "innate" response) that can interfere with viral replication.
4. The cell translates some of the remaining viral RNA into viral protein.
5. An autophagosome delivers viral protein to another kind of endosome, where the protein is broken up.
6. A fragment of viral protein is loaded onto an MHC II molecule and presented at the cell surface, triggering a specific "adaptive" immune response against cells infected with the virus.

disease, is a bacterium that readily gets inside a cell. But if *L. pneumophila* bacteria are engulfed by an autophagosome, they can delay or even prevent the autophagosome from fusing with a lysosome. Thus instead of serving as a vehicle that helps to rid the cell of a pathogen, the infected organelle becomes a niche where the bacteria can replicate, using the sequestered cytoplasm as a nutrient supply.

Some micro-organisms have learned to subvert autophagy. HIV can even accelerate the process in neighboring immune system cells, causing them to commit suicide.

The very existence of such clever evolutionary tactics is good evidence that autophagy has long functioned as a major barrier to invasion by pathogens and their replication in human cells—a barrier that disease-causing agents must overcome to survive. Not surprisingly, HIV is another good example of a pathogen that can harness autophagy for its own purposes. Two groups in France, one led by Martine Biard-Piechaczyk of the Center for Studies of Pathogenic Agents and Biotechnologies for Health and the other by Patrice Codogno of INSERM, have jointly shown that HIV, which infects immune system cells known as CD4$^+$ T cells, can increase cell death in uninfected "bystander" cells of the same kind. As HIV enters a cell, it sheds its outer envelope, and the protein that makes up the envelope induces uncontrolled, excess autophagy and then apoptosis in cells that surround the HIV-infected cell. Thus by activating autophagy in "innocent" bystander cells, HIV further reduces the number of healthy CD4$^+$ T cells in the body. Eventually the catastrophic loss of immune system cells brings about full-blown AIDS.

The Immune Connection

Autophagy not only eliminates pathogens directly; investigators have also found that it takes part in immune responses [see box "Repelling Invaders"]. For example, autophagosomes help to

deliver pathogens or pathogen products to membrane molecules called toll-like receptors (TLRs), a subset of the regulators that control the so-called innate immune response. The role of autophagosomes in the process is to make a clever "topological" inversion. A pathogen in the cytoplasm can hide from TLRs because TLR binding sites for pathogens face away from the cytoplasm. The binding sites point either toward the space outside the cell or toward the inside of an endosome, or intracellular compartment. But autophagosomes can fix this topological problem by scooping up pathogens or their parts from the cytoplasm and delivering them to an endosome that embeds TLRs in its membrane. There the pathogen molecules meet TLRs at last. Their encounter signals the cell to produce chemicals called interferons, which act, for instance, to suppress the replication of the pathogen. This innate immune response is generated to combat infection as soon as it starts—no time is needed for the cell to build a highly specific response to the pathogen.

But autophagosomes can also help build that highly specific immune response, known as adaptive immunity. For example, when a virus invades the cytoplasm and tricks the cell into making viral protein, an autophagosome engulfs some of the viral protein and ushers it into another kind of endosome that embeds so-called MHC class II molecules in its membrane. Once inside that endosome, the viral protein is partly broken up, and a piece of it is loaded onto a part of an MHC class II molecule that faces

Autophagy in Medicine
New Weapons against Disease

Intensifying, suppressing or otherwise manipulating autophagy in specific kinds of cells could become a powerful part of the medical arsenal. Here are just a few examples of the potential treatment options.

Disease	Strategy	Goals
Cancer	Inhibit autophagy in cells of cancerous tumors	Help to prevent tumor cells from consuming the contents of their own cytoplasm, thereby surviving in oxygen- or nutrient-starved environments
Cancer	Enhance autophagy in cells at risk of cancer	Lower the chances that mutations and secondary tumors will arise when too little autophagy enables DNA-damaging molecules to accumulate in the cell
Huntington's disease	Enhance autophagy with drug rapamycin (sirolimus)	Help to remove toxic microaggregates of proteins that accumulate in nerve cells
Tuberculosis	Enhance autophagy	Kill disease-causing agents that hide in the cytoplasm, both in people who are sick and in carriers who are symptom-free

the inside of the endosome. (Just as with the TLR, the MHC class II molecule would not meet properly with the pathogen molecule if the autophagosome did not bring the pathogen molecule inside the endosome.) Once the MHC class II molecule is bound to the pathogen fragment and the assemblage is transported to the surface of the cell, the immune system begins mounting an adaptive immune response, a slower but far more specific and more efficient response than innate immunity can muster.

Long Life?

Remarkably, autophagy may also play a role in determining the human life span. Most people take it for granted that many diseases become more frequent with age, including cancer and the degeneration of neurons. The reason, in part, may be a decline in the efficiency of autophagy. According to Ana Maria Cuervo of the Albert Einstein College of Medicine, the current thinking is that cellular

systems, including autophagy, undergo a steady loss of function with age. In particular, the systems that remove aberrant or dysfunctional proteins and organelles begin to work less efficiently, and the resulting buildup of damaged cellular components leads to disease.

If inefficient autophagy is to blame, Cuervo says, that could help explain why caloric restriction has been found to extend average life spans in several kinds of experimental animals. The less food such animals eat (provided they get an adequate supply of essential nutrients), the longer they live, and the same may be the case for people. Recall that a restricted food supply—incipient starvation—speeds up autophagy. Hence, caloric restriction as one ages might offset the natural age-related decline of autophagy and so prolong the essential housekeeping function of the process in cells. Furthermore, Cuervo adds, recent research shows that if you can prevent the decline of autophagy in experimental animals, you can often avoid the usual age-related buildup of proteins damaged by reactions with oxygen compounds.

What was once seen primarily as a hedge against cellular starvation has come to be recognized as central to a broad range of factors affecting human health and disease. Research into autophagy is expanding in new and unexpected directions, generating an exponentially increasing body of scientific knowledge. But we have only begun. Learning to promote or inhibit autophagy at will holds great promise for the treatment of disease and perhaps even for slowing down the natural process of aging. But whether autophagy can be harnessed to benefit health, much less to become the elusive fountain of youth, will depend on gaining a fuller understanding of its mechanisms and of the intricate biochemical signals on which it depends.

More to Explore

Cell Suicide in Health and Disease. Richard C. Duke, David M. Ojcius and John Ding-E Young in *Scientific American*, Vol. 275, pages 80–87; December 1996.

Autophagy in Health and Disease: A Double-Edged Sword. T. Shintani and D. J. Klionsky in *Science*, Vol. 306, pages 990–995; November 5, 2004.

Autophagy in Immunity and Infection: A Novel Immune Effector. Edited by Vojo Deretic. Wiley-VCH, 2006.

Potential Therapeutic Applications of Autophagy. D. C. Rubinsztein, J. E. Gestwicki, L. O. Murphy and D. J. Klionsky in *Nature Reviews Drug Discovery*, Vol. 6, pages 304–312; April 2007.

VOJO DERETIC is a professor and chair of the molecular genetics and microbiology department at the University of New Mexico Health Sciences Center; he also holds a joint appointment there as a professor of cell biology and physiology. He was educated in Belgrade, Paris, and Chicago. Deretic is fascinated with autophagy both as a fundamental biological process and as an effector of innate and adaptive immunity.
DANIEL J. KLIONSKY is Alexander G. Ruthven Professor of Life Sciences at the University of Michigan Life Sciences Institute. He is a former fellow of the John Simon Guggenheim Memorial Foundation, a National Science Foundation Distinguished Teaching Scholar, and editor in chief of the journal *Autophagy*.

Start Early to Prevent Genital HPV Infection—And Cervical Cancer

ELIZABETH HEAVEY, RN, CNM, PhD

In June 2006, the Food and Drug Administration (FDA) approved the human papillomavirus (HPV) vaccine (Gardasil), which protects against four strains of HPV. Although most HPV infections are asymptomatic and transient, four HPV strains, 6, 11, 16, and 18, together are responsible for 70% of cervical cancers and 90% of genital warts.[1] Transmitted by direct or indirect genital contact with an infected person, genital HPV infection also rarely causes respiratory tract warts in children.[1]

The Gardasil vaccine can be given to girls as young as age 9. In this article, I'll discuss what to teach patients and their parents about HPV and vaccination.

Common Threat

Genital HPV infection is the most common sexually transmitted infection in the United States, with approximately 20 million people currently infected. Approximately 6.2 million Americans become newly infected each year, and more than half of sexually active adults will be infected during their lifetime.[1] Over 99% of cervical cancers have HPV DNA isolated in the tumor cells. In 2008, over 11,000 women are expected to be diagnosed with invasive cervical cancer, and nearly 3,900 American women are expected to die of cervical cancer.[2]

Women who are from minority or low socioeconomic groups and who lack insurance are disproportionately affected; they also have the lowest rates of screening with Pap tests. Of the women diagnosed with cervical cancer, an estimated 50% were never screened and an additional 10% hadn't been screened for at least 5 years before diagnosis.[1]

Primary Prevention Pointers

The first line of primary prevention is patient education. Teach both female and male patients about the risks of sexually transmitted diseases and about protective sexual behaviors, including abstinence, monogamy, and condom use. Encourage your patients to talk about these issues with their partner and to negotiate sexual decision making before they become sexually active. Keep in mind that as many as half of HPV-infected males and females are between the ages of 15 and 24.[1]

Another Vaccine in the Pipeline

Another cervical cancer vaccine, Cervarix, is currently undergoing clinical trials in the United States to determine its safety and effectiveness. This vaccine targets HPV types 16 and 18. Like Gardasil, this vaccine can't give anyone the infection. It's also given by I.M. injection, and the recommended schedule is a three-dose series. It's been approved in the European Union, Australia, the Philippines, Mexico, and Singapore, among other countries.

Educate your patients about these HPV facts:

- Genital HPV is usually transmitted through direct skin-to-skin contact, most often during penetrative genital contact (vaginal or anal sex).[3]
- Women whose partners consistently use latex condoms have lower rates of cervical cancer.[3,4] However, using latex condoms doesn't eliminate the risk of transmitting HPV and other sexually transmitted diseases.

Teach patients these points about the vaccine:

- It doesn't protect against all types of HPV that cause cervical cancer or genital warts.
- It's not intended for the treatment of cervical cancer.
- It's not a "live" vaccine and it contains no viral DNA, so no one can get infected with HPV from the vaccine.
- After receiving it, women need to continue regular cervical cancer screening.

The Centers for Disease Control and Prevention (CDC) recently adopted the recommendations of its Advisory Committee on Immunization Practices, which recommends administering the HPV vaccine to girls ages 11 and 12. The vaccine has been studied and safely administered to girls as young as 9 and women up to age 26. Current studies predict immunity will last for 5 years and the efficacy level for the four strains of HPV is almost 100%.[1]

Administering the Vaccine

Gardasil is administered in three I.M. injections over a 6-month period. The second and third doses should be given 2 and 6 months after the first dose. According to the CDC's latest guidelines, there must be a minimum of 4 weeks between doses one and two and a minimum of 12 weeks between doses two and three, *and* a minimum of 24 weeks between doses one and three for adolescents on a catch-up schedule.[5] However, adhering to any schedule can be a problem for adolescents. Missing doses or getting them at the wrong intervals may result in a lower level of immunity or none at all.

Very few adverse reactions to Gardasil have been reported; the most common is local discomfort at the injection site. The vaccine contains no thimerosal or mercury. At this time, it's indicated for girls and women 9 to 26 years of age but is being studied for male patients. It isn't recommended during pregnancy.

Pap Tests Still Needed

Secondary prevention for cervical cancer includes regular screening with a conventional or liquid-based Pap test.[3] Screening should begin approximately 3 years after having vaginal intercourse for the first time, but no later than 21 years of age. For additional screening guidelines, see the American Cancer Society Guidelines for the Early Detection of Cancer.[6]

A Pap test (also called a Pap smear) is used to collect cervical cells for cytology testing. A Pap test can identify cancerous or precancerous abnormalities, as well as cellular changes related to infection.[6]

In March 2000, the FDA approved the HPV DNA test for use in women who had abnormal Pap test results. In 2003, the FDA approved expanded use of the test, allowing it to be used for screening in conjunction with the Pap test in women over age 30.[7] The HPV DNA test, like the Pap test, is performed by collecting cervical cells for cytology testing. lt's important to remember that the HPV DNA test isn't a substitute for regular Pap screening.

Challenges to Overcome

The Gardasil vaccine currently costs approximately $120 per injection, making the entire series $360.[1] Children and adolescents under age 19 who are eligible for Medicaid, who don't have health insurance, or who are American Indian or Alaska natives can get it for free through the federal Vaccines for Children program.[1]

We must increase the use of this vaccine in pediatric settings, school health clinics, and family practice settings. One study found that more than 42% of urban adolescents had engaged in vaginal intercourse before the age of 14.[8] A recent study of girls ages 14 to 19 found that 18% of them were infected with HPV. Of teenage girls in this study who admitted having had sex, 40% had a sexually transmitted disease.[9]

To obtain a sexual history and discuss sexual behavior with young women, you need to collect information openly and with great sensitivity. To improve your comfort level and that of your colleagues, try role playing. Practice asking open-ended questions and identify personal barriers to these discussions.

Why So Young?

Giving the Gardasil vaccine to a minor requires parental consent. Parents may ask you why a preadolescent girl needs a vaccine that's intended to prevent sexually transmitted infection. Inform them that the vaccine is two to three times more effective when administered by age 11.[1] If it's given before she engages in sexual activity, it helps protect her against four strains of the virus, including two that cause cervical cancer. Reassure parents that vaccination doesn't encourage sexual activity and reinforce the fact that abstinence from sexual activity is the only sure way to avoid sexually transmitted diseases.

If a girl is vaccinated when she's young, she may need to have titers drawn and receive boosters during later adolescence, Data are still pending and guidelines may change, but currently there aren't any indications that a booster dose wilt be needed. Advise parents to follow their health care provider's recommendations.

Getting Parents on Board

Teach patients and their parents about the vaccine during routine checkups. In my practice, I've found that many parents are uncomfortable addressing their child's sexuality, so it's imperative that you initiate the discussion. Encourage parents to talk openly with their child about responsible sexual behavior and their family's values.

References

1. Centers for Disease Control and Prevention. HPV and HPV vaccine—information for healthcare providers. http://www.cdc .gov/std/HPV/STDfactHPV-vaccine-hcp.htm. Accessed March 13, 2008.
2. American Cancer Society. *Cancer Facts and Figures 2008*. http://www.cancer.org/docroot/stt/stt_0.asp. Accessed March 13, 2008.
3. *Human Papillomavirus; HPV lnformation for Clinicians* brochure. Centers for Disease Control and Prevention, April 2007. http://www.cdc.gov/STD/Hpv/hpv-clinicians-brochure .htm. Accessed March 12, 2008.
4. Centers for Disease Control and Prevention. Sexually transmitted treatment guidelines, 2006. *MMWR Recommendations and Reports.* 55(RR11): l-94, August 4, 2006.
5. Centers for Disease Control and Prevention. Catch-up immunization schedule for persons aged 4 months–18 years who start late or who are more than one month behind. http://www.cdc.gov/vaccines/recs/schedules/downloads/ child/2008/08_catch-up_schedule_pr.pdf. Accessed March 25, 2008.

6. American Cancer Society Guidelines for the Early Detection of Cancer. http://www.cancer.org/docroot/ped/content/Ped_2_3x_acs_cancer_detection_guidelines_36.asp. Accessed March 11, 2008.

7. Food and Drug Administration. FDA approves expanded use of HPV test. *FDA News.* March 31, 2003, http://www.fda.gov/bbs/topics/NEWS/2003/NEW00890.htm1. Accessed March 13, 2008

8. Ampad DC, et al. Predictors of early initiation of vaginal and oral sex among urban young adults in Baltimore, Maryland. *Archives of Sexual Behavior.* 35(I):53–65, February 2006.

9. Centers for Disease Control and Prevention, 2008 National STD Prevention Conference. Nationally representative CDC study finds 1 in 4 teenage girls has a sexually transmitted disease. March 11, 2008. http://www.cdc. gov/stdconference/2008/media/release-11 march2OO8.htm. Accessed March 19, 2008.

ELIZABETH HEAVEY is an assistant professor of nursing at State University of New York College at Brockport.

How Safe Are Vaccines?

Parents worried that vaccines trigger autism are increasingly declining the shots for their kids. That's raising fears that long-dormant diseases could return. What the science says about the real risks—and what you should do.

ALICE PARK

L ife, if you're a bacterium or virus, boils down to this: finding a pristine human home to provide for your every need, from food and nutrients to shelter against biological storms. As a microbial drifter, you can literally travel the world, hopping from host to host when the opportunity presents itself or when conditions at your temporary residence start heading south. There's no worry about taking along life's necessities either—viruses in particular are adept at traveling light; incapable of reproducing on their own, they think nothing of co-opting the reproductive machinery of their cellular sponsors to help them spawn generation after generation of freeloading progeny.

But ever since Edward Jenner, a country doctor in England, inoculated his son and a handful of other children against smallpox in 1796 by exposing them to cowpox pus, things have been tougher on humans' most unwelcome intruders. In the past century, vaccines against diphtheria, polio, pertussis, measles, mumps and rubella, not to mention the more recent additions of hepatitis B and chicken pox, have wired humans with powerful immune sentries to ward off uninvited invasions. And thanks to state laws requiring vaccinations for youngsters enrolling in kindergarten, the U.S. currently enjoys the highest immunization rate ever; 77% of children embarking on the first day of school are completely up to date on their recommended doses and most of the remaining children are missing just a few shots.

Yet simmering beneath these national numbers is a trend that's working in the microbes' favor—and against ours. Spurred by claims that vaccinations can be linked to autism, increasing numbers of parents are raising questions about whether vaccines, far from panaceas, are actually harmful to children. When the immune system of a baby or young child is just coming online, is it such a good idea to challenge it with antigens to so many bugs? Have the safety, efficacy and side effects of this flood of inoculations really been worked through? Just last month the U.S. government, which has always stood by the safety of vaccines, acknowledged that a 9-year-old Georgia girl with a pre-existing cellular disease had been made worse by inoculations she had received as an infant, which "significantly

Vaccine Tally 28
Number of doses of vaccines American children receive by age 2 if they get the complete schedule of immunizations recommended by the Centers for Disease Control and Prevention.

Staying Protected 77%
Percentage of kindergartners in the U.S. who are completely up to date on their vaccinations, in part because schools require it. This is the country's highest rate of immunization ever.

Opting out 2%–3%
Percentage of school-age children in the U.S. whose parents have received a religious or philosophical exemption from state vaccination requirements.

Autism 1 in 150
The prevalence of autism among 8-year-olds in the U.S. Autism rates have not declined, even though thimerosal, which some believe contributes to the disease, was removed from vaccines in 2001.

Polio 888
Number of polio cases in Nigeria in 2006, after religious leaders convinced parents they should not allow their children to be vaccinated. The country reported fewer than 30 cases in 2000.

aggravated" the condition, resulting in a brain disorder with autism-like symptoms.

Though the government stressed that the case was an exceptional one, it provided exactly the smoking gun that vaccine detractors had been looking for and vaccine proponents had been dreading. More and more, all this wrangling over risks and benefits is leading confused parents simply to opt out of vaccines altogether. Despite the rules requiring students to be vaccinated, doctors can issue waivers to kids whose compromised immune system might make vaccines risky. Additionally, all but two states allow waivers for children whose parents object to vaccines on religious grounds; 20 allow parents to opt out on philosophical grounds. Currently, nearly one-half of 1% of kids enrolled in school are unvaccinated under a medical waiver; 2% to 3% have a nonmedical one, and the numbers appear to be rising.

Parents of these unimmunized kids know that as long as nearly all the other children get their shots, there should not be enough pathogen around to sicken anyone. But that's a fragile shield. Infectious-disease bugs continue to travel the globe, always ready to launch the next big public-health threat. Pockets of intentionally unvaccinated children provide a perfect place for a disease to squat, leading to outbreaks that spread to other unprotected kids, infants and the elderly. Ongoing measles outbreaks in four states are centered in such communities; one originated with an unimmunized boy from San Diego who contracted the virus while traveling in Europe—where the bug was thriving among intentionally unimmunized people in Switzerland. Dr. Anne Schuchat, director of the National Center for Immunization and Respiratory Diseases at the Centers for Disease Control and Prevention (CDC), says, "We are seeing more outbreaks that look different, concentrated among intentionally unimmunized people. I hope they are not the beginning of a worse trend."

If they are, it's possible that once rampant diseases such as measles, mumps and whooping cough will storm back, even in developed nations with robust public-health programs. That is forcing both policymakers and parents to wrestle with a dilemma that goes to the heart of democracy: whether the common welfare should trump the individual's right to choose. Parents torn between what's good for the world and what's good for their child will—no surprise—choose the child. But even then, they wonder if that means to opt for the vaccines and face the potential perils of errant chemistry or to decline the vaccines and face the dangers of the bugs. There is, as yet, no simple solution, but answers are emerging.

The Autism Riddle

More than any other issue, the question of autism has fueled the battle over vaccines. Since the 1980s, the number of vaccinations children receive has doubled, and in that same time, autism diagnoses have soared threefold. In 1998, British gastroenterologist Dr. Andrew Wakefield of London's Royal Free Hospital published a paper in the journal the *Lancet* in which he reported on a dozen young patients who were suffering from both autism-like developmental disorders and intestinal symptoms that included inflammation, pain and bloating. Eight of the kids began exhibiting signs of autism days after receiving the MMR vaccine against measles, mumps and rubella. While Wakefield and his co-authors were careful not to suggest that these cases proved a connection between vaccines and autism, they did imply, provocatively, that exposure to the measles virus could be a contributing factor to the children's autism. Wakefield later went on to speculate that virus from the vaccine led to inflammation in the gut that affected the brain development of the children.

Like the initial tremor that triggers a massive earthquake, Wakefield's theories resonated throughout the autism community, where vaccines had been regarded with suspicion for another reason as well. Ever since the 1930s, a mercury compound known as thimerosal had been included in some vaccines—though not the measles inoculation—as a preservative to keep them free of fungi and bacteria. Thimerosal can do serious damage to brain tissue, especially in children, whose brains are still developing. It was perhaps inevitable that parents would make a connection between the chemical and autism, since symptoms typically appear around age 2, by which time babies have already received a fair number of vaccines. That link could be merely temporal, of course; babies also get their first teeth after they get their first vaccines, but that doesn't mean one causes the other.

In 2001, however, a U.S. Food and Drug Administration study revealed that a 6-month-old receiving the recommended complement of childhood vaccinations was exposed to total levels of vaccine-based mercury twice as high as the amount the EPA considers safe in a diet that includes fish. By the end of that year, thimerosal-free formulations of the five inoculations that included it—hepatitis B, diphtheria, tetanus and pertussis and some versions of *Haemophilus influenzae* type b (Hib)—had replaced the older versions. The result was a drop in mercury exposure in fully immunized 6-month-old babies from 187.5 micrograms to just trace amounts still found in some flu vaccines. Yet there's been no effect on autism rates. In the seven years since the cleaned-up vaccines were introduced, new cases of autism continue to climb, reaching a rate of 1 in every 150 8-year-olds today. That trend suggests that other factors, including heightened awareness of the condition and possible genetic anomalies or environmental exposures, are behind the climbing rates. What's more, in the decade since Wakefield's watershed paper, 10 of its 13 authors have retracted their hypothesis, admitting that the study did not produce solid enough evidence to support a connection between the measles virus in the MMR vaccine and autism.

But the damage had been done. Parents, already uneasy about immunizations, now felt betrayed by government health authorities and a vaccine industry that simply kept the shots coming, with today's kids receiving up to 28 injections for 14 diseases, more than double the number of shots required in the 1970s. "There is no doubt in my mind that my child's first cause of autism is the mercury in vaccines," says Ginny DeLeo, a New York science teacher whose son Evan, born in 1993, was developing normally until he was a year old. The day the boy received his fourth dose of Hib vaccine, DeLeo had to rush him to the hospital with tremors and a 104°F (40°C) fever, which later led to seizures. Evan recovered, and several months later he received the first of two MMR shots. Within months, he stopped talking, and autism was diagnosed.

So, is there a link? In 2003, a 15-person committee impaneled by the CDC and the National Institutes of Health analyzed the available studies on thimerosal and its possible connections to autism and concluded that there was no scientific evidence to support the link. In a further show of confidence, the committee noted that it did "not consider a significant investment in studies of the theoretical vaccine-autism connection to be useful." Instead, the panel recommended that studies focus on less explored genetic or biological explanations for the disease.

There is also little evidence to support the claim made by antivaccine activists that the battery of shots kids receive can damage the immune system rather than strengthen it. Experts stress that it's not the number of inoculations that matters but the number of immune-stimulating antigens—or proteins—in them. Thanks to a better understanding of which viral or bacterial proteins are best at activating the immune system, that number has plummeted. The original smallpox injection alone packed 200 different immune-alerting antigens in a single shot. Today there are only 150 antigens in all 15 or so shots babies get before they are 6 months old. "The notion that too many vaccines can overwhelm the immune system is just not based on good science," says Dr. Paul Offit, chief of infectious diseases at Children's Hospital in Philadelphia.

My Child, My Choice

If the push-back against vaccines were only about the science, doctors might have an easier time making their case. But there's more going on than that. Parents object to the mandatory nature of the shots—and the fact that their child's access to education hinges on compliance with the immunization regulations. There's also the simple reality that the illnesses kids are being inoculated against are rarely seen anymore. When diseases like polio ran free in the early 1900s, the clamor was less about why we needed vaccines than about why there weren't more of them. Once you've seen your neighbor's toddler become paralyzed, you're a lot more likely to worry that the same thing will happen to yours. "The fact is," says Offit, "young mothers today never grew up with the disease."

What worries him and others is that young mothers of tomorrow will—and that could be disastrous. CDC officials estimate that fully vaccinating all U.S. children born in a given year from birth to adolescence saves 33,000 lives, prevents 14 million infections and saves $10 billion in medical costs. Part of the reason is that the vaccinations protect not only the kids who receive the shots but also those who can't receive them—such as newborns and cancer patients with suppressed immune systems. These vulnerable folks depend on riding the so-called herd-immunity effect. The higher the immunization rate in any population, the less likely that a pathogen will penetrate the group and find a susceptible person inside. As immunization rates drop, that protection grows thinner. That's what happened in the current measles outbreaks in the western U.S., and that's what happened in Nigeria in 2001, when religious and political leaders convinced parents that polio vaccines were dangerous and their kids should not receive them. Over the next six years, not only did Nigerian infection rates increase 30-fold, but the disease also broke free and ranged out to 10 other countries, many of which had previously been polio-free.

As long ago as 1905, the U.S. Supreme Court recognized the power of the herd and ruled that states have the right to mandate immunizations, not for the individual's health but for the

5 Questions
What You Need to Know

Uncertainty over the need for and safety of vaccines is fueling fear and confusion. Here are answers.

Are Vaccinations Necessary?

Absolutely. Immunizing all babies born in the U.S. in a given year prevents 14 million infections and saves 33,000 lives and $10 billion in medical costs by the time the children reach adolescence.

Do Vaccines Cause Autism?

The best scientific evidence says no. Experts are instead focusing on genetic and environmental factors.

Will My Child React Badly to Immunization?

The vast majority will not. Genetic-screening advances may help doctors identify the few who might.

Is Mercury Still Used in Vaccines?

Only in the flu vaccine. By 2001 thimerosal-free formulations of vaccines were introduced, which dramatically cut mercury exposure for 6-month-olds but had no discernible effect on autism rates.

Must I Vaccinate My Children?

Yes. All but two states allow exemptions if families object to vaccines on religious grounds; 20 allow them for philosophical grounds.

community's. That principle, say vaccine proponents, should still apply. "The decision to vaccinate is a decision for your child," says Dr. Jane Seward, deputy director of viral diseases at the CDC, "but also a decision for society."

Some parents have taken to cherry-picking vaccines, leaving out only the shots they believe their children don't need—such as those for chicken pox and hepatitis B—and keeping up with what they see as the life-or-death ones. But that can be a high-stakes game, as Kelly Lacek, a Pennsylvania mother of three, learned. She stopped vaccinating her 2-month-old son Matthew when her chiropractor raised questions about mercury in the shots. Three years later, she came home to find the little boy feverish and gasping for breath. Emergency-room doctors couldn't find the cause—until one experienced physician finally asked the right question. "He took one look at Matthew and asked me if he was fully vaccinated," says Lacek. "I said no." It turned out Matthew had been infected with Hib, a virus that causes meningitis, swelling of the airway and, in severe cases, swelling of the brain tissue. After relying on a breathing tube for several days, Matthew recovered without any neurological effects, and a grateful Lacek immediately got him and his siblings up to date on their immunizations. "I am angry that people are promoting not getting vaccinated and messing with people's lives like that," she now says.

Health officials are angry too. Encouraged in part by the government report that seemed to clear vaccines of the autism charges, they are beginning to take a harder line with parents who submit vaccine exemptions for nonmedical reasons. In Maryland, where unvaccinated students are not permitted in school, officials last November threatened to take parents to court for truancy violations if their kids did not get all their shots so that they could be cleared for class. On Long Island, N.Y., vaccine objectors are called in for what some parents call "sincerity" interviews with school officials and school-board attorneys to determine how genuinely the vaccines conflict with religious convictions.

Even in cities where such interviews are not required, the tensions are palpable. Says Sue Collins, a New Jersey mother who has not had either of her two sons vaccinated: "Things are getting so nasty. People are calling us bad parents, saying it's child abuse if we don't vaccinate our children." In an effort to avoid potential conflicts, some parents are bypassing the school system altogether, preferring to homeschool their kids so they won't be forced to vaccinate them.

Common Ground

That still leaves the broader community at risk. So, is there room between public health and personal choice? Science may eventually provide a way out. Most people agree that there may be kids with genetic predispositions or other underlying conditions that make them susceptible to being harmed by vaccines. The Georgia girl in the recent vaccine case is the first such documented child, but her story suggests there could be others. Though CDC director Julie Gerberding was quick to insist that the case should not be considered an admission that vaccines can cause autism, some parents will surely take it as just that. "In rare instances, there could be some gene-vs.-exposure interaction that in theory could lead from the vaccine to autism," says Dr. Tracy Lieu, director of the center for child-health-care studies at Harvard Medical School. "The future of vaccine-safety research lies in trying to answer questions of genomic contributions to responses to vaccines." Screening for genetic profiles that are most commonly associated with immune disorders, for example, would be a good place to start.

Whether tests like these, combined with detailed family histories, will make a difference in the rates of developmental disorders like autism isn't yet clear. But such a strategy could reveal new avenues of research and lead to safer inoculations overall. Parents concerned about vaccine safety would then have stronger answers to their questions about how their child might be affected by the shots. Vaccines may be a medical marvel, but they are only one salvo in our fight against disease-causing bugs. It's worth remembering that viruses and bacteria have had millions of years to perfect their host-finding skills; our abilities to rebuff them are only two centuries old. And in that journey, both parents and public-health officials want the same thing—to protect future generations from harm.

Caution: Killing Germs May Be Hazardous to Your Health

Our war on microbes has toughened them. Now, new science tells us we should embrace bacteria.

JERRY ADLER AND JENEEN INTERLANDI

Behold yourself, for a moment, as an organism. A trillion cells stuck together, arrayed into tissues and organs and harnessed by your DNA to the elemental goals of survival and propagation. But is that all? An electron microscope would reveal that you are teeming with other life-forms. Any part of your body that comes into contact with the outside world—your skin, mouth, nose and (especially) digestive tract—is home to bacteria, fungi and protozoa that outnumber the cells you call your own by 10, or perhaps a hundred, to one.

Their ancestors began colonizing you the moment you came into the world, inches from the least sanitary part of your mother's body, and their descendants will have their final feast on your corpse, and join you in death. There are thousands of different species, found in combinations "as unique as our DNA or our fingerprints," says Stanford biologist David Relman, who is investigating the complex web of interactions microbes maintain with our digestive, immune and nervous systems. Where do you leave off, and they begin? Microbes, Relman holds, are "a part of who we are."

Relman is a leader in rethinking our relationship to bacteria, which for most of the last century was dominated by the paradigm of Total Warfare. "It's awful the way we treat our microbes," he says, not intending a joke; "people still think the only good microbe is a dead one." We try to kill them off with antibiotics and hand sanitizers. But bacteria never surrender; if there were one salmonella left in the world, doubling every 30 minutes, it would take less than a week to give everyone alive diarrhea. In the early years of antibiotics, doctors dreamed of eliminating infectious disease. Instead, a new paper in The Journal of the American Medical Association reports on the prevalence of Methicillin-resistant Staphylococcus aureus (MRSA), which was responsible for almost 19,000 deaths in the United States in 2005—about twice as many as previously thought, and more than AIDS. Elizabeth Bancroft, a leading epidemiologist, called this finding "astounding."

As antibiotics lose their effectiveness, researchers are returning to an idea that dates back to Pasteur, that the body's natural microbial flora aren't just an incidental fact of our biology, but crucial components of our health, intimate companions on an evolutionary journey that began millions of years ago. The science writer Jessica Snyder Sachs summarizes this view in four words in the title of her ground-breaking new book: *Good Germs, Bad Germs*. Our microbes do us the favor of synthesizing vitamins right in our guts; they regulate our immune systems and even our serotonin levels: germs, it seems, can make us happy. They influence how we digest our food, how much we eat and even what we crave. The genetic factors in weight control might reside partly in their genes, not ours. Regrettably, it turns out that bacteria exhibit a strong preference for making us fat.

Microbes synthesize vitamins, and regulate immune systems and even serotonin levels. Germs, it seems, can make you happy.

Our well-meaning war on microbes has, by the relentless process of selection, toughened them instead. When penicillin began to lose its effectiveness against staph, doctors turned to methicillin, but then MRSA appeared—first as an opportunistic infection among people already hospitalized, now increasingly a wide-ranging threat that can strike almost anyone. The strain most commonly contracted outside hospitals, dubbed USA300, comes armed with the alarming ability to attack immune-system cells. Football players seem to be especially vulnerable: they get scraped and bruised and share equipment while engaging in prolonged exercise, which some researchers believe temporarily lowers immunity. In the last five years outbreaks have plagued the Cleveland Browns, the University of Texas and the

University of Southern California, where trainers now disinfect equipment almost hourly. The JAMA article was a boon to makers of antimicrobial products, of which about 200 have been introduced in the United States so far this year. Press releases began deluging newsrooms, touting the benefits of antibacterial miracle compounds ranging from silver to honey. Charles Gerba, a professor of environmental microbiology at the University of Arizona, issued an ominous warning that teenagers were catching MRSA by sharing cell phones. Gerba is a consultant to the makers of Purell hand sanitizer, Clorox bleach and the Oreck antibacterial vacuum cleaner, which uses ultraviolet light to kill germs on your rug.

To be sure, MRSA is a scary infection, fast-moving and tricky to diagnose. Hunter Spence, a 12-year-old cheerleader from Victoria, Texas, woke up one Sunday in May with pain in her left leg. "I think I pulled a calf muscle," she told her mother, Peyton. By the next day, the pain was much worse and she was running a low-grade fever, but there was no other sign of infection. A doctor thought she might have the flu. By Wednesday her fever was 103 and the leg pain was unbearable. But doctors at two different community hospitals couldn't figure out what was wrong until Friday, when a blood culture came up positive for MRSA. By the time she arrived at Driscoll Children's Hospital in Corpus Christi—by helicopter—her temperature was 107 and her pulse 220. Doctors put her chance of survival at 20 percent.

Hunter needed eight operations over the next week to drain her infections, and an intravenous drip of two powerful new antibiotics, Zyvox and Cubicin. She did survive, and is home now, but her lung capacity is at 35 percent of normal. "We are seeing more infections, and more severe infections" with the USA300 strain, says Dr. Jaime Fergie, who treated her at Driscoll. In many cases, there's no clue as to how the infection was contracted, but a study Fergie did in 2005 of 350 children who were seen at Driscoll for unrelated conditions found that 21 percent of them were carrying MRSA, mostly in their noses. Then all it may take is a cut . . . and an unwashed hand.

And there are plenty of unwashed hands out there; Gerba claims that only one in five of us does the job properly, getting in all the spaces between the fingers and under the nails and rubbing for at least 20 seconds. Americans have been obsessed with eradicating germs ever since their role in disease was discovered in the 19th century, but they've been partial to technological fixes like antibiotics or sanitizers rather than the dirty work of cleanliness. Nancy Tomes, author of *The Gospel of Germs,* believes the obsession waxes and wanes in response to social anxiety—about diseases such as anthrax, SARS or avian flu, naturally, but also about issues like terrorism or immigration that bear a metaphoric relationship to infection. "I can't protect myself from bin Laden, but I can rid myself of germs," she says. "Guarding against microbes is something Americans turn to when they're stressed." The plastic squeeze bottle of alcohol gel, which was introduced by Purell in 1997, is a powerful talisman of security. Sharon Morrison, a Dallas real-estate broker with three young daughters, estimates she has as many as 10 going at any time, in her house, her car, her purse, her office and her kids' backpacks. She swabs her grocery cart with sanitizing wipes and, when her children were younger, she would bring

Bacteria's Base

By the time you turn 2, mircobes have colonized every inch of your body. Some regions are more densely populated than others.

Sharing Intelligence

How resistant bacteria spreads:

1. **Armed:** Some bacteria—but not all—carry extra genes that make them resistant to certain antibiotics.
2. **Dangerous Liaisons:** They can pass on these resistant genes by connecting to their nonresistant neighbors through a protein tube.
3. **Building Ranks:** As antibiotics kill off the vulnerable bacteria, the resistant ones thrive, and continue to pass their genes along.

The Hiding Places

- **The Mouth** is made up of dozens of distinct microbial neighborhoods. Each tooth has its own species and strains.
- **The Appendix** is now thought to be stockpiling gut microbes to replenish your intestine in the event of an illness.
- **The Gut** houses more microbes than all other body parts combined. These bugs aid digestion and produce nutrients.
- **The Groin** supplies most people with their first microbial residents, acquired as you pass through the birth canal.
- **The Skin** is covered in different species of friendly staphylococcus that may help keep infectious strains from getting in.

Hospitalized by Staph

Methicillin-resistant Staphylococcus aureus hospital stays, 1993–2005.

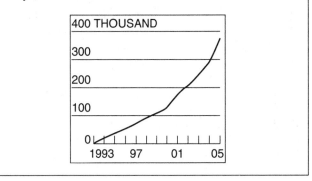

her own baby-seat cover from home and her own place mats to restaurants. Sales of Purell last year were $90 million, so she's clearly not alone. There's no question it kills germs, although it's not a substitute for washing; the Centers for Disease Control Web site notes that alcohol can't reach germs through a layer of dirt. Alcohol gels, which kill germs by drying them out, don't cause the kind of resistance that gives rise to superbugs like

MRSA. But they're part of the culture of cleanliness that's led to a different set of problems.

In terms of infectious disease, the environment of the American suburb is unquestionably a far healthier place than most of the rest of the world. But we've made a Faustian bargain with our antibiotics, because most researchers now believe that our supersanitized world exacts a unique price in allergies, asthma and autoimmune diseases, most of which were unknown to our ancestors. Sachs warns that many people drew precisely the wrong conclusion from this, that contracting a lot of diseases in childhood is somehow beneficial. What we need is more exposure to the good microbes, and the job of medicine in the years to come will be sorting out the good microbes from the bad.

That's the goal of the Human Microbiome Project, a five-year multinational study that its advocates say could tell us almost as much about life as the recently completed work of sequencing the human genome. One puzzling result of the Human Genome Project was the paltry number of genes it found—about 20,000, which is only as many as it takes to make a fruit fly. Now some researchers think some of the "missing" genes may be found in the teeming populations of microbes we host.

And the microbe project—which as a first step requires sampling every crevice and orifice of 100 people of varying ages from a variety of climates and cultures—is "infinitely more complex and problematic than the genome," laments (or boasts) one of its lead researchers, Martin Blaser of NYU Medical School. Each part of the body is a separate ecosystem, and even two teeth in the same mouth can be colonized by different bacteria. In general, researchers know what they'll find—*Escherechia* (including the ubiquitous microbial Everyman, *E. coli*) in the bowel, lactobacilli in the vagina and staphylococcus on the skin. But the mix of particular species and strains will probably turn out to be unique to each individual, a product of chance, gender (men and women have different microbes on their skin but are similar in their intestines) and socioeconomic status and culture. (Race seems not to matter much.) Once the microbes establish themselves they stay for life and fight off newcomers; a broad-spectrum antibiotic may kill most of them but the same kinds usually come back after a few weeks. The most intriguing question is how microbes interact with each other and with our own cells. "There is a three-way conversation going on throughout our bodies," says Jane Peterson of the National Human Genome Research Institute. "We want to listen in because we think it will fill in a lot of blanks about human health—and human disease."

The vast majority of human microbes live in the digestive tract; they get there by way of the mouth in the first few months of life, before stomach acid builds to levels that are intended to kill most invaders. The roiling, fetid and apparently useless contents of the large intestine were a moral affront to doctors in the early years of modern medicine, who sought to cleanse them from the body with high-powered enemas. But to microbiologists, the intestinal bacteria are a marvel, a virtual organ of the body which just happens to have its own DNA. Researchers at Duke University claim it explains the persistence of the human appendix. It serves, they say, as a reservoir of beneficial microbes which can recolonize the gut after it's emptied by diseases such as cholera or dysentery.

Microbes play an important role in digestion, especially of polysaccharides, starch molecules found in foods such as potatoes or rice that may be hundreds or thousands of atoms long. The stomach and intestines secrete 99 different enzymes for breaking these down into usable 6-carbon sugars, but the humble gut-dwelling *Bacterioides theta* produces almost 250, substantially increasing the energy we can extract from a given meal.

Of course, "energy" is another way of saying "calories." Jeffrey Gordon of the University of Washington raised a colony of mice in sterile conditions, with no gut microbes at all, and although they ate 30 percent more food than normal mice they had less than half the body fat. When they were later inoculated with normal bacteria, they quickly gained back up to normal weight. "We are finding that the nutritional value of food is pretty individualized," Gordon says. "And a big part of what determines it is our microbial composition."

We can't raise humans in sterile labs, of course, but there's evidence that variations between people in their intestinal microbes correspond to differences in body composition. And other factors appear to be at work besides the ability to extract calories from starch. Bacteria seem able to adjust levels of the hormones ghrelin and leptin, which regulate appetite and metabolism. Certain microbes even seem to be associated with a desire for chocolate, according to research by the Nestlé Research Center. And a tiny study suggests that severe emotional stress in some people triggers an explosion in the population of B. theta, the starch-digesting bacteria associated with weight gain. That corresponds to folk wisdom about "stress eating," but it is also a profoundly disturbing and counterintuitive observation that something as intimate as our choice between a carrot and a candy bar is somehow mediated by creatures that are not us.

But these are the closest of aliens, so familiar that the immune system, which ordinarily attacks any outside organism, tolerates them by the trillions—a seeming paradox with profound implications for health. The microbes we have all our lives are the ones that colonize us in the first weeks and months after birth, while our immune system is still undeveloped; in effect, they become part of the landscape. "Dendritic" (treelike) immune cells send branches into the respiratory and digestive tracts, where they sample all the microbes we inhale or swallow. When they see the same ones over and over, they secrete an anti-inflammatory substance called interleukin-10, which signals the microbe-killing T-cells: stand down.

And that's an essential step in the development of a healthy immune system. The immune reaction relies on a network of positive and negative feedback loops, poised on a knife edge between the dangers of ignoring a deadly invader and overreacting to a harmless stimulus. But to develop properly it must be exposed to a wide range of harmless microbes early in life. This was the normal condition of most human infants until a few generations ago. Cover the dirt on the floor of the hut, banish the farm animals to a distant feedlot, treat an ear infection with penicillin, and the inflammation-calming interleukin-10 reaction may fail to develop properly. "Modern sanitation is a good thing, and pavement is a good thing," says Sachs, "but they keep kids at a distance from microbes." The effect is to tip the immune system in the direction of overreaction, either to outside stimuli or even to the body's own cells. If the former, the result is allergies or asthma. Sachs writes that "children who receive antibiotics in the first year of life have more than double the rate of allergies and asthma in later childhood." But if the immune system turns on the body itself, you see irritable bowel syndrome, lupus or multiple sclerosis, among the many autoimmune diseases that were virtually unknown to our ancestors but are increasingly common in the developed world.

That is the modern understanding of the "Hygiene Hypothesis," first formulated by David Strachan in 1989. In Strachan's original version, which has unfortunately lodged in the minds of many parents, actual childhood illness was believed to exert a protective effect. There was a brief vogue for intentionally exposing youngsters to disease. But researchers now believe the key is exposure to a wide range of harmless germs, such as might be found in a playground or a park.

The task is complicated, in part because some bacteria seem to be both good and bad. The best-known is *Helicobacter pylori,* a microbe that has evolved to live in the acid environment of the stomach. It survives by burrowing into the stomach's mucous lining and secreting enzymes that reduce acidity. Nobel laureates Barry Marshall and Robin Warren showed it could cause gastric ulcers and stomach cancer. But then further studies discovered that infection with H. pylori was protective against esophageal reflux and cancer of the esophagus, and may also reduce the incidence of asthma. H. pylori, which is spread in drinking water and direct contact among family members, was virtually universal a few generations ago but is now on the verge of extinction in the developed world. The result is fewer ulcers and stomach cancer, but more cancer of the esophagus—which is increasing faster than any other form of cancer in America—more asthma, and . . . what else? We don't know. "H. pylori has colonized our guts since before humans migrated out of Africa," says Blaser. "You can't get rid of it and not expect consequences."

Blaser questions whether eliminating H. pylori is a good idea. Someday, conceivably, we might intentionally inoculate children with a bioengineered version of H. pylori that keeps its benefits without running the risk of stomach cancer. There is already a burgeoning market for "probiotics," bacteria with supposed health benefits, either in pill form or as food. Consumers last year slurped down more than $100 million worth of Dannon's Activia, a yogurt containing what the Web site impressively calls "billions" of beneficial microbes in every container. The microbes are a strain of *Bifidobacterium animalis,* which helps improve what advertisers delicately call "regularity," a fact Dannon has underscored by rechristening the species with its trademarked name "Bifidus regularis." Other products contain *Lactobacillus casei,* which is supposed to stimulate production of infection-fighting lymphocytes. Many others on the market are untested and of dubious value. Labels that claim ANTIBIOTIC RESISTANT ought to be considered a warning, not a boast. Bacteria swap genetic material among themselves, and the last thing you want to do is introduce a resistant strain, even of a beneficial microbe, into your body.

And there's one more thing that microbes can do, perhaps the most remarkable of all. *Mycobacterium vaccae,* a soil microbe found in East Africa that has powerful effects on the immune system, was tested at the University of Bristol as a cancer therapy. The results were equivocal, but researchers made the startling observation that patients receiving it felt better regardless of whether their cancer was actually improving. Neuroscientist Chris Lowry injected mice with it, and found, to his amazement, that it activated the serotonin receptors in the prefrontal cortex—in other words, it worked like an antidepressant, only without the side effects of insomnia and anxiety. Researchers believe M. vaccae works through the interleukin-10 pathway, although the precise mechanism is uncertain. But there is at least the tantalizing, if disconcerting, suggestion that microbes may be able to manipulate our happiness. Could the hygiene hypothesis help explain the rise in, of all things, depression? We're a long way from being able to say that, much less use that insight to treat people. But at least we are asking the right questions: not how to kill bacteria, but how to live with them.

With Matthew Philips, Raina Kelley and Karen Springen.

Why We're Sicker

Scientists blame modern living for increases in allergies and autoimmune diseases.

ROB STEIN

First, asthma cases shot up, along with hay fever and other common allergic reactions, such as eczema. Then, pediatricians started seeing more children with food allergies. Now, experts are increasingly convinced that a suspected jump in lupus, multiple sclerosis and other afflictions caused by misfiring immune systems is real.

Though the data are stronger for some diseases than others, and part of the increase may reflect better diagnoses, experts estimate that many allergies and immune-system diseases have doubled, tripled or even quadrupled in the last few decades, depending on the ailment and country. Some studies now indicate that more than half of the U.S. population has at least one allergy.

The cause remains the focus of intense debate and study, but some researchers suspect the concurrent trends all may have a common explanation rooted in aspects of modern living—including the "hygiene hypothesis" that blames growing up in increasingly sterile homes, changes in diet, air pollution, and possibly even obesity and increasingly sedentary lifestyles.

"We have dramatically change our lives in the last 50 years," says Fernando Martinez, who studies allergies at the University of Arizona. "We are exposed to more products. We have people with different backgrounds being exposed to different environments. We have made our lives more antiseptic, especially early in life. Our immune systems may grow differently as a result. And we may be paying a price for that."

Along with a flurry of research to confirm and explain the trends, scientists have also begun testing possible remedies. Some are feeding high-risk children gradually larger amounts of allergy-inducing foods, hoping to train the immune system not to overreact. Others are testing benign bacteria or parts of bacteria. Still others have patients with MS, colitis and related ailments swallow harmless parasitic worms to try to calm their bodies' misdirected defenses.

"If you look at the incidence of these diseases, a lot of them began to emerge and become much more common after parasitic worm diseases were eliminated from our environment," says Robert Summers of the University of Iowa, who is experimenting with whipworms. "We believe they have a profound symbiotic effect on developing and maintaining the immune system."

Although hay fever, eczema, asthma and food allergies seem quite different, they are all "allergic diseases" because they are caused by the immune system responding to substances that are ordinarily benign, such as pollen or peanuts. Autoimmune diseases also result from the body's defense mechanisms malfunctioning. But in these diseases, which include lupus, MS, Type 1 diabetes and inflammatory bowel disease, the immune system attacks parts of the body such as nerves, the pancreas or digestive tract.

"Overall, there is very little doubt that we have seen significant increases," says Syed Hasan Arshad of the David Hide Asthma and Allergy Centre in England, who focuses on food allergies. "You can call it an epidemic. We're talking about millions of people and huge implications, both for health costs and quality of life. People miss work. Severe asthma can kill. Peanut allergies can kill. It does have huge implications all around. If it keeps increasing, where will it end?"

One reason that many researchers suspect something about modern living is to blame is that the increases show up largely in highly developed countries in Europe, North America and elsewhere, and have only started to rise in other countries as they have become more developed. "It's striking," says William Cookson of the Imperial College in London.

The leading theory to explain the phenomenon holds that as modern medicine beats back bacterial, viral and parasitic diseases that have long plagued humanity, immune systems may fail to learn how to differentiate between real threats and benign invaders, such as ragweed pollen or food. Or perhaps because they are not busy fighting real threats, they overreact or even turn on the body's own tissues.

"Our immune systems are much less busy," says Jean-Francois Bach of the French Academy of Sciences, "and so have much more strong responses to much weaker stimuli, triggering allergies and autoimmune diseases."

Several lines of evidence support the theory. Children raised with pets or older siblings are less likely to develop allergies, possibly because they are exposed to more microbes. But perhaps the strongest evidence comes from studies comparing thousands of people who grew up on farms in Europe to those who lived in less rural settings. Those reared on farms were one-tenth as likely to develop diseases such as asthma and hay fever.

"The data are very strong," says Erika von Mutius of the Ludwig-Maximilians University in Munich. "If kids have all sorts of exposures on the farm by being in the stables a lot, close to the animals and the grasses, and drinking cow's milk from their own farm, that seems to confer protection."

The theory has also gained support from a variety of animal studies. One, for example, found that rats bred in a sterile laboratory had far more sensitive immune systems than those reared in the wild, where they were exposed to infections, microorganisms and

Overactive Immunity

A variety of ailments appear to be increasing because of malfunctions of the immune system. Here are some:

Ailment	Cause	Effect
Asthma	An allergic reaction in which the immune system reacts to something in the environment, such as pollen.	Passages that carry air into and out of the lungs become inflamed, making it difficult to breathe.
Hay fever	An allergic reaction in which the immune system reacts to pollen.	Tissues in the nasal passages and upper airways become inflamed.
Eczema	An allergic reaction to something, such as food or dust mites.	Red, swollen and itchy skin.
Type 1 diabetes	The immune system attacks cells in the pancreas.	Blood sugar levels rise to dangerous levels.
Multiple sclerosis	The immune system attacks the nerves' myelin sheath.	Signals that control muscles, sensation and vision are blocked.
Inflammatory bowel diseases	The immune system attacks the digestive system.	Cramps, diarrhea, nausea and other symptoms: Crohn's disease and ulcerative colitis.
Food allergies	An immune reaction in the digestive system.	Nausea, cramps and vomiting. In serious cases, chemicals released by the immune system cause anaphylaxis, which can be life-threatening.

parasites. "It's sort of a smoking gun of the hygiene hypothesis," says William Parker of Duke University.

Researchers believe the lack of exposure to potential threats early in life leaves the immune system with fewer command-and-control cells known as regulatory T cells, making the system more likely to overreact or run wild. "If you live in a very clean society, you're not going to have a lot of regulatory T cells," Parker says.

While the evidence for the hygiene theory is accumulating, many say it remains far from proven.

"That theory is so full of holes that it's clearly not the whole story," says Robert Wood of the Johns Hopkins School of Medicine.

It does not explain, for example, the rise in asthma, since that disease occurs much more commonly in poor, inner-city areas where children are exposed to more cockroaches and rodents that may trigger it, Wood and others say.

Several alternative theories have been presented. Some researchers blame exposure to fine particles in air pollution, which may give the immune system more of a hair trigger, especially in genetically pre-disposed individuals. Others say obesity and a sedentary lifestyle may play a role. Still others wonder whether eating more processed food or foods processed in different ways, or changes in the balance of certain vitamins that can affect the immune system, such as vitamins C and E and fish oil, are a factor.

"Cleaning up the food we eat has actually changed what we're eating," says Thomas Platts-Mills of the University of Virginia.

But many researchers believe the hygiene hypothesis is the strongest, and that the reason one person develops asthma instead of hay fever or eczema or lupus or MS is because of a genetic predisposition. "We believe it's about half and half," Cookson says. "You need environmental factors and you need genetic susceptibility as well."

Some researchers have begun to try to identify specific genes that may be involved, as well as specific components of bacteria or other pathogens that might be used to train immune systems to respond appropriately.

"If we could mimic what is happening in these farm environments, we could protect children and prevent asthma, allergies and other diseases," von Mutius says.

Some researchers are trying to help people who are at risk for allergies or already ill with autoimmune diseases.

With new research suggesting that food allergies may be occurring earlier in life and lasting longer, several small studies have been done or are underway in which children at risk for milk, egg and peanut allergies are given increasing amounts of those foods, beginning with tiny doses, to try to train the immune system.

"I'm very encouraged," says Wesley Burks, a professor of pediatrics at Duke who has done some of the studies. "I'm hopeful that in five years, there may be some type of therapy from this."

Another promising line of research involves giving patients microscopic parasitic worms to try to tamp down the immune system. "We've seen rather dramatic improvements in patients' conditions," says Summers of the University of Iowa, who has treated more than 100 people with Crohn's disease or ulcerative colitis by giving them parasitic worms that infect pigs but are harmless to humans. "We're not claiming that this is a cure, but we saw a very dramatic improvement. Some patients went into complete remission."

Doctors in Argentina reported last year that MS patients who had intestinal parasites fared better than those who did not, and researchers at the University of Wisconsin are planning to launch another study as early as next month testing pig worms in 20 patients with the disease.

"We hope to show whether this treatment has promise and is worth exploring further in a larger study," says John O. Fleming, a professor of neurology who is leading the effort.

Novel Anti-Infectives: Is Host Defence the Answer?

PAMELA HAMILL ET AL.

Resistance to antimicrobial agents and the limited development of novel agents are threatening to worsen the burden of infections that are already a leading cause of morbidity and mortality. This has increased interest in the development of novel strategies such as selective modulation of our natural immune defences. Innate immunity is a complex, evolutionarily conserved, multi-facetted response to defeating infection that is naturally stimulated by pathogenic organisms through pattern recognition receptors on host cells. It is amplifiable and broad spectrum but if overstimulated can lead to the potential for harmful inflammatory responses. A broad variety of therapies are already available or increasingly under development, to stimulate protective innate immunity without overtly stimulating harmful inflammation or even suppressing such damaging pro-inflammatory responses.

Introduction

Despite the enormous positive impact that the development of antibiotic, antiviral and antifungal drugs have made on human health in recent decades, infectious diseases remain a major contributor to morbidity and mortality and a considerable burden to healthcare systems globally. The current WHO global burden of disease report indicates that infectious diseases account for nearly a third of global deaths while HIV, malaria, tuberculosis and lower respiratory tract infections were among the top eight leading causes of death in 2004.[1] The alarming increase in the prevalence of antibiotic-resistant bacteria together with the threat of new and variant pathogens, exemplified by the emergence of HIV, SARS and avian influenza, highlights the urgent need for new strategies to combat infectious diseases.

Innate immunity represents a conserved, complex and multipronged response to overcoming infection that is present in all complex host species of life. Natural stimulation of innate immunity by pathogens results in an amplifiable, broad spectrum and protective immune response but if this response is too vigorous or prolonged it can lead to the potential for harmful inflammatory responses. Selective modulation of innate immunity as an anti-infective strategy is an emerging concept driven by the huge advances in our understanding of this crucial host defence

system. The discovery of key pathogen recognition receptors (PRRs) such as the Toll-like receptors (TLRs), and intracellular sensors of microbial components such as the Nod-like receptors (NLRs) and RIG-I-like receptors (RLRs),[2] has stimulated a rapid expansion in information regarding pathogen-sensing mechanisms and intracellular signaling pathways and effector strategies that lead to a rapid, highly effective clearance of pathogens. This has provided valuable insights into the role of host immunity in the pathogenesis of infectious diseases and revealed possible targets for therapeutic intervention. Here we discuss the potential for this new approach in developing urgently needed novel anti-infective therapies and the progress being made towards this goal. There is also considerable activity currently within the biotech community focused on development of immunomodulators as treatments for inflammatory conditions and as vaccine adjuvants (see [3–6] for recent reviews); however, here we discuss recent developments and potential for the use of immunomodulators in the direct treatment of infectious diseases.

Why Target Host Innate Immunity Defence Systems as an Anti-Infectious Strategy?

Innate immunity is a highly effective defence system, considering the relative infrequency with which infectious diseases occur, despite our constant daily exposure to pathogens. Symptomatic diseases can progress either through damage caused directly by microbial factors or as a consequence of the immune response itself; some pathogens stimulate an overtly powerful pro-inflammatory response, while the response to other pathogens may be insufficient.[7] Hence immunomodulation offers the potential to tip the balance back in the favour of the host, either by boosting or inhibiting selected elements of the immune response as well as exploiting the powerful and multi-faceted effector mechanisms that have evolved specifically for the purpose of pathogen clearance. There are several advantages to modulating host innate immunity as an anti-infective strategy. Since the pathogen itself is not targeted, there is no selective pressure and hence a very small possibility of development of resistance to treatment. Another attractive prospect is that, since host innate immunity utilizes effector mechanisms that are effective against a diverse array of pathogens, immunomodulation could form the basis for broad spectrum therapeutics to treat infections of bacterial, viral, fungal or parasitic origin. Further, since innate immunity is highly instrumental in directing subsequent adaptive responses, modulation of innate immunity could be used to initiate or reinforce immune responses or 'skew' them to either a Type I or Type II antigen-specific response, thereby encouraging the appropriate adaptive response for either intracellular or extracellular pathogens, respectively. However it is important to note that there are many potential disadvantages including the inappropriate dysregulation of immunity causing, for example sepsis or what is known as the cytokine storm, possible unfavourable interactions between infectious agents and the immunomodulators in combination, the induction of aggressive and damaging activated cells such as inflammatory macrophages and neutrophils, a range of immunotoxicities including histamine release, apoptosis, complement hyperactivation, among others, and the induction of autoimmunity or chronic inflammation. A simple example of the potential perils of immunotherapy is that a side effect of immune suppressive chemicals and irradiation is that the body becomes extremely vulnerable to infections.

Existing and Potential Innate Immune Targets for Development of Anti-Infectives

There is a wealth of potential targets for therapeutic intervention that capitalize on the complexities of the innate immune response and these include soluble mediators, membrane-bound receptors as well as intracellular signaling molecules. Currently approved immunomodulators are predominantly cytokine-based and exploit their natural role in potentiating antimicrobial responses, particularly against intracellular viral infections. The best-established cytokine therapies are recombinant and modified forms of IFN-α and IFN-β (e.g. pegylated IFN-α), which are effective treatments for several different viral diseases. They are most widely used in chronic HCV and HBV infections but are also administered for severe herpesvirus-associated disease in immunocompromised patients. Alternatively colony stimulating factors (CSF) are licensed to treat neutropenia, a depletion of neutrophils common in patients receiving medication or chemotherapy that damages the bone marrow. For example, GM-CSF restores and stimulates neutrophil functions that are crucial to fight bacterial and fungal functions, and is licensed for use in chemotherapy and HIV-infected patients to diminish susceptibility to infection. There are also examples of approved immunomodulators that act upon other innate immune targets. Imiquimod, a synthetic small molecule Toll-like receptor (TLR)-7 agonist, is currently used in the topical treatment of papillomavirus-associated genital warts and is believed to work through TLR-7-mediated secretion of cytokines such as IFN-α,

IL-6 and TNF-α as well as activation of NK cells and macrophages.[8] Isoprinosine is an immunostimulant that enhances T-cell proliferation and activity and is approved for the treatment of HSV, EBV and viral hepatitis. Microbes themselves are also licensed for use as immunomodulators in Europe. Products such as Bronchomunal® (Lek) and Luivac® (Daiichi Sankyo Co.) consist of lyophilized bacteria and bacterial lysates, respectively, and are used as preventative and/or therapeutic treatments for respiratory tract infections. In addition, intravenous pooled human immunoglobulin is FDA approved and used as a general immunomodulator for pediatric HIV with off-label uses for several infection-related issues including sepsis and *C. difficile* colitis.

Future Prospects for Development of Anti-Infective Immunomodulators

The case for increased research and development of immunomodulators as anti-infective therapies is bolstered by the success of therapeutic immunomodulators already in clinical use. At present, the most viable drugs are based on pathogen signature molecules (agonists of PRRs), cytokines and antimicrobial (host defence) peptides, and include TLR agonists/antagonists and agents targeting chemokines and cytokines. Many of these are aimed, although not exclusively, at viral infections (Table 1). However their potential in bacterial infections and the associated inflammatory sequelae seems strong, especially since we are running out of novel treatment options for bacteria.

Antivirals

Although the majority of antiviral research remains focused on viral targets such as protease and polymerase enzymes, immunomodulation as an antiviral strategy is also being actively pursued. The switch to alternative approaches in antiviral development reflects the difficulties associated with successful targeting of crucial points in the virus life cycle and exploits an ever-increasing understanding of viral life cycles and immune evasion strategies. Many viruses establish chronic disease through their ability to subvert host immune responses,[9,10] hence therapy based on immunomodulation to counteract viral immune evasion could be of great therapeutic value. The development of antiviral immunomodulators has been dominated to date by therapies targeting chronic viral infections such as HIV, HBV and HCV, for which there is a large unmet clinical need.

General strategies for therapeutic intervention in viral infection include boosting the host's natural antiviral effector mechanisms by inducing release of cytokines such as IFN-α, TNF-α and IL-12, NK cell activation and strong CD8$^+$ T cell responses, all of which are hallmarks of Th1-like immune responses that are crucial for the clearance of intracellular pathogens. The main approach to eliciting Th1-type responses is through stimulation of intracellular TLRs involved in the recognition of conserved microbial molecular signature molecules, using TLR agonists,

Table 1 A Selection of Immunomodulatory Agents Currently in Clinical Trials or Developmental Stages

Drug	Description	Company	Status/Results	Reference
Chronic HCV-Directed Therapies				
Zadaxin®	Thymosin α1 (thymalfasin)	SciClone/Sigma-Tau	**Phase III:** Synthetic peptide. Promotes MHC class I expression, IL-2 and IFNγ secretion, proliferation and activation of CD4 Th1, CD8, and NK cells. Decreases Th2 cytokines IL-4 and IL-10 that are counter productive to viral infections.	www.hcvadvocate.org
Oglufanide disodium	Dipeptide	Implicit Bioscience	**Phase II:** Intranasal synthetic formulation of a natural dipeptide of L-glutamic acid and L-tryptophan. Reverses suppression of the immune system.	www.hcvadvocate.org
SCV-07	Dipeptide	SciClone	**Phase II:** γ-D-glutamyl-L-tryptophan, a synthetic dipeptide that activates Th1 cells	www.hcvadvocate.org
ANA773	TLR-7 agonist	Anadys	**Phase 1:** Induced secretion of IFN-a from human PBMC, increased NK cell cytotoxicity and cytokine secretion *in vitro*	July 2008, Anadys press release
IMO-2125	TLR-9 agonist	Idera Pharma	**Phase I:** TLR-9 agonist that induces IFN	www.hcvadvocate.org
Locteron	IFN-α	Biolex Therapeutics Inc.	**Phase II complete:** Controlled-release drug delivery technology of IFN-α	www.hcvadvocate.org
Albuferon	IFN-α2b	Human Genome Sciences/Hospira	**Phase III complete:** Bioengineered, long acting IFN-α conjugated to albumin	August 2007, HGS Press Release
IL-29	IFN-λ	ZymoGenetics	**Phase I complete:** PEG-IFN-λ (long acting)	www.hcvadvocate.org
Omega IFN	Omega IFN	Intarcia Therapeutics	**Phase II complete:** Implantable infusion pump releases Omega interferon	www.hcvadvocate.org
Bavituximab (Tarvacin)	Monoclonal Ab against phospholipids	Peregrine	**Phase I complete:** Binding, for example, to phosphatidyl serine on the surface of virally infected cells will alert the immune system.	November 2007, Peregrine press release
Civacir	HCV Immune Globulin	NABI/Biotest AG	**Phase II:** Plasma-derived polyclonal antibody	www.hcvadvocate.org
Ceplene	Histamine dihydrochloride	Maxim	**Phase II (with PEG-IFNα-2b and Ribavirin):** Histamine inhibits HCV NS3-induced oxidative stress and apoptosis in T cells, NK and NKT cells.	www.hcvadvocate.org
PeviPRO™/ PeviTER™	Therapeutic vaccine*	Pevion Biotech	**Phase I:** Virosome-based synthetic cocktail that targets CD8 and CD4 helper T lymphocytes	December 2006, Pevion press release
ChronVac-C	Therapeutic vaccine*	Inovio/Tripep AB	**Phase I/II:** DNA-based, immune boosting vaccine with unique pulse-delivery method. Activates T cells to kill HCV-infected liver cells.	www.hcvadvocate.org
TG4040	Therapeutic vaccine*	Transgene	**Phase I:** Attenuated strain of vaccinia virus (MVA), expressing non-structural proteins (NS3, NS4 and NS5B) of HCV	www.hcvadvocate.org
GI-5005 (Tarmogen)	Therapeutic vaccine*	Globe Immune	**Phase II:** Delivered in combination with pegylated interferon and ribavirin. Targets two conserved HCV replication proteins.	www.hcvadvocate.org
IC41	Therapeutic vaccine*	Intercell/Novartis	**Phase II:** Combination vaccine of five synthetic peptides with HCV, CD4 and CD8 T-cell epitopes	www.hcvadvocate.org

(continued)

Table 1 A Selection of Immunomodulatory Agents Currently in Clinical Trials or Developmental Stages *(continued)*

Name	Type	Company	Phase / Description	Website
NOV-205	Immune modulator	Novelos Therapeutics	**Phase Ib complete:** Hepatoprotective agent with immunomodulatory/anti-inflammatory properties	www.novelos.com
CTS-1027	MMP inhibitor	Conatus	**Phase II:** MMP inhibitor that reduces aminotransferase (ALT) activity and protects liver cells from damage during a viral infection	www.conatuspharma.com
LGD-4665	Receptor agonist	Ligand Pharmaceuticals	**Phase I complete:** Thrombopoeitin receptor agonist. Stimulates platelet production. For use before or with other HCV treatments.	www.hcvadvocate.org
Eltrombopag (Promacta)	Receptor agonist	Glaxo SmithKline	**Phase II complete:** Thrombopoeitin receptor agonist. Stimulates platelet production. For use before or with other HCV treatments.	www.hcvadvocate.org
HIV-Directed Therapies				
Ampligen	TLR-3 Agonist	Hemispherx Biopharma, Inc.	**Phase I complete:** Poly IC-12U is a dsRNA agonist of TLR-3 that stimulates the immune system and inhibits HIV replication *in vitro*.	www.hemispherx.net
Vacc-4x	Therapeutic Vaccine	Bionor Immuno AS	**Phase II complete:** Therapeutic vaccine consisting of modified HIV peptides with increased immunogenicity	www.bionorimmuno.com
Leflunomide	Pyrimidine synthesis inhibitor	Sanofi-Aventis	**Phase I (Licensed for arthritis):** Immunosuppressive agent that blocks cell division in activated (virally infected) T cells thereby maintaining a sufficient population of T cells for defence while limiting the number of cells in which HIV can reproduce.	products.sanofi-aventis.us/arava/arava.pdf
Alpha Lipoic Acid	Antioxidant glutathione	NIAID (NCCAM)	**Phase II** (in use as an over the counter medicine; Phase IIIs for neuropathy): Prevents oxidative stress and HIV-induced apoptosis of T cells	http://nccam.nih.gov/
Antibacterial Therapies				
Soluble beta-1, 3/1,6 glucan	Poly-saccharide	Biotec Pharmacon	**Phase I studies:** For gingivitis and periodontitis and general immune modulation. Beta 1,3-D glucans are polysaccharides that occur in cereal bran, yeast and fungi, and are biological response modifiers that activate the immune system.	www.biotec.no
IMX942	Peptide modeled on host defence peptides	Inimex	**Preclinical:** Parent molecule IDR-1 was demonstrated to work through selective stimulation of innate immunity, upregulating protective immunity while suppressing pro-inflammatory cytokine production in response to bacterial TLR agonists.	www.inimexpharma.com
CLS001	12-mer analog of antimicrobial peptide indolicidin	Migenix/Cutanea	**Phase III:** Developed as an antimicrobial peptide, Omiganan (CLS001/CP-226/MX-226) has been demonstrated in Phase II trials to have anti-inflammatory properties vs. Rosacea and Acne.	www.migenix.com www.cutanealife.com
hLF-1-11	Small peptide derived from human lactoferrin	AM-Pharma	**Phase II:** While originally developed as an antimicrobial peptide this peptide has very weak antibiotic activity and may protect via immunomodulatory activity. Trials address allogeneic bone marrow stem cell transplantation-associated infections.	www.am-pharma.com

*Unlike traditional prophylactics, therapeutic vaccines are administered post-infection.

157

and currently synthetic agonists directed at TLR-3 (dsRNA), TLR-7/8 (ssRNA) and TLR-9 (unmethylated CpG DNA motifs) are all in pre-clinical or clinical development Phases (Table 1).

Ampligen, a synthetic TLR-3 agonist consisting of PolyI: $PolyC_{12}U$ RNA, is being developed by Hemispherx (Philadelphia, USA) and is in clinical trials as an HIV therapy, having had success in earlier trials.[11] Following on from the success of imiquimod in the treatment of HPV-associated warts, other TLR-7 agonists are being developed. Takeda Pharmaceuticals (Japan) has compound R851 in Phase II clinical trials in the US also for the treatment of HPV, while Anadys (San Diego, USA) has ANA773, an oral TLR-7 agonist prodrug, at a preclinical stage of development with current indications being Hepatitis C and cancer. Synthetic agonists of TLR-9 are currently the focus of much research and development since TLR-9 activation induces both innate and adaptive immune responses, making it an attractive target for the development of both anti-infective treatments and vaccine adjuvants. Currently, Dynavax (San Francisco, USA) has a candidate Type C TLR-9 agonist in the development for HCV therapy and similarly, Idera (Cambridge, USA) has a TLR-9 agonist, IMO-2125, as its lead candidate for the treatment of HCV in Phase I trials for patients not responding to standard interferon/ribavirin dual therapy.[12]

One potential caution with such strategies is that the combination of a virus manipulating innate immunity and a TLR agonist doing the same could potentially yield surprising results, with an exacerbation of pro-inflammatory responses being the most concerning. Another issue is that TLRs have been found to be required not only for antiviral defence but also for viral infectivity that further adds to the complexity of such treatments.[13,14] TLR stimulation boosts antiviral defences by stimulating natural induction of IFN and other cytokines required to initiate a Th1 type response, but there are also ongoing efforts into developing improved IFNs for exogenous administration through increasing stability, strategies for oral delivery and increasing binding affinities for IFN receptors (Table 1).

Alternative approaches to stimulating Th1-type responses to treat viral infections are also in development. The concept of therapeutic vaccines is a growing trend in developing treatments for chronic viral infections such as HCV and HIV. Like prophylactic vaccines they contain pathogen-specific epitopes to elicit an immune response; however they are administered post-infection and frequently have additional components to encourage the appropriate Th1 responses necessary for effective clearance of intracellular pathogens. Pevion Biotech (Berne, Switzerland) is developing a therapeutic vaccine that incorporates HCV antigen peptides encapsulated in proprietary 'virosomes', a carrier system designed to induce specific CTL and T helper cell responses in patients with chronic HCV infection.[15,16] Bionor AS (Skien, Norway) has HIV therapeutic vaccines in clinical trials that are based on novel peptide fragments with modified sequences to enhance immunogenicity.

In addition to strategies aimed at boosting general antiviral defence mechanisms, the possibility to develop custom immunomodulators tailored to treat specific viruses also exists, provided there is suitably detailed knowledge of the virus life cycle and pathogenesis.

Host Defence Peptides

Other classes of immunomodulatory agents with the potential for development as anti-infectives include the host defence peptides (HDPs). HDPs are important innate immune effector molecules that are conserved in virtually all life forms.[17] While some possess direct antimicrobial activity (the ability to lyse or destabilize bacterial membranes or viral envelopes; often weak in natural peptides), others, such as the endogenous human HDP LL-37, exert potent and pleiotropic immunomodulatory effects.[18–20] Synthetic peptides that retain many of the immunomodulatory properties of naturally occurring HDPs are currently being explored for their therapeutic potential[21] owing to their unique ability to promote protective innate immunity while suppressing potentially harmful inflammatory responses. Innate Defence Regulator-1 (IDR-1), an anti-infective peptide that selectively modulates the innate immune response, represents the first proof of principle that synthetic immunomodulatory peptides offer therapeutic potential.[22••] Despite possessing no direct antimicrobial activity, IDR-1 confers protection against multiple bacterial pathogens, including multiply antibiotic resistant strains (methicillin resistant *S. aureus,* vancomycin resistant *Enterococcus*), in mouse infection models. Mechanism of action studies have shown that IDR-1 stimulates the production of monocyte chemokines, dampens pro-inflammatory cytokine responses and activates monocyte-macrophage cells, without inducing toxic side effects. A 5 amino acid immunomodulatory peptide, IMX942, is in pre-clinical development by Inimex Pharmaceuticals (Vancouver, Canada) with likely indications for future clinical trials being hospital-treated pneumonia, surgical site infections and chemotherapy-induced neutropenia. Intriguingly the peptide Omeganan, developed by Migenix Inc. (Vancouver, Canada) as an antimicrobial peptide, has demonstrated efficacy as an anti-inflammatory in suppressing the effects of acute acne and rosacea in Phase II clinical trials. The development of immunomodulatory HDPs is particularly attractive owing to their ability to resolve infections by antibiotic resistant bacteria and their ability to stimulate natural host defence effector mechanisms without inducing potentially harmful excessive pro-inflammatory responses. Since they also exhibit multi-faceted immunomodulatory capabilities, they may also circumvent the problems associated with stimulation or inhibition of one individual process that could affect TLR-targeted agonists/antagonists.

Other immunomodulatory peptides in development for infectious disease include SCV-07 (gamma-D-glutamyl-L-tryptophan), a synthetic peptide with proven immune stimulating properties, which is already licensed in Russia for the treatment of tuberculosis. SciClone (California, USA) has introduced this compound into Phase II trials for HCV treatment.[23] Implicit Biosciences (Brisbane, Australia) is developing IM862 (oglufanide disodium), a dipeptide of L-glutamyl-L-tryptophan with known anti-angiogenic and immunomodulatory properties initially isolated from the thymus,[24] for HCV therapy, and has recently embarked on a Phase II trial in the USA.

Challenges in the Development of Immunomodulatory Anti-Infectives

The development of immunomodulatory therapeutics is subject to unique difficulties arising due to interspecies and intraspecies variation and redundancy within the innate immune system. The potential lack of correlation between outcomes observed in animal models and human subjects can occur owing to fundamental interspecies differences in innate immunity (e.g. between human and mouse chemokine systems[25]) and is a problem that necessitates greater thought in establishing appropriate animal models as well as improved orthology predictions for innate immune-related genes. Genetic variation between individuals is another important yet unpredictable feature that can impact upon the effectiveness of therapies and has particular relevance for the development of immune-targeted therapies, since genes encoding immune-related proteins are among the fastest evolving within mammalian genomes. Given its hugely important role, it makes sense that redundancy is a feature of the innate immune system. The duplication of crucial features creates an important safety net, but also potentially undermines the effectiveness of therapies that target a single immune component, such as chemokines or their receptors. Despite their enormous potential, many new immunomodulator candidates have not progressed through clinical trials. This can be attributed partly to our incomplete understanding of the complex nature of the immune system and interindividual variability.[26] Certainly, continued interrogation of innate immunity using functional genomic and proteomic approaches at the systems biology level together with sophisticated bioinformatics analysis is necessary for the progression of novel immunomodulatory candidates into clinical use.

Knowledge and Technology Advances That Will Accelerate the Development of Novel Anti-Infective Immunomodulators

The application of new systems biology approaches and techniques such as siRNA gene silencing and transcriptional network profiling to study innate immunity is already underway and will undoubtedly improve our understanding of innate immune responses and host–pathogen interactions and thus hopefully expedite the development of novel immunomodulators.[27•–29] New bioinformatics resources have recently become available with the launches of Innate DB (www.innatedb .com)[30•] and IIDB (http://db.systemsbiology.net/IIDB),[31] which are innate immunity-specific databases that include data analysis resources to facilitate the functional analysis of innate immunity responses, as well as the IIPGA program (www .innateimmunity.net), which is a collaborative effort to analyze polymorphisms in human innate immunity genes.

It is noteworthy that the majority of existing and prospective immunomodulators are proteins, peptides or nucleic acids, which renders them more prone to difficulties in drug stability and delivery than conventional small molecule drugs. The development of pegylated forms of IFN, that display greatly enhanced stability, demonstrates how technological advances can improve the clinical usefulness of biologic therapies. Idera Pharmaceuticals (Cambridge, MA, USA) has recently generated chemically modified RNA compounds with increased resistance to nuclease activity that stimulate TLR-7 and TLR-8 in vitro and in vivo[32] as part of its development of nucleic acid-based TLR agonists. Advances in delivery systems for cytokines have also been made, such as the use of biodegradable microparticles constructed from poly-(lactideco-glycolide) (PLGA), designed for ex vivo T-cell expansion, that permit the sustained release of IL-2.[33] Such advances in formulation and delivery systems will also help expedite the progression of immunomodulators into clinical use.[34]

Conclusion

In recent years it has become evident that we are entering a 'post-antibiotic era' in which many previously successful drug regimes are becoming ineffective and it is only a matter of time before others follow suit. Pathogen-directed treatments will always be subject to the risk of the emergence of resistance; consequently the time has come to vigorously pursue alternative potential treatment approaches to infectious diseases. The targeting of innate immunity represents an intuitive new approach, and one that should be explored since there are few practical options available. Since disease is a manifestation of the pathogen's ability to overcome or subvert host immune responses, we argue that the development of novel anti-infectives that target the host immune system should warrant high priority. While this approach is not without its risks, as with conventional, pathogen-directed drug therapies, the therapeutic targeting of innate immunity is a concept in its infancy and hence the risks associated may be mitigated by more extensive research into the field, which is already underway, together with the implementation of new technology platforms and methodologies. Greater understanding of the processes and regulation of innate responses will intuitively lead to improved strategies and solutions to overcome the problems that accompany the development of any new treatment.

So far, the concept of harnessing innate immunity to treat infectious diseases has been adopted primarily by the biotech community to target viral infections, despite the urgent clinical need also for treatments of antibiotic resistant bacterial strains. This perhaps reflects the recent emergence of viral diseases of global concern such as SARS and avian influenza, but also the fact that since the generation of pathogen targeting antiviral therapies has traditionally been more challenging, the prospect of alternate approaches to treating viral disease has been received with greater enthusiasm. In addition, because antibiotics have been so successful for many decades, most ongoing anti-infective development has remained focused on pathogen-directed therapies. It is well documented that the extent of antibacterial research is woefully mismatched to the need for new treatments and this is also due to financial obstacles in the development of new treatments, since anti-infectives tend to be less profitable then ventures targeting other types of diseases and are therefore less attractive to large pharmaceutical companies.[35,36]

The existence of immunomodulators already in clinical use highlights that intelligent targeting of components of innate immunity is an achievable and efficacious route, and one that offers great potential for the future since it circumvents the problems of resistance that blight current treatments for infectious diseases. Immunomodulators may also be useful as adjunct therapies in conjunction with current anti-infectives, such that the treatment regime might consist of dual or multiple drugs that conjointly directly target the microbe itself in addition to providing an immunomodulator that boosts, suppresses or subtly adapts the immune response in such a way as to enhance host defences.

References and Recommended Reading

Papers of particular interest, published within the period of review, have been highlighted as:

• of special interest
•• of outstanding interest

1. World Health Statistics on World Wide Web URL: www.who.int/whosis/whostat/2008/en/index.html.
2. Creagh EM, O'Neill LA: TLRs, NLRs and RLRs: a trinity of pathogen sensors that co-operate in innate immunity. *Trends Immunol* 2006, **27**:352–357.
3. O'Neill LA: Targeting signal transduction as a strategy to treat inflammatory diseases. *Nat Rev Drug Discov* 2006, **5**:549–563.
4. Kanzler H, Barrat FJ, Hessel EM, Coffman RL: Therapeutic targeting of innate immunity with Toll-like receptor agonists and antagonists. *Nat Med* 2007, **13**:552–559.
5. Romagne F: Current and future drugs targeting one class of innate immunity receptors: the Toll-like receptors. *Drug Discov Today* 2007, **12**:80–87.
6. Wales J, Andreakos E, Feldmann M, Foxwell B: Targeting intracellular mediators of pattern-recognition receptor signalling to adjuvant vaccination. *Biochem Soc Trans* 2007, **35**:1501–1503.
7. Casadevall A, Pirofski LA: The damage-response framework of microbial pathogenesis. *Nat Rev Microbiol* 2003, **1**:17–24.
8. Bilu D, Sauder DN: Imiquimod: modes of action. *Br J Dermatol* 2003, **149(Suppl. 66)**:5–8.
9. Finlay BB, McFadden G: Anti-immunology: evasion of the host immune system by bacterial and viral pathogens. *Cell* 2006, **124**:767–782.
10. Schwegmann A, Brombacher F: Host-directed drug targeting of factors hijacked by pathogens. *Sci Signal* 2008, **1**:re8.
11. Thompson KA, Strayer DR, Salvato PD, Thompson CE, Klimas N, Molavi A, Hamill AK, Zheng Z, Ventura D, Carter WA: Results of a double-blind placebo-controlled study of the double-stranded RNA drug polyI:polyC12U in the treatment of HIV infection. *Eur J Clin Microbiol Infect Dis* 1996, **15**:580–587.
12. Agrawal S, Kandimalla ER: Synthetic agonists of Toll-like receptors 7, 8 and 9. *Biochemical Society Transactions* 2007, **035**:1461–1467.
13. Wang T, Town T, Alexopoulou L, Anderson JF, Fikrig E, Flavell RA: Toll-like receptor 3 mediates West Nile virus entry into the brain causing lethal encephalitis. *Nat Med* 2004, **10**:1366–1373.
14. de Jong MA, de Witte L, Oudhoff MJ, Gringhuis SI, Gallay P, Geijtenbeek TB: TNF-alpha and TLR agonists increase susceptibility to HIV-1 transmission by human Langerhans cells *ex vivo, J Clin Invest* 2008, **118**:3440–3452.
15. Kammer AR, Amacker M, Rasi S, Westerfeld N, Gremion
• C, Neuhaus D, Zurbriggen R: A new and versatile virosomal antigen delivery system to induce cellular and humoral immune responses. *Vaccine* 2007, **25**:7065–7074.

This paper describes innovations in the development of immunopotentiating influenza virosomes, which increase stability while inducing potent cytotoxic T-cell responses, thus improving their suitability as a carrier adjuvant system for therapeutic and prophylactic vaccines.

16. Moser C, Amacker M, Kammer AR, Rasi S, Westerfeld N, Zurbriggen R: Influenza virosomes as a combined vaccine carrier and adjuvant system for prophylactic and therapeutic immunizations. *Expert Rev Vaccines* 2007, **6**:711–721.
17. Jenssen H, Hamill P, Hancock REW: Peptide antimicrobial agents. *Clin Microbiol Rev* 2006, **19**:491–511.
18. Yu J, Mookherjee N, Wee K, Bowdish DM, Pistolic J, Li Y, Rehaume L, Hancock REW: Host defense peptide LL-37, in synergy with inflammatory mediator IL-1beta, augments immune responses by multiple pathways. *J Immunol* 2007, **179**:7684–7691.
19. Bowdish DM, Davidson DJ, Hancock REW: Immunomodulatory properties of defensins and cathelicidins. *Curr Top Microbiol Immunol* 2006, **306**:27–66.
20. Tjabringa GS, Rabe KF, Hiemstra PS: The human cathelicidin LL-37: a multifunctional peptide involved in infection and inflammation in the lung. *Pulm Pharmacol Ther* 2005, **18**:321–327.
21. Hancock REW, Sahl HG: Antimicrobial and host-defense peptides as new anti-infective therapeutic strategies. *Nat Biotechnol* 2006, **24**:1551–1557.
22. Scott MG, Dullaghan E, Mookherjee N, Glavas N, Waldbrook M,
•• Thompson A, Wang A, Lee K, Doria S, Hamill P *et al.*: An anti-infective peptide that selectively modulates the innate immune response. *Nat Biotechnol* 2007, **25**:465–472.

This paper presents the first proof of principle study showing that a synthetic immunomodulatory peptide can confer protection against infection with multiple bacterial pathogens, including multiple antibiotic resistant strains, without toxicity and while suppressing harmful pro-inflammatory responses.

23. Aspinall RJ, Pockros PJ: SCV-07 (SciClone Pharmaceuticals/ Verta). *Curr Opin Investig Drugs* 2006, **7**:180–185.
24. Tulpule A, Scadden DT, Espina BM, Cabriales S, Howard W, Shea K, Gill PS: Results of a randomized study of IM862 nasal solution in the treatment of AIDS-related Kaposi's sarcoma. *J Clin Oncol* 2000, **18**:716–723.
25. Viola A, Luster AD: Chemokines and their receptors: drug targets in immunity and inflammation. *Annu Rev Pharmacol Toxicol* 2008, **48**:171–197.
26. Brown KL, Cosseau C, Gardy JL, Hancock REW: Complexities of targeting innate immunity to treat infection. *Trends Immunol* 2007, **28**:260–266.
27. Alper S, Laws R, Lackford B, Boyd WA, Dunlap P, Freedman JH,
• Schwartz DA: Identification of innate immunity genes and pathways using a comparative genomics approach. *Proc Natl Acad Sci U S A* 2008, **105**:7016–7021.

This is an elegant study of the application of siRNA knockdown methods and comparative genomics to identify novel genes involved in the regulation of innate immunity, in both *C. elegans* and murine systems, which may also represent novel therapeutic targets.

28. Katze MG, Fornek JL, Palermo RE, Walters KA, Korth MJ: Innate immune modulation by RNA viruses: emerging insights from functional genomics. *Nat Rev Immunol* 2008, **8**:644–654.

29. Tegner J, Nilsson R, Bajic VB, Bjorkegren J, Ravasi T: Systems biology of innate immunity. *Cell Immunol* 2006, **244**:105–109.

30. Lynn DJ, Winsor GL, Chan C, Richard N, Laird MR, Barsky A,
• Gardy JL, Roche FM, Chan TH, Shah N *et al.*: InnateDB: facilitating systems-level analyses of the mammalian innate immune response. *Mol Syst Biol* 2008, **4:** 218.

Introduces a novel, highly detailed, and manually curated, database of genes and cellular pathways involved in innate immunity in human and murine systems together with new analysis and visualization tools to facilitate innate immunity research.

31. Korb M, Rust AG, Thorsson V, Battail C, Li B, Hwang D, Kennedy KA, Roach JC, Rosenberger CM, Gilchrist M *et al.*: The innate immune database (IIDB). *BMC Immunol* 2008, **9**:7.

32. Lan T, Kandimalla ER, Yu D, Bhagat L, Li Y, Wang D, Zhu F, Tang JX, Putta MR, Cong Y *et al.*: Stabilized immune modulatory RNA compounds as agonists of Toll-like receptors 7 and 8. *Proc Natl Acad Sci U S A* 2007, **104**:13750–13755.

33. Steenblock ER, Fahmy TM: A comprehensive platform for *ex vivo* T-cell expansion based on biodegradable polymeric artificial antigen-presenting cells. *Mol Ther* 2008, **16**:765–772.

34. Kobsa S, Saltzman WM: Bioengineering approaches to controlled protein delivery. *Pediatr Res* 2008, **63**:513–519.

35. Talbot GH, Bradley J, Edwards JE Jr, Gilbert D, Scheld M, Bartlett JG: Bad bugs need drugs: an update on the development pipeline from the Antimicrobial Availability Task Force of the Infectious Diseases Society of America. *Clin Infect Dis* 2006, **42**:657–668.

36. Bradley JS, Guidos R, Baragona S, Bartlett JG, Rubinstein E, Zhanel GG, Tino MD, Pompliano DL, Tally F, Tipirneni P *et al.*: Anti-infective research and development—problems, challenges, and solutions. *Lancet Infect Dis* 2007, **7**:68–78.

Acknowledgments—The author's research in this area of work was supported by the Foundation for the National Institutes of Health, Gates Foundation and Canadian Institutes for Health Research through two separate Grand Challenges in Global Health Initiatives and by Genome British Columbia for the Pathogenomics of Innate Immunity Research Program. REWH is the recipient of a Canada Research Chair.

Malaria Vaccines and Their Potential Role in the Elimination of Malaria

Geoffrey A. Targett and Brian M. Greenwood

Background

Development of a malaria vaccine has been difficult. Greatly expanded investment in malaria vaccine research and development in recent years has resulted in the identification of a substantial number of vaccine candidates that are now in clinical trials or in the late stages of pre-clinical development. Now the malaria vaccine community is faced with a new challenge. Do the vaccine development plans developed several years ago, when the main target of malaria vaccine development was reduction in the burden of clinical malaria, fit with the new and ambitious aim of achieving malaria elimination. Here the current situation with respect to malaria control, the particular challenges elimination strategies present, and the progress being made in vaccine development are considered. An assessment is made of what vaccines are needed and how they could be used most effectively as part of a malaria elimination programme.

The much quoted figures for malaria deaths and clinical cases—around 1 million deaths and 300–500 million clinical cases per annum, are still the best estimates available. The majority of these deaths are due to *Plasmodium falciparum* malaria and occur in sub-Saharan Africa.[1]

The importance of *Plasmodium vivax* infection, in particular in South-East Asia, and the severity of some infections caused by this malaria parasite have been underestimated but are now receiving more attention.[2]

There are, however, encouraging recent reports that show that a very significant improvement in the malaria situation is possible using existing control tools. Effective malaria control in high transmission areas seemed a remote possibility, even just a few years ago, but, with the substantial increase in political commitment and financial investment in control measures over the past 5–6 years, some dramatic results have been obtained. Some of the reported successes have occurred in countries or regions where malaria transmission was already low[3] but, in other cases, a significant downward trend has been achieved in places where transmission is stable; Zanzibar,[4] Eritrea,[5] The Gambia[6] and Kenya[7,8] are good examples.

These successes have involved scaling up of existing control measures, notably treatment with artimisinin-based combination therapy (ACT) or other effective drug combinations,

insecticide-treated nets (ITNs) and, increasingly, a return to insecticide-residual spraying (IRS). There has also been increased use of intermittent preventive treatment in pregnancy,[9] and this approach to malaria control is being explored in infants and older children. A high level of commitment to the discovery of new drugs and insecticides is essential to ensure that these gains are not lost when the drugs and insecticides in current use lose their effectiveness.

Effective malaria control is defined as a reduction in cases of clinical malaria and mortality to a level at which malaria ceases to be a major problem. The malaria parasite still persists in the community, country or region and, if the control measures are not sustained, there is every likelihood that transmission and numbers of cases will increase rapidly again. However, the somewhat surprising impact of scaling-up the use of existing control measures has prompted the call, first by the Bill and Melinda Gates Foundation, quickly endorsed by WHO, and then by the Roll Back Malaria partnership, for malaria elimination to become the new goal. This has a very different and far more challenging aim of stopping transmission completely within a defined region, so that the only cases of malaria that occur are through importation from outside the region.[10] Elimination by this definition can be achieved only by killing all of the parasites within the target population.

It is clear that, despite the successes achieved by scaling-up use of existing tools, additional or alternative strategies will be needed if malaria elimination is to be achieved—a possible exception being some island situations.[11]

Persistence of Infection

The focus of enhanced research and malaria control has, understandably, been primarily on *P. falciparum* malaria, given the mortality and severity of disease associated with this species. However, there is increasing recognition that the risk of infection and the burden of disease due to *P. vivax* malaria is substantial and, although of limited importance in sub-Saharan Africa, this parasite is often the dominant one in the other major endemic regions of the world. Frequently, *P. vivax* and *P. falciparum* occur sympatrically and elimination programmes, in such cases, must take account of the different challenges presented by the two species, especially in terms of persistence of infection, and the complex interactions between the species. This complex balance may be disturbed by vaccination against either *P. falciparum* or *P. vivax* in areas where both parasites are prevalent. Eliminating *P. falciparum* but not *P. vivax* would be a step forward, but if this was all that was achieved could damage the reputation of a malaria elimination programme.

Malaria elimination means stopping infection. This may be achieved if the measures directed against asexual parasites are fully effective, or by targeting sexual stages directly with drugs or vaccines.

Transmission of P. falciparum

In areas of high transmission, asexual parasite densities are highest in young children, and it is in this age group that microscopic detection of gametocytes is most common. Both asexual blood stage and gametocyte densities then decline with age, though the patterns of decline are somewhat different.[12] Epidemiological studies show, however, that transmission of *P. falciparum* is as dependent on the parasites not detected by routine blood screening as on those that are readily seen. The cumulative evidence for the importance of low-grade asymptomatic infections as a reservoir for infection of mosquitoes is strong. In areas of highly seasonal malaria, where there is often a long dry season in which little or no transmission occurs, persisting very low gametocytaemias are the source from which transmission occurs at the onset of the subsequent rainy season.[13] In areas where the endemicity of malaria allows acquired immunity to develop, there are many asymptomatic individuals, particularly adults, who are an important source of infection for mosquitoes.[14,15]

Detection of gametocytes using molecular techniques, such as reverse transcriptase polymerase chain reaction (RT-PCR)[13] and quantitative nucleic acid sequence-based amplification (QT-NASBA) has shown that gametocytaemias can persist at sub-microscopic levels for months and that the prevalence of gametocytaemic individuals is much higher than was assumed from blood film examinations.[16] Of particular relevance to consideration of malaria elimination, Shekalaghe *et al*,[17] in a study of parasite prevalence in an area of low, seasonal transmission in Tanzania, showed that, while microscopically the parasite rates were low (1.9% asexual parasites, 0.4% gametocytes), QT-NASBA revealed much higher prevalence rates (32.5% asexual parasites and 15.0% gametocytes).

These observations indicate the need to adopt a community-based approach to elimination; any selective interventions used should be based more on the focality of malaria,[18] rather than on particular groups especially at risk from the clinical consequences of malaria infection, such as young children or pregnant women. Those at low risk clinically may still be important transmitters of infection.

Transmission of P. vivax

A key factor in the persistence of *P. vivax* is the fundamental difference in the life cycle shown by *P. vivax* and *Plasmodium ovale,* compared with *P-. falciparum* and *Plasmodium.*

malariae, namely the occurrence of dormant hypnozoites in the liver, from which relapse infections can emerge, weeks, months or years after infection.

There is some evidence from use of PCR techniques, that, as for *P. falciparum,* the number of *P. vivax* infections in an affected community is generally significantly underestimated.[19] Though the pattern of gametocyte production in *P. vivax* is different from that of *P. falciparum,* low grade asymptomatic infections are equally likely to give rise to infectious gametocytes.

Progress with Vaccine Development

The challenges set to vaccine developers[20] by those who drew up the malaria vaccine road map are first, by 2015, to produce a licensed vaccine that has a protective efficacy of more than 50% against severe disease and death from malaria which lasts longer than one year. Secondly, by 2025, to develop a vaccine with a protective efficacy greater than 80% against clinical disease and death that lasts longer than four years.

These targets focus on the prevention of clinical disease, especially its severe and life-threatening form, valid objectives for vaccines that are to be introduced into areas of medium or high transmission. A vaccine that conformed to the 2015 objective, providing protection to half of those vaccinated, would be valuable as part of an integrated control programme alongside vector control and chemo-preventative measures. That level of efficacy would not justify its use alone as an alternative to other means of malaria control.

The expanding programme of experimental vaccine-related research has two broad but overlapping approaches. One is to achieve a much needed better understanding of the nature of protective immune mechanisms against malaria, thus providing a basis for rational vaccine design.[21] The other approach is more empirical and involves issues of design and presentation as vaccines of antigens that have been recognised for a long time and which are known to have an important role in the parasite's life cycle.

Pre-Erythrocytic Stage Vaccines

The primary objective of pre-erythrocytic stage vaccines is provision of a level of protection that prevents any invasion of the blood and hence any clinical malaria. This has been achieved readily in experimental malarias by vaccination with radiation-attenuated sporozoites; this formed the starting point for the extensive investigations into pre-erythrocytic vaccines. Complete protection was also achieved in humans against *P. falciparum* and *P. vivax* by exposing them to the bites of mosquitoes that had been irradiated to attenuate their sporozoites. However, this delivery procedure was totally impractical and the research emphasis shifted to the synthetic design and genetic-engineering of sub-unit vaccines.[22]

There has, however, been a renewal of interest in the development of attenuated-sporozoite vaccines. Given the high level of sterile protection these whole organism vaccines can induce, this is a very welcome development. Stephen Hoffman established and directs Sanaria Inc. specifically to produce radiation-attenuated sporozoites of *P. falciparum* from infected mosquitoes in sufficient quantity and in a way that meets the regulatory standards required for their use as a vaccine.[23] This has been achieved and phase I/2a clinical trials will begin shortly using irradiation-attenuated sporozoites delivered by intradermal or subcutaneous injection.

Genetic attenuation of sporozoites is also being investigated. Mueller *et al*[24] produced sporozoites from *Plasmodium berghei* deficient in the vis3 gene. These gave complete protection when used for vaccination. Sporozoites lacking 6-cysteine secretory proteins required for parasitophorous vacuole formation are equally effective vaccines.[25] The immune responses of importance following vaccination with attenuated sporozoites are CD8 + T cell-mediated with production of interferon gamma.[26–28]

Synthetic and genetically-engineered sub-unit vaccines have generally been based on the two surface proteins, circumsporozoite protein (CSP) and thrombospondin-related anonymous protein (TRAP) involved in sporozoite motility and invasion of liver cells. Most of the phase 2 trials of vaccines based on these antigens, using a variety of vaccine constructs, have given poor results and it is perhaps surprising that the RTS, S vaccine candidate has proved to be much more promising. This vaccine is a hybrid molecule expressed in yeast, that consists of the tandem repeat tetra-peptide (R) and C-terminal T-cell epitope containing (T) regions of CSP fused to the hepatitis B surface antigen (S), plus unfused S antigen. The adjuvant ASO2, which consists of an oil in water emulsion containing immunostimulants monophosphoryl lipid A and QS-21, a fraction of *Quillaia saponaria,* is an essential component of the vaccine. Variants of ASO2 have been tested successfully,[29] and an alternative adjuvant, ASO1, where a liposomal formulation replaces the oil in water emulsion, has given very encouraging results.[30,31]

The first challenge trials with RTS, S/ASO2 gave impressive, but short-lived, protection in naïve adults.[32] Similarly, in a trial in Gambian adults, there was greater than 70% protection against parasitaemia in the nine weeks post-vaccination, but the immunity waned rapidly.[33] The most comprehensive study of RTS, S/AS02 has been in children in Mozambique, who have been followed up for two years. Over this period, there was 30% protection against clinical malaria and close to 50% protection against severe malaria.[34] Most recently, in a small-scale trial in infants designed primarily to test safety and immunogenicity, RTS, S/AS02D had a vaccine efficacy of 65.9% against new infections in the six months of follow-up.[35] A series of phase 2 studies preparatory to a large phase 3 trial and potential licensure of this vaccine are in progress.

RTS, S is several years ahead of any other vaccine in terms of assessment of its efficacy in clinical trials. Trials of other pre-erythrocytic stage vaccines, based on CSP, TRAP and other liver stage antigens, several of which have used viral vector delivery systems, have shown some initial promise, but are not sufficiently advanced or effective to be considered yet for evaluation in phase 3 trials.[36,37]

Asexual Blood-Stage Candidate Vaccines

A range of blood stage antigens have progressed to phase 1 and phase 2 trials. Most are molecules that have been identified as being involved in the process of invasion of erythrocytes by merozoites.[38] Promising results from studies with rodent malarias are proving difficult to transfer to human infections,[39–43] though some early encouraging results with MSP-3 antigen have been reported.[44] The expectation is that such candidate vaccines might give protection against disease, but not against infection. This was the case with one phase 2 clinical trial of a vaccine containing MSP-1, MSP-2 and RESA antigens which reduced parasite density, but not prevalence of infection.[45] It was also strain-specific in its effect and the polymorphism of these antigens, coupled with the variability in invasion pathways *P. falciparum* can adopt[46,47] are a severe challenge to the design of this kind of vaccine. Blocking invasion of reticulocytes by *P. vivax* merozoites might be a more hopeful strategy, given that the Duffy antigen is thought to be the obligatory receptor on the erythrocyte surface[48] yet, even here an alternative invasion strategy has been proposed.[49]

Plasmodium falciparum-infected erythrocytes express highly variable parasite molecules on the red blood cell surface. Naturally-acquired immunity involves variant-specific responses to these antigens and the very complexity of these responses may limit the potential of these antigens as vaccine candidates. However, this variability might be exploitable for specifically targeted vaccines. The *P. falciparum*-infected erythrocytes that sequester in the placenta of pregnant women have a very selected sub-set of variant surface antigens, notably one coded VAR2CSA, through which they bind to chondroitin sulphate A (CSA) in the placenta. This opens the possibility of designing a vaccine that would be beneficial to pregnant women,[50] but which would probably not affect the other variants that circulate and sequester elsewhere using different receptor-ligand interactions. Another postulated approach to vaccination, much less studied, is to block parasite molecules that mediate disease by inducing pro-inflammatory responses. Glycosylphosphotidyl-inositol (GPI) is strongly implicated,[51] but any vaccine effect would alleviate symptoms of disease without affecting infection. The immunity induced by such a vaccine might prevent some of the severe complications of malaria mediated by cytokine-induced response to infection. Use of vaccines of this kind, however, has the potential disadvantage of damping down the early clinical features of malaria, such as fever, which could delay the time before a patient sought treatment whilst still allowing parasite multiplication to occur.

Transmission-Blocking Vaccines

The concept of a malaria vaccine that could provide an effective immune response when the antibodies induced had been ingested by blood-feeding mosquitoes, was developed thirty years ago.[52] The target antigens of the passively transferred antibodies that blocked transmission were shown to be sexual-stage specific surface molecules (Pfs 48/45 and Pfs 230 in *P. falciparum*), that are involved in the process of fertilization of macrogametocytes by microgametes. Subsequently, other antigens (P25 and P28 proteins), that are uniquely expressed in the mosquito by zygotes and ookinetes (i.e. after fertilization), were shown to be equally good for induction of transmission-blocking immunity (TBI). The end result in each case is to prevent sporogonic development in the vector.

Experimental studies with animal models have shown that it is possible to induce a highly effective TBI.[22,53] The most extensive studies have focused on Pfs 48/45, Pfs 230, Pfs 25, Pfs 28 of *P. falciparum* and orthologues in other *Plasmodium* species, but more potential vaccine candidates have been identified.[54]

The gamete surface molecules (48/45 and 230) are also expressed in gametocytes circulating in the blood. This has made it possible to study the nature and duration of naturally-acquired sexual-stage specific immunity, and has contributed importantly to an understanding of the epidemiology of gametocytes in comparison with that of the much more comprehensively studied asexual blood stages.[12] The P25 and P28 proteins are not expressed in gametocytes and hence there is no natural infection-related immunity to them. However, the mRNA that encodes these proteins is measurable in gametocytes and is used as the basis of highly sensitive means of detecting gametocytes.[16]

The transmission-blocking activity of sera from vaccinated animals or humans, or from individuals naturally infected, has mostly been assessed by an *ex vivo* assay, during which mosquitoes feed through a membrane on blood containing gametocytes and a serum under test. Comparison with controls of the numbers of mosquitoes that become infected, and the number of oocysts they carry, gives a measure of the potency of the serum under test.

Experimentally, sera from rabbits, monkeys and mice vaccinated with vaccine candidates Pfs25 from *P. falciparum* and Pvs25 from *P. vivax* contained antibodies with transmission-blocking activity. The antibody levels measured by ELISA correlated with both oocyst reduction and the number of mosquitoes that failed to become infected;[55] antibody levels persisted at a high level for months after a second or third injection in mice.[56] Phase 1 human vaccine trials were also effective,[57] though not yet at the level that will be required and can be achieved experimentally.

Alternatives to the membrane-feeding assay are also being assessed for evaluation of transmission-blocking activity. Transmission of the transgenic *P. berghei* expressing the P25 antigens of either *P. falciparum*[58] or *P. vivax*[59] was blocked by antibodies obtained from animal vaccination and phase 1 clinical trials.

Though Pfs48/45 has been clearly shown to induce antibodies that prevent fertilization and correlate with transmission-blocking activity, the conformational nature of the epitopes, and the cysteine-rich nature of the protein has made production of a correctly folded recombinant molecule difficult. Encouragingly, production of a stable, properly folded C terminal portion of the molecule that induces transmission-blocking antibodies has recently been described.[60,61]

Use of Vaccines in an Elimination Programme

Since elimination of malaria requires complete removal of all parasites, the focus of measures used to accomplish this goal is quite different from that for disease control. Whether the measures used for parasite elimination involve drugs[9] or vaccines, or a combination of the two, what is required are tools that prevent production of gametocytes or that render them non-infective to mosquitoes.

The ideal vaccine would be one that induces complete sterilizing immunity or which is fully effective at blocking transmission. Nothing approaching induction of a sterile immunity has been shown so far and it seems unlikely that this will be achieved with sub-unit vaccines. This does not rule out the use of such vaccines as part of an integrated approach to malaria-elimination, but they are unlikely to induce an anti-parasite immunity of sufficient efficacy to eliminate all parasites. It remains to be seen whether attenuated sporozoites, which, in small-scale studies with a demanding and unusable vaccination regime, did give full protection, will be as effective in the trials now beginning.

The transmission-blocking vaccines currently being tested induce good, but certainly not complete, transmission-blocking immunity. Since the membrane-feeding and the transgenic rodent malaria assays allow sera from clinical trials to be tested for efficacy, it is possible to set up a series of small-scale phase 1/2a trials with different vaccine constructs and vaccination regimes. It should then be possible to make speedy progress towards improving vaccine efficacy and selecting the best vaccine for development. There is an urgency to do this.

There are biological and clinical features of infection that need to be taken into account in designing elimination measures that include vaccines. These are listed in Table 1 and, in particular, involve the interaction between the acquired immune response and the gametocyte infectious reservoir. As discussed earlier in this review, the gametocyte reservoir is grossly underestimated when assessed by conventional means, even in areas of low transmission. The same is true of asexual parasitaemias. In areas of stable transmission, numbers of gametocytes decrease with age more rapidly than do numbers of asexual parasites, but the proportion of individuals who are gametocytaemic is actually higher at lower transmission intensities.[12] An increase in gametocyte prevalence has also been seen following control studies.[62] For this and other reasons adult carriage becomes as important as that of children in terms of the source of infection for mosquitoes.

Naturally acquired immune responses to the sexual stages of the malaria parasite develop rapidly (in individuals with little or no previous exposure to malaria) and have transmission-blocking activity.[63] However, in endemic areas, the ability of sera to reduce transmission decreases in older age groups, corresponding to a reduction in gametocyte numbers.[64] The biological significance of this loss of immunity has been reviewed[12] and further supports evidence that adults with few parasites are nevertheless important as a reservoir of infection. There is evidence too, particularly for *P. vivax*, that a low level of immunity can enhance transmission to mosquitoes.[65]

Table 1 Acquired Immunity, Persistence of Gametocytes, and Transmission

1. Acquired immunity to asexual blood stages increases with exposure but allows the persistence of low level parasitaemias from which gametocytes develop.

2. Anti-parasitic immunity will persist after interruption of transmission and may allow the occurrence of asymptomatic infections and hence gametocytaemias for a number of years.

3. Transmission-blocking immunity to gametocytes develops rapidly then wanes so that adults carrying small numbers of gametocytes are less likely to have antibodies that render them non-infective.

4. Low levels of transmission-blocking antibodies have been shown to enhance transmission.

5. The gametocyte reservoir is much larger than that determined by blood film examination even in areas where transmission is low. Transmission of infection can occur from individuals with very low numbers of gametocytes.

6. Gametocyte infectivity is broadly related to gametocyte density but small numbers of infectious gametocytes in adults are as important in maintaining transmission as larger numbers of gametocytes in children susceptible to clinical attacks of malaria.

7. A vaccine for elimination of malaria must ensure everyone susceptible to malaria infection is protected completely or that all gametocytes are made non-infective.

It is frequently stated that, as malaria control and elimination programmes progress, the population no longer has any immunity and becomes highly susceptible to epidemic infections. While this is true to some extent, especially for younger age groups, there is another aspect of immunity to consider. It is well known that those who had immunity to malaria quite rapidly lose their clinical immunity if no longer exposed and may develop a symptomatic infection at low parasite densities on subsequent exposure. However, the anti-parasite immunity they had acquired can persist for many years.[66–68] In other words, there is immunological memory directed against the infection. The relevance of this to malaria elimination is that, for many years, there may be a proportion of the population capable of supporting low grade infections from which gametocytes can be derived.

Conclusion

Elimination programmes are focused on populations, not individuals, and, optimally, a herd immunity that is sustained and prevents transmission is required. The current vaccine development programmes are largely concerned with disease control and must be sustained, but, from the results obtained so far in clinical trials of the RTS, S vaccine, the impact is not dissimilar to that of ITNs, i.e. there is a greater impact on severe disease than on infection. It is difficult to imagine partially effective asexual blood stage vaccines being very useful for parasite elimination. Given that all age groups can potentially provide a source of infection to mosquitoes, a high proportion of the population will need to be given a vaccine. However, there is much evidence to show that malaria is a highly focal disease and, initially at least, it might be beneficial to vaccinate those at greatest risk of being bitten by vector mosquitoes. Whatever vaccine is employed, pre-erythrocytic or transmission-blocking, its efficacy would need to be very high to achieve elimination.[9]

A transmission-blocking vaccine has no direct effect on clinical malaria, but would break the life-cycle between human and malaria. It might be used as a stand-alone vaccine, but, more appropriately, as part of an integrated programme involving drug-treatment and vector control. The concept of a multi-component, multi-stage vaccine[69] with an effect from one component mainly on disease and from another on infection is intuitively appealing though the type of vaccine design required to kill parasites or render gametocytes non-infective to mosquitoes might be quite different from vaccines whose beneficial effect is mainly against disease severity. What is needed is an expanded malaria vaccine programme targeting the particular requirements of malaria elimination.

Competing Interests

The authors declare that they have no competing interests.

Authors' Contributions

GT prepared the first draft of the manuscript. BG and GT determined the content of the review and the reference sources, and agreed the final form of the manuscript.

References

1. Guerra CA, Gikandi PW, Tatem AJ, Noor AM, Smith DL, Hay SI, Snow RW: The limits and intensity of *Plasmodium falciparum* transmission: implications for malaria control and elimination worldwide. *PLoS Med* 2008, **5:**e38.

2. Baird JK: Neglect of *Plasmodium vivax* malaria. *Trends Parasitol* 2007, **23:**533–539.

3. Barnes KI, Durrheim DN, Little F, Jackson A, Mehta U, Allen E, Dlamini SS, Tsoka J, Bredenkamp B, Mthembu DJ, White NJ, Sharp BL: Effect of artemether-lumefantrine policy and improved vector control on malaria burden in KwaZulu-Natal, South Africa. *PLoS Med* 2005, **2:**e330.

4. Bhattarai A, Ali AS, Kachur SP, Martensson A, Abbas AK, Khatib R, Al-Mafazy AW, Ramsan M, Rotllant G, Gerstenmaier JF, Molteni F, Abdulla S, Montgomery SM, Kaneko A, Björkman A: Impact of artemisinin-based combination therapy and insecticide-treated nets on malaria burden in Zanzibar. *PLoS Med* 2007, **4:**e309.

5. Nyarango PM, Gebremeskel T, Mebrahtu G, Mufunda J, Abdulmumini U, Ogbamariam A, Kosia A, Gebremichael A, Gunawardena D, Ghe-brat Y, Okbaldet Y: A steep decline of malaria morbidity and mortality trends in Eritrea between 2000 and 2004: the effect of combination of control methods. *Malar J* 2006, **5**:33.

6. Ceesay SJ, Casals-Pascual C, Erskine J, Anya SE, Duah NO, Fulford AJ, Sesay SS, Abubakar I, Dunyo S, Sey O, Palmer A, Fofana M, Corrah T, Bojang KA, Whittle HC, Greenwood BM, Conway DJ: Changes in malaria indices between 1999 and 2007 in The Gambia: a retrospective analysis. *Lancet* 2008, **372**:1545–1554.

7. Okiro EA, Hay SI, Gikandi PW, Sharif SK, Noor AM, Peshu N, Marsh K, Snow RW: The decline in paediatric malaria admissions on the coast of Kenya. *Malar J* 2007, **6**:151.

8. O'Meara WP, Bejon P, Mwangi TW, Okiro EA, Peshu N, Snow RW, Newton CR, Marsh K: Effect of a fall in malaria transmission on morbidity and mortality in Kilifi, Kenya. *Lancet* 2008, **372**:1555–1562.

9. Greenwood BM: From malaria control to elimination: implications for the research agenda. *Trends Parasitol* 2008, **24**:449–454.

10. Feachem R, Sabot O: A new global malaria eradication strategy. *Lancet* 2008, **371**:1633–1635.

11. Kaneko A, Taleo G, Kalkoa M, Yamar S, Kobayakawa T, Bjorkman A: Malaria eradication on islands. *Lancet* 2000, **356**:1560–1564.

12. Drakeley C, Sutherland C, Bousema JT, Sauerwein RW, Targett GA: The epidemiology of *Plasmodium falciparum* gametocytes: weapons of mass dispersion. *Trends Parasitol* 2006, **22**:424–430.

13. Abdel-Wahab A, Abdel-Muhsin AM, Ali E, Suleiman S, Ahmed S, Walliker D, Babiker HA: Dynamics of gametocytes among *Plasmodium falciparum* clones in natural infections in an area of highly seasonal transmission. *J Infect Dis* 2002, **185**:1838–1842.

14. Drakeley CJ, Akim NI, Sauerwein RW, Greenwood BM, Targett GA: Estimates of the infectious reservoir of *Plasmodium falciparum* malaria in The Gambia and in Tanzania. *Trans R Soc Trop Med Hyg* 2000, **94**:472–476.

15. Schneider P, Bousema JT, Gouagna LC, Otieno S, Vegte-Bolmer M van de, Omar SA, Sauerwein RW: Submicroscopic *Plasmodium falciparum* gametocyte densities frequently result in mosquito infection. *Am J Trop Med Hyg* 2007, **76**:470–474.

16. Bousema JT, Schneider P, Gouagna LC, Drakeley CJ, Tostmann A, Houben R, Githure JI, Ord R, Sutherland CJ, Omar SA, Sauerwein RW: Moderate effect of artemisinin-based combination therapy on transmission of *Plasmodium falciparum*. *J Infect Dis* 2006, **193**:1151–1159.

17. Shekalaghe SA, Bousema JT, Kunei KK, Lushino P, Masokoto A, Wolters LR, Mwakalinga S, Mosha FW, Sauerwein RW, Drakeley CJ: Sub-microscopic *Plasmodium falciparum* gametocyte carriage is common in an area of low and seasonal transmission in Tanzania. *Trop Med Int Health* 2007, **12**:547–553.

18. Woolhouse ME, Dye C, Etard JF, Smith T, Charlwood JD, Garnett GP, Hagan P, Hii JL, Ndhlovu PD, Quinnell RJ, *et al.*: Heterogeneities in the transmission of infectious agents: implications for the design of control programs. *Proc Natl Acad Sci USA* 1997, **94**:338–342.

19. Kasehagen LJ, Mueller I, McNamara DT, Bockarie MJ, Kiniboro B, Rare L, Lorry K, Kastens W, Reeder JC, Kazura JW, Zimmerman PA: Changing patterns of Plasmodium blood-stage infections in the Wosera region of Papua New Guinea monitored by light microscopy and high throughput PCR diagnosis. *Am J Trop Med Hyg* 2006, **75**:588–596.

20. Malaria Vaccine Technology Roadmap [http://www.malariavaccineroadmap.net]

21. Stevenson MM, Zavala F: Immunology of malaria infections–implications for the design and development of malaria vaccines. *Parasite Immunol* 2006, **28**:1–60.

22. Targett GA: Malaria vaccines 1985–2005: a full circle? *Trends Parasitol* 2005, **21**:499–503.

23. Sanaria [http://www.sanaria.com]

24. Mueller AK, Labaied M, Kappe SH, Matuschewski K: Genetically modified Plasmodium parasites as a protective experimental malaria vaccine. *Nature* 2005, **433**:164–167.

25. Labaied M, Harupa A, Dumpit RF, Coppens I, Mikolajczak SA, Kappe SH: *Plasmodium yoelii* sporozoites with simultaneous deletion of P52 and P36 are completely attenuated and confer sterile immunity against infection. *Infect Immun* 2007, **75**:3758–3768.

26. Jobe O, Lumsden J, Mueller AK, Williams J, Silva-Rivera H, Kappe SH, Schwenk RJ, Matuschewski K, Krzych U: Genetically attenuated *Plasmodium berghei* liver stages induce sterile protracted protection that is mediated by major histocompatibility complex Class I-dependent interferon-gamma-producing CD8 + T cells. *J Infect Dis* 2007, **196**:599–607.

27. Mueller AK, Deckert M, Heiss K, Goetz K, Matuschewski K, Schluter D: Genetically attenuated *Plasmodium berghei* liver stages persist and elicit sterile protection primarily via CD8 T cells. *Am J Pathol* 2007, **171**:107–115.

28. Tarun AS, Dumpit RF, Camargo N, Labaied M, Liu P, Takagi A, Wang R, Kappe SH: Protracted sterile protection with *Plasmodium yoelii* pre-erythrocytic genetically attenuated parasite malaria vaccines is independent of significant liver-stage persistence and is mediated by CD8 + T cells. *J Infect Dis* 2007, **196**:608–616.

29. Barbosa A, Naniche D, Manaca MN, Aponte JJ, Mandomando I, Aide P, Renom M, Sacarlal J, Ballou WR, Moris P, Cohen J, Dubovsky F, Millman J, Alonso PL: Assessment of cellular immune responses in infants participating in a RTS, S/ASO2D phase I/IIB trial in Mozambique. *Am J Trop Med Hyg* 2007, **77(Abstr 9)**:2–3.

30. Lell B, Agnandji S, von Glasenapp I, Oyakhiromen S, Haertle S, Kremsner PG, Ramboer I, Lievens M, Ballou WR, Vekemans J, Dubois MC, Demoitie M-A, Cohen J, Villafana T, Carter T, Petersen T: A randomized, observer-blind trial to compare safety and immunogenicity of two adjuvanted RTS, S anti-malaria vaccine candidates in Gabonese children. *Am J Trop Med Hyg* 2007, **77(Abstr 10)**:3.

31. Anyona SB, Hunja CW, Kifude CM, Polhemus ME, Heppner DG, Leach A, Lievens M, Ballou WR, Cohen J, Sutherland C: Impact of RTS, S/ASO2A and RTS, s/ASO1B on multiplicity of infections and CSP T-cell epitopes of *Plasmodium falciparum* in adults participating in a malaria vaccine clinical trial. *Am J Trop Med Hyg* 2007, **77(Abstr 578)**:166.

32. Kester KE, McKinney DA, Tornieporth N, Ockenhouse CF, Heppner DG, Hall T, Krzych U, Delchambre M, Voss G, Dowler MG, Palensky J, Wittes J, Cohen J, Ballou WR, RTS S Malaria Vaccine Evaluation Group: Efficacy of recombinant circumsporozoite protein vaccine regimens against experimental *Plasmodium falciparum* malaria. *J Infect Dis* 2001, **183**:640–647.

33. Bojang KA, Milligan PJ, Pinder M, Vigneron L, Alloueche A, Kester KE, Ballou WR, Conway DJ, Reece WH, Gothard P, Yamuah L, Delchambre M, Voss G, Greenwood BM, Hill A, McAdam KP, Tornieporth N, Cohen JD, Doherty T, RTS S Malaria Vaccine Trial Team: Efficacy of RTS, S/AS02 malaria vaccine against *Plasmodium falciparum* infection in semi-immune adult men in The Gambia: a randomised trial. *Lancet* 2001, **358**:1927–1934.

34. Alonso PL, Sacarlal J, Aponte JJ, Leach A, Macete E, Aide P, Sigauque B, Milman J, Mandomando I, Bassat Q, Guinovart C,

Espasa M, Corachan S, Lievens M, Navia MM, Dubois MC, Menendez C, Dubovsky F, Cohen J, Thompson R, Ballou WR: Duration of protection with RTS, S/AS02A malaria vaccine in prevention of *Plasmodium falciparum* disease in Mozambican children: single-blind extended follow-up of a randomised controlled trial. *Lancet* 2005, **366:**2012–2018.

35. Aponte JJ, Aide P, Renom M, Mandomando I, Bassat Q, Sacarlal J, Manaca MN, Lafuente S, Barbosa A, Leach A, Lievens M, Vekemans J, Sigauque B, Dubois MC, Demoitié MA, Sillman M, Savarese B, McNeil JG, Macete E, Ballou WR, Cohen J, Alonso PL: Safety of the RTS, S/AS02D candidate malaria vaccine in infants living in a highly endemic area of Mozambique: a double blind randomised controlled phase I/IIb trial. *Lancet* 2007, **370:**1543–1551.

36. Hill AV: Pre-erythrocytic malaria vaccines: towards greater efficacy. *Nat Rev Immunol* 2006, **6:**21–32.

37. Walther M: Advances in vaccine development against the pre-erythrocytic stage of *Plasmodium falciparum* malaria. *Expert Rev Vaccines* 2006, **5:**81–93.

38. Hu J, Chen Z, Gu J, Wan M, Shen Q, Kieny MP, He J, Li Z, Zhang Q, Reed ZH, Zhu Y, Li W, Cao Y, Qu L, Cao Z, Wang Q, Liu H, Pan X, Huang X, Zhang D, Xue X, Pan W: Safety and immunogenicity of a malaria vaccine, *Plasmodium falciparum* AMA-1/MSP-1 chimeric protein formulated in montanide ISA 720 in healthy adults. *PLoS ONE* 2008, **3:**e1952.

39. Greenwood BM, Fidock DA, Kyle DE, Kappe SH, Alonso PL, Collins FH, Duffy PE: Malaria: progress, perils, and prospects for eradication. *J Clin Invest* 2008, **118:**1266–1276.

40. Dicko A, Sagara I, Ellis RD, Miura K, Guindo O, Kamate B, Sogoba M, Niambele MB, Sissoko M, Baby M, Dolo A, Mullen GE, Fay MP, Pierce M, Diallo DA, Saul A, Miller LH, Doumbo OK: Phase 1 study of a combination AMA1 blood stage malaria vaccine in Malian children. *PLoS ONE* 2008, **3:**e1563.

41. Malkin E, Long CA, Stowers AW, Zou L, Singh S, MacDonald NJ, Narum DL, Miles AP, Orcutt AC, Muratova O, Moretz SE, Zhou H, Diouf A, Fay M, Tierney E, Leese P, Mahanty S, Miller LH, Saul A, Martin LB: Phase 1 study of two merozoite surface protein 1 (MSP1(42)) vaccines for *Plasmodium falciparum* malaria. *PLoS Clin Trials* 2007, **2:**e12.

42. Richie T: High road, low road? Choices and challenges on the pathway to a malaria vaccine. *Parasitology* 2006, **133(Suppl S1):**13–144.

43. Remarque EJ, Faber BW, Kocken CH, Thomas AW: Apical membrane antigen 1: a malaria vaccine candidate in review. *Trends Parasitol* 2008, **24:**74–84.

44. Druilhe P, Spertini F, Soesoe D, Corradin G, Mejia P, Singh S, Audran R, Bouzidi A, Oeuvray C, Roussilhon C: A malaria vaccine that elicits in humans antibodies able to kill *Plasmodium falciparum*. *PLoS Med* 2005, **2:**e344.

45. Genton B, Betuela I, Felger I, Al-Yaman F, Anders RF, Saul A, Rare L, Baisor M, Lorry K, Brown GV, Pye D, Irving DO, Smith TA, Beck HP, Alpers MP: A recombinant blood-stage malaria vaccine reduces *Plasmodium falciparum* density and exerts selective pressure on parasite populations in a phase 1–2b trial in Papua New Guinea. *J Infect Dis* 2002, **185:**820–827.

46. Baum J, Maier AG, Good RT, Simpson KM, Cowman AF: Invasion by *P. falciparum* merozoites suggests a hierarchy of molecular interactions. *PLoS Pathog* 2005, **1:**e37.

47. Takala SL, Coulibaly D, Thera MA, Dicko A, Smith DL, Guindo AB, Kone AK, Traore K, Ouattara A, Djimde AA, Sehdev PS, Lyke KE, Diallo DA, Doumbo OK, Plowe CV: Dynamics of polymorphism in a malaria vaccine antigen at a vaccine-testing site in Mali. *PLoS Med* 2007, **4:**e93.

48. Devi YS, Mukherjee P, Yazdani SS, Shakri AR, Mazumdar S, Pandey S, Chitnis CE, Chauhan VS: Immunogenicity of *Plasmodium vivax* combination subunit vaccine formulated with human compatible adjuvants in mice. *Vaccine* 2007, **25:**5166–5174.

49. Cavasini CE, Mattos LC, Couto AA, Bonini-Domingos CR, Valencia SH, Neiras WC, Alves RT, Rossit AR, Castilho L, Machado RL: *Plasmodium vivax* infection among Duffy antigen-negative individuals from the Brazilian Amazon region: an exception? *Trans R Soc Trop Med Hyg* 2007, **101:**1042–1044.

50. Duffy PE: Plasmodium in the placenta: parasites, parity, protection, prevention and possibly preeclampsia. *Parasitology* 2007, **134:**1877–1881.

51. Schofield L, Hewitt MC, Evans K, Siomos MA, Seeberger PH: Synthetic GPI as a candidate anti-toxic vaccine in a model of malaria. *Nature* 2002, **418:**785–789.

52. Carter R: Transmission blocking malaria vaccines. *Vaccine* 2001, **19:**2309–2314.

53. Saul A: Efficacy model for mosquito stage transmission blocking vaccines for malaria. *Parasitology* 2008, **135:** 1497–1506.

54. Saul A: Mosquito stage, transmission blocking vaccines for malaria. *Curr Opin Infect Dis* 2007, **20:**476–481.

55. Miura K, Keister DB, Muratova OV, Sattabongkot J, Long CA, Saul A: Transmission-blocking activity induced by malaria vaccine candidates Pfs25/Pvs25 is a direct and predictable function of antibody titer. *Malar J* 2007, **6:**107.

56. Kubler-Kielb J, Majadly F, Wu Y, Narum DL, Guo C, Miller LH, Shiloach J, Robbins JB, Schneerson R: Long-lasting and transmission-blocking activity of antibodies to *Plasmodium falciparum* elicited in mice by protein conjugates of Pfs25. *Proc Natl Acad Sci USA* 2007, **104:**293–298.

57. Malkin EM, Durbin AP, Diemert DJ, Sattabongkot J, Wu Y, Miura K, Long CA, Lambert L, Miles AP, Wang J, Stowers A, Miller LH, Saul A: Phase 1 vaccine trial of Pvs25H: a transmission blocking vaccine for *Plasmodium vivax* malaria. *Vaccine* 2005, **23:**3131–3138.

58. Mlambo G, Maciel J, Kumar N: Murine model for assessment of *Plasmodium falciparum* transmission-blocking vaccine using transgenic *Plasmodium berghei* parasites expressing the target antigen Pfs25. *Infect Immun* 2008, **76:**2018–2024.

59. Ramjanee S, Robertson JS, Franke-Fayard B, Sinha R, Waters AP, Janse CJ, Wu Y, Blagborough AM, Saul A, Sinden RE: The use of transgenic *Plasmodium berghei* expressing the *Plasmodium vivax* antigen P25 to determine the transmission-blocking activity of sera from malaria vaccine trials. *Vaccine* 2007, **25:**886–894.

60. Outchkourov N, Vermunt A, Jansen J, Kaan A, Roeffen W, Teelen K, Lasonder E, Braks A, Vegte-Bolmer M van de, Qiu LY, Sauerwein R, Stunnenberg HG: Epitope analysis of the malaria surface antigen pfs48/45 identifies a subdomain that elicits transmission blocking antibodies. *J Biol Chem* 2007, **282:**17148–17156.

61. Outchkourov NS, Roeffen W, Kaan A, Jansen J, Luty A, Schuiffel D, van Gemert GJ, Vegte-Bolmer M van de, Sauerwein RW, Stunnenberg HG: Correctly folded Pfs48/45 protein of *Plasmodium falciparum* elicits malaria transmission-blocking immunity in mice. *Proc Natl Acad Sci USA* 2008, **105:**4301–4305.

62. Molineaux L, Gramiccio G: *The Garki project: research on the epidemiology and control of malaria in the Sudan savanna of West Africa WHO Geneva;* 1980.

63. Ong CS, Zhang KY, Eida SJ, Graves PM, Dow C, Looker M, Rogers NC, Chiodini PL, Targett GA: The primary antibody response of malaria patients to *Plasmodium falciparum* sexual stage antigens which are potential transmission blocking vaccine candidates. *Parasite Immunol* 1990, **12:**447–456.

64. Bousema JT, Drakeley CJ, Sauerwein RW: Sexual-stage antibody responses to *P. falciparum* in endemic populations. *Curr Mol Med* 2006, **6:**223–229.

65. Peiris JS, Premawansa S, Ranawaka MB, Udagama PV, Munasinghe YD, Nanayakkara MV, Gamage CP, Carter R, David PH, Mendis KN: Monoclonal and polyclonal antibodies both block and enhance transmission of human *Plasmodium vivax* malaria. *Am J Trop Med Hyg* 1988, **39:**26–32.

66. Deloron P, Chougnet C: Is immunity to malaria really short-lived? *Parasitol Today* 1992, **8:**375–378.

67. Struik SS, Riley EM: Does malaria suffer from lack of memory? *Immunol Rev* 2004, **201:**268–290.

68. Filipe JA, Riley EM, Drakeley CJ, Sutherland CJ, Ghani AC: Determination of the processes driving the acquisition of immunity to malaria using a mathematical transmission model. *PLoS Comput Biol* 2007, **3:**e255.

69. Saul A, Fay MP: Human immunity and the design of multi-component, single target vaccines. *PLoS ONE* 2007, **2:**e850.

Acknowledgements—This article has been published as part of *Malaria Journal* Volume 7 Supplement 1, 2008: Towards a research agenda for global malaria elimination. The full contents of the supplement are available online at http://www.malariajournal.com/supplements/7/S1

UNIT 6

Emerging Infectious Diseases and Public Health

Unit Selections

Learning Objectives

- Describe three factors that have contributed to the rise in emerging infectious diseases.

- How does climate change impact human health?

- What are some challenges faced in developing vaccines against infectious diseases?

- Why is epidemiological surveillance of critical importance to global health?

Student Website

www.mhcls.com

Internet References

World Health Organization: Emerging Infectious Diseases
http://www.who.int/topics/infectious_diseases/en

National Center for Preparedness, Detection and Control of Infectious Disease
http://www.cdc.gov/ncpdcid/

Infectious Diseases Society of America
http://www.idsociety.org

The Johns Hopkins University HIV Guide Q&A
http://www.hopkins-hivguide.org/q_a/index.html?categoryId=9352&siteId=7151#patient_forum

Pandemic Flu
http://www.pandemicflu.gov

Doctors Without Borders
http://www.doctorswithoutborders.org

In recent years we have seen the number of new infectious disease outbreaks increasing around the world. An important consequence of globalization has been the rapid redistribution of microbes to new habitats; when we travel, our microbes do as well. In addition, global climate change has upset the delicate balance of ecosystems on a macroscopic and microscopic scale. Rising temperatures in the world's oceans have led to increased occurrences of "red tides"—algal blooms that can poison fish and humans who eat those fish. Exploration of previously uninhabited jungles, caves, and rainforests provide new human hosts for animal viruses. Many emerging infectious diseases spring from animal reservoirs to humans. Increased global travel has made it easier than ever for infectious diseases to spread silently on airplanes as we have seen in the case of swine flu, SARS, and extremely drug-resistant (XDR) tuberculosis. It is essential to have global infection control measures in place to thwart the spread of emerging infectious diseases. Cooperation and open communication between public health agencies are of paramount importance—microbes don't recognize international borders. In this unit, we examine several examples of infectious disease transmission, discuss the role of climate change and human behavior in spreading disease, and learn about new methods for microbial surveillance to prevent future outbreaks.

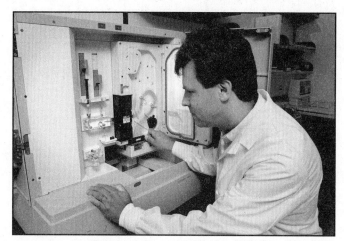

© Centers for Disease Control and Prevention /James Gathany

Most of the emerging infectious diseases affecting the world today originated in wild animals. SARS, Ebola, dengue fever, and influenza (swine and avian) are a few examples of diseases caused by viruses found in wild animals. In "Preventing the Next Pandemic," Nathan Wolfe describes a new program established to track microbes associated with wild animals in order to study the migration of these pathogens to human reservoirs. Careful monitoring may enable the early detection of these microbes before they cause pandemics. Wolfe enumerates the five-step process by which an animal pathogen may turn into a human pathogen. Scientists from the Global Viral Forecasting Initiative (GVFI) have been working to identify viral, bacterial, and parasitic pathogens from animal sources and track their movement through human populations. The science of "epidemic forecasting" has never been more relevant than it is today, and hopefully the GVFI will expand to encompass more nations.

The seasonal outbreaks of meningococcal meningitis that occur in Africa are described in "An Ill Wind, Bringing Meningitis." The vaccine treatment historically used to halt these epidemics is portrayed as a bandaid at best because it does not prevent the transmission of the disease and its effects are short-lived. For these reasons the World Health Organization (WHO) does not endorse this vaccine for preventing outbreaks but encourages its use for stopping epidemics once they have begun. This article discusses the work of the Meningitis Vaccine Project (MVP), a public-private partnership established to develop a more long-lasting, effective vaccine for meningitis that can be used for disease prevention. This new vaccine is expected to reach nine million people in clinical trials in Burkina Faso later this year. Especially in developing nations, cheap effective vaccines are essential for preventing the spread of disease in a population. In addition to developing a malaria vaccine (discussed in Unit 5),

researchers in Africa are exploring two different strategies for combating malaria; in one case, malaria-resistant mosquitoes are being engineered and in the other, the genetic basis for natural immunity against malaria in mosquitoes is being studied. Hopefully a combination of these approaches will result in alleviating the suffering of the millions of people affected by malaria.

Whenever an outbreak occurs anywhere in the world, we rely on the scientists and epidemiologists who go to the ends of the earth to trace the pathogen's reservoir, investigate the chain of transmission and put measures into place to halt its spread. *Smithsonian* magazine profiles avian influenza expert Dr. Robert Webster in "The Flu Hunter." This is a gripping narrative that highlights the importance of early intervention in preventing a flu pandemic. Webster compares the current H5N1 strain of bird flu to the strain from the Spanish flu pandemic of 1918 that killed around 100 million people across the globe. Because of his insistent warnings, some governments now have plans underway to deal with a potential pandemic, if and when the avian influenza virus mutates into a form capable of efficient human-to-human transmission. We have recently seen the emergence of swine flu (H1N1), which is transmitted among humans and is thought to contain a genetic combination of human, avian, and swine influenza viruses. This epidemic has already cost the government of Mexico billions of dollars in lost revenue, after the shutdown of businesses, restaurants, and tourist attractions during the peak of the outbreak. We have not seen the last of this virus, and new cases continue to arise in the United States and around the world. This latest outbreak is testing the public health communications and containment systems put into place after the SARS outbreak. Masked travelers at airports, temperature checks, and warnings to the public about prevention measures signal the constant threat of emerging infectious diseases.

As mentioned earlier, climate change is one factor that has contributed to the rise of emerging infectious diseases. The 2008 American Institute for Biological Sciences meeting addressed this topic and a review of their findings is presented in "Climate, Environment, and Infectious Disease." "The White Plague" details the re-emergence of an infectious disease that has been around since the time of the ancient Egyptians—tuberculosis. Unfortunately, *Mycobacterium tuberculosis* has evolved into an extremely drug-resistant (XDR) form that resists all of the antibacterial agents used against it. This is extremely frightening, especially given the fact that tuberculosis can be spread through just a cough, and one-third of the world's population is infected with this pathogen. Organizations such as Partners in Health pioneered the use of DOTS (direct observation treatment system) for treating resistant cases of tuberculosis by directly monitoring patient compliance with medications and using novel drug combinations to treat them. Although their strategy has now been accepted and the World Health Organization agrees that resistant TB must be treated and not ignored as it has been in the past, more work needs to be done in efforts to combat this deadly disease. This unit closes with an examination of infectious diseases caused by viruses and fungi—norovirus and Valley fever cases are increasing in numbers and the reasons for this are explored. Finally, another fungal infection is spreading through the air, literally! Cryptococcal infections are on the rise in Canada, and researchers are learning more about how this unique fungus lives and travels on air currents in "They Came from Above."

Preventing the Next Pandemic

An international network for monitoring the flow of viruses from animals to humans might help scientists head off global epidemics.

NATHAN WOLFE

Sweat streamed down my back, thorny shrubs cut my arms, and we were losing them again. The wild chimpanzees my colleagues and I had been following for nearly five hours had stopped their grunting, hooting and screeching. Usually these calls helped us follow the animals through Uganda's Kibale Forest. For three large males to quiet abruptly surely meant trouble. Suddenly, as we approached a small clearing, we spotted them standing below a massive fig tree and looking up at a troop of red colobus monkeys eating and playing in the treetop.

The monkeys carried on with their morning meal, oblivious to the three apes below. After appearing for a moment to confer with one another, the chimps split up. While the leader crept toward the fig tree, his compatriots made their way up two neighboring trees in silence. Then, in an instant, the leader rushed up his tree screaming. Leaves showered down as the monkeys frantically tried to evade their attacker. But the chimp had calculated his bluster well: although he failed to capture a monkey himself, one of his partners grabbed a juvenile and made his way down to the forest floor with the young monkey in tow, ready to share his catch.

As the chimps feasted on the monkey's raw flesh and entrails, I thought about how this scene contained all the elements of a perfect storm for allowing microorganisms to jump from one species to the next, akin to space travelers leaping at warp speed from one galaxy to another. Any disease-causing agent present in that monkey now had the ideal conditions under which to enter a new type of host: the chimps were handling and consuming fresh organs; their hands were covered with blood, saliva and feces, all of which can carry pathogens; blood and other fluids splattered into their eyes and noses. Any sores or cuts on the hunters' bodies could provide a bug with direct entry into the bloodstream. Indeed, work conducted by my group and others has shown that hunting, by animals such as chimpanzees as well as by humans, does provide a bridge allowing viruses to jump from prey to predator. The pandemic form of HIV began in this way, by moving from monkeys into chimpanzees and, later, from chimpanzees into humans.

Today HIV is so pervasive that it is hard to imagine the world without it. But a global pandemic was not inevitable: If

Key Concepts

- Most human infectious diseases originated in animals.
- Historically, epidemiologists have focused on domestic animals as the source of these scourges. But wild animals, too, have transmitted many diseases to us, including HIV.
- To address the threat posed by wild animals, researchers are studying the microbes of these creatures and the people who come into frequent contact with them.
- Such monitoring may enable scientists to spot emerging infectious diseases early enough to prevent them from becoming pandemics.

—The Editors

scientists had been looking for signs of new kinds of infections in Africans back in the 1960s and 1970s, they could have known about it long before it had afflicted millions of people. With a head start like that, epidemiologists might well have been able to intervene and mitigate the virus's spread. HIV is not alone in having emerged from an animal reservoir. More than half of human infectious diseases, past and present, originated in animals, including influenza, SARS, dengue and Ebola, to name a few. And today the vast inter-connectedness of human populations, linked so extensively by road and air travel, allows new diseases to become pandemic more quickly, whether they come directly from wild animals, as did HIV, or indirectly, by passing from wild animals to domestic ones and then to us, as in the case of Japanese encephalitis virus and some strains of influenza. In response to these threats, my colleagues and I recently developed a bold new plan to monitor wild animals and the people who come into frequent contact with them for signs of new microorganisms or changes in the bugs' activity. We believe such eavesdropping may provide the early warning needed to stop pandemics before they start.

Stalking Viruses

Our surveillance vision grew out of research we began 10 years ago, when we initiated a study of viruses in rural villagers in the Central African country of Cameroon who hunt and butcher wild animals, as well as keep them as pets. We were trying to determine whether new strains of HIV were entering into human populations, and we suspected that these people would be at particularly high risk of infection.

To understand why we thought these Central African populations are vulnerable, consider a typical bushmeat hunter going about his day. The hunter wears only simple cotton shorts as he walks barefoot along a forest path, carrying on his back a 50-pound baboon. He has transported the animal for some miles and still has more to go before he reaches his village. As the hunter travels, the blood from his prey mingles with his own sweat and drips down his leg, flowing into open cuts along the way. Any infectious agents in the baboon's blood now have access to the hunter's circulatory system and tissues.

If the hunter had his choice, he and his fellow villagers might very well prefer pork or beef to monkey. But those forms of animal protein are rare here. And so he does what humans throughout the world have done for millennia: he hunts the local fauna, just as my friends in New Jersey do on their farm during deer season, in preparation for their annual venison dinner party. The only differences, perhaps, are that the Central African hunter relies on this food for his own survival and that of his family and that his primate quarry is more likely to transfer its viruses and other microorganisms to the hunter than is a deer, which is related to humans much more distantly.

Persuading the villagers to cooperate with us on this project was not easy. Many feared that we were going to seize their game. Only after gaining their trust could we begin collecting data. Their cooperation was essential: in addition to drawing samples of their blood for study and peppering them with questions about their health and hunting activities, we needed blood samples from their prey. We relied on them to obtain these samples by using pieces of filter paper we gave them.

Our analyses of the blood from the hunters and the hunted revealed several animal viruses not previously seen in humans. One agent, which we first reported in a paper published in 2004 in *Lancet,* is known as simian foamy virus (SFV), and it is a member of the same family of viruses—the so-called retroviruses—to which HIV belongs. SFV is native to most primates, including guenon monkeys, mandrills and gorillas, and each of these primate species harbors its own genetically distinctive variant of the bug. We found that all three variants had entered the hunter populations. In one particularly telling example, a 45-year-old man who reported having hunted and butchered gorillas—animals rarely pursued by subsistence hunters—had contracted gorilla SFV.

In those same Central African populations we also found a variety of retroviruses known as human T lymphotropic viruses (HTLVs), so named because of their propensity for infecting immune cells called T lymphocytes. Two of the HTLVs, HTLV-1 and HTLV-2, were already well known to affect millions of people around the world and contribute to cancer and neurological disease in some infected individuals. But HTLV-3 and HTLV-4, which we described in 2005 in the *Proceedings of the National*

Context
Infectious Diseases from Wild Animals

Many of the major infectious diseases of humans are believed to have come from wild animals. This fact underscores the need to monitor the microbes of wild creatures, in addition to those of livestock. The table at the bottom lists 10 such diseases and the animals from which they likely emerged.

Disease	Source
AIDS	Chimpanzees
Hepatitis B	Apes
Influenza A	Wild birds
Plague	Rodents
Dengue fever	Old World primates
East African sleeping sickness	Wild and domestic ruminants
Vivax malaria	Asian macaques
West African sleeping sickness	Wild and domestic ruminants
Yellow fever	African primates
Chagas' disease	Many wild and domestic mammals

Academy of Sciences USA, were new to science. Given the high degree of genetic similarity between HTLV-3 and its simian counterpart, STLV-3, it appears as if this virus was picked up through hunting STLV-3-infected monkeys. The origin of HTLV-4 remains unclear, but perhaps we will find its primate ancestor as we continue to explore these viruses in monkeys. We do not yet know whether SFV or the new HTLVs cause illness in people. Viruses do not necessarily make their hosts sick, and viruses that do sicken people and even spread from person to person do not always cause pandemics; often they retreat spontaneously. But the fact that SFV and HTLV are in the same family as HIV, which did spawn a global epidemic, means that epidemiologists must keep a close eye on them.

My colleagues and I have outlined five stages in the transformation of a pathogen of animals into one that specializes on humans. In stage 1, the agent lives only in animals. In stage 2, it can be transmitted to a human only from an animal. A stage 3 germ is transmitted primarily from animals to humans, but it can also spread among humans for a short time before dying out. Once the agent reaches stage 4, it can sustain longer outbreaks among humans. By the time it attains stage 5, it has become an exclusive pathogen of humans and no longer utilizes an animal host. Pathogens in stage 4 or stage 5 have the potential to cause massive human die-offs.

Forecasting the Next Pandemic

Had we been watching hunters 30 years ago, we might have been able to catch HIV early, before it reached the pandemic state. But that moment has passed. The question now is, How

Stages to Watch
From Animal Microbe
to Human Pathogen

The process by which a pathogen of animals evolves into one exclusive to humans occurs in five stages. Agents can become stuck in any of these stages. Those in early stages may be very deadly (Ebola, for example), but they claim few lives overall because they cannot spread freely among humans. The better able a virus is to propagate in humans, the more likely it is to become a pandemic.

Disease Examples

Reichenowi malaria

Stage 1: Pathogen is present in animals but has not been detected in humans under natural conditions.

Rabies

Stage 2: Animal pathogen has been transmitted to humans but not between humans.

Ebola

Stage 3: Animal pathogen can be transmitted between humans causes an outbreak of disease but only for a short period before dying out.

Dengue

Stage 4: Pathogen exists in animals and undergoes a regular cycle of animal-to-human transmission but also sustains long outbreaks arising from human-to-human transmission.

HIV

Stage 5: Pathogen has become exclusive to humans.

Source: *Origins of Major Human Infectious Diseases,* by Nathan D. Wolfe, Claire Panosian Dunavan and Jared Diamond, in *Nature,* Vol. 447; May 17, 2007.

Two-Way Street

Pathogens do not jump only from animals into humans—they can also travel in the other direction. Some infectious diseases that people have transmitted, and continue to transmit, to animals include:

- Tuberculosis (cattle)
- Yellow fever (South American monkeys)
- Measles (mountain gorillas)
- Poliomyelitis (chimpanzees)

Though still a fledgling effort, the GVFI now has around 100 scientists following sentinel populations or animals in Cameroon, China, the Democratic Republic of the Congo, Laos, Madagascar and Malaysia—all hotspots for emerging infectious diseases. Many of the sentinels are hunters, but we are also screening other populations at high risk of contracting diseases from wildlife, such as individuals who work in Asia's "wet markets," where live animals are sold for food.

Finding a new microorganism in a hunter is only the first step in tracking an emerging pathogen, however. We must then determine whether it causes disease, whether it is transmissible from person to person, and whether it has penetrated urban centers, where the high density of occupants could fuel its spread. The appearance in an urban center, away from the original source, would be a particularly worrisome sign of pandemic potential.

In the cases of HTLV-3 and HTLV-4, we are beginning to study high-risk populations in cities near hotspots for emerging infectious disease, regularly testing them for these viruses. Individuals with sickle cell disease who receive routine blood transfusions for their condition are one such population that could become infected early on. If we find people in these populations who are infected, we would work to initiate worldwide monitoring of blood supplies, to protect blood recipients. To that end, we are working with our long-term collaborator Bill Switzer and our colleagues at the U.S. Centers for Disease Control and Prevention to develop new diagnostic tests to check for the presence of viruses in the blood supply. Another urgent priority would be to determine the agent's mode of transmission, which would inform tactics for blocking its spread. If an agent were sexually transmitted, for example, public health officials could launch awareness campaigns urging condom use, among other precautions.

Governments can also take measures to keep new viruses from entering the blood banks in the first place. In fact, following our discoveries concerning the relation between exposure to primates and these new viruses, the Canadian government modified its blood donation policies to exclude donors who have had contact with nonhuman primates.

In addition to our forecasting efforts, the new science of pandemic prevention includes programs such as HealthMap and ProMED, which compile daily reports on outbreaks around the world, and cutting-edge cyberwarning systems such as those piloted by Google.org to use patterns in search engine data to successfully forecast influenza. Likewise, national

can we prevent the next big killers? Once my colleagues and I had determined that we could study remote populations effectively, we knew we could extend our work more broadly to listen in on viral "chatter"—the pattern of transfer of animal viruses to humans. With global surveillance, we realized, we might be able to sound the alarm about an emerging infectious disease before it boils over.

Fortunately, through partnership with Google.org and the Skoll Foundation we were able to launch the Global Viral Forecasting Initiative (GVFI), a program in which epidemiologists, public health workers and conservation biologists the world over collaborate to identify infectious agents at their point of origin and to monitor those organisms as they bubble up from animals into humans and flow outward from there. Instead of focusing narrowly on just viruses or a particular disease du jour, the GVFI works to document the full range of viruses, bacteria and parasites that are crossing over from animals into humans.

Prevention Proposal
Building a Surveillance Network

By monitoring microorganisms in wild animals and the people who are frequently exposed to them, scientists may be able to spot an emerging infectious disease before it becomes widespread. To that end, the author recently organized the Global Viral Forecasting Initiative (GVFI), a network of 100 scientists and public officials in six countries who are working to track potentially dangerous agents as they move from animals into human populations. The GVFI focuses on tropical regions in particular, because they are home to a wide variety of animal species and because humans there commonly come into contact with them through hunting and other activities. Eventually the GVFI hopes to expand the network to include more countries with high levels of biodiversity.

Country: Cameroon
Viruses previously spawned: HIV
Sentinel population under study for new pathogens: People who hunt and butcher wild animals

Country: Democratic Republic of the Congo
Viruses previously spawned:
Marburg, monkeypox, Ebola
Sentinel population: People who hunt and butcher wild animals

Country: China
Viruses previously spawned: SARS, H5N1
Sentinel population: "Wet market" workers

Country: Malaysia
Viruses previously spawned: Nipah
Sentinel population: Wildlife hunters

Taking Action

If investigators find signs that an emerging pathogen has spread beyond humans who have direct contact with animals into the mainstream population, they will sound an alarm. Protecting the blood supply will be one important step toward preventing a pandemic. This measure will require rapid development and deployment of a diagnostic test for the germ.

The Threat from Pets

Wild animals and farm animals are not the only potential sources for the next major pandemic. Fido and Fluffy—and other pets, too—could harbor pathogens devastating to humans. This possibility arises when pets come into contact with germ-carrying wild animals. The germs can jump into pets, which can then transmit these agents to their owners.

diversity of animal species that could transmit pathogens to humans. Fuller development of the GVFI will be expensive: building out our network so that we have adequate staff and lab facilities for testing the sentinel populations every six months and testing the animals with which these people are in contact will cost around $30 million, and keeping it running will cost another $10 million a year. But if it succeeds in averting even a single pandemic within the next 50 years, it will more than pay for itself. Even just mitigating such an event would justify the cost.

Humans work to forecast a variety of very complex natural threats. We rarely question the logic behind trying to predict hurricanes, tsunamis, earthquakes and volcanoes. Yet we really have no reason to believe that predicting pandemics is inherently harder then predicting tsunamis. Given the enormous sums of money required to stop pandemics once they have already been established, it only makes sense to spend a portion of those public health dollars on stopping them in the first place. The ounce of prevention principle has never been more apt.

More to Explore

Naturally Acquired Simian Retrovirus Infections in Central African Hunters. Nathan D. Wolfe et al. in *Lancet,* Vol. 363, No. 9413, pages 932–937; March 20, 2004.

Emergence of Unique Primate T-Lymphotropic Viruses among Central African Bushmeat Hunters. Nathan D. Wolfe et al. in *Proceedings of the National Academy of Sciences USA,* Vol. 102, No. 22, pages 7994–7999; May 31, 2005.

Bushmeat Hunting, Deforestation, and Prediction of Zoonotic Disease Emergence. Nathan D. Wolfe et al. in *Emerging Infectious Diseases,* Vol. 11, pages 1822–1827; December 2005.

Origins of Major Human Infectious Diseases. Nathan D. Wolfe, Claire Panosian Dunavan and Jared Diamond in *Nature,* Vol. 447, pages 279–283; May 17, 2007.

The Global Viral Forecasting Initiative Web site: www.gvfi.org

NATHAN WOLFE is Lorry I. Lokey Visiting Professor in Human Biology at Stanford University and director of the Global Viral Forecasting Initiative. He earned a doctorate in immunology and infectious diseases from Harvard University in 1998. A recipient of the National Institutes of Health Director's Pioneer Award and the National Geographic Society's Emerging Explorer Award, Wolfe currently has active research and public health projects in 11 countries in Africa and Asia.

and international surveillance and response systems of local governments and the World Health Organization will play an important role in stopping the next plague.

For our part, we would ultimately like to expand our surveillance network to more countries around the world, including such nations as Brazil and Indonesia, which have a tremendous

An Ill Wind, Bringing Meningitis

Crippling epidemics of meningococcal meningitis sweep across Africa with the onset of the dry season and harsh harmattan winds. An affordable, effective vaccine in the works could change that.

LESLIE ROBERTS

The dust is inescapable, burning your eyes, clogging your nose, penetrating into your lungs, and making breathing ragged. In March, on the road to Koudougou, some 100 km west of Ouagadougou, the landscape is moonlike. In the cratered bottom of a lakebed, dust-caked men, barely distinguishable from their surroundings, fashion bricks from the mud. The bricks will dry quickly in the baking heat, which tops 45°C each day.

It is the dry season in Burkina Faso. And with the dust and the hot, dry wind, known as the harmattan, that blasts across the Sahel come meningococcal meningitis epidemics, caused by the bacterium *Neisseria meningitidis*. What, exactly, about these conditions triggers the epidemics remains mysterious, but they come like clockwork, hitting Burkina Faso every year and engulfing the entire "meningitis belt," which runs from Ethiopia in the east to Senegal and The Gambia in the west, every 6 to 12 years.

The last big one, in 1996–97, sickened hundreds of thousands and killed more than 25,000 in 10 countries. In 2007, the death toll climbed alarmingly high again, prompting the World Health Organization (WHO) to warn that another huge epidemic was likely in 2008. But this season turned out to be relatively quiet, with some 9400 cases in Burkina Faso and 27,000 across the entire belt. As always, the epidemics in Burkina Faso stopped suddenly with the first rains in May, as the population in this country, one of the poorest in the world, braced for the inevitable onslaught next year.

Koudougou district officially passed the epidemic threshold in mid-March, and scarce supplies of the meningitis vaccine were made available to try to curb the epidemic's spread. At a rudimentary health center there, hundreds of people—mostly women and children—queue up for vaccinations, seeking shade by the buildings or under a scrawny tree. Most have been waiting patiently for hours, but some occasionally surge to the front of the line only to be pushed back by the men in charge of crowd control.

At best, this reactive vaccination strategy, as it is called, is a "Band-Aid," says Rosamund Lewis, a physician and meningitis expert at the GAVI Alliance (formerly the Global Alliance for Vaccines and Immunization). The reason is that the vaccine being used, a 1960s design using a polysaccharide from the bacterium's coat and still the only affordable one in Africa, doesn't work very well. Although this vaccine prevents those carrying the bacterium from getting sick, it doesn't stop them from passing it on to others; immunity lasts only a few years; and the vaccine has minimal effect on children under age 2. Because of these limitations, WHO has long recommended that it be used only to control epidemics, not to prevent them—a strategy that has its critics. "The epidemic is sometimes over by the time vaccine arrives," concedes William Perea, a Colombian-born epidemiologist who leads Epidemic Readiness and Interventions at WHO and who nonetheless supports the strategy for lack of a cost-effective alternative.

F. Marc LaForce wants to change all that. He is heading an innovative public-private partnership known as the Meningitis Vaccine Project (MVP) to develop an affordable, effective, long-lasting vaccine for African meningitis—a conjugate vaccine that includes a protein to boost the immune reaction. The conjugate will cost roughly 50 cents a dose—a price many African governments say they can afford—and is already being tested in clinical trials at several African sites. Barring any further delays, it will be introduced in a massive trial of some 9 million people in Burkina Faso in late 2009. LaForce hopes the new vaccine will eventually be used in preventive campaigns across the entire meningitis belt and spell the end of these devastating epidemics.

"It will change completely the approach to meningitis, and that will be great," says Myriam Henkens, a physician at Médecins Sans Frontières (MSF) in Brussels, Belgium. "We are counting on the conjugate," she says. "It will be so much cheaper and so much better."

And all indications are that countries will be clamoring for it. Although other diseases exact a bigger toll, they are not as feared as meningitis, which can kill within 24 hours and often leaves survivors deaf or otherwise disabled. "Boy, do they want this vaccine," agrees Emil Gotschlich of Rockefeller University in New York City, who developed the existing polysaccharide vaccine.

But the road between here and there is littered with potholes. MVP, a joint project of WHO and the Seattle, Washington–based health nonprofit PATH, has already encountered unexpected obstacles. Originally promised for 2007, the vaccine may not be ready for the 2009 rollout in Burkina Faso, LaForce admits. And questions remain about just how long immunity will last, whether a booster will be needed—which would affect the overall cost of a vaccination strategy—and perhaps more worrisome, whether once *N. meningitidis* group A—which causes the majority of epidemic disease in Africa and is the target of this vaccine—has been beaten down, other strains will arise to replace it.

"Vaccine-making is not for the faint of heart," concedes LaForce.

Mysterious Cycles

Meningitis is an infection of the meninges, the thin membrane that surrounds the brain and spinal column. Several different bacteria can cause meningitis, including *Haemophilus influenzae* type B (*Hib*) and *Streptococcus pneumoniae*—and also a few viruses—but only *N. meningitidis,* or meningococcus, spawns the huge epidemics that sweep across the belt. Of the dozen or so meningococcal groups, A is by far the worst, causing roughly 80% to 90% of epidemic disease most years. (Groups C and B, by contrast, fell adolescents and college students in Europe and the United States.)

Roughly 1 in 20 people "carry" the bacterium asymptomatically in the back of their throats and can transmit it to others; what triggers invasive disease remains mysterious. Once the disease begins, with its characteristic sudden fever, headache, and stiff neck, it progresses rapidly. Even with prompt treatment with antibiotics, which is often impossible in remote villages, about 10% die, and up to 25% of survivors are left with permanent disabilities such as deafness or mental retardation.

For about 100 years now, experts have puzzled over the remarkable seasonality of the epidemics, which begin with the dry season in December, peak after a few months, and disappear with the first rains in May or June. One hypothesis is that dust and wind increase transmission of the disease, which is spread person to person through respiratory droplets. The other, which seems more likely, is that the harsh environmental conditions irritate the mucus membranes, enabling the bacterium to more easily penetrate and enter the spinal fluid, where it causes invasive disease.

Similarly, although epidemics are inevitable, almost all efforts to predict exactly when and where they will strike have failed miserably. It has defied the best scientific minds, says Perea, who also works with MVP.

Failed Policy?

MVP was born of what everyone refers to as the "terrible epidemic" in 1996–97. African leaders and global health officials watched in horror as the largest meningococcal meningitis epidemic ever recorded swept across the belt. More than 250,000

people fell ill, and the death toll soared beyond 25,000. Clearly, the epidemic response system wasn't working.

Shortly thereafter, WHO, UNICEF, MSF, and the International Federation of Red Cross and Red Crescent Societies banded together to create the International Coordinating Group (ICG) on Vaccine Provision for Epidemic Meningitis Control. Headquartered at WHO in Geneva, Switzerland, ICG tries to ensure that limited supplies of the polysaccharide vaccine are rapidly sent to where they can do the most good. Following the recommendations of a pivotal 2000 *Lancet* paper by Lewis—then at MSF's research institute, Epicentre—and colleagues, WHO lowered the epidemic threshold to 10 cases among 100,000 people in 1 week, shaving more than a week off the usual response time. Perea says that if a country can launch a reactive vaccination campaign within 3 to 4 weeks of an epidemic's onset, it can prevent 70% of the cases. Response times have improved substantially, but campaigns start within 3 weeks only about 60% of the time, he says.

These improvements were just a stopgap, all conceded. What was needed, recommended a WHO advisory group in 2000, was an entirely new type of meningitis vaccine, a conjugate vaccine that would confer lasting immunity and could be used preventively, modeled on the Men C vaccine that had all but eliminated the disease in the United Kingdom.

A conjugate vaccine uses the same polysaccharide but links, or conjugates, it to a protein to increase its immunogenicity. Conjugate vaccines developed to date, including those for *Hib* and *S. pneumoniae,* confer longer lasting immunity than polysaccharide vaccines, work in infants, and, perhaps more important, reduce transmission of the bacteria, thereby providing herd immunity and protecting even those who are not immunized.

The newly established Bill and Melinda Gates Foundation didn't need much convincing; in May 2001, it sprang for $70 million over 10 years to establish MVP to develop, test, license, and introduce an affordable conjugate vaccine for Africa.

From the outset, some questioned the approach, worrying that it would take too long and arguing that alternatives already existed. Gotschlich and vaccinologist John B. Robbins of the U.S. National Institute of Child Health and Human Development, who both won Lasker Awards for their pioneering work, had long railed against what they considered WHO's failed policy of waiting for an epidemic before starting mass vaccination. In a series of papers, including a roundtable in the *WHO Bulletin* in 2003, they argued that WHO should urge countries to use the polysaccharide vaccine preventively.

True, the polysaccharide vaccine is by no means perfect, says Gotschlich—although he says it is more effective than others now acknowledge. And without question, he says, the conjugate vaccine will be far superior: "I am all for it," says Gotschlich, who serves on MVP's advisory committee. But the polysaccharide vaccine was available and could save lives right away for just pennies a dose, says Gotschlich, who kicks himself for not urging WHO to be proactive back in the 1970s.

"In theory, he is right," says Perea, but he doubts the strategy would work. And it would be hugely expensive. Because

Costs of Meningitis Outbreaks Are Crippling, Too

The government of Burkina Faso prides itself on providing free health care to anyone affected by meningitis. Try telling that to the family from Koudougou gathered at Yalgado Hospital here in the capital city.

Last year, during a "reactive" vaccination campaign, their daughter hid from the vaccinators, afraid of the needle, her mother explains. That may be why she got sick this March when another epidemic hit the same district. Now the 14-year-old is lying on a bare cot in a sparse, concrete-floored room, naked in the heat except for her underpants and an IV dripping into her thin arm. She has been here 2 weeks and still has a stiff neck and seems listless. Rigobert Thiombiano, the head of the infectious disease ward, suspects she may have septicemia.

Both her parents are with her in the hospital room, as is the custom in Burkina Faso. Hospitals here do not provide the services Western patients take for granted. The family must buy the medicine, provide and cook the food, wash the laundry, and otherwise care for their kin.

One illness in a family can exact a huge toll on household income, says Anaïs Colombini, a health economist at the French aid group Agence de Médecine Préventive (AMP) in Ouagadougou, who, with her colleagues, recently completed a detailed socioeconomic study, supported by the Meningitis Vaccine Project (MVP) and the World Health Organization, of the burden of meningitis in Burkina Faso.

Government claims aside, AMP found that most families pay on average $25 in direct medical costs, which includes medicine, testing, and lab analysis. Other nonmedical direct costs, say, for food, soap, transportation, and telephone, run another $15.

The indirect costs, from loss of income and property such as cattle and crops, are even higher, running $50 per episode. That adds up to almost half of a family's average annual income of roughly $220. Worse still, says Colombini, just one episode can throw a family into "a downward spiral of poverty" from which it can be impossible to recover.

"Most people don't realize that meningitis is such an important contributor to poverty," she says, with both short- and long-term costs. "For instance, if you can't be home to tend to the crops, they die, and future income is lost." Colombini is eager for MVP's new meningitis vaccine, which she hopes will help eliminate poverty along with this feared disease in Africa.

—L.R.

polysaccharide as a preventive vaccine," says Perea. With limited resources, he asks, why not focus instead on a "real solution": a conjugate vaccine.

Trials and Tribulations

Two months after the Gates money came through, LaForce was on the job, setting up shop in July with a small, energetic staff in Ferney-Voltaire, across the French border from WHO headquarters in Geneva. An infectious disease expert and former meningitis officer for the U.S. Centers for Disease Control and Prevention (CDC), LaForce had recently quit academic medicine, fed up with the paperwork and the ever-increasing administrative demands. "I wanted to make an impact on global public health before I retired," he says.

In August, LaForce and WHO's Luis Jodar, an early member of MVP, were on a plane to Africa where they began informal consultations with African leaders and health officials about what they wanted in a vaccine. He distinctly remembers a discussion with Hassane Adamou, the secretary general for Niger's Ministry of Health, who said: "Please don't give us a vaccine we can't afford. That is worse than no vaccine." When he asked leaders what was affordable, the answer was 50 cents a dose or less, significantly lower than the $2 to $3 a dose the collaborators had originally envisioned. Won over, LaForce immediately began pushing for the best and least expensive vaccine possible: a First World vaccine at a Third World price.

Vaccine expert Rino Rappuoli, now head of global vaccine research at Novartis, argued vociferously that waiting for the cheapest vaccine was actually more expensive, in terms of lives lost in the interim. While at Chiron in the 1990s, Rappuoli had developed the first conjugate vaccine against meningitis, targeted against groups A and C. It was a "beautiful vaccine," he laments, that performed well in field tests in Niger and The Gambia. But as the group C component was developed for the lucrative market in the United Kingdom, "we were asked to remove A," which was needed only for Africa, he recalls. "The market was only for C." Instead of reinventing the wheel, urged Rappuoli, MVP should dust off his vaccine and make it quickly available.

MVP stuck to its plan for an affordable vaccine. But LaForce says he could find no major vaccine manufacturer willing to produce a Men A conjugate for $2 a dose, much less the price he thought was needed. So in an unusual strategy, LaForce insisted on a guaranteed selling price—50 cents or less—and then found a developing country manufacturer, Serum Institute of India Limited (SIIL) in Pune, willing to take it on. (Although the company has agreed to a fixed price for Africa, it is free to sell the vaccine elsewhere at higher prices.)

Next, MVP lined up suppliers for the raw ingredients: the group A polysaccharide and the protein it would be conjugated with, a tetanus toxoid. And they contracted with a European research group to develop a new conjugation technology and then transfer it to SIIL.

From the outset, MVP decided to concentrate on a monovalent vaccine against group A. Creating a multivalent vaccine

epidemics are so unpredictable, he says, this approach would require vaccinating the entire population of the meningitis belt—roughly 400 million people—every 3 years for an estimated $400 million a campaign. "To arrange a countrywide vaccination every 3 years is nuts. The number of resources, from my point of view, is totally unjustified to introduce the

Clinical Trials: Dispelling Suspicions, Building Trust in Mali

Across West Africa, suspicions of Western medicine—and in particular the fear of being used as a guinea pig in clinical trials—run high. That is true here in Bamako, where Samba Sow of the Center for Vaccine Development (CVD) is doing clinical testing of a new conjugate vaccine developed by the Meningitis Vaccine Project (MVP) (see main text).

Before trials can begin, Sow, like his counterparts leading trials in The Gambia and Senegal, has to win the trust of the local community so he can enlist participants and ensure that consent is truly informed—a task that can't be undertaken lightly, Sow says. He does it by convincing the poor, largely Muslim and illiterate or semiliterate population that the vaccine being tested is designed to help break the cycle of deadly meningitis epidemics rather than make them sterile or infect them with the AIDS virus, as is widely believed.

At the same time, he and MVP have to design a study protocol and consent process that will pass muster with the four institutional review boards (IRBs) that oversee the studies: one at each of MVP's partners, PATH in Seattle, Washington, and the World Health Organization; one at the University of Bamako; and another at the University of Maryland (UMD), with which CVD-Mali is affiliated. A Malian and Muslim physician who went to medical school in Mali and then was trained in epidemiology at the London School of Hygiene and Tropical Medicine—and who now also serves on the faculty of UMD—Sow manages to bridge both worlds.

Sow starts with the all-important "chef de village"—Bamako, with a population of 1.5 million to 2 million, has six districts, which in turn contain five to 12 "souscartiers," or local jurisdictions, each with its own leader. Sow explains the trial to the village head and his advisers, showing them the regulatory approvals from the Minister of Health and the Bamako regional health director, and ethical approvals from all four IRBs, and asks him to convene a community meeting.

"If the leader is not convinced, they will say no, and no one is allowed to sign up. That is the way it works in Mali," Sow says. On the flip side, he says, once one leader says yes, others are also likely to. Sow chairs every single community meeting, during which he describes the study and how the vaccine "acts like a soldier in the body" to protect against meningitis; then the community can ask questions.

The information sheet and consent forms are available in print and on audiotape in English, French, and the most common spoken language here, Bambara. Some questions concern the motives of his group and the Western collaborators. He answers: "Why would I hurt the people where I grew up and who paid for my education? I am here to make my country proud." Other questions are specific, such as "Why do you have to take blood, and why twice?"

The process is repeated again when individuals sign up. Each participant—or the mother if the participant is a child—meets privately with a physician at CVD's clinical center, where they again review the forms and listen to the audiotape. Sometimes the mothers go home to confer with family and community before deciding, says Sow. But so far, the refusal rate is "very, very low."

–L.R.

that could protect against the other groups in Africa would jack up the cost, not to mention the risk of failure, and stretch out the time frame of development. "If we could [address] 85% of the burden with the simplest approach, . . . to me, that was a completely acceptable wager with public money," says LaForce. The unexpected emergence of a new epidemic strain, W135, in 2000 caused considerable soul-searching. Even though W135 appeared to be a significant new threat, in the interest of speed, MVP decided against a midcourse change. "History will tell if that was the right decision," says Lewis.

Manufacturing didn't go smoothly. The conjugation technology proved finicky, causing some delays. But still, the partners thought they were roughly on track until, in spring 2003, the European research group MVP was collaborating with announced it was unwilling to transfer the technology to SIIL. That was the absolute low point, says LaForce. Even friends of the project said it was doomed, he recalls, and several called for his ouster. But others lobbied to give him more time.

MVP found a solution at the U.S. Food and Drug Administration, where longtime vaccine experts Robert Lee and Carl Frasch at the Center for Biologics Evaluation and Research had already developed an alternative conjugation technology and quickly transferred it to SIIL with no strings attached. "These guys are heroes," says LaForce. Preclinical animal studies began soon after. Since then, LaForce has been traveling around the world selling his vision—a continent free of deadly meningitis outbreaks—while generally greasing the wheels for the vaccine's introduction.

LaForce, a 69-year-old American, is a big man with a big voice. "*Mon ami, mon ami, comment ça va?*" he boomed on a recent trip to Ouagadougou, as he clasped hands and patted shoulders of collaborators and hotel clerks alike. "*Le vieux blanc*," or old white man, as he sometimes refers to himself, is invariably upbeat, even as he delivers the bad news that, because of a regulatory snafu in India, the vaccine's introduction in Burkina Faso will be delayed from 2008 until 2009. But the vaccine will come, he assures, and it will be great.

Thinking Locally

Phase II and II/III clinical trials of MVP's Men A vaccine candidate are under way in Mali, The Gambia, Senegal, and India, where SIIL plans to license the vaccine to protect against that country's occasional outbreaks.

Samba Sow, a Malian physician and epidemiologist, is running the trials in Bamako, Mali, at the Center for Vaccine Development, a partner lab of the University of Maryland's CVD. With MVP support, a former leprosarium—a small group of former patients still lives on the grounds—has been converted into a cheery clinical center for testing the new conjugate and other vaccine candidates.

The challenge in setting up the trials in some of the poorest countries in the world, says Simonetta Viviani, an Italian physician who heads MVP's vaccine development from Ferney-Voltaire, was to create a "functional but minimalist" system that would meet all international standards for ethics and good clinical practices but could also be continued with local experts once MVP is gone. That means hiring and training local staff, working closely with the community, and respecting local traditions, she says (see sidebar).

In Mali, explains Sow, the initial contacts with the community—and also the first person parents see when bringing their children to the clinic—should be older and preferably religious, which implies a certain wisdom and trustworthiness. Even in this predominantly Muslim nation, says Sow, a Catholic nun still carries significant clout, whereas bright, young doctors, no matter how prestigious their degrees, or white people will not pass muster, he says.

Results from the "pivotal" phase II trial of the Men A conjugate vaccine, conducted here and in Basse, The Gambia, and announced in June 2007, "put us on the map in Africa," says LaForce. The trial of 600 healthy toddlers age 12 to 23 months, half at each site, showed that the conjugate vaccine produced antibody titers almost 20-fold higher than the current polysaccharide vaccine.

MVP's clinical team recently unblinded the results from the second arm of the study, in which the same cohort of 12- to 23-month-olds were randomized to receive a booster dose of the conjugate or the polysaccharide or the control vaccines 8 to 12 months later. The as-yet-unpublished data are "fantastic," raves Viviani. She and LaForce suspect that the vaccine will protect for at least 10 years, although Gotschlich and others caution that the duration of protection won't be known until the vaccine is used in real-world conditions.

A phase II/III trial of 900 participants and controls, under way in Bamako, Basse, and Dakar, Senegal, is testing the safety and immunogenicity of a single dose in 2- to 29-year-olds and will also look at its effect on "carriage"—that is, whether it actually does reduce the load of bacteria carried in the back of the throat. Next up is a study of safety and immunogenicity of different dose schedules in infants, expected to start later this year in Ghana.

Optimism, Tempered

If the remaining trials go as expected, if the lot-consistency studies under way in Pune, India, go without a hitch, if production can be scaled up, if India licenses the vaccine and WHO "prequalifies" it, if funding comes through, and tens of other details go right, MVP will introduce the vaccine in Burkina Faso in 2009 or perhaps 2010. And that, LaForce hopes, will be the beginning of the end of meningitis epidemics in Africa.

The vaccine will be given to Burkina Faso's entire population of 1- to 29-year-olds, roughly 9 million people. Kader Konde, director of WHO's Multi-Disease Surveillance Centre

(MDSC) in Ouagadougou, who is also MVP's general troubleshooter for Africa, says the president and senior health officials are on board and are pushing MVP to move faster. LaForce says he wishes they could but adds that "it's important that all the regulatory steps are taken so no one feels they have a substandard product."

MVP has already lined up partners to conduct follow-on studies to measure the vaccine's impact. Surveillance will be critical. CDC will help MDSC look for changes in circulating strains. "We must be able to document any case to see if it is a failure of the vaccine or another strain," like W135, rearing its head, says MDSC epidemiologist Mamoudou Harouna Djingarey. That requires strengthening surveillance across the entire belt.

The hub of these efforts is MDSC in Ouagadougou. Housed in a building that still bears the name Onchocerciasis, the revamped facility boasts state-of-the-art equipment, including a real-time polymerase chain reaction machine, a recent gift from CDC, for analyzing cerebrospinal fluid samples to determine the bug and the group. Meanwhile, a half-dozen epidemiologists, microbiologists, and data experts track the bug's every move in 14 countries.

Before MDSC, there was "not much," recalls Perea, who notes that surveillance is now "pretty good" but is still spotty in some countries, such as Chad and Nigeria. Even in Burkina Faso, adds Djingarey, cases are still missed, cerebrospinal samples are degraded in transport, and data dribble in late from some districts. All that must be fixed, he says.

Following the planned 2009 introduction, MVP, MDSC, the Burkina Faso Ministry of Health, and CDC, in collaboration with the Norwegian Institute of Public Health and the Centre for Prevention of Global Infections at the University of Oslo, will conduct carriage studies. Other academic, governmental, and nongovernmental partners, coordinated by Brian Greenwood at the London School of Hygiene and Tropical Medicine, will monitor how long immunity lasts and whether a booster shot is needed, as turned out to be the case with the Men C conjugate in the United Kingdom.

Provided no major problems surface, WHO and its AFRO bureau and UNICEF will introduce the vaccine, first in the three hyperendemic countries: Burkina Faso, Mali, and Niger. Because production will be limited to about 45 million doses for the first few years, the partners are trying to allocate it to the populations at highest risk. By 2016, there should be enough vaccine for the most vulnerable population of the meningitis belt, roughly 250 million people, says LaForce. He envisions that countries will do "catch-up" vaccination campaigns every 5 years or so until the vaccine is approved for infants and can be integrated into routine childhood immunizations. As *Science* went to press, the GAVI secretariat recommended that its board approve $370 million to cover the vaccine introduction in Burkina Faso and subsidize the vaccine's rollout across the belt. Eventually, GAVI's support would wane and countries would pick up the tab themselves.

In the interim, stresses Perea, it will be essential to keep up supplies of the polysaccharide vaccine, which plummeted because of a production decline after MVP was announced, leaving a global shortfall that still persists. It was an unpredictable market to begin with, and manufacturers thought the conjugate would "put them out of business," says LaForce, so "most moved on."

They stopped 10 years too soon," says Lewis. Since then, two manufacturers have agreed to produce the vaccine if WHO guarantees to purchase it.

The next priority, all agree, is an affordable multivalent conjugate vaccine that would offer even wider protection from all meningococcal groups in Africa. (The Menactra quadrivalent conjugate vaccine licensed in the United States sells for about $100 a dose.) Right now, it's not clear who will take the lead on the multivalent. MVP won't, says LaForce—although he hopes it has shown what is possible—as the project will shut its doors and he will retire in 2011.

The rains started in late May in Burkina Faso, and the epidemic, which had affected more than 9000 and killed 900 there, waned. Although that's a much lower toll than everyone had feared for this year, "it's still 9000 cases too many," says LaForce. He won't hazard a guess about how bad next year's epidemic will be, but he is hoping for a quiet season. "We need another year's breathing room," he says, before the next big one hits.

The Flu Hunter

For years, Robert Webster has been warning of a global influenza outbreak. Now governments worldwide are finally listening to him.

Michael Rosenwald

Robert Webster was in the backyard of his home in Memphis doing some landscaping. This was in the early winter of 1997, a Saturday. He was mixing compost, a chore he finds enchanting. He grew up on a farm in New Zealand, where his family raised ducks called Khaki Campbells. Nothing pleases him more than mucking around in the earth. He grows his own corn, then picks it himself. Some of his friends call him Farmer Webster, and although he is one of the world's most noted virologists, he finds the moniker distinguishing. He was going about his mixing when his wife, Marjorie, poked her head out the back door and said, "Rob, Nancy Cox is on the phone." Cox is the chief of the influenza division at the Centers for Disease Control and Prevention, in Atlanta. Webster went to the phone. He has a deep voice and a thick accent, which people sometimes confuse with pomposity. "Hello, Nancy," he said.

Cox sounded distressed. She told him there had been a frightening development in Hong Kong—more cases, and another death.

Oh my God, Webster recalls thinking. *This is happening. It's really happening this time.*

Some months before, a 3-year-old boy in Hong Kong had developed a fever, a sore throat and a cough. The flu, his parents thought. But the boy grew sicker. Respiratory arrest set in, and he died. The case alarmed doctors. They could not recall seeing such a nasty case of the flu, particularly in a child so young. They sent off samples of his lung fluid for testing, and the results showed that he did indeed have the flu, but it was a strain that had previously appeared only in birds. H5N1, it's called. Webster is the world's preeminent expert on avian influenza, and it was only a matter of time before the test results made their way to him. But he was not yet troubled. He thought there must have been some sort of contamination in the lab. H5N1 had never crossed over into humans. Had to be a mistake, he thought.

That was until Cox interrupted his gardening to tell him about the new cases.

It immediately occurred to Webster that he should be on an airplane. "I had to go into the markets," he told me recently. "I had to get into the markets as fast I could." He meant the poultry markets, where chickens are bought and sold by the hundreds of thousands. The little boy who died a few months before had been around some chickens, as have most little boys in that part of the world, where families often live side by side with their chickens, pigs, ducks and dogs. If H5N1 was, in fact, in the markets, as Webster suspected, that was the beginning of his worst-case scenario: the virus could mutate in the chickens and perhaps other animals, and then acquire the know-how to pass from person to person, possibly initiating a pandemic that, he thought, might kill as many as 20 million people.

Webster has been predicting and preparing for such an event for his entire career as a scientist. His lab at St. Jude Children's Research Hospital in Memphis is the world's only laboratory that studies the human-animal interface of influenza. It was Webster who discovered that birds were likely responsible for past flu pandemics, including the one in Asia in 1957 that killed about two million people. He has spent a good part of his life collecting bird droppings and testing them for signs of influenza. Some of that collecting has taken place while he and his family were on vacation. One evening in Cape May, New Jersey, his school-age granddaughter ran toward him on the way to dinner saying that she had discovered some poop for him. He was so pleased.

A couple of days after Cox's phone call, Webster stepped off a plane in Hong Kong. He stopped at the University of Hong Kong to drum up some help to sample chicken droppings in the market. He also phoned his lab in Memphis and some scientists in Japan whom he had trained. He told them to pack their bags.

It occurred to Webster that there was a problem. The problem was H5N1. Neither he nor any members of his staff had ever been exposed to the virus strain, meaning they did not have any antibodies to it, meaning they had no defense against it. If they became infected, they would likely meet the same fate as the little boy who died.

They needed a vaccine. Four decades before, Webster had helped create the first widespread commercial flu vaccine. Until he came along, flu vaccines were given whole—the entire virus was inactivated and then injected. This caused numerous side effects, some of which were worse than the flu. Webster and his colleagues had the idea to break up the virus with detergents, so that only the immunity-producing particles need be injected to spur an immune response. Most standard flu shots still work like this today.

Before they went to work in Hong Kong, Webster and his colleagues created a sort of crude vaccine from a sample containing the H5N1 virus. They declined to discuss the matter in detail, but they treated the sample to inactivate the virus. Webster arranged for a pathologist in Hong Kong to drip the vaccine into his nose and the noses of his staff. In theory, antibodies to the virus would soon form.

"Are you sure this is inactivated?" the pathologist said.

Webster pondered the question for a moment.

"Yes it is. I hope."

And the fluid began dripping.

"It's very important to do things for yourself," Webster told me recently. "Scientists these days want other people to do things for them. But I think you have to be there, to be in the field, to see interactions." In many ways, Webster's remarkable career can be traced to a walk along an Australian beach in the 1960s, when he was a microbiology research fellow at Australian National University.

He was strolling along with his research partner Graeme Laver. Webster was in his 30s then, Laver a little older. Every 10 or 15 yards they came across a dead mutton bird that apparently had been washed up on the beach. By that time, the two men had been studying influenza for several years. They knew that in 1961, terns in South Africa had been killed by an influenza virus. Webster asked Laver: "What if the flu killed these birds?"

It was a tantalizing question. They decided to investigate further, arranging a trip to a deserted coral island off Queensland. Their boss was not entirely supportive of the adventure. "Laver is hallucinating," the boss told a colleague. They were undeterred. "Why there?" Laver once wrote of the trip. "Beautiful islands in an azure sea, hot sand, a baking sun, and warm coral lagoon. What better place to do flu research!" They snorkeled during the day. At night, they swabbed the throats of hundreds of birds. Back at their lab, they had a eureka moment: 18 birds had antibodies to a human flu virus that had circulated among people in 1957. Of course this meant only that the birds had been exposed to the virus, not that they were carrying or transmitting it.

To figure out if they were, Webster and Laver took subsequent trips to the Great Barrier Reef, Phillip Island and Tryon Island. More swimming during the day, sherry parties at dusk, and then a few hours of swabbing birds. They took the material back to their lab at Australian National University, in Canberra. It is standard procedure to grow flu viruses in chicken eggs. So they injected the material from the swabs into chicken eggs, to see if the influenza virus would grow. Two days later the fluid was harvested. In most of the eggs, the virus had not grown.

But in one of the eggs, it had grown. That could mean only one thing: the virus was in the birds.

Webster wanted to know more. Specifically, he wanted to know whether birds might have played a role in the influenza pandemic of 1957. He traveled to the World Influenza Center, in London, which has a large collection of influenza virus strains from birds and also antibody samples from flu victims. His experiment there was rather simple. He gathered antibody samples from victims of the 1957 flu pandemic. He also gathered samples of several avian flu strains. Then he mixed the samples. What did the antibodies do? They attacked the bird flu strains, meaning the human flu virus had some of the same molecular features as avian flu viruses.

How could that be? The answer is something now known as reassortment. The influenza virus, whether it's carried by birds or humans, has ten genes, which are arranged on eight separate gene segments. When two different influenza viruses infect the same cell, their genes may become reassorted—shuffled, mixed up. The net effect is that a new strain of flu virus forms, one that people have never been exposed to before. Webster refers to the mixing process as "virus sex." Perhaps Webster's greatest contribution to science is the idea that pandemics begin when avian and human flu viruses combine to form a new strain, one that people lack the ability to fight off.

After he entered the Hong Kong poultry markets, Webster needed only a few days to turn up enough chicken droppings to show that the H5N1 strain was indeed circulating. Along with many of his colleagues, he recommended that all the chickens in the market area be killed, to prevent spread of the virus. About 1.5 million chickens in Hong Kong met their maker. And that seemed to do the trick. The virus was gone.

But Webster had a hunch it would be back. The reason was ducks. Webster thinks the most dangerous animal in the world is the duck. His research has shown that ducks can transmit flu viruses quite easily to chickens. But while chickens that come down with bird flu die at rates approaching 100 percent, many ducks don't get sick at all. So they fly off to other parts of the world carrying the virus. "The duck is the Trojan horse," Webster says.

After the chickens in Hong Kong were killed, wild ducks probably relocated the virus to other parts of Asia, where it continued to infect chickens and shuffle its genetic makeup. When the strain emerged from hiding again, in Thailand and Vietnam in late 2003, it was even stronger. The virus passed directly from birds to people, killing dozens in what the World Health Organization has described as the worst outbreak of purely avian influenza ever to strike human beings.

Webster says the world is teetering on the edge of a knife blade. He thinks that H5N1 poses the most serious public health threat since the Spanish flu pandemic of 1918, which killed an estimated 40 million to 100 million people worldwide. Though the H5N1 strain has so

A Century of Influenza

1889–90: Influenza spreads from Central Asia to Russia, Europe and North America. The so-called "Russian flu" kills roughly one million people.

1918–1919: At least 40 million die of the "Spanish flu," the most deadly disease episode in history.

1925

1934: Microbiologist Thomas Francis Jr. confirms identification of the influenza virus.

1941: Francis and Jonas Salk develop a flu vaccine, used during World War II.

1950

1955: Avian flu is shown to be caused by the most virulent of the three forms of the virus, Influenza A.

1957: A pandemic of "Hong Kong flu" kills more than two million people worldwide.

1968: An influenza pandemic kills about one million. Many deaths are blamed on an inadequate supply of vaccine.

1975

1997: A highly pathogenic avian influenza virus, called H5N1, is discovered in Hong Kong. Authorities slaughter 1.5 million area poultry to prevent an epidemic.

2000

2003: H5N1 virus influenza surfaces at a few chicken farms near Seoul, South Korea.

2004: H5N1 virus is transmitted from birds to people in Thailand and Vietnam, killing dozens.

2005: In August, U. S. officials say an experimental vaccine for one strain of the H5N1 virus shows promise for preventing infection.

2005: In October, U.S. scientists report that the H5N1 virus shares features with the virus that caused the 1918 pandemic.

2005: In November, President Bush announces a pandemic flu plan, requesting $7.1 billion for vaccine development, drug stockpiling, surveillance and other measures.

Also, H5N1 virus surfaces in several locations in China, killing at least three people, though some say hundreds died.

for one, with hundreds of millions of dollars to be spent on further developing a new vaccine that was recently hatched in Webster's lab.

Webster has been advising federal health officials every step of the way. He does so out of fear of this virus and also because it is his job. When the H5N1 strain emerged in the late 1990s, the National Institute of Allergy and Infectious Diseases awarded Webster a major contract to establish a surveillance center in Hong Kong, to determine the molecular basis of transmission of avian flu viruses and isolate strains that would be suitable to develop vaccines. "He's certainly one of those people in this field who have been way ahead of the curve in bringing attention to this issue," Anthony Fauci, the institute's director, told me. "He was out ahead of the pack. He's one of the handful of people who have not only been sounding the alarm, but working to prevent this thing from turning into something that nobody wants to see happen."

Webster's job keeps him out of the country two to three weeks a month. Back in Memphis, his lab analyzes samples of influenza virus strains from around the world, to see how they are mutating. Recently, health officials have reported finding H5N1 avian flu in birds in Turkey, Romania, Croatia and Kuwait. It has not yet been found in birds in North America. If H5N1 makes its way here, Webster will likely be among the first to know.

This past June, I caught up with Webster at a meeting of the American Society for Microbiology, in Atlanta, where he was scheduled to deliver a speech about the threat of bird flu. There were more than 5,000 microbiologists in attendance, which, because I am a recovering hypochondriac, I found strangely comforting. Walking around with Webster at a meeting of scientists is an experience that must be similar to walking around with Yo-Yo Ma at a meeting of cellists. When Webster walked by, people suddenly stopped speaking, a fact to which he seemed oblivious.

He opened his talk by asking a series of intriguing questions: "Will the H5N1 currently circulating in Vietnam learn to transmit, reproduce, from human to human? Why hasn't it done so already? It's had three years to learn how, and so what's it waiting for? Why can't it finish the job? We hope it doesn't."

He paused. "Is it the pig that's missing in the story?"

Webster explained that the strain is still not capable of acquiring the final ingredient needed to fuel a pandemic: the ability to transmit from person to person. For that to happen, Webster and others believe that a version of the human flu virus, which is easily transmittable between people, and the H5N1 avian virus have to infect the same mammalian cell at the same time and have virus sex. If H5N1 picks up those genes from the human flu virus that enable it to spread from person to person, Webster says that virtually nobody will have immunity to it. If an effective vaccine based specifically on that newly emerged virus isn't quickly available, and if antiviral drugs aren't also, many deaths will ensue.

Watching Webster speak, I couldn't help thinking that animals are not always our friends. It turns out that animals are a frequent source of what ails us. University of Edinburgh

far shown no signs that it will acquire the ability to transmit easily from person to person—all evidence is that flu victims in Vietnam and Thailand acquired the virus from direct contact with infected poultry—that has provided Webster no comfort. It's only a matter of time before this virus, as he puts it, "goes off." He has been saying this for several years. The world is finally taking notice. Elaborate plans are now being created in dozens of countries to deal with a pandemic. In November, President Bush requested that $7.1 billion be set aside to prepare researchers recently compiled a rather frightening list of 1,415 microbes that cause diseases in humans. Sixty-one percent of those microbes are carried by animals and transmitted to humans. Cats and dogs are responsible for 43 percent of those microbes, according to the Edinburgh researchers; horses, cattle, sheep, goats and pigs transmit 39 percent; rodents, 23 percent; birds, 10 percent. Primates originally transmitted AIDS to humans. Cows transmit bovine spongiform encephalopathy, or mad cow disease. In their 2004 book, *Beasts of the Earth: Animals, Humans and Disease,* the physicians E. Fuller Torrey and Robert Yolken cite evidence suggesting that a parasite transmitted by cats, *Toxoplasma gondii,* causes schizophrenia. A couple of years ago, the monkeypox virus broke out among several people in the Midwest who had recently had close contact with pet prairie dogs.

And then there are pigs. For many years, Webster has theorized that pigs are the mixing bowls for pandemic flu outbreaks. He has actually enshrined the theory in his house. He has a stained-glass window next to his front door that depicts what he perceives to be the natural evolution of flu pandemics. At the top of the glass, birds fly. Below them, a pig grazes. Man stands off to the left. Below all of them are circles that represent viruses and seem to be in motion. They are set in a backdrop of fever red.

The pig is in the picture because its genome, perhaps surprisingly, shares certain key features with the human genome. Pigs readily catch human flu strains. Pigs are also susceptible to picking up avian flu strains, mostly because they often live so close to poultry. If a human flu strain and an avian flu strain infect a pig cell at the same time, and the two different viruses exchange genetic material inside a pig cell, it's possible that the virulent avian strain will pick up human flu virus genes that control transmission between people. If that happens with H5N1, that will almost certainly mean that the virus will be able to pass easily from person to person. A pandemic may not be far behind.

During his talk in Atlanta, Webster pointed out that this H5N1 virus was so crafty that it has already learned to infect tigers and other cats, something no avian flu has ever done. "The pig may or may not be necessary" for a pandemic to go off, Webster said. "Anyway, this virus has a chance at being successful." He said he hoped world health officials "would keep making their plans because they may face it this winter. We hope not."

I went hunting with Webster. Hunting for corn. His cornfield is on a patch of land he owns about five miles from his home on the outskirts of Memphis. He grows genetically modified corn that he gets from Illinois. An extra gene component known for increasing sweetness has been inserted into the corn's DNA, producing some of the sweetest corn in the United States. Three of his grandchildren were with us, visiting from North Carolina. They had come, among other reasons, for Webster's annual Corn Fest, where members of the virology department at St. Jude Hospital gather in his backyard to sit around eating corn on the cob. The record for the most ears of corn eaten in one sitting at the Corn Fest is 17. The record holder is the teenage son of one of Webster's protégés. Webster reports the prize was a three-day stomachache. He encouraged me not to beat this record.

"There's a good one," Webster said, bending down to pull off an ear. He was wearing long shorts, a plaid blue shirt and a wide-brimmed canvas hat. He had been fussing around among the stalks for a few minutes before he found an ear he liked. He seemed unhappy with the quality of the corn, muttering into his chest. In between picking some ears, I asked why he was down on the crop. "I believe I planted too soon," he said. "The ground was still too damp." This caused many of the ears to bloom improperly. I asked why he had planted so early. He said, "I had to be in Asia." It occurred to me that attempting to stop a global epidemic was a reasonable excuse for a so-so batch of corn.

Webster was home this weekend for the first time in many weeks. He had been to Asia and back nearly a dozen times in the past year. I asked Marjorie Webster how often she sees him, and she replied, "Not much these days." It is a sacrifice she seems willing to make; Webster has told her plenty about the bug and what it can do.

We picked corn for about half an hour, then went back to Webster's home to do some shucking. He shucked at a pace nearly double mine. We must have shucked 250 ears of corn. We placed the shucked ears in a cooler of ice. By noon we had finished, so I decided to go do some sightseeing. Beale Street, Elvis impersonators, several barbecue joints. A little before 5 P.M., I wandered into the lobby of the Peabody Hotel, a landmark. I wanted to see the ducks. Since the 1930s, ducks have swum in a fountain in the hotel's lobby. The ducks live upstairs in a sort of duck mansion. In the morning, they ride down in an elevator. When the elevator doors open in the lobby, the ducks wobble down a red carpet, single file, about 30 yards, in front of hundreds of people who snap photographs as if they were duck *paparazzi*. When the ducks plop into the fountain, people cheer. At 5 P.M., the ducks are done for the day; they wobble back along the carpet to the elevator, then ride back to their mansion for dinner. One generally has to witness the occasion to believe it.

I wondered whether Webster had ever tested these ducks. That evening, at the corn party, after my third ear, and Webster's second, I told him that I had gone to see the ducks. "Oh, the Peabody ducks," he said, the first time I'd seen him visibly happy in days. "The kids loved the ducks when they were little." I asked whether he liked the ducks too. "Why not? I enjoy the ducks," he said. I said, "Have you ever swabbed them?" He answered: "No. Sometimes you just don't want to know. There are some ducks I won't swab."

MICHAEL ROSENWALD is a staff writer for the *Washington Post*. This is his first article for Smithsonian.

Climate, Environment, and Infectious Diseases: A Report from the AIBS 2008 Annual Meeting

The American Institute of Biological Sciences dedicated this year's annual meeting to the challenge of Earth's changing climate and its effects on the environment, and the spread of infectious diseases. The conference, held in May in Arlington, Virginia, attracted more than 250 biologists, climatologists, and other scientists, as well as physicians and public health officials.

CHERYL LYN DYBAS

Malaria. Dengue fever. West Nile virus. Lyme disease. These and other infectious diseases are on the rise in humans. Is our interaction with Earth's environment somehow responsible? What factors are involved in the emergence and transmission of infectious diseases? How are climate change and other environmental parameters involved? These complexities have created a tangled web that scientists must soon unravel.

If biologists, climatologists, physicians, and other researchers work closely together, said Richard O'Grady, executive director of the American Institute of Biological Sciences (AIBS), "we have the potential to predict disease outbreaks and to mitigate their effects. Accelerating climate change and threats to health are now two [sides] of the same coin."

Links among the environment and infectious diseases were demonstrated in many ways and for many locations by AIBS conference speakers; they made connections between microbes, environmental events such as El Niño, and disease incidence in people. Large-scale environmental events—global climate change, land-use change and habitat destruction, and human population growth and urbanization—alter the risks of viral, parasitic, and bacterial diseases. Agricultural intensification, deforestation and reforestation, increased precipitation, and ocean warming all play a role.

For example, in a case that might be called "CSI: Infectious Diseases and Climate Change," otherwise healthy people began dying of a mysterious respiratory disease in 1993 in the US Southwest. Tests yielded surprising results: The victims had hantavirus pulmonary syndrome, a result of a previously undetected type of hantavirus.

Research by biologist Terry Yates of the University of New Mexico, honored posthumously at this year's meeting with the AIBS Distinguished Scientist Award, showed that the outbreak was connected to the El Niño climate phenomenon, a pattern of changes in ocean circulation and atmospheric weather. Increased rainfall led to more plants, more mice, and more opportunities for people to come into contact with the rodents' droppings and urine, which contain the virus. Named for the Hantaan River in Korea, hantaviruses were known to spread from mice to humans in Asia and Europe, but until the 1993 outbreak, hantaviruses had been seen only outside America.

"We can no longer discuss infectious diseases simply as those that affect the US, or that affect New York City, or Washington, DC," said scientist and cholera expert Rita Colwell, of the University of Maryland and Johns Hopkins University Bloomberg School of Public Health. "Infectious diseases that arise in Africa or in Asia reach the United States, as evidenced by the SARS [severe acute respiratory syndrome] episodes," she said. "The connections to the environment are very dramatic. For example, the origin of SARS was tracked to bats in caves in Asia."

Colwell, who is also the president of AIBS, spoke of cholera as a disease not just restricted to Bangladesh or India but prevalent also in Latin America and Africa. The environmental burden of disease falls on those least prepared to deal with it, she said. Countries in Africa and in Asia, where populations are large and poverty is high, are particularly affected, as shown in studies by the Centers for Disease Control and Prevention (CDC) in Atlanta, Georgia. Dealing with the effects of climate change on infectious disease falls—so far—mostly on these populations.

How long will it be before the rest of the world follows suit? That depends, Colwell believes, on decisions we make now about our environmental future. Climate change poses risks to ecosystems and the life support they provide to people, and to all animals and plants on Earth. The bottom line, AIBS meeting participants agreed, is that to safeguard human health, we must live within our planet's environmental limits.

The Environmental Century: A World Beyond Fossil Fuels?

Terry Maple, president of the Palm Beach Zoo in Florida, spoke of the premise of the book *A Contract with the Earth,* which he wrote with Newt Gingrich. Maple called ours "the environmental century." Green investments are growing, he said, and we need to address what he called the overwhelming issues facing life on Earth in the near future. Through their book, Maple and Gingrich hope to appeal to the public to open up dialogues and debate issues.

Looking at innovation as a solution to climate change is a promising path, Maple said. "If we get serious about innovation, we're going down the road America is good at. We'll be selling this technology to the people we're now buying oil from." For example, he discovered that almost no solar power projects exist in the state of Florida. Through his efforts, the Palm Beach Zoo is joining with Florida Power and Light to build a facility that's nearly half solar powered.

An engaged and informed populace and finding new energy sources are the only answers to the challenges we face, Maple and Gingrich believe. Answers will come not a moment too soon, many at the meeting said. Global warming, say climate and health scientists, is beginning to set off a worldwide "domino effect."

"In the story of Earth's recent climate, we have all the characteristics of 'the perfect storm,'" said climatologist James Hansen, director of the NASA Goddard Institute for Space Studies. "Among them is that the primary greenhouse gas we're putting into the atmosphere by burning fossil fuels, carbon dioxide, lasts a very long time. A large fraction will stay in the air for several centuries."

What level of carbon dioxide is an imminent threat to life on Earth? "The dangerous amount of carbon dioxide is less than what's there now," Hansen said. "We've increased carbon dioxide to 385 parts per million. We're going to have to reduce it to at least 350 parts per million."

A third of the warming in the last century has occurred over the past 30 years. Isotherms are moving forward at a rate of some 60 kilometers with each passing decade. If we continue business as usual, Hansen believes, that rate will double in the next century. "The most important thing in the short term," Hansen said, "is to deal with coal, the biggest source of carbon dioxide. I think if we addressed this, and got the process going in the right direction, there would be positive feedbacks in the social system that could, in fact, take us to a world beyond fossil fuels." We can't double or triple the amount of carbon dioxide in the atmosphere, he said, "or we're going to produce a completely different planet."

Infectious Disease Risks and Public Health: The Climate Factor

"Unlike directly transmissible disease between humans, zoonotic and vectorborne diseases are very much dependent on the environment, especially climate," said biologist Durland Fish, director of the Yale University Center for EcoEpidemiology. Fish focused on Lyme disease as an example.

Lyme disease was discovered in the early 1970s in Lyme, Connecticut, but it occurs throughout much of the United States and in Europe and Asia. Lyme disease risk remains highest in northeastern US states like Connecticut and Rhode Island. "There's been a lot of speculation about what's going to happen with Lyme disease and climate change," Fish said. "We could have more severe disease at more northerly latitudes."

Environmental change and reforestation have resulted in the epidemic of Lyme disease, scientists believe. A century ago, there were no deer—nor was there Lyme disease—in southern New England. In the absence of deer, deer ticks can't survive. It's only since reforestation and the subsequent proliferation of deer, Fish said, "that Lyme disease has been able to expand its range." The changing environment, he said, has everything to do with changing human disease risk.

"We're witnessing the emergence of a different world than the one in which our grandparents grew up,"said Howard Frumkin, a physician and director of the National Center for Environmental Health at the CDC."The world is changing almost before our eyes, including the health burden of climate."

Heat waves in cities, said Frumkin, have become more common as the climate warms. "We know who in cities is the most vulnerable: the elderly, people who are shut-ins, people who live on upper floors, people without air conditioning, people on certain medications that impair the ability to dissipate heat." Heat waves are an example of a public health problem related to climate change that physicians know how to handle. Through simple buddy systems, people are brought to safety in refuge centers.

Infectious diseases are much harder to deal with. As tropical climates expand their ranges, tropical diseases may march along in lockstep."Sometimes this is predictable," Frumkin said,"and sometimes not." He cited dengue fever as an example of predictability. Climate modeling has been used to forecast an extension of dengue fever in various parts of the world."Sure enough, a decade after that climate model was run," Frumkin said, "we're seeing headlines about the expansion of dengue fever."

Scientists need to take existing models that describe climate system changes and Earth system changes, then scale down and extend them so they can forecast health outcomes and implications of climate change, Frumkin believes. "Then we need to develop plans to protect the public, test those plans using research protocols so we know what works best, and ultimately implement steps to protect public health."

"If everything is connected," asked ecologist David Rogers, of the University of Oxford, "how can we sort out cause and effect? . . . How many 'inconvenient truths' are there? Is there something else we should know about?" Rogers questioned whether climate change may already have had an effect on some diseases. For that, he said, the disease must have changed "in the right place, at the right time, and in the right direction, according to our understanding of the relationship between climate and disease." In the case of malaria in East Africa, he believes, most highland areas will become climatically suitable for the disease in the near future. Evidence exists that some of those regions already harbor it.

Rogers also cited a rise in the number of human cases of tick-borne encephalitis in the Baltic states from 1970 to 2005. "Are they due to global warming?" he asked. The answer lies in many complex variables, he believes, not in climate change alone. In Lithuania, for example, economic changes in total employment, agricultural changes in number of cattle, environmental changes in tick abundance, and other factors play an important role. We have a distance to go, Rogers said, in understanding the factors involved in the spread of infectious diseases. Only when we do can we say how much of that advance is a result of climate change.

The Challenge Ahead: Predicting the Emergence of Infectious Diseases

Ecological changes are important for providing new opportunities for pathogens to reach the human population or to cross species, said scientist Stephen Morse of Columbia University's Mailman School of Public Health. "A problem that has bedeviled all people who attempt to predict the effect of changing climate on infectious diseases," he said, "is that the specific scenarios—what diseases will move where—depend greatly on exact changes in climate."

Avian flu, Morse said, may or may not be the next pandemic strain, but it's of concern because there have been human cases and limited human-to-human transmission, almost all of them associated with domestic poultry. What effects might climate have on avian influenza? Migratory birds that interact with infected poultry follow age-old flyways for now, but will these birds continue to use those flyways when climate changes? Or will they change their pattern? "I suspect they will adapt," Morse said, "just like other living things, to find their 'comfort zone.'"

To answer questions about climate and disease, he said, "we need to understand a great deal more about seasonality and related factors. Specific prediction will depend on knowing specific conditions."

As the world gets wetter or warmer, how is that going to affect disease predictions? asked biologist Andrew Dobson, of Princeton University. "For most diseases, there is likely to be a seasonal driver." It gets complicated, Dobson said, "because seasonal cycles interact with the natural dynamics of disease."

Cholera and malaria are the poster children for looking at the effects of climate change on disease, he said. "With cholera, we can develop models showing that climate variability, particularly El Niño, drives cholera dynamics in Bangladesh." If scientists ignore El Niño, he continued, they won't do a

very good job of predicting diseases; if they include it, they still may miscalculate because they forgot to include all the details of immunity. "Unless you know how many susceptible people are out there, you don't know how the population will respond."

Scientists looking at climate and malaria in Africa have noted a small change in temperature from 1950 to 2000, an increase of about 3 percent. "If you use that temperature change to drive a model for the dynamics of the [mosquito] vectors, then that 3 percent change results in a 30 percent increase in the number of vectors," Dobson said. "A 30 percent increase in vector density is likely to give you a significant increase in malaria."

More than one variable needs to be considered, he believes. There's the evolution of drug resistance. There's biodiversity. "The more biodiversity, the more it's protecting us from, for example, mosquito-transmitted diseases," Dobson said. "That's relevant to climate change, because as you go from the tropics to the temperate zone, you have less and less biodiversity, which means that diseases can focus more on humans."

The dynamics of all these interactions are complex mathematical problems. "It's not rocket science," Dobson said. "It's much harder."

And it all started long ago. Sometime in the past, "humans working in the forest brought the dengue virus into villages, where epidemics first began," said scientist Duane Gubler, of the University of Hawaii John A. Burns School of Medicine. But because human populations were small then, the epidemics didn't persist. "The viruses burned through the villages and disappeared," Gubler said. Then as port cities around the world developed, and as mosquitoes were transported and infested those port cities, an urban epidemic cycle began.

The mosquito *Aedes aegypti* is the principal vector of this disease. It's a "highly domesticated mosquito," Gubler said. "It prefers to feed on humans. It prefers to live in houses with humans—and not anywhere else."

In the early 1980s, scientists saw a dramatic geographic expansion of both the viruses and the vectors. "As we go into the 21st century," Gubler said, "dengue has become one of the most important emerging tropical diseases."

Urbanization, along with modern transportation patterns, is behind the global reemergence of dengue hemorrhagic fever. Dengue viruses and their mosquito vector (*Aedes aegypti*) have completely adapted to an urbanized lifestyle, needing only humans to complete their life cycles.

The combination of urbanization, modern transportation, and general globalization, along with increased movement of people, animals, and commodities, is perfect for moving these pathogens around, Gubler said. "Our single biggest challenge is to do something about the movement of vectors and pathogens by modern transportation. And we need a better understanding of disease ecology because most of these diseases won't have a drug or a vaccine."

New Approaches for a Different World

The malaria world is changing dramatically and rapidly, said Stephen Hoffman, chief executive officer of Sanaria, Inc., in Rockville, Maryland. Hoffman is a physician and tropical infectious disease specialist who, with colleagues at Sanaria, is working to develop the first successful malaria vaccine.

"We go to work every day with one dream and one dream only: to develop a malaria vaccine to prevent millions of deaths from malaria," Hoffman said. "Malaria is responsible for more deaths in children than any other infectious agent. More than 300 million people will suffer clinical attacks of malaria during the next year. I've heard estimates recently that it's up to one billion." At least 70 percent of the illnesses will occur in sub-Saharan Africa, where 90 percent of malaria-related deaths occur, most of which are children under five years old.

"Malaria not only has a huge health impact," Hoffman said, "it also has a huge economic impact. It's the leading health-care expenditure in most of Africa. It's a major cause of poverty, and poverty enhances disease. It's been estimated that at least 10 [billion] to 12 billion dollars are lost annually to malaria in Africa."

Malaria is not only a disease of children in remote villages in Africa. Tens of millions of travelers from North America, Europe, Australia, the Middle East, and East Asia visit areas of the world with malaria every year; some 30,000 contract the disease. "Five to ten million American travelers are at serious risk annually, with equivalent numbers for Europe," Hoffman said.

Sanaria was formed five years ago with the mission to develop, license, and release a vaccine based on an attenuated version of the malaria-causing parasite *Plasmodium falciparum.* "The vaccine, we hope, will prevent infection with this parasite in greater than 90 percent of the recipients for at least six months," Hoffman said.

The vaccine development method at Sanaria is different from other approaches. The company is using a live, attenuated, whole-organism vaccine; other efforts have been "subunit recombinant" (a subunit apart from the entire infectious agent is used, and the vaccine is made recombinantly in the laboratory). Sanaria is now in the process of manufacturing its malaria vaccine for clinical trials, first in the United States, then in Africa.

"The road to success has lots of potential potholes," Hoffman said, but he believes that there will be a vaccine for malaria in five years. "I can hear footsteps behind me, so we all work every day to try to prevent the 3000 children who are dying from malaria today from dying."

In her AIBS annual meeting closing remarks, Colwell stated: "Interdisciplinary approaches will be required to address these critical problems of the 21st century. They're international, in that we will need to work together to develop a global understanding of these diseases, and also of global public health measures to be taken in a time of climate change and of societal change."

Billions of people are depending on the answers.

CHERYL LYN DYBAS (e-mail: cldybas@nasw.org) is a biologist and science journalist who specializes in the environment and health.

From *BioScience,* October 2008, pp. 792–797. Copyright © 2008 by American Institute of Biological Sciences. Reprinted by permission.

Mosquito Modifications:
New Approaches to Controlling Malaria

Malaria kills about one million people each year, but efforts to destroy disease-carrying mosquitoes have succeeded only in breeding tougher bugs. Researchers have begun to look for ways to create malaria-resistant mosquitoes. One approach is to bioengineer transgenic mosquitoes that, when released into the wild, would lead to a new race of malaria-proof young. Another approach uses mosquitoes' natural resistance to *Plasmodium* infection.

SHARON LEVY

As daylight wanes on the island of São Tomé, a team of biologists heads out to spy on one of the most important, but least studied, bits of natural history in Africa: the sex life of the mosquito *Anopheles gambiae*, the most widespread vector of malaria on the continent. The researchers, led by J. D. Charlwood of the Danish Bilharziasis Laboratory, have scoped out likely spots scattered on the outskirts of a village, places where a footpath intersects grassland or a swath of dark soil meets the bleached wood of a tree stump. At dusk, male mosquitoes gather over these areas of color contrast, form swarms, and await the arrival of potential mates.

Looking into the columns of hovering insects in the fading light, the researchers watch and count as more and more mosquitoes begin to couple, their hind ends interlocking as they fall out of the swarm. The mating process peaks and drops off within 15 to 20 minutes, and then, in the darkness, females fly to nearby homes to bite the villagers. Each female needs a series of blood meals for sustenance as she incubates the eggs that will form a new generation.

Malaria, a parasitic disease transmitted by infected mosquitoes, threatens an estimated three billion people in 106 nations. Most of the fatalities caused by malaria are young African children. In recent years, a new global effort to control the disease has risen from the ashes of a failed campaign that once tried to eradicate it.

Failed Attempts

That first major antimalaria drive was based on two chemicals that seemed to hold the promise of destroying both *Plasmodium falciparum*, the most virulent of the protozoan parasites that can attack human liver and red blood cells during malaria infection, and the mosquito vectors that carry the disease. Chloroquine, a cheap, effective equivalent of the plant extract quinine, long the most successful antimalaria drug in the world, was first synthesized in the 1940s.

In 1939, chemist Paul Muller discovered that an organochlorine compound known as DDT worked as a powerful insecticide. His achievement was widely celebrated, and DDT was used for disease control worldwide during and after World War II. Louse-infested refugees were doused with it, and the chemical was dropped in many areas where mosquitoes were thought to breed. When Muller was awarded the Nobel Prize for his work in 1948, many still cherished the hope that DDT would wipe out malaria-bearing mosquitoes forever. Yet the first wild mosquitoes to evolve resistance to DDT had already been identified two years earlier, in 1946. Excessive use of DDT in agriculture accelerated the evolution of insect resistance. By the early 1960s, about 400,000 metric tons of DDT were used annually, 70 to 80 percent of which was for control of crop pests.

Malaria has been effectively wiped out in the United States and many other developed nations, but both *Plasmodium* and its mosquito vectors still flourish in many poorer, hotter countries. The malaria parasite has evolved resistance to chloroquine and to subsequent generations of drugs. Today the only reliable malaria treatment is a cocktail of drugs that hit the parasite in several different ways at once. Likewise, mosquitoes and other insects have shown a great facility for detoxifying DDT and several forms of alternative insecticide. Recent studies of DDT resistance in the fruit fly *Drosophila melanogaster*, which is used as an experimental template by many insect researchers, show that a mutation at a single gene locus confers resistance to DDT and an array of other pesticides, and it is likely that a similar mutation occurs in DDT-resistant mosquitoes.

By 1972, when the United States banned DDT because of its long-lived toxic impacts on wildlife and human health, 19 species of malaria-transmitting mosquito were resistant to the chemical. When the World Health Organization recently reiterated its support for limited use of DDT inside the homes of rural people living in malaria-affected regions, there was a flurry of passionate responses in the North American press, including claims that environmentalists who supported the ban on DDT had the blood of millions of African malaria victims on their hands. Those claims ignore political facts—DDT has remained available in many countries—as well as a basic biological reality."Genes for DDT resistance can persist in populations for decades," writes entomologist May Berenbaum, of the University of Illinois. "Spraying DDT in the interior walls of houses, the form of chemical use now advocated as the solution to Africa's malaria problem, led to the evolution of resistance 40 years ago, and will almost certainly lead to it again unless resistance monitoring and management strategies are put into place."

Berenbaum points out that modern-day pockets of mosquito resistance to DDT are already well documented in Africa. Mosquitoes can also quickly evolve resistance to alternative poisons: research on Bioko Island, off the coast of Cameroon, recently found that a new pyrethroid insecticide lost its punch in less than two years. For now, indoor spraying of DDT to help control the raging epidemic may be the best tool at hand in some parts of Africa, but the threat of mosquitoes developing resistance remains—and a less toxic alternative, pyrethroid-laden mosquito bed nets, can be just as effective.

Building a Better Mosquito

With the insecticide arms race doomed to fail, researchers have begun to explore an intriguing new strategy. Instead of wiping out winged vectors with poisons, they hope to build a better mosquito, one that is immune to *Plasmodium* infection. The goal is to someday neutralize the deadly threat of malaria by making mosquitoes healthier, leaving the victims of *Anopheles* bites at risk of nothing worse than an itchy bump.

Laboratory work on ways to manipulate the mosquito genome to confer malaria resistance is in some ways surprisingly advanced. Marcelo Jacobs-Lorena, of Johns Hopkins University, and his colleagues have inserted an extra gene into *Anopheles stephensi,* a mosquito that transmits malaria in India; the gene makes the insects resistant to mouse malaria, *Plasmodium berghei.* (*P. berghei* is the commonly used laboratory malaria model, because working with *P. falciparum* requires expensive, sophisticated biohazard facilities.) Several different research groups in the United States and Europe are working with different varieties of transgenic mosquitoes that have been made immune not only to malaria but also to dengue fever, another deadly mosquito-borne illness. Still, transferring such a trait into wild insect populations presents a formidable challenge.

A major problem is that lab-reared mosquitoes are likely to have trouble competing with their wild relatives. Many details of mosquito life histories remain mysterious, and it's unlikely that humans can manufacture mosquitoes whose immune responses have been engineered to thwart malaria without inadvertently changing other important traits along the way. In the wild, mosquitoes must adapt to local conditions, which are

sometimes harsh. In some parts of the world, they must survive a long dry spell each year. In others, mosquitoes can breed year-round in continuously wet habitats but are targeted by a multitude of predators. The great majority of wild larvae—more than 90 percent—don't survive.

For those that do live, size can be a critical factor. Often, the smallest adults die before they have a chance to mate or eat. If adults make it to a sunset swarm to hunt for a mate, subtle factors can affect their success. Males use a complex sensory organ to track and amplify the sound of a female's whine. Both a female's ability to produce the right tones and a male's capacity to track them are crucial to successful pairing.

Despite these complexities, wild mosquito populations are often exceedingly dense. Millions of mosquitoes can hatch daily in a single village. Pushing a bio-engineered trait into such a vast wild population would be an uphill struggle. In a recent review of the existing studies of mosquito reproduction and survival, Charlwood estimates that, assuming a highly fit malaria-resistant mosquito can be produced, it would take many decades for the resistance trait to come to dominate a wild population through normal genetic inheritance. And complete population replacement is the goal: if even a small proportion of vector mosquitoes live, they'll continue to spread malaria to people.

In any human-designed mosquito hatchery, the insects are bound to be subject to adaptive pressures that differ from those in the outside world. Any tweaks to their innate timing systems could render transgenic mosquitoes useless in the wild. In nature, mating takes place during a precise 20-minute window at dusk. Colonies that breed indoors, where the lights are either on or off, are likely to undergo selection for insects that will mate after dark. Even a slight delay in the biological clocks of human-reared mosquitoes could leave them unable to find wild mates.

Antimalarial transgenes themselves make mosquitoes less likely to survive in the wild."Bioengineered *Plasmodium*-resistant mosquitoes have so far all had a fitness disadvantage compared to wild strains," says Willem Takken, an entomologist with Wageningen University in the Netherlands. In order to replace malaria carriers in nature, "genetically modified mosquitoes would need to overcome that handicap and demonstrate strong behavioral and biological advantages over wild mosquitoes."

Selfish Transgenes

Molecular biologists are aware of the problem, and for several years they have been talking about possible ways to force a transgene into a wild insect population at rates much faster than those produced by normal Mendelian inheritance. In a paper published in *Science* in April 2007, Bruce Hay, Chun-Hong Chen, and their colleagues at the California Institute of Technology describe a genetic trick that accomplishes this task in *Drosophila,* and they hope it will translate with relative ease into the mosquito genome.

Hay's group has designed a genetic element they've dubbed *Medea,* a set of genes that spreads quickly through a population not because it makes individuals more fit, but because it kills the competition. "You can think of it as someone running a race," says Hay. "One way to win is to be better than your competitor. The other way is to whack them in the knees. That's the way *Medea* works, through everyone's favorite behavior: spite."

The inspiration for this new twist in bioengineering came from the discovery by Richard Beeman, an entomologist at Kansas State University, of a naturally occurring gene in the flour beetle that dominated populations by killing competing genotypes rather than by conferring a fitness advantage. Beeman hypothesized that this trait was actually a pair of genes, one a toxin that poisoned every egg cell a mother beetle produced, the other an antidote that rescued only the offspring carrying the selfish gene set. Hay and his group read Beeman's studies and began to brainstorm ways to design a system in *Drosophila* that could replicate this pattern. In theory, such a set of selfish genes, once linked to a disease-resistance trait, could be an invaluable "gene driver" able to rapidly push engineered characteristics into wild insect populations.

Hay's group tinkered with the installation of various toxin genes but could not find a workable system. Either the toxins were so powerful they killed every oocyte before it could be fertilized, or so weak they had no significant effect at all. While trying to formulate an alternative approach, says Hay, "We realized there's an entirely different way to make a toxin. Instead of adding the expression of a poisonous protein, we could silence a gene needed for normal embryonic development."His lab has since been able to produce a selfish gene set, coding for RNA sequences that silence the expression of *Myd88,* a gene essential for the normal early development of fly embryos. This is linked with an "antidote" gene, a spare copy of *Myd88* that is expressed by the embryo soon after fertilization, just in time to rescue the newly conceived fly. So a female with the Medea trait—"Medea"stands for maternal effect dominant embryo arrest"—dooms every one of her own oocytes by failing to express *Myd88* herself, and only the embryos that have inherited the *Medea* genes have the ability to save themselves.

The next step is to apply this work from *Drosophila* to *A. gambiae.* The mosquito's entire genome was recently sequenced, but it remains little known compared with *Drosophila*'s genetic code, which has been explored and tinkered with by a multitude of researchers. Still, early oogenesis and embryogenesis in the mosquito is similar to that in *Drosophila*. The hope is that bioengineers will be able to use the counterparts of the genes identified in *Drosophila* to build a *Medea* element in mosquitoes, then link it to malaria-resistance genes.

"Building *Medea* in *Drosophila* is the equivalent of building a model airplane as a test for seeing how you'd build a real plane," says Hay."We know we can build something that flies in a lab, in a model insect. Now we need to go do it in a real insect, the mosquito, and we want it to fly not only in a nice simple lab environment but [also] in the wild, where it's exposed to uncontrolled temperature, humidity, predators, and genetic diversity."

Hay envisions factories that will someday churn out disease-resistant, *Medea*-bearing male mosquitoes for release to areas of endemic malaria. Any releases of transgenic mosquitoes are expected to be limited to males, which don't feed on blood and so have no direct contact with people. If the plan succeeds, the right gene driver could replace an entire population of wild mosquitoes with offspring bearing the antimalaria transgene in the course of a year or two.

He acknowledges that any factory environment would probably select for traits that reduce mosquitoes' fitness in the wild, but he believes an effective gene driver can overcome the problem."Some proportion of them are going to make it," he says. "Once you produce a first generation of hybrid mosquitoes in the wild, the selfish genetic element can go do its thing."

Naturally Resistant Mosquitoes

Wild mosquitoes may already have evolved the most effective immune defense against malaria, beating bioengineers to the punch. Recent work by geneticists Ken Vernick and Michelle Riehle, of the University of Minnesota, in collaboration with field researchers in Mali and Kenya, shows, for the first time, that the majority of *A. gambiae* in Africa possess innate resistance to the disease. Biologists in Mali captured wild female mosquitoes resting on the walls of huts in a malaria-affected village. The mosquitoes had mated in the wild, and their young were reared in a laboratory, then fed on blood from malaria-infected people in the same village. A few days later, the researchers dissected the mosquitoes and found that a majority had been able to kill the parasites they had ingested.

"The prevailing notion had been that it was a large proportion of the mosquito population that was transmitting malaria," says Riehle. "Now it seems the reverse is true: the majority of wild mosquitoes are resistant, and it's a minority that actually cause infection in humans." Sifting through the mosquito genome in the Minnesota lab, the Vernick group has identified a cluster of genes—which they've dubbed the *Plasmodium* resistance island, or PRI—responsible for innate immunity. They've homed in on a single gene locus coding for a leucine-rich protein, APL1, similar to molecules known to work in antipathogen responses of plants and mammals.

The study, published in *Science* in 2006, is unique in that it examines malaria resistance in mosquitoes that are the offspring of natural matings by wild parents in a malaria-endemic area. Most work on transgenic malaria resistance uses colonies of mosquitoes not only removed from the selective pressures of the wild but also tested by their response to rodent, rather than human, malaria. "I believe starting in nature is most important," says Riehle."There the mosquitoes have to fend for themselves. Anything you develop in the lab, where all the insect's needs are taken care of, is very artificial."

In a study published in *Malaria Journal* in July 2007, the same researchers show that mosquitoes in Kenya, on the opposite side of Africa from Mali, possess innate malaria resistance that can be mapped to the same gene cluster. The same mechanism of immunity seems to exist in *A. gambiae* across Africa, suggesting it is an ancient, well-established trait that might be exploited to help stop the spread of the disease.

Riehle believes building a malaria control strategy around this natural immune response is a much safer bet than relying on transgenes."We need to find a way to tip the balance in nature, to increase the fitness cost of being susceptible to malaria so that resistant mosquitoes outnumber and eventually wipe out the disease carriers," she says. The finding that resistant mosquitoes dominate wild populations is both encouraging and daunting: there must be some fitness benefits for malaria-susceptible insects, or they would naturally die out.

One potential weapon for tipping the balance against malaria-bearing mosquitoes is a fungus, endemic to Africa, that attacks adult insects. The fungus weakens and kills more malaria-infected mosquitoes than malaria-free mosquitoes. Willem Takken is one of a group of researchers exploring the use of the fungus, already commercially produced for use against agricultural insect pests, as a way of knocking down mosquito populations in African villages. Fungal spores mixed in oil, sprayed on sheets of fabric, and then hung on the walls of homes killed a significant number of mosquitoes within days of their first blood meal. Even if the insects were susceptible to malaria, they died before they'd had time to incubate the contagious form of the *Plasmodium* parasite in their bodies. "We see a strong fitness effect of the fungus on the mosquitoes," says Takken. "The process is slow, which may slow down the eventual development of resistance mechanisms against the fungus. The advantage is that the fungus is naturally present in Africa, and mosquitoes may be regularly exposed to it in the wild."

Takken shares Riehle's belief that using innate immunity and naturally occurring pesticides is a better idea than attempting to bioengineer a solution. "Making use of natural resistance mechanisms has in principle a much brighter future, as the evolutionary selection processes have already taken place," he says. Resistance traits evolved in nature are less likely to cause unforeseen problems than completely novel, laboratory-built genes.

Other researchers disagree. "Most of the people I know who are trying to engineer resistance in mosquitoes are using human-designed genes, rather than trying to tweak the innate immune response," says entomologist Thomas Scott, of the University of California–Davis. "Natural immune systems might more quickly select for resistance in the malaria parasite, because it's something the organism is already dealing with."

Hay suggests an eventual combination of the two approaches: someday, bioengineers may be able to link a human-designed selfish gene driver to the naturally occurring malaria resistance genes identified by Vernick's lab. That kind of artificial boost may be able to wipe out the minority of the *A. gambiae* population that remain susceptible to malaria.

If transgenic mosquitoes are ever released into the wild, it will only be after a long series of experiments. Researchers will have to slowly move their bioengineered insects from the lab to carefully sealed outdoor cages to study the effects of natural temperature, humidity, and light regimes. They'll need to find ways to mass-produce designer insects. Bioengineered mosquitoes will face many legal and ethical hurdles: people might choose not to eat genetically modified foods, but no one in Africa will have a choice about living with genetically modified mosquitoes once they're released. Some scientists working in the field acknowledge that African health officials are wary of the idea of any genetically modified organism being set loose on their turf.

But as Scott points out, there are parts of Africa where families don't name their children until they're two years old because so many of them don't survive. Young children, whose immune systems are not fully developed, are more susceptible than adults to the loss of red blood cells caused by malaria, and they often die from it. "Around 30 percent of children between the ages of one and five were dying of malaria when I worked in western Kenya," he recalls. "How do the local people feel about this? They'd like it to stop. If you've got a good idea, they'll be open to it."

There's yet another layer of complication. While *A. gambiae* is the most populous malaria vector in Africa, other species of anopheline mosquitoes also carry the disease. Even if the entire population of *A. gambiae* was made malaria resistant, other species would continue to transmit malaria to people. "Taking *gambiae* out of the equation would be a fantastic contribution," says Scott, "but it won't eliminate the problem."

For now, practical malaria-control efforts continue to focus on distributing pyrethroid-laden bed nets, spraying DDT and other pesticides indoors, cleaning up mosquito breeding habitats near villages, and delivering effective anti-malaria drug cocktails to the impoverished communities that need them most. Overcoming malaria will have to involve somehow controlling its passage through insect vectors, but the old vision of wiping mosquitoes off the face of the planet with mass pesticide spraying is history. Understanding mosquitoes and their complex relationship with the malaria parasite is now the critical challenge.

SHARON LEVY (e-mail: levyscan@sbcglobal.net) is a freelance writer based in Arcata, California.

The White Plague

A disease that fends off nearly every drug you throw at it could already be taking hold in a town near you.

Debora MacKenzie

You've heard of bird flu, and you probably remember the SARS panic of 2003. Now there is a new bug on the loose, right here in your backyard, and it is one that the World Health Organization is declaring just as big a threat as those two diseases—although few column inches have been devoted to it so far.

The disease is an old enemy, but it has a deadly new twist. The bacteria that cause TB have mutated to resist most of the weapons we have against it. Dubbed extremely or extensively drug-resistant TB (XDR-TB), the strain is now spreading to every part of the globe, with cases discovered last year in the UK, US and Canada. And *New Scientist* can reveal that last month, Italy reported a case of TB that was invulnerable to every single drug we have—the first-ever known incidence of completely drug-resistant TB.

The advent of this "extreme TB" is prompting some radical countermeasures. Because the disease can be passed on through mere coughing, countries such as the US have resorted to locking up patients who violate quarantine. The WHO is holding crisis meetings and doctors are experimenting with drastic surgery for those who have lost all other hope. Bob Swanepoel, a virologist at South Africa's National Institute for Communicable Diseases, has battled plagues of Ebola and HIV, and knows a formidable enemy when he sees one. "XDR," he says, "is a frightening disease."

TB has been a scourge of humanity for millennia—even Egyptian mummies have been found with its deadly fingerprints. Four centuries ago, for reasons that are still unclear, the disease seems to have gone on the rampage in Europe, and was responsible for one in five deaths from the 1600s to the 1900s. Sometimes called the "white plague", in reference to the deathly pallor of those affected and the only slightly deadlier black death, TB spawned the familiar literary character from that period: the "consumptive"—feeble, coughing up blood and slowly but inexorably wasting away.

Today about a third of the world's population are thought to carry the TB bacterium, *Mycobacterium tuberculosis,* in their lungs. In most it remains dormant, or latent, but one in 10 of those infected develop the active disease at some point, usually after some stress or setback to the immune system. That is when TB kills, and when it can spread.

The disease is most common in poor countries, and poor communities within rich countries, because dirt, malnutrition and overcrowding prompt the switch to active disease and help the infection spread. It typically passes between people in crowded sleeping areas, though it is possible that simply sharing a long-haul flight may be enough.

TB mainly attacks the lungs, killing tissue and creating holes and abscesses. People become gradually weaker, develop fever and night sweats, and usually die from respiratory failure or lung haemorrhage. Untreated, about a third of people with active TB recover, a third die relatively quickly and the rest linger, spreading infection before they die.

The first antibiotics against TB appeared in the 1940s. For a few decades, it seemed as though the white plague would be eradicated, at least among the wealthy, but now TB is on the rise once more.

Initially, it started spiralling among the poor, and the HIV-positive, whose ravaged immune systems cannot suppress active infection. Now the rise in the west is mainly among immigrants, whose latent infections turn active with the stress of emigration and poverty in a new land. It is this group that accounted for the UK's record 10 per cent rise in cases between 2004 and 2005, and for most new cases in the US, although the total number of cases there is stable.

What is really frightening doctors, though, is the growing number of infections that are resistant to antibiotics. Resistance develops because TB is a tough bacterium to kill: it has a slow growth rate and a waxy coat that is hard for drugs to penetrate. Added to that is its habit of living inside the immune cells that should destroy it. So people with the infection must take the standard or "first-line" antibiotics for up to nine months, and during this long siege any mutant strains that can withstand the drugs have time to flourish. In an attempt to pre-empt this, treatment begins with four antibiotics and continues with two of them, to ensure that any bacteria resisting one drug will fall prey to the others in the end. However, if patients stop their treatment too soon, either because they feel better, or because the medicines are unavailable or unaffordable, drug-resistant bacteria may remain.

Resistant from the Start

Some people end up with drug-resistant TB not because they have failed to keep to their treatment regimes, but because they contracted a resistant strain in the first place. Until recently, this was not thought to be a big problem. In theory microbes pay a price for carrying genes for drug resistance: in the drugs' absence, they are weaker than their normal drug-sensitive cousins and so spread less effectively.

That's the theory. But Gary Schoolnik, a microbiologist at Stanford University in California, took samples of bacteria from individual patients as their infections changed from drug sensitive to resistant, then pitted them against each other in the test tube. In research published last year, he showed that if anything, the resistant strains were stronger.

This fits with the few field observations available to date. Half the people at Tugela Ferry, where extremely drug-resistant TB (XDR-TB) was first reported in South Africa, had had no prior TB treatment, showing that resistance was already present in the bacteria rather than developing through poor treatment adherence.

Even people who are receiving treatment may be getting "super-infected" with circulating resistant strains. That's what Qian Gao and colleagues at Shanghai Medical College in China reported this month in *The Journal of Infectious Diseases.* They tracked 35 people whose TB became drug-resistant. In only five did their initial bacteria evolve resistance—the other 27 were super-infected with new multi-drug-resistant (MDR) strains during treatment.

If MDR and XDR circulate that readily, they could pose a much greater risk to people than ordinary TB. Gao says his work shows the importance of strict quarantine. "We cannot wait to implement measures to block transmission," he warns.

The Doctor

"I realised we had something very serious on our hands," says Tony Moll, head of the 350-bed Church of Scotland Hospital in Tugela Ferry, a market town in the dusty, impoverished hills of South Africa's KwaZulu-Natal province.

He is referring to the day in 2005 when test results came back for a group of patients whose TB was not responding to treatment. "I thought we might have multi-drug-resistant TB. When the results came back, 10 were extremely drug-resistant (XDR-TB), with resistance to six drugs—virtually untreatable."

Part of the problem, says Moll, was that the hospital lacked the resources to ensure patients took their full course of medicine and thus prevent the development of drug resistance. "We had two nurses, with no vehicle, who were supposed to track patients defaulting on treatment and do home visits for 7000 people in the community. It just wasn't possible."

At least, since the Tugela Ferry cases caused a sensation in the TB world, the hospital now has more equipment.

But they also have 222 XDR-TB patients. "Are we a hotspot or just the tail of the elephant?" asks Moil. "I think when they look for more XDR-TB, they'll find it."

Since 2001, 53 people at one clinic had been diagnosed with XDR-TB, and only one had survived, with the rest dying within a few weeks. "That surprised us," says Mario Raviglione, head of TB at the WHO. "We didn't think there was much XDR-TB in Africa."

By October Raviglione had called a crisis meeting at the WHO headquarters in Geneva, Switzerland. The WHO now has response plans in place, and is trying to improve basic treatment overall.

It is also trying to assess the scale of the problem. Early figures seem to show that XDR is present across South Africa. Public-health labs around the world have pooled data on nearly 20,000 samples from people with TB in 49 countries taken between 2000 and 2004. In February these labs announced that nearly 2 per cent of the samples were XDR. The WHO has new figures due to be announced this month showing that in the US, the proportion of MDR cases that were in fact XDR grew from 3 per cent in 2000 to 11 per cent in 2004.

"We are as helpless against completely drug-resistant TB as we were in the 19th century, before antibiotics."

—Mario Raviglione

The 1980s saw growing numbers of cases of multi-drug-resistant TB (MDR-TB), defined as bacteria that can withstand the two main first-line drugs. After that, only four antibiotic classes remain. They are much more expensive and must be taken for up to two years. Some have to be given intravenously, and all have nasty side effects, ranging from nausea and diarrhoea to convulsions and kidney failure.

A year ago this month, US doctors highlighted the first cases of XDR-TB, defined as resistance to the two main first-line drugs, and a specific duo of the second-line ones. This combination means the bacteria tend to be left susceptible to fewer than four other kinds of drugs, the minimum needed for a reliable treatment regimen.

Initially XDR-TB was thought to be a limited problem, primarily in countries such as Russia and China where enough people are taking the second-line drugs under less-than-optimal conditions to promote resistance. Then came a report from South Africa, which shocked delegates at the 26th International AIDS Conference in Toronto last August.

Now *New Scientist* has learned of the first known case of completely drug-resistant TB, in Italy. Raviglione points out that we are as helpless against this strain as we were before the

The Quarantine Officer

The TB world erupted in January after three doctors wrote about extremely drug-resistant TB (XDR-TB) in *PLoS Medicine* (DOI: 10.1371/journal.pmed.00h0050). They warned that authorities in South Africa and elsewhere might have to get tough with those who refuse quarantine. "We didn't write that article lightly," says one of the authors, Ross Upshur, a Toronto doctor and ethicist, "but we had to say something."

You cannot accuse Upshur of ignorance of his subject. He acted as a quarantine officer during the SARS epidemic that hit Toronto in 2003. "I want to treat people and cure them, not quarantine them. No one likes you for that," he says.

Ultimately, says Upshur, the answer is better TB treatment. "But we have to ask what we can do now to protect the community. Otherwise, we might as well just say, 'here's kindling, kerosene and a match'. The kindling is the high prevalence of HIV, the kerosene is XDR-TB, and the match is doing nothing."

Before the development of antibiotics, he points out, western nations had sanatoriums where TB patients were sent—not always on a voluntary basis—until they either recovered or died. "We need to consider that again," says Upshur. "Or we need to admit that we are going to tolerate the transmission of an incurable disease to innocent people."

development of antibiotics. "We are back in the 19th century," he says.

With XDR-TB, doctors can still attempt to treat people with the remaining drugs, in the hope that they will kill the bacteria before further resistance mutations arise. But for people with HIV in the developing world, XDR-TB is an almost certain death sentence. Even for those who don't have HIV and live in the west, the prognosis seems grim; there are few statistics because this disease is so new, but a group of 64 US patients who received treatment had a death rate of one in three. Some US doctors are resorting to an approach reminiscent of the days before antibiotics—surgically removing infected lung tissue, or in some cases, a whole lung.

Jail Wards

Doctors face another difficulty: how to cope with uncooperative patients. Being admitted to an isolation unit is a daunting prospect—patients must take nasty drugs, often for many months, possibly until their death. Unsurprisingly, some simply discharge themselves and go home, where they infect others. Another group that presents a dilemma are those who have

a track record of stopping taking their TB medicines before the course is complete. Even if such patients are theoretically non-infectious, because they are responding to the antibiotics, doctors may strongly suspect that sending them home risks the life of the patient and those around them.

In the US, one such patient in Arizona is in a hospital jail ward, without visiting privileges, a phone or showers, allegedly because he stopped taking the antibiotics and visited food stores without wearing a mask. The patient, 27-year-old Robert Daniels, who contracted XDR-TB while visiting his native Russia, says he was afraid he would be mistaken for a masked robber. Another person with XDR-TB, in Toronto, Canada, was forced to return to hospital under a court order.

Authorities around the world are debating how to treat such cases. The WHO has issued a statement saying that forced quarantine may be needed as a last resort. But Raviglione warns: "If people think they are going to be locked up they will not come for treatment. We will drive [TB] underground."

In the longer term, a key goal is to develop new antibiotics effective against TB. The pharmaceutical industry has long neglected the disease, as it has mainly affected the poor. Progress is being made, however, by the Stop TB consortium, funded by the WHO, donors and governments. It has nine drugs in development with two in clinical trials (*New Scientist*, 17 February, p 10), although none may be ready for at least another five years.

We also need to be able to diagnose cases of XDR-TB much more quickly than the present time of two months. The bacteria multiply so slowly that a sputum sample must be cultured for several weeks before getting enough bacteria to test for resistance, which is done by adminstering the drugs and seeing if the growth rate of the bacteria slows. By that time the patient could have infected many others—and patients with HIV could be dead.

The UK Health Protection Agency's TB reference lab has developed better tests, which it will use in southern Africa later this year. One uses an improved culturing method to speed the bacteria's growth, allowing XDR to be diagnosed within two weeks. Another can identify MDR within two days using DNA analysis to test for the mutations that cause resistance to the first-line drugs. The tests will be made available to the developing world at an affordable price. Researchers are now trying to identify the mutations that cause resistance to the second-line drugs so they can identify XDR-TB in the same way.

The WHO says $650 million a year is needed to treat the estimated 1.5 million with MDR or XDR-TB worldwide—of which so far it only has $250 million.

Perhaps the first known case of completely drug-resistant TB will help galvanise global concern. There must be others—perhaps the person sitting next to you on that plane. The Italian case, says Raviglione, "is probably just the tip of the iceberg".

Rising Incidence of Valley Fever and Norovirus

Known infections seem to be on the rise.

Carol Potera and Shawn Kennedy, MA, RN

While influenza and community-acquired methicillin-resistant *Staphylococcus aureus* have been making headlines, illnesses such as valley fever and norovirus have received less attention but have had a considerable impact on health.

Coccidioidomycosis

Valley fever—the common name for coccidioidomycosis, a fungal lung infection that mimics influenza and pneumonia—has been appearing more frequently in western and southwestern states (mostly in Arizona and California but also in Nevada, New Mexico, Utah, and Texas) and Mexico.[1] Although the disease has been known for a century, its incidence started soaring in the 1990s. In Arizona there were about 4,900 cases last year,[2] and "more than 5,500 cases were reported in 2006, including 33 deaths."[3] Another outbreak occurred at a prison near Fresno, California, where according to a December 30 *New York Times* story, 900 inmates and 80 workers have contracted the illness since 2004.[3]

The fungus that causes valley fever, *Coccidioides immitis,* thrives in dry, alkaline soil; spores are inhaled when dusty desert soil is stirred up, and symptoms, if they do appear, generally do so in seven to 21 days.[1] Experts blame the rising infection rate on several factors, including an influx of new arrivals and visitors to the Southwest who lack a natural immunity to *Coccidioides,* as well as the recent boom in new construction. Workers with an occupational exposure to dusty soil, such as agricultural and construction workers, are especially at risk, as are smokers and older adults. Those at risk of more serious infection or complications include Filipinos, Asians, Hispanics, and blacks (for unknown reasons); pregnant women; and people with compromised immune systems. Animals, including dogs, cats, and horses, are also susceptible to valley fever.[1]

Symptoms may include fever, cough, headache, rash, and myalgia. Although only about 40% of infected people develop symptoms and most recover unscathed, the infection can advance to severe pneumonia or spread to the central nervous system (including the meninges), bones, joints, or skin.[4] Azole antifungals (fluconazole [Diflucan] and itraconazole [Sporanox]) are used to treat acute, chronic, and disseminated infections, Rapidly progressing infections or infections in pregnant women are treated with amphotericin B. To protect against infection, the Centers for Disease Control and Prevention (CDC) suggests controlling dust exposure by wearing masks that filter particles as small as 0.4 microns, using air conditioning, and wetting soil before digging.[1]

Even in endemic regions, medical personnel don't always recognize valley fever. Researchers at the Valley Fever Center for Excellence (VFCE) at the University of Arizona College of Medicine found that primary care physicians at three clinics in Tucson misdiagnosed almost one in three patients with valley fever as having pneumonia.[5] Furthermore, 81% of patients with valley fever were prescribed ineffective antibiotics. The study's authors recommend that primary care-givers obtain a travel history from patients as soon as pneumonia is suspected. In a recent interview conducted by e-mail, John Gagliani, MD, director of the VFCE and a coauthor of the study, suggested that travelers who contract a pneumonia-like illness within a month of returning home from an endemic area should let their physicians know about their travel and ask whether their illness could be valley fever.

For more information, see www.cdc.gov/ncidod/dbmd/diseaseinfo/coccidioidomycosis_t.htm.

Norovirus

The so-called "stomach flu"—diarrhea, vomiting, abdominal cramps, nausea, and mild fever—is most often caused by norovirus, a highly contagious organism that accounts for more than half of all foodborne outbreaks of gastroenteritis. The CDC estimates that 23 million cases occur in the United States yearly.[6] Last year, the CDC found that outbreaks of acute gastroenteritis had increased in 2006 and 2007 and that two new strains of norovirus were to blame.[7]

Previously known as Norwalk virus (after the Ohio town that suffered a 1968 outbreak of gastroenteritis that was linked to the newly identified microbe), norovirus spreads rapidly in environments like schools, hospitals, cruise ships, and nursing homes. Incubation takes 24 to 48 hours, symptoms last one to three days, and the main therapy consists of replacing lost fluids.[6] The very young, people with weakened immune systems, and the elderly are especially at risk because of fluid loss.

The virus spreads primarily via the fecal-oral route, either through the consumption of contaminated food or water or through person-to-person contact. According to the CDC, there is "good evidence" of transmission resulting from the "aerosolization of vomitus that presumably results in droplets contaminating surfaces or entering the oral mucosa and being swallowed."[6] The CDC also says there is no evidence that infection "occurs through the respiratory system."[6] Outbreaks often begin with unsanitary handling of foods such as salads, sandwiches, and bakery products. Foods like oysters that come from contaminated waters or prepackaged salad greens that may have been irrigated with contaminated water have caused widespread out-breaks. The virus is present in both stool and stomach contents from the onset of symptoms to at least three days after recovery; viral shedding may continue for as long as two weeks after recovery. Those who are symptomatic should stay home, wash hands frequently avoid handling food meant for others, and flush vomit and stool down the toilet. Those caring for the sick should wash hands frequently, disinfect household surfaces with a bleach solution, and launder the clothing and linen of the sick person.

A norovirus outbreak in February 2007 at a grade school in Washington, D.C., was traced to a contaminated computer mouse and keyboard shared by students and staff.[8] Public health officials who investigated the outbreak said that the source of infection highlighted "the difficulty in identifying and properly disinfecting all possible environmental sources of norovirus during outbreaks."[8] For further information about norovirus, see www.cdc.gov/ncidod/dvrd/revb/gastro/norovirus-factsheet.htm.

References

1. Mayo Clinic. *Infectious disease: Valley fever.* 2008. http://www.mayoclinic.com/health/valley-fever/DS00695.
2. Arizona Department of Health Services. *Arizona—Valley fever report.* Phoenix, AZ; 2007 Dec. http://azdhs.gov/phs/oids/epi/disease/cocci/cocci_report_dec_2007.pdf.
3. McKinley J. Infection hits a California prison hard. *New York Times.* Dec. 30, 2002. http://www.nytimes.com/2007/12/30/US/30inmates.html.
4. Centers for Disease Control and Prevention. Division of Bacterial and Mycotic Diseases. *Coccidioidomycosis.* Atlanta, GA; 2005 Oct 6. http://www.cdc.gov/ncidod/dbmd/diseaseinfo/coccidioidomycosis_t.htm.
5. Valdivia L, et al. Coccidioidomycosis as a common cause of community-acquired pneumonia. *Emerg Infect Dis* 2006;12(6):958–62.
6. National Center for Infectious Diseases. Respiratory and Enteric Viruses Branch. *Norovirus: technical fact sheet.* Atlanta: Centers for Disease Control and Prevention; 2006 Aug 3. http://www.cdc.gov/ncidod/dvrd/revb/gastro/norovirus-factsheet.htm.
7. Centers for Disease Control and Prevention. Norovirus activity-United States, 2006-2007. *MMWR Morb Mortal Wkly Rep* 2007;56(33):842–6.
8. Centers for Disease Control and Prevention. Norovirus outbreak in an elementary school-District of Columbia, February 2007. *MMWR Morb Mortal Wkly Rep* 2008; 56(51-52):1340–3.

CAROL POTERA AND SHAWN KENNEDY, MA, RN, editorial director; Emerging Infections is coordinated by Pancy Leung-Chen, MPA, RN, CIC: pancyl@gmail.com.

They Came from Above

Opportunistic infections seem to pop up out of nowhere, but new strains are appearing in new places, striking otherwise healthy animals—including humans. A few microbiologists go hunting.

BRENDAN BORRELL

I n the spring of 2000, veterinarian Craig Stephen walked up to the biology department at Vancouver Island University in Nanaimo for what he thought would be a routine autopsy of a dead porpoise. "In my experience of doing stranded marine mammals, the vast majority of them, you don't get anything," says Stephen, who runs the university's Center for Coastal Health, "They've died, they've sunk, they've started to rot, they float back up, they get on the beach and then somebody finds them."

Stretched out in the department's backyard the sleek, four-foot long corpse was remarkably fresh, and Stephen saw nothing unusual as he scanned the eyes, mouth, and blowhole for hints of what led to its demise. Then he slit open the abdomen, removed each organ, and placed pieces into sterile plastic bags for laboratory tests. When he cracked open the chest cavity, Stephen was immediately struck by the lungs. "Usually you feel them and they are spongy, right?" he says. "You feel these and they are like liver." They were dense and heavy, swollen with pneumonia. Under a microscope, the sectioned tissue was dotted with tumor-like cysts. Stephen was shocked to learn the diagnosis: Cryptococcosis, a fungal disease best known for ravaging AIDS patients with weakened immune systems, had killed a seemingly healthy marine mammal.

That was just the start. By the summer of that year, local veterinarians had recorded 12 cases of cryptococcosis in domestic dogs, cats, and even llamas—up from an average of about five per year. In 2001, scientists noticed that relatively healthy people were getting sick—and even dying from cryptococcal infection.

The infectious yeast looks like a fried egg, a sphere surrounded by a cloudy sheath. By shedding this sheath in a process experts jokingly call "vomitcytosis," Cryptococcus absconds from the bacteria in the decaying plant matter, soil, and pigeon droppings where it lives. This tactic works equally well against human and animal phagocytes, thus enabling Cryptococcus to evade the immune system. The disease proceeds from shortness of breath to fever and headaches. If it is diagnosed soon enough, several months of intravenous anti-fungal Amphotericin B can knock it out. But not everyone is so lucky: This past April, 45-year-old Sandra Agostini became Vancouver Island's latest fatality due to Cryptococcus.

O n a recent visit to the British Columbia Center for Disease Control in Vancouver, medical microbiologist Linda Hoang opens a humming incubator to reveal a small sample of the isolates the center has amassed over the last eight years. Today, the reference lab she supervises diagnoses 25–30 human cases of this virulent strain of Cryptococcus each year using restriction fragment length polymorphisms (RFLP), and is testing a PCR-based strategy to quicken the turnaround time.

Unlike diseases transmitted strictly via an animal vector or from person-to-person, the ubiquity of opportunistic pathogens like Cryptococcus presents a new and daunting set of challenges for scientists and medical professionals. Any organism small enough to be lifted into the air has the potential to achieve a cosmopolitan distribution, provided it can survive where it lands. In the cool climates of a temperate zone, untold numbers of potentially pathogenic bacteria and fungi may subsist—but fail to thrive—just below the level of detection. Microbial ecologists have a saying for this: Everything is everywhere, but the environment selects. [1]

Biologists now recognize that this dogma is only partly true, particularly in the face of Earth's warming climate. For example, over 50% of Kazakhstan's croplands have been sucked dry, while the Sahara expands into Nigeria and Ghana at a rate of 3,500 km^2 per year. This global process of desertification is increasing the number of dust storms that ferry microbes across continents and oceans. Meanwhile, in the temperate zone, rising temperatures have rendered some regions more hospitable to colonization by microbial hitchhikers arriving on soils from tropical climes. This new fungal strain cropping up in people and other animals with healthy immune systems may have been a new arrival to Vancouver Island, or it may have always been tucked away in some hidden valley for many years, until one balmy summer triggered its unfortunate bloom. And there are hints that it is

Same-Sex Mating in *Cryptococcus?*

Since fungi must mate to produce airborne spores, *Cryptococcus* researchers have been faced with a mystery. "The unusual thing about *C. neoformans* and *C. gattii* is for the vast majority of the world you can only find [one sex]," says molecular geneticist James Fraser of the University of Queensland. The trouble is that, without mating, the yeast reproduces clonally, and individual cells typically are not tiny enough to infiltrate the lungs. So how can only a single sex of *Cryptococcus* infect people in the open air?

There's some surprising evidence to suggest that two strains of a single sex can, in fact, mate—a process which could produce spores small enough to become airborne. A 2005 study in *Nature* by Joseph Heitman's group at Duke University demonstrated that a unisexual population of *C. neoformans* could mate to produce airborne spores in the laboratory.[1] Later that year, Fraser, Heitman, and colleagues published a genealogical analysis suggesting that the Vancouver Island fungus, *C. gattii*, had also undergone same-sex mating, a finding they suggested was key to its ability to expand into its new habitat on Vancouver Island.[2]

Fraser and colleagues used multilocus sequence typing (MLST) at as many as 30 polymorphic loci to identify four discrete molecular types within *C. gattii* worldwide. Vancouver Island contained two of those four types, which the authors term a major form and a minor form, both of which were only present in a single sex. The major form, it turned out, was identical to a rare strain isolated from a human sputum sample in Seattle 30 years earlier, and to a sample collected from a San Francisco *Eucalyptus camaldulensis* tree in 1992. The minor form is identical to a widespread *C. gattii* strain from Australia and South America. The two strains were also closely related to each other, sharing 14 out of 30 alleles.

Fraser believes that the minor strain was one of two parents that begat the major strain and, possibly, its airborne spores. When they sequenced the sex-determining alleles in the two strains, they found that they were closely related but clearly different, suggesting that the major strain had inherited its sex-determining allele from an unknown parent of the same sex. Finding the second parent, Fraser admits, would have been the smoking gun, but he says it may no longer exist.

Karen Bartlett of the University of British Columbia remains skeptical. "It's a good theory," she says, "I'm not dissing it, I'm just not of the same opinion." Instead, she believes that *C. gattii* clones are hardy and small enough to become airborne and are able to infiltrate human lungs. For instance, she argues that the melanin-like pigments in *Cryptococcus* cells suggest that they have evolved to resist the high UV radiation encountered during airborne dispersal. Furthermore, her airborne samples have mostly fallen out in the upper three stages of the Andersen airborne sampler, indicating that the propagules range in size from 3.3μ up to 7μ or higher, far larger than the estimated $1-2\mu$ spore size.[3]

It's certainly plausible for desiccated cells to evade the lung's ciliary defenses, Fraser says. Researchers have infected mice with *Cryptococcus* cells rather than spores in the lab, and Bartlett points out that spores are not required for airborne infection in other species: hyphae of *Coccidioides immitis,* which causes San Joaquin valley fever, dry out and break into fine particles which are readily inhaled. At least for now, this yeast's sex life will remain a mystery.

—B.B.

References

1. X. Lin et al., "Sexual reproduction between partners of the same mating type in *Cryptococcus neoformans,"* *Nature,* 434:1017–21, 2005.

2. J.A. Fraser et al., "Same-sex mating and the origin of the Vancouver Island *Cryptococcus gattii* outbreak," Nature, 437:1360–4, 2005.

3. S.E. Kidd et al., "Characterization of environmental sources of the human and animal pathogen *Cryptococcus gattii* in British Columbia, Canada and the Pacific Northwest of the United States," *Appl Environ Microbiol,* 73:1433–43, 2007.

steadily furthering its progress. "The question we've been asking over the last 10 years," Hoang says, "Is it going to get to the mainland and will it spread across the Pacific Northwest?"

A key moment in aeromicrobiology, or the study of airborne microbes, came in 1933, when Fred Meier of the US Department of Agriculture convinced Charles Lindberg to collect samples during an arctic flight from Maine to Denmark. Upon finding everything from fungal spores to algae and diatoms, Meier wrote, "the potentialities of worldwide distribution of spores of fungi and other organisms caught up and carried abroad by transcontinental winds may be of tremendous consequence." We now know that particles of dust, organic matter, and aerosolized water droplets support hardy communities of bacteria, fungi, and viruses—a mere 0.08% of which have ever been cultured.[2]

Some 10,000 bacteria are present in every gram of airborne sediment, and the atmosphere contains at least one billion metric tons of dust.[3] That translates to a quintillion dust-borne bacteria—enough, according to Dale Griffin of the USDA office in St. Petersburg, Fla., "to form a microbial bridge between Earth and Jupiter." Over the course of five days in 2001, NASA tracked a large dust cloud that originated in the Gobi Desert as it moved east across the Pacific, North America, and the Atlantic, before petering out over Europe. Frequently, during African dust storms, a smoke-like strand is visible in satellite photos swirling off the continent, and looming over Italy, Spain, and southern France.

One of the most surprising new findings about airborne microbes is that far from being passive passengers of the wind, some are truly adapted to life in the mesosphere—70 km above the earth's surface—where they must constantly repair their DNA following bombardment by direct UV radiation. Or take a

2008 study that found that airborne microbes haunting Singapore shopping malls are not a random sample of what's outside, but are specialized for survival in the indoor air environment.[2]

One morning this summer, microbial ecologist Emilio Casamayor from the Center for Advanced Studies in Blanes, Spain, and visiting postdoc Maria Vila-Costa hike past hordes of tourists and backpackers to Lake Redo, one of 1,200 alpine lakes in the Pyrenees Mountains. After finding a comfortable rock to squat on, the pair pull out metal mesh screens about one meter long and bound by a wooden frame. Vila-Costa presses her screen to the water's mirrored surface and lifts it into the air, a thin film bridging the wire grid. She tilts the screen to the neck of her plastic bottle and fills it one drip at a time. "How long does it take usually?" she asks Casamayor. "One liter, one hour," he responds.

The clock starts ticking, and Casamayor takes time to explain his work, which is aimed at understanding the effect of dust-borne microbes on the native communities in the water. Most of the lake's biodiversity, he says, is restricted to this hydrophilic layer on the surface, where native microbes make use of nutrients coming from dust drifting in from Africa. He says that humic acids and other nutrients in atmospheric dust are easier for bacteria to break down than those from local soils, because they have been exposed to intense UV radiation. The dust particles also contain living microbes, and he wants to know how long they survive on the lake's cool surface. In addition to collecting and sequencing samples from 30 lakes in the region, Casamayor has set up a system to collect dust before it hits the lake and sequence organisms within it. These data will help him understand the impact of dust storms on microbial ecology.

Casamayor first noticed the dust in the Pyrenees back in June 2004, when his research focused on microbes living in the water. A late-season snowstorm had dumped fresh snow around Lake Redon near the town of Vielha and which sits at a slightly lower elevation than Redo. But when Casamayor came out to sample under the ice, he noticed the snow was already covered with a fine layer of brown grit. The phenomenon was not new—he had seen it in photos dating back to the 1860s—but with the drying of North Africa over the last 50 years, he reasoned that larger storms may be bringing more foreign microbes.

Indeed, the microbial community may be transforming in ways that scientists are only beginning to notice.

The microbial community may be transforming in ways that scientists are only beginning to notice.

Around the same time the *Cryptococcus* outbreak began in Canada, scientists finally discovered what was killing sea fan coral in the Caribbean. Since the early 1980s, massive numbers of *Gorgonia ventalina* in the West Indies have been smothered by four highly virulent strains of *Aspergillus sydowii,*[4] a common,

cosmopolitan fungus that had never caused widespread disease in plants or animals. The virulent *Aspergillus* strains were later shown to have originated in Africa.[5] The dust also brings crop pathogens such as sugar cane rust, *Puccinia melanocephala,* and banana leaf spot, *Mycosphaerella musicola.* During African dust storms, USDA microbiologist Dale Griffin says the number of culturable colonies of microorganisms airborne over the US Virgin Islands rises by a factor of 10. Since scientists began monitoring in 1973, an increase in African dust arriving on the island of Barbados has coincided with a 17-fold increase in asthma levels. In Spain, these dust clouds can have fatal consequences: A study published in the November issue of *Epidemiology* found that daily mortality in Barcelona increased by 8.4% during periods when African dust clouds were present over the region, possibly due to biological irritants and allergens.[6]

Casamayor began monitoring Pyrenean lakes last year, and one of the first organisms he found was *Acinetobacter,* an opportunistic pathogen that is highly resistant to antibiotics. A PhD student is currently developing molecular probes to track what happens when *Acinetobacter* lands in Lake Redo and what conditions would allow it to proliferate. Presently the team is only monitoring the effect of dust on microbial populations, but Casamayor plans to develop collaborations with other biologists to look for effects in plants and wildlife.

For instance, an unknown pestivirus has recently been ravaging populations of the endangered chamois, a goat-like ungulate that lives high in the mountains, and Casamayor wonders whether it could have been brought by the wind. Chamois researcher Emmanuelle Gilot-Fromont at the University of Lyon doubts that the pestivirus could survive over long distances, but agrees that airborne dispersal is a potential threat to humans and wildlife in the region. Q-fever, she notes, is caused by the bacteria *Coxiella burnetii,* and has traveled by wind from rural regions to infect humans in population centers in southern France.

Wandering through Vancouver Island's Rathtrevor Beach Provincial Park this summer, there is little evidence of the panic that gripped this coastal campground in June 2002, when campers cancelled their reservations in droves. Today, children fly kites on the windy beach and young couples pitch their tents beneath stands of towering Douglas Fir trees. No one pays much notice to the turquoise signs posted at every parking lot warning of the ever-present threat in the air.

In late 2001, as *Cryptococcus* infections spiked, Murray Fyfe of the British Columbia Center for Disease Control was just beginning to assemble the province's historic records of cryptococcal disease, when he received results from a Cryptochek serotyping test. The fungus infecting porpoises and people was not the cosmopolitan strain of *C. neoformans* that infected immunocompromised patients, but was *C. gattii,* a species known primarily from Australian eucalyptus groves and other tropical regions that could infect relatively healthy people. Fyfe's team tested the few groves of eucalyptus present on Vancouver Island, but failed to isolate the fungus. "We were a bit stumped," he says.

So Fyfe contacted Karen Bartlett of the University of British Columbia to find where *C. gattii* was lurking on the island. Bartlett is an expert in environmental health who had spent years studying occupational hygiene, publishing papers on endotoxin exposure in metal shops and the spread of pathogens in schoolrooms. Fyfe believed Bartlett's expertise in airborne pathogens would help the agency pinpoint the fungus in the wild. Bartlett agreed, and in March 2002 took the 90-minute ferry out to the island to join the team as they traveled the east coast from Victoria to Parksville, taking sterile swabs in the cracks, scars, and bark of native trees and transferring it to an agar derived from birdseeds.

Most of the dishes incubated back in Bartlett's lab turned up nothing, except for one sample from a Douglas fir tree in Rathtrevor Park that was peppered with smooth, brown dots. This melanin-like pigment is a tell-tale sign of *Cryptococcus* and protects it from damaging UV radiation when it is floating in the air. After discovering the positive sample, Bartlett returned to Rathtrevor with her Andersen air sampler in tow. Invented by army bacteriologist Ariel Andersen in the 1950s, the sampler contains six stages with progressively narrower holes to mimic human bronchial passages. Back at Rathtrevor, Bartlett flipped on an air pump for 10 minutes and let it suck 283 liters of outside air through the sampler. Then, she diligently counted dots on petri dishes. One sample contained 1,081 colonies of *Cryptococcus* in every cubic meter of air, practically a pure culture and far higher than any environmental samples Bartlett had cultured in the past. "We don't normally see airborne organisms in that kind of concentration," she says, with characteristic understatement, during an interview at her lab this summer.

On June 6, 2002, BC CDC issued a public advisory, warnings were posted at Rathtrevor, and cryptococcal infection became a reportable disease. By 2003, the infection rate had soared to 37 cases per million residents per year, far higher than the 0.94 cases per million residents in endemic regions of Australia. Over the summer and fall, Bartlett, postdoctoral researcher Sarah Kidd, and collaborators, continued to sample areas where human or animal cases had occurred, taking hundreds of soil, bark, and leaf samples along with additional air samples in the vicinity. *C. gattii*, they found, was turning up in small clusters all along the coast. These patches, they now know, blink on and off like Christmas lights, while others remain permanently colonized. All of these patches fall within two similar biogeoclimatic zones, the Coastal Douglas Fir zone and the Coastal Western Hemlock zone, which are drier than other parts of the Pacific Northwest.

Using PCR-fingerprinting and amplified fragment length polymorphism (AFLP), the team identified most of the fungal samples as the rare VGIIa genotype of *C. gattii*.[7] But the presence of small numbers of the VGIIb subtype suggested that there had been a second introduction of the fungus. Both fungal strains were a single mating type, which later led Duke University researchers to suggest that the outbreak was the result of an unusual same-sex mating event (see sidebar "Same-Sex Mating in *Cryptococcus?*").

Today, many questions about the ongoing outbreak remain unanswered. The Duke lab, led by Joseph Heitman, believes that *C. gattii* arrived in the Pacific Northwest with shipments of *Eucalyptus camaldulensis* in the early 20th century. But no Eucalyptus trees on Vancouver Island have turned up positive for the fungus, and VGII strains of *C. gattii* have not been found on *E. camaldulensis* in Australia. For her part, Bartlett believes that the hardy fungus, which can endure months in seawater, has probably been around much longer than anyone realizes, but was only been able to propagate after a series of warm summer seasons. Her colleague James Kronstad has been using comparative genome hybridization to try to understand the basis for differences in virulence in strains of *C. gattii* isolated from both the clinic and the environment.

As rainy areas of the Pacific Northwest grow warmer and drier with the warming climate, Bartlett suspects *C. gattii* will continue to show up. Her recent work demonstrates that it doesn't just hitch rides on air currents, but also inside the wheel wells of vehicles and the bottoms of shoes. So there's practically no limit to where it could go from here.

Karen Bartlett's recent work demonstrates that *C. gattii* doesn't just hitch rides on air currents, but also inside the wheel wells of vehicles and the bottoms of shoes.

The first animal and human cases of disease have already begun to show up on mainland British Columbia and in Washington and Oregon State. Although Bartlett has obtained positive air samples in the city of Vancouver, she has yet to pinpoint an infected tree. "If it does become colonized on the lower mainland," she says, "That's two million potential cases of exposure."

References

1. M.A. O'Malley. "The nineteenth century roots of 'everything is everywhere', *Nat Rev Microbiol*, 5:647–51, 2007.
2. S.G. Tringe et al., "The airborne metagenome in an indoor urban environment," *PLOS One*, 3:e1862, 2008.
3. D.W. Griffin, "Atmospheric movement of microorganisms in clouds of desert dust and implications for human health," *Clin Microbiol Rev*, 20:459–77, 2007.
4. K.B. Ritchie and G.W. Smith, "Cause of sea fan death in the West Indies," *Nature*, 394:137–8, 1998.
5. J.R. Weir-Bush et al., "The relationship between gorgonian coral (*Cnideria: Gorgonacea*) diseases and African dust storms," *Aerobiologia*, 20:119–26, 2004.
6. L. Perez et al., "Coarse particles from Saharan dust and daily mortality," *Epidemiology*, 19:800–7, 2008.
7. S.E. Kidd et al., "A rare genotype of *Cryptococcus gattii* caused the cryptococcosis outbreak on Vancouver Island (British Columbia, Canada)," *Proc Natl Acad Sci*, 101:17258–63, 2004.

From *The Scientist*, December 2008, pp. 36–42. Copyright © 2008 by The Scientist. Reprinted by permission.

UNIT 7

Microbes and Food: Friend and Foe

Unit Selections

Learning Objectives

- Why has there been an increase in the numbers of foodborne illnesses reported in the U.S. in recent years?

- What are three steps you can follow to keep your family's food safe?

- What are prions? How can the agricultural industry prevent the spread of "Mad Cow" disease?

- Compare the use of yeasts and bacteria in food production.

Student Website

www.mhcls.com

Internet References

Center for Food Safety and Applied Nutrition
http://www.foodsafety.gov/list.html
American Dietetics Association: Home Food Safety
http://www.homefoodsafety.org/index.jsp
Government Food Safety Information
http://www.foodsafety.gov
USDA Food Safety and Inspection Service (FSIS)
http://www.fsis.usda.gov
Fankhauser's Cheese Page
http://biology.clc.uc.edu/fankhauser/cheese/CHEESE.html

Throughout history the products of microbial fermentations have been used in making food and drink, from alcoholic beverages such as wine and beer to various types of breads, cheeses, yogurt, and pickles. At the same time, however, humans have always taken measures to prevent the microbial spoilage and contamination of food. Using salt to preserve meats is an ancient practice that exploits the low tolerance of most microbes to high osmotic pressure. Similarly, the age-old practice of using sugar as a preservative (for example, in the preparation of jellies and jams) also uses the principle of high osmotic pressure to prevent microbial spoilage. Lemon juice and vinegar were commonly used as preservatives because their low pH inhibits the growth of microbes. Thus, even in the days before refrigeration or industrial canning, people had "low-tech" methods of preserving foods from microbial contamination. This unit further illustrates the theme of this anthology—that microbes are both friend and foe to humans.

Foodborne illnesses are a growing public health concern in the United States. It is hard to escape the media attention given to those microbes on our "Most Wanted" list: *Salmonella, E. coli, Campylobacter, Listeria,* and many others. The U.S. Centers for Disease Control (CDC) estimates that there are 80 million cases of foodborne illness every year, with approximately 9000 deaths. With our food supply becoming increasingly globalized, and due to the fact that the FDA is underfunded, it is estimated that less than one percent of imported food is inspected every year. Consumers are learning that knowing where food comes from is important to maintaining one's health. In addition to animal products like chicken and beef being contaminated, there has been an increase in the number of cases of foodborne illness arising from contaminated fruits and vegetables.

The first half of this unit focuses on foodborne illness caused by various types of microbes including bacteria, prions, and algae. In particular, the causes of major outbreaks are examined. Farming practices such as irrigation need to be scrutinized for links to contamination of fruits and vegetables, such as in the case of the *E. coli O157:H7* outbreak in spinach discussed in "Toxic Salad." Throughout history humans have battled against marine based microbial toxins as well. "Protecting Ourselves from Shellfish Poisoning" provides an excellent guide to the practices in place for toxin detection and preventing its spread in seafood. Prions are infectious protein particles that cause "Mad Cow" disease. An excellent *Scientific American* article by Dr. Stanley Prusiner, the researcher who first discovered prions, describes new methods for detecting this infectious agent in the beef industry. What can consumers do to protect themselves from microbial contamination of food? Safe food handling practices at home are of utmost importance, and since most illness-causing microbes can be killed at high temperatures, cooking food thoroughly is essential. "Fear of Fresh, How to

© M. Lamotte/Cole Group/Getty Images

Avoid Foodborne Illness from Fruits and Vegetables" consists of a Q&A with Robert Tauxe, the Acting Deputy Director of the Division of Foodborne, Bacterial and Mycotic Diseases at the U.S. Centers for Disease Control. He provides practical advice for consumers in terms of how to handle produce, symptoms of foodborne illness, and emerging foodborne illnesses. An article in *U.S. News and World Report* informs readers how to shop smart and use appropriate safety measures in the kitchen to avoid microbial contamination of food.

Finally, the last section of this unit comprises two articles that illustrate how microbes are essential for food production. An article from *Current Opinion in Biotechnology* describes new research into our understanding about the role of the unicellular eukaryotic microbe, yeast, in the industrial production of both food and beverages. The last article in this unit focuses on the microbial communities used in making quality chocolate. The latest research on cocoa fermentation ends this chapter on a sweet note, reminding us that microbes contribute as much to our pleasure as they do to our discomfort!

Toxic Salad

What are fecal bacteria doing on our dinner plates?

JOSIE GLAUSIUSZ

A pig, a bug, a bag of spinach: It sounds like the answer to a riddle, and in a way, it is. Sometime late last summer, a wandering band of wild pigs trampled a fence and trotted into a spinach field in California's Salinas Valley. As they rooted around in the leafy greens, they most likely left behind feces infested with a virulent strain of the intestinal bacterium *Escherichia coli*—picked up, it appears, from a nearby cattle pasture. The hardy breed of *E. coli,* dubbed O157:H7, normally lives harmlessly by the billions at the rear end of a cow's gut. But in this case, the bacteria nestled craftily in the crevices of the spinach leaves, sticking to the vegetables as they were harvested, chopped, washed, bagged, and then transported across the country to states from Oregon to Maine. In September and October, nearly 200 people who dined on the infested spinach became ill with bloody diarrhea. Thirty-one developed severe kidney disease, and three people died.

Scientists at the Food and Drug Administration traced the toxic strain of *E. coli* to the California ranch after testing bacteria-tainted spinach leaves and matching them to fecal samples from a wild boar that had been killed on the farm—the first time an outbreak of *E. coli* poisoning from produce had been tracked directly to its source. This spate of illness was just one of at least three major produce-linked outbreaks of *E. coli* O157 that occurred in 2006, triggering a nationwide fear of fresh vegetables. In November and December, 71 people in five states became sick after eating at Taco Bell outlets, an epidemic that was later traced to contaminated iceberg lettuce, and another 81 in three states became sick after eating tainted lettuce at a fast-food chain called Taco John's.

The wave of salad-linked sickness highlights an alarming twist in the history of an infection originally known as "hamburger disease" when it popped up in 1982. Now a bug that is famous for lurking in undercooked beef is hitching rides on greens like lettuce and spinach. The question is, why? And how can we stop it?

C ontamination of fruits and vegetables by rogue *E. coli* is not new: Since 1991, small outbreaks have been linked to melons, grapes, coleslaw, sprouts, and apple juice.

By 1998, mandatory warning labels on unpasteurized apple juice led to a decline in drink-related outbreaks. At the same time, though, the craze for bagged greens had begun. Between 1997 and 2005, prewashed packaged salad sales nearly tripled, creating a $3-billion-a-year industry.

The nationwide distribution of mixed leafy greens makes it far more likely that any *E. coli* they harbor will be spread far afield. "When lettuce was shipped as entire heads, a contaminated head of lettuce was unlikely to contaminate another head of lettuce," says James Kaper, a microbiologist at the University of Maryland School of Medicine in Baltimore. "But when lettuce is chopped up and washed together, then contamination from one head of lettuce could be spread much further." Although processors triple-wash the leaves in chlorinated water before bagging, a minute amount of surviving bacteria—just 50 organisms—is enough to make a person ill.

E. coli O157 can wreak such havoc because it secretes one of the most potent toxins ever described, second only to that released by botulism bacteria. Known as shiga, this toxin is harmless to cattle. In humans, however, it sparks a cascade of symptoms that begin when the bacterium injects a sort of syringe into the cells of the intestinal wall. First the bugs secrete a protein that helps them adhere to the gut epithelium; then they shoot out shiga toxin, which can reach the bloodstream after destroying intestinal cells. As it circulates, the toxin can attack the kidneys, invade the brain, and—in the most extreme cases—bring on multiple organ failure and death. Those who survive may suffer paralysis, blindness or chronic kidney failure. Antibiotics are worse than useless; because they break up bacterial cells, they can trigger the release of yet more toxin.

According to the Centers for Disease Control, *E. coli* O157 causes an average of 73,000 cases of infection and 61 deaths in the United States each year. The number of outbreaks has actually declined since reaching a peak in 2000. Nevertheless, "it is a worldwide problem," says microbiologist John Fairbrother of the University of Montreal, "and the ideal solution has not been found yet." Although the number of cases from contaminated meat may be declining—the result, he says, of factors like more rigorous sanitary controls in slaughterhouses and thorough heating of hamburgers by fast-food chains—the bug is spreading farther in foods that are less likely to be cooked (and therefore sterilized), such as fruit and lettuce. Infections from other, rarer toxic *E. coli* strains, dubbed O26 and O111, are also being diagnosed more often.

The bacteria have a worrisome ability to survive in many different environments outside the colon, Fairbrother says. Experiments carried out by Michael Doyle, director of the University of Georgia Center for Food Safety in Griffin, show that *E. coli* O157 can survive in the field for more than 77 days on lettuce, on carrots for at least 175 days, and on onions for at least 85 days. The question is how a bacterium from a cow's gut makes its way onto a carrot. One obvious way is through the spreading of animal manure or recycled, treated human sewage on produce fields. "There is no prohibition to using raw manure on crops, so long as the crops aren't organic. Not very appealing," says Caroline Smith DeWaal, director of food safety at the Center for Science in the Public Interest in Washington, D.C. "Treated sewage may be used in certain water systems that then are used for irrigation. Even tertiary treatment of sewage may not be enough to totally eliminate the hazards linked to *E. coli*." Contaminated runoff from dairies can seep into adjacent fields—as probably happened in the Taco John lettuce outbreak of 2006—and in summer, dust from cattle stalls can drift onto nearby crops. Other animal feces can also carry the bacteria.

If the bagged salad scare continues, produce farmers could face the same kind of financial trouble that threatened cattle ranchers in the 1990s. A 2003 article in *Meat & Poultry,* an industry magazine, concludes that *E. coli* fears cost the beef industry as much as $2.7 billion in lost sales over the previous decade. Fortunately for lettuce producers, they can benefit from the science being pursued by beef farmers to save their industry. One such line of research is a vaccine to prevent cows from shedding virulent *E. coli* in their feces. Veterinarian David Smith at the University of Nebraska at Lincoln has spent five years testing this vaccine, which stimulates the production of antibodies against the proteins that the bacteria secrete to help them attach to the gut wall. Smith's results show that the vaccine reduces the number of cattle shedding virulent *E. coli* in their feces by 65 percent. In December 2006, Bioniche Life Sciences, a Canadian pharmaceutical company, was granted a license by the Canadian Food Inspection Agency to distribute the vaccine to cattle veterinarians there. Approval is still pending in the United States.

Other approaches are more experimental. Todd Callaway, a microbiologist at the USDA Agricultural Research Service in College Station, Texas, has poked around in cattle feces to find an array of bacteriophages—viruses that attack bacteria—that target *E. coli* O157. In a recent study, he fed a cocktail of these viruses to sheep infected with the O157 bacteria and found a sharp reduction in the quantity of that strain of *E. coli* in their intestines. He is now investigating whether spraying the viruses onto cattle hides—a major source of infection in slaughterhouses—could also reduce their bacterial load.

Even if these prevention methods succeed, they are unlikely to be foolproof, so scientists are also developing new ways to attack rogue *E. coli* if they reach humans. One approach is to target the genes that make *E. coli* O157 so virulent. James Kaper has discovered that when the bacterial population reaches a critical mass in the large intestine, the microbes secrete hormone-like compounds. These compounds, in turn, switch on the genes that enable the bacteria to colonize the gut wall and to exude their toxin. Disrupting this bacterial groupthink could provide a way to prevent invasion. Vanessa Sperandio, Kaper's former postdoc who now heads her own lab at the University of Texas Southwestern Medical Center in Dallas, has discovered three molecules that can block interbacterial signaling among *E. coli.* She plans shortly to begin animal testing of a drug derived from one of the three molecules.

Another option is to defang the shiga toxin itself, a tactic adopted by microbiologist Alison O'Brien of the Uniformed Services University in Bethesda, Maryland. She has created toxin-smothering antibodies that have been shown to be safe in humans by Caprion Pharmaceuticals, paving the way for efficacy trials. Such an antibody-based treatment would be most effective if administered soon after the onset of infection.

Ultimately, quashing the bacteria at their source—the cows—is probably the most effective way to prevent *E. coli* O157 infections in humans.

Ultimately, though, quashing the bacteria at their source—the cows—is probably the most effective way to prevent *E. coli* O157 infections in humans. "This isn't rocket science," says DeWaal. "It's critically important that both human and animal waste be kept off crops. It's appropriate to have restrictions on the use of raw manure, to have monitoring of composting processes to make sure they're effective, and to restrict the use of water that may contain treated sewage." But Kaper adds a caveat. "There are so many different bugs, there are so many different types of foods, so many different ways in which food is prepared and processed," he says. "I don't think the food supply can be totally safe."

Fear of Fresh

How to Avoid Foodborne Illness from Fruits & Vegetables

ROBERT TAUXE

Kyle Allgood, of Chubbuck, Idaho, would have been three in December. He died on September 20 after *E. coli* O157:H7 damaged his kidneys. Kyle got the *E. coli* from the fresh spinach smoothie that his mother made for him.

Jillian Kohi, of Milwaukee, was luckier. The graduate student, who used to run marathons, suffered stomach cramps, muscle aches, fever, and bloody diarrhea after eating spinach. After 2½ weeks in the hospital, Jillian was discharged with less than 10 percent of her normal kidney function.

Since when do we have to worry that even the healthiest foods could harm or kill us?

Q: Is it riskier to eat produce these days?

A: Yes. The outbreaks are bigger and more frequent than they were 20 or 30 or even 15 years ago. Even though we can identify and control outbreaks better than we used to, when contamination occurs with lettuce, spinach, cantaloupe, or tomatoes, we can have a big problem on our hands.

The headlines about *E. coli* O157:H7 in spinach tell the story. Since some produce is very conveniently packed in a bag and prewashed, there is nothing consumers can do to lower their risk. In many cases, you don't cook it. You don't blanch it. You don't do much except eat it. So it's critically important that it not be contaminated from the beginning.

Q: Why are outbreaks in produce on the rise?

A: One reason is that people are eating more fresh fruits and vegetables than they did 30 years ago. That's a good thing. We want people to eat fresh produce. Before fresh spinach went off the market, something like a third of people we called had eaten fresh spinach in the previous week. That's great news.

Q: Are greens in bags causing more outbreaks?

A: Conceivably. Rather than just one head of lettuce or a bunch of spinach, there could be leaves from many different plants in one bag. Bacteria from one contaminated leaf in a bag, you can just imagine, would be over all the leaves in the bag by the end of the distribution chain.

It's a pooling issue, like ground beef. How many cows are in one patty of ground beef? How many cows contribute to one glass of milk? It means that procedures have to be in place to make sure that none of it is contaminated.

Q: Does the bacteria come from nearby animals?

A: Possibly. We know that feedlots have E. coli O157:H7. How wise is it to grow spinach or lettuce plants, which are very close to the ground, just downwind or downstream or down the hill from a feedlot or cow pasture? Doesn't sound like a good idea to me.

The produce industry has to figure out how to prevent contamination. Maybe we need a half-mile buffer between feedlots and produce farms.

Q: Does it help to wash produce?

A: We recommend washing produce in general, even if you plan to peel it. When you slice a melon, for example, the knife can transfer bacteria from the surface to the inside.

But it's tough to get bacteria off greens. Those germs are very sticky. This triple-washed stuff that comes out of the bag—if it's got contamination on it, there's no way to wash it off, even if you use bleach or detergent.

And you can't wash off germs if they're inside the melon, mango, or apple. For example, bacteria can creep in through the apple core. Germs can go in the hole at the very bottom of the apple where the flower was—it's called the calyx.

Q: Is organic produce less likely to have *E. coli?*

A: No. I don't see this as an organic versus conventional issue, just like it's not a domestic versus imported issue. There's room for improvement with both kinds of production and on both sides of the border.

Q: Which produce is most likely to have *E. coli* O157:H7?

A: The recurrent outbreaks have come from leafy greens—especially lettuce—sprouts, and unpasteurized juices and cider. It's not all fruits and vegetables. And apple cider might be a special case because it's sometimes made from apples that have fallen from the tree.

What those fruits and vegetables have in common is that they're grown fairly close to the ground, they're not cooked, and they're not acidic. In general, the bacteria that cause

Bugs Are Breaking out All Over

The Centers for Disease Control and Prevention (CDC) estimates that 76 million Americans become sick, more than 325,000 are hospitalized, and 5,000 die from foodborne illness each year. Experts estimate that foodborne illness costs the nation $10 billion to $83 billion each year in pain and suffering, reduced productivity, and medical expenses.

Here are some of the foods that are most likely to cause foodborne illness. The list is drawn from a database of some 5,000 outbreaks that sickened more than 150,000 people from 1990 through 2004. The database is maintained by the Center for Science in the Public Interest (publisher of *Nutrition Action*). A few caveats from CSPI director of food safety Caroline Smith DeWaal:

- Most foodborne illnesses are isolated cases, so outbreaks (two or more people sickened by the same food) are the exception.
- Many outbreaks and illnesses are never investigated or reported to the CDC.
- The outbreak data cover 15 years, so they don't necessarily reflect current trends. For example, eggs caused a much higher percentage of outbreaks in the early 1990s than they do today.
- Foods with multiple ingredients (like tacos, lasagna, cheeseburgers, chili, egg salad, stuffing, and some sandwiches) are in the database but most aren't listed here.

All Bugs (includes *E. coli* O157:H7)

Food	Outbreaks	Cases*
Greens-based salad	199	7,555
Turkey	109	5,832
Chicken	215	3,979
Ground beef	171	3,425
Shellfish	155	3,399
Berries	20	3,330

Food	Outbreaks	Cases*
Tomatoes	19	2,852
Lettuce	56	2,380
Eggs	74	2,117
Ham	46	2,107
Sprouts	30	2,018
Ice cream	44	1,807
Cheese	50	1,791
Melon	29	1,683
Juice & cider	22	1,514
Unpasteurized milk	56	1,457
Scallions	9	1,221
Potatoes	40	1,099
Luncheon meat	50	1,014
Bread	32	980
Fresh tuna	193	824
Fruit salad	17	538
Mahi-mahi	71	422
Fresh salmon	14	195

E. coli O157:H7

Food	Outbreaks	Cases*
Ground beef	107	2,028
Lettuce	12	336
Unpasteurized milk	4	222
Greens-based salad	9	216
Coleslaw	3	194
Apple cider/juice	6	142
Alfalfa sprouts	3	120
Potato salad	2	49
Melon	2	36

*Number of people sickened. (For the latest report from the CSPI outbreak database, see www.cspinet.org/foodsafety/outbreak_report.html.)
Source: *Outbreak Alert 2005,* Center for Science in the Public Interest.

---

foodborne disease don't like acid. But Americans have a phenomenal sweet tooth, and my understanding is that apples and tomatoes are getting sweeter and less tart and acidic.

Q: Is all foodborne disease rising?

A: No. Over the last ten years, we've witnessed important decreases, 29 to 32 percent, in infections that are related to meat and poultry—like *E. coli* O157:H7 linked to ground beef; or *Listeria* infections, which are often linked to processed meats; or *Campylobacter,* which is linked to poultry. We haven't eliminated them, but we're certainly headed in the right direction.

Q: What about Salmonella?

A: In the same ten-year period, it's down 9 percent, which is not a brilliant success. *Salmonella* is complicated, because it can come from a number of foods and even non-food sources like pet turtles or lizards. And antibiotic-resistant *Salmonella*—much of it from ground beef—has become more of a problem over the last decade.

When to Call the Doctor

Q: Which symptoms should alert people to call the doctor?

A: The red flag is diarrheal illness that's not resolved in three days or that's accompanied by a fever over 101.5°F or by blood in the stools.

The Dirty Dozen

Bug	Major Symptoms	Some Foods That Have Caused Outbreaks	How Soon It Typically Strikes	How Soon It Typically Ends
Campylobacter (bacteria)	diarrhea (can be bloody), cramps, fever, vomiting	undercooked poultry, unpasteurized (raw) milk, contaminated water	2 to 5 days	2 to 10 days
Ciguatera (toxin)	*within 2 to 6 hours:* abdominal pain, diarrhea, general pain and weakness, nausea, temperature reversal (hot things feel cold and cold things feel hot), tingling, vomiting *within 2 to 5 days:* slow heartbeat, low blood pressure	large reef fish like barracuda, grouper, red snapper, and amberjack	2 hours to 5 days	days to months
Clostridium botulinum (bacteria)	vomiting, diarrhea, blurred vision, double vision, difficulty swallowing, muscle weakness that spreads from the upper to the lower body	home-canned foods, improperly canned commercial foods, herb-infused oils, potatoes baked in aluminum foil, bottled garlic	12 to 72 hours	days to months (get treatment immediately)
Cyclospora (parasite)	diarrhea (usually watery), loss of appetite, substantial weight loss, stomach cramps, nausea, vomiting	imported berries, lettuce	1 to 14 days (usually at least 1 week)	weeks to months
E. coli O157:H7 (bacteria)	severe diarrhea that is often bloody, abdominal pain, vomiting (usually accompanied by little or no fever)	undercooked beef, unpasteurized (raw) milk or juice, raw produce, salami, contaminated water	1 to 8 days	5 to 10 days (get treatment immediately, especially for a child or elderly person)
Hepatitis A (virus)	diarrhea, dark urine, jaundice (yellow "whites" of the eyes), flu-like symptoms	shellfish, raw produce, foods that are not reheated after coming into contact with an infected food handler	15 to 50 days	2 weeks to 3 months
Listeria (bacteria)	fever, muscle aches, nausea, diarrhea (pregnant women may have mild flu-like symptoms; can lead to premature delivery or stillbirth)	fresh soft cheeses, unpasteurized (raw) or inadequately pasteurized milk, ready-to-eat deli meats and hot dogs	9 to 48 hours for gastrointestinal symptoms, 2 to 6 weeks for infections in the blood, brain, or uterus	days to months (get treatment immediately)
Noroviruses (virus)	nausea, vomiting (more common in children), abdominal cramping, diarrhea (more common in adults), fever	poorly cooked shellfish, ready-to-eat foods touched by infected food handlers, salads, sandwiches	12 to 48 hours	12 to 60 hours
Salmonella (bacteria)	diarrhea, fever, abdominal cramps, vomiting	eggs, poultry, unpasteurized (raw) milk or juice, cheese, raw produce	1 to 3 days	4 to 7 days
Scombrotoxin (toxin)	flushing; rash; burning sensation in skin, mouth, and throat; dizziness; hives; tingling	fresh tuna, bluefish, mackerel, marlin, mahi-mahi	1 minute to 3 hours	3 to 6 hours
Vibrio parahaemolyticus (bacteria)	watery diarrhea, abdominal cramps, nausea, vomiting	undercooked or raw seafood	2 to 48 hours	2 to 5 days
Vibrio vulnificus (bacteria)	vomiting, diarrhea, abdominal pain, bacteria in the blood, wounds that become infected	undercooked or raw shellfish (especially oysters), other contaminated seafood	1 to 7 days	2 to 8 days (get treatment immediately)

Source: Adapted from *Diagnosis and Management of Foodborne Illnesses: A Primer for Physicians and Other Health Care Professionals* (www.cdc.gov/mmwr/preview/mmwrhtml/rr5304a1.htm), by the American Medical Association, American Nurses Association, Centers for Disease Control and Prevention, Food and Drug Administration, and U.S. Department of Agriculture.

If a very young child seems lethargic or doesn't seem to be making much urine or tears, that could be a sign of dehydration and is another reason to seek medical attention. Any young child with diarrhea should start drinking pediatric electrolyte solution—it's in all the drugstores—to prevent dehydration.

Q: How soon do the symptoms of E. coli O157:H7 show up?

A: They usually appear within 3 to 4 days, but it could be anywhere from 1 to 10 days. Most people have bloody diarrhea and severe abdominal cramps, but sometimes the infection causes non-bloody diarrhea or no symptoms at all. Usually the person gets little or no fever, and the illness resolves in 5 to 10 days.

Q: How many people end up with life-threatening complications?

A: About 3 to 8 percent get hemolytic uremic syndrome, or HUS. Some don't have complete kidney failure, but most do.

There are two parts to HUS. One is the kidney failure—that's uremia, the 'U.' The other is hemolysis, the 'H.' That's when red blood cells look like they've been through a blender.

The *E. coli* toxin damages blood vessels by creating small strands across the insides, so when the red blood cells go through them, it's like they're going through a cheese cutter. It just slices up the red cells.

So people may need transfusions and dialysis before the blood vessels get better. They're destroying their own red blood cells. Even with intensive care, 3 to 5 percent of these patients die.

Q: Should doctors send a stool sample to the local health department?

A: That's absolutely vital to our tracing the outbreaks. But the doctor may wait to see if the diarrhea goes away in a few days rather than sock the patient with a bill for the lab. We recommend a stool specimen if it's a severe illness, bloody diarrhea, high fever, or if the illness is lasting.

A stool sample can also tell the doctor whether the infection will respond to an antibiotic and to which antibiotic. With so much resistant bacteria around, that could be especially important. But often the patient will get better with no antibiotics.

Q: Can antibiotics make E. coli O157:H7 infections worse?

A: There's a real paradox. The *E. coli* harms people by producing a toxin that destroys blood vessels in the gut, kidney, and brain. An antibiotic drug kills the *E. coli,* but it also can provoke them to make a lot more toxin. So it may make the patient worse. Antidiarrheal agents like Imodium should also be avoided.

What's Next?

Q: What new foodborne illnesses are emerging?

A: Some new and highly resistant strains of *Salmonella* have appeared in recent years. In Japan, they've seen cases of hepatitis E from pork sushi.

Safe at Home

How you handle food matters. With enough warmth, moisture, and nutrients, one bacterium that divides every half hour can produce 17 million progeny in 12 hours.

Putting food in the fridge or freezer stops most bacteria from growing. Exceptions: *Listeria* (typically found in soft cheese, lunch meats, and hot dogs) and *Yersinia enterocolitica* (typically found in undercooked pork and unpasteurized milk) grow at refrigerator temperatures.

Rules for Leftovers
2 Hours—2 Inches—4 Days
2 Hours from Oven to Refrigerator

Refrigerate or freeze leftovers within 2 hours of cooking. Otherwise throw them away.

2 Inches Thick to Cool It Quick

Store food at a shallow depth—about 2 inches—to speed chilling.

4 Days in the Refrigerator—
Otherwise Freeze It

Use leftovers from the refrigerator within 4 days.

Exception: use stuffing and gravy within 2 days. Reheat solid leftovers to 165°F and liquid leftovers to a rolling boil. Toss what you don't finish.

- Buy **fresh-cut produce** like half a watermelon or bagged salad greens only if it's refrigerated or surrounded by ice.
- Store **perishable fresh fruits and vegetables** (like strawberries, lettuce, herbs, and mushrooms) or **pre-cut or peeled produce** in a clean refrigerator at a temperature of 40°F or below.
- Wash your hands for 20 seconds with warm water and soap before and after preparing **any food.**
- Wash **fruits and vegetables** under running water just before eating, cutting, or cooking, even if you plan to peel them. Don't use soap (it leaves a residue). Produce washes are okay. (Exception: triple-washed bagged lettuce or other produce needs no further washing.)
- Scrub **firm produce,** like melons and cucumbers, with a clean produce brush. Let them air dry before cutting.
- Remove the outer leaves of **heads of leafy vegetables** like cabbage and lettuce.
- Don't eat **raw sprouts** (alfalfa, bean, clover, or radish).
- Cooking a food at 160° F will kill any *E. coli* O157:H7.
- Neither **processed spinach** (frozen or canned) nor **other fresh or processed leafy greens** (like lettuce or kale) were implicated in the recent *E. coli* outbreak.
- Drink only pasteurized **milk, juice, or cider.**
- For more information on handling produce safely: www.cfsan.fda.gov/~dms/prodsafe.html.
- For information on *E. coli* O157:H7: www.cdc.gov/ncidod/dbmd/diseaseinfo/escherichiacoli_g.htm.

Sources: Centers for Disease Control and Prevention, U.S. Department of Agriculture, Food and Drug Administration, Center for Science in the Public Interest.

And in Finland, they're seeing outbreaks of *Yersinia pseudotuberculosis*. It's a second cousin to the bug that causes the plague. Finland recently realized that the outbreaks of what looks like appendicitis were due to this bug, which was traced to eating local lettuce.

Q: So people there may have unnecessary surgery for appendicitis?

A: Yes. Doctors perform surgery and find no problem with the appendix, but they see big swollen nodes all over the intestines. That's what the gastrointestinal tuberculosis looks like.

The working assumption is that *Yersinia* causes disease in deer and rabbits, which are getting into the lettuce and carrot fields and contaminating them. It's not a problem in this country, but there are other problems elsewhere in the world, and I expect that we'll be finding more of them in our food supply.

Q: What about in the United States?

A: Some investigators have raised the possibility that the *E. coli* that causes urinary tract infections comes from the animals we eat, but the link is by no means proven.

Q: Have antibiotics made some Salmonella resistant?

A: Yes. Anytime an antibiotic is used in a hospital, in a child with an earache, or in animals on a farm, there are winners and losers in the local bacterial population. You hope that the losers are the bacteria that were making the patient, child, or animal sick. But if there's a bacterium there that's resistant, it's a winner.

The bacteria in food that make us ill—like *Salmonella* or *Campylobacter*—have their natural home in animals, not people. So antibiotic use in those animals can make those bacteria resistant.

Q: Are bacteria that live in people also becoming resistant?

A: Yes. The antibiotics we use in people cause resistance in the bacteria that cause pneumonia and tuberculosis. We have a program here at the CDC called Get Smart, which tries to reduce unnecessary antibiotic use for people who have colds. Antibiotics aren't going to make them better. Likewise, we have a program called Get Smart on the Farm to promote prudent antibiotic use in animals.

Q: And the farm is where contamination starts?

A: Right. Most of the progress we've made with *E. coli, Listeria,* and *Campylobacter* has been at the slaughterhouse. The process is cleaner than it was before. But many of the animals coming in off ranches and farms are still contaminated.

And I'm concerned that back on the farm, bugs are transferring from animals to plants or cycling back and forth from animals to plants. You can't slaughter a spinach plant in a way that guarantees that it comes out clean. So the effort has to be focused on the farm, and that includes the animal farm.

Q: Should people stop eating leafy greens?

A: No. We want to encourage people to eat fresh fruit and vegetables. But obviously, the spinach problem—and previous problems with lettuce, tomatoes, and other fresh produce—show us that contamination is not under control.

ROBERT TAUXE is Acting Deputy Director of the Division of Foodborne, Bacterial and Mycotic Diseases at the Centers for Disease Control and Prevention. He spoke to *Nutrition Action*'s Bonnie Liebman by phone from his office in Atlanta.

Protecting Ourselves from Shellfish Poisoning

Molecular probes deployed by California scientists are just the latest weapons in our species' long battle with harmful algae.

MARY WILCOX SILVER

As the sun set over San Francisco Bay on July 15, 1927, area residents had plenty to talk about: Aviators Ernie Smith and Emory Bronte had just become the first to fly a single-engine aircraft, *City of Oakland,* 2,100 miles nonstop from Oakland to Hawaii.

But the next day, a panic began to grip the area. Residents who had eaten mussels gathered along the beaches around San Francisco were falling gravely ill. That day the *San Francisco Examiner* reported the first two deaths on its front page. An alarm went out. Signs were posted along the beaches, and scientists and public-health officials got to work to understand what was happening.

Thanks to the research that followed the San Francisco scare, shellfish poisoning is now rare in California. Indeed, a monitoring strategy developed in the state in response to the incident has saved countless lives around the world over the past eight decades. Scientists have learned a great deal about what can make shellfish and other aquatic organisms dangerous to eat, and this knowledge has been put to practical use in harvesting regulations, monitoring and food-testing programs and public education.

At the same time, we've also learned that people have been protecting themselves from ingesting marine toxins for millennia and perhaps much longer. Today, even as modern technology is being harnessed to tackle this daunting and persistent problem, we find that ways of protecting ourselves from toxins in seafood likely have been a part of maritime cultures for thousands of years of human history and may even have roots in prehistoric culture.

Shellfish and Ancient Diets

What was happening on the beaches of San Francisco that warm July day, then, was anything but new. People living around the Pacific Ocean have always eaten shellfish, and consuming these filter-feeders has probably always posed certain risks. The San Francisco scare was one of the occasions that expanded our understanding of those risks.

Shellfish are easily harvested from shallow aquatic environments in much of the world, where they are protein-rich food for predators, including people. It is not surprising that evidence of their use is found throughout the archaeological record left in Africa by earlier hominids, *Homo erectus* and *H. habilis,* and by modern human beings. Along the waterways of the world's continents can be found the remains of shell mounds that, in association with the artifacts found with them, have led anthropologists to speculate that hominids ate shellfish, including mussels and other bivalve mollusks, possibly as early as one million years ago. Anatomically modern *H. sapiens* left sizable middens on coastal sites in Africa and at various locations in Eurasia, supplying good evidence that shellfish were collected more than 100,000 years ago.

The middens found up and down the West Coast of North America are much younger, mostly less than 10,000 years old. Between the new and ancient middens, our record of coastal dietary culture thins out, owing to the fact that the world's sea level rose and fell multiple times during ice ages. Our ancestors would have moved out onto continental shelves to collect fish during cooler periods, leaving behind evidence that is underwater today. Underwater archaeology has located some of these sites, but wave disturbance and wave action have destroyed much of the record. About 10,000 years ago, sea-level rise slowed, so that coastal archaeologists have a good record of shellfish use in locations such as the islands off southern California.

Was eating shellfish risky in ancient times? There is intriguing evidence that it was. Consider, from the Jewish tradition, the laws of kosher eating set out in the book of Leviticus (here, the King James Bible version):

> These shall ye eat of all that are in the waters: whatsoever hath fins and scales in the waters, in the seas, and in the

rivers, them shall ye eat. . . . Whatsoever hath no fins nor scales in the waters, that shall be an abomination unto you.

In the views of the rabbis who dictated the Kashrut, filter-feeding shellfish, crustaceans, gastropods and cephalopods were scavengers, indiscriminate eaters, unclean animals. References to avoiding shellfish can also be found in certain Islamic and Christian traditions. Although other explanations have been offered, knowledge of the special hazards of eating shellfish might have informed the makers of these laws.

The traditions of the indigenous peoples of the Pacific Coast of North America also hint at ancient knowledge of shellfish dangers. Anthropologists have found that some of these tribes customarily watch for the ripening of elderberries, a sign of summer, as a signal that it is time to stop harvesting shellfish. And sentinels were traditionally posted on cliffs in the Pacific Northwest to watch the sea for bioluminescence—an indicator, we now know, of a particularly dense bloom that might make shellfish dangerous to eat. Today scientists and physicians are working with the Quileute Tribe in the Olympic region in a collaboration intended both to develop new monitoring tools to protect human health and to learn about the effect of long-term exposure to shellfish toxins.

To understand these intriguing connections between human culture and shellfish poisoning, it is useful to take a close look at the biology behind shellfish poisoning.

Toxins in the Marine Food Web

As the ancient rabbis recognized, a mussel will consume just about anything. As collectors of tiny particles floating in the water, the bivalves, cemented to the seafloor, continuously siphon great volumes of water, straining out diatoms and other nutritious tidbits. A mussel harvested on any given day will contain whatever was in the water around it that day, sometimes along with traces of items it's picked up on other days. These can include viruses, bacteria, pollutants, parasites and toxic algae. Even if these don't affect the mussel, they can spell trouble for a susceptible animal that eats it.

Viruses and bacteria are perhaps the best-known risks associated with eating shellfish. These risks grew in Europe during medieval times as settlements around estuaries and bays grew larger, polluting the enclosed waters with coliform bacteria from human waste. *Salmonella* bacteria including *S. typhii*, the cause of typhoid fever, began to be found in shellfish with regularity. In the Middle Ages, shellfish were also implicated in cholera outbreaks, since they can harbor *Vibrio* species including *V. cholerae*.

Fortunately bacteria and viruses such as hepatitis A and poliovirus can be destroyed by cooking, and so measures to prevent shellfish-borne disease have included cooking practices along with closings of contaminated areas during disease outbreaks. Before the San Francisco scare, new immigrants along the California coast were unaware of the dangers of eating shellfish, either cooked or raw—dangers known to some of the native inhabitants of the region.

Potential Dangers from Shellfish Consumption

Hazard	Destroyed by Cooking?
Viral Diseases	
Hepatitis A	yes (but not by steaming)
Norwalk virus	yes (but not by steaming)
poliovirus	yes
Bacterial Disease	
Listeria monocytogenes	yes
Vibrio (including *V. cholerae*)	yes
Salmonella (including *S. typhii*)	yes
Algal Toxins	
Paralytic shellfish poisoning	no
Amnesic shellfish poisoning	no
Other Dangers	
Pollutants (organic and inorganic)	no
parasites (including *Schistosoma*)	yes

But in 1927 San Francisco beachgoers learned that even cooked shellfish from a clean environment could sicken and kill. The reason is that the organisms I study—certain of the tiny unicellular marine organisms known as dinoflagellates and diatoms—produce potent toxins, poisons that are not destroyed even by cooking. Mollusks feeding on these toxic algae are generally thought to be unaffected by these substances, but the toxin acquired through filtering its food may, like the other shellfish contaminants mentioned above, be enough to kill a susceptible predator that eats the mussel.

Recent research has shown that knowledge of the effects of these toxins exists in many aboriginal cultures. But the San Francisco incident came as a surprise to an urbanizing nation that knew only of the shellfish hazards associated with pathogens. Even after newspaper warnings were issued and signs posted on the beaches, some people continued to eat mussels. The worst affected were those who ate large numbers of mussels on an empty stomach, feasting on these especially tasty mollusks on a warm summer's weekend with a good low tide that allowed easy access to the mussel beds. The neurological effects set in soon, usually within an hour or so. Ultimately more than 100 people were affected, and a handful died.

Discovering *paralytic shellfish poisoning* or PSP, as this syndrome is now known, took some work. Scientists first suspected backwash from local sewage, materials leaking from garbage barges, or even copper from a disintegrating tanker. But three men—physician-scientists Herman Sommer and Karl Meyer of

Human Symptoms of Paralytic Paralytic Shellfish Poisoning

- Tingling and numbness of lips, fingers, face and extremities
- Vomiting, diarrhea
- Blurred vision
- Shortness of breath
- Dry mouth, difficulty swallowing
- Slurred speech
- Headache
- Rapid heartbeat and sweating
- Floating sensation
- Death from respiratory paralysis

Human Symptoms of Amnesic Shellfish Poisoning

- Nausea
- Vomiting, diarrhea
- Abdominal cramps
- Disorientation, confusion
- Incapacitating headache
- Hallucinations
- Seizures
- Memory loss
- Death (uncommon and only in elderly)

the Hooper Foundation (later part of the University of California, San Francisco) and Charles Kofoid, a microbiologist at UC Berkeley—pursued the case further as the early leads appeared inadequate and the geographic extent of the poisoning became known. They began to look closely at water samples taken on San Francisco's open coast and speculated that there might be something of natural origin making the water itself toxic.

After several more years of study, more cases of shellfish poisoning in the region and elegant scientific detective work, Sommer, Meyer and Kofoid began to focus on a microorganism— a photosynthetic swimming alga, a dinoflagellate—and came up with their prime suspect. They turned out to be right: An organism we now call *Alexandrium catanella* was the culprit revealed in an article published in 1937. This was the first time that a waterborne alga was found to be responsible for shellfish poisoning.

The substance produced by *A. catanella,* which sometimes can cause a luminescent "red tide" when the cells become especially abundant, was first synthesized in 1977. Saxitoxin is tasteless, odorless and water-soluble, with a toxicity similar to that of the biological-weapon poison ricin. When you eat a mussel containing saxitoxin, you may first experience tingling in your fingers, lips, face and extremities as if someone is poking you with pins. Then your lips go numb, along with your arms, legs and neck. It's time to get to the doctor! You're in the first stages of paralysis, and without medical assistance you may die, because you will be unable to breathe. There is no antidote, and just a few milligrams can kill you. Patients survive with help from intravenous fluids and a respirator.

The algal cells capable of wreaking this havoc are almost spherical and about 40 micrometers long; that is, a line of about 25 would be a millimeter long. The reddish cells form colonies, an important fact because they thus become food for organisms that cannot capture smaller particles. Since the discovery of PSP, *A. catanella* and saxitoxin (also known as STX) have turned up in other contexts. Hong Kong's harbor has been known to turn

red; indeed, toxic algal events have been observed along the mid- and upper-latitude coasts of every continent except Antarctica. And pilots who flew U-2 spy missions over the Soviet Union were given tiny pellets of saxitoxin extracted from the algae and reportedly were instructed to take the suicide capsules if they were shot down.

The Pelican Peril

In 1991, a new kind of incident in California caught the monitoring community by surprise. Fortunately for the state's human population, birds and marine mammals served as sentinels in this case. This may be the only case in which the local presence of a poisonous alga was first recognized because of animal deaths.

To Californians familiar with PSP, it was soon clear that a different poison was now at work. This time, brown pelicans and cormorants began washing ashore in the Monterey Bay area. Not all the birds were dead, and the survivors exhibited very odd behavior. Working with veterinarians, we biologists proceeded to examine their stomach contents and found anchovies full of a different algal species: a common coastal diatom now known as *Pseudo-nitzschia.* These slender cells are about one-tenth of a millimeter long, over twice the length of *A. catanella,* and they also form long chains.

As we studied what was known about this organism, we realized that our local "species" had long been misidentified in the region, so we had to use a newly revised taxonomy of the *Pseudo-nitzschia* genus. It turns out that within this genus are species that are poisonous in some places and not poisonous in others, along with species that are mostly or rarely poisonous. The poisonous species—which today can be distinguished using electron microscopy or molecular tools—make a neurotoxin quite different from saxitoxin, one called domoic acid. This is a naturally occurring amino acid that affects glutamate receptors in the brain and particularly damages memory centers. Our waters in Monterey Bay contained two toxic and a number of nontoxic species of *Pseudo-nitzschia.*

The pelicans and cormorants' behavior was a manifestation of a toxin that causes *amnesic shellfish poisoning* in humans. The phenomenon acquired this name from an earlier incident, this one involving people who ate a crop of farmed mussels from Prince Edward Island, on Canada's east coast, in 1987. The victims suffered gastrointestinal symptoms and terrible headaches, but also hallucinations, seizures and memory loss; several died. In the Canadian event, the source of the toxin was *Pseudo-nitzschia,* a toxin that strangely turned out to be an anti-worming agent, previously known in Japan, as I'll describe below.

How had the birds come down with domoic acid poisoning? One lesson clearly demonstrated by the 1991 incident is that animals other than shellfish can be poisonous. Pelicans and cormorants forage on schooling fish—anchovies in particular. Off the California coast, schools of anchovies appear as large shadows on the sea surface. A brown pelican in search of a meal dives into a school and scoops up a fish in its expandable pouch.

For their part, anchovies acquire food with the aid of *gill rakers,* comblike structures that allow them to sieve from the water microorganisms of millimeter size. The anchovy concentrates this food in its stomach. Algae thus end up in an anchovy's digestive tract, which can be a condensed packet of partially digested toxic cells. But the fish itself is normally unaffected by the domoic acid and typically contains little toxin in its tissue. When a pelican dives and scoops up an anchovy in its pouch, it gets the fish's stomach contents as well. Fish like anchovies and sardines turn out to be responsible for many of the poisonings of marine vertebrates in our area.

Since the 1991 incident we've found evidence that a great variety of organisms are affected by domoic acid in the marine food web. We've found domoic acid and the tiny skeletons of *Pseudo-nitzschia* in the feces of humpback whales, in mole crabs and fish such as white croakers, a favorite of pier-anglers. My UC Santa Cruz colleagues Sibel Bargu and Kathi Lefebvre examined the feces of the largest mammal on Earth, the blue whale, and found fragments of the toxic species *Pseudo-nitzschia australis* along with concentrations of domoic acid that were 10 times the level considered dangerous to humans. In the case of the blue whale, neither shellfish nor anchovies are the source; the domoic acid is being passed along by krill, the whale's chief food source.

And among the animals most obviously affected are sea lions. This year stranded sea lions are turning up on California beaches along with disoriented pelicans. Sea lions have indeed become the sentinels of the central California coast. Rescuers from the Marine Mammal Center at Sausalito look for animals exhibiting seizures, head-weaving, strange arching postures, lethargy and unresponsiveness. They restrain the animals and transport them to rehabilitation centers, where their symptoms can be treated with anticonvulsant agents and hydration.

Is domoic acid, in fact, to blame for the strandings of marine mammals that are becoming so frequent in California? It turns out that the connection is becoming less clear the more data we get. In 1998, Christopher Scholin and his colleagues at the Monterey Bay Aquarium Research Institute (MBARI) began taking measurements of the abundance of *Pseudo-nitzschia* cells in the bay's waters and studied how they correlated with marine-mammal strandings observed by Frances Gulland of the Marine Mammal Center. During their initial study period, the correlation appeared strong and positive, but as I've added data over the following years, the apparent correlation has become muddied.

The links between blooms of toxic *Pseudo-nitzschia* and strandings of sea lions that show symptoms of domoic acid poisoning are not clear. We are beginning to notice, however, that the diatom cells seem to be more toxic offshore than close to shore, where we monitor the algae. The more-toxic offshore populations, unfortunately, are located in the sea lions' foraging range, and thus we may be underestimating their exposure in more dangerous waters than those we monitor close to shore.

Domoic acid was synthesized by chemists in 1982, well before it was known to occur in phytoplankton. It was first isolated in Japan in 1953, not from a diatom but from *Chondria,* a red seaweed on the Japanese coast. It turns out that the toxin has been known for a very long time in Japan, where in small amounts it has a traditional use in the home. *Chondria* is known to kill flies that settle on the seaweed; this evident toxicity has been put to long use by Japanese parents as an antihelminth, a treatment for internal parasites in children.

But What to Do?

I've left out what may be the most important part of the story of shellfish (and anchovy) poisoning: how people have responded to this hazard in our natural environment and managed to keep it largely at bay. Ever since the San Francisco incident, science has played a major part in protecting us from shellfish poisoning, a battle in which we're now deploying 21st-century tools. For the most part, though, we continue to rely chiefly on a 20th-century tool: the mouse.

When the Berkeley and Hooper Foundation investigators discovered what had killed the people of San Mateo and San Francisco in 1927, they set to work developing a way to prevent future deaths. They developed a monitoring technique that still is, remarkably, in widespread global use today.

The concept is fairly simple. Mussel tissue is ground up, and the toxins extracted using an acidic agent. After filtering, the liquid phase is injected into a mouse. A low concentration of toxin may have little effect; a high concentration will kill the mouse quickly. Thus the toxicity of shellfish is routinely measured in "mouse units," an indication of how quickly it kills a 20-gram white mouse. This bioassay, the very test that proved that a waterborne alga was the culprit in the San Francisco poisonings and that was first used in 1927, remains the international "gold standard" for detecting PSP. The impressive field observations and clever laboratory work of Sommer and his Hooper Foundation colleague W. Forest Whedon in the 1930s gave us the convincing evidence that the abundance of *Alexandrium catanella* cells correlates well with mouse units of PSP toxicity.

As a result, California and now other states and countries around the world have monitoring programs and are able to prevent deaths by closing shellfishing beds when there is a dangerous increase in toxicity. Commercial shellfish growers do their own testing before preparing their harvest. Because the monitoring was able to show seasonal patterns of toxicity, California could set a seasonal quarantine, routinely forbidding sport harvesting of shellfish during late spring, summer and early fall.

Scientists would like to find a more rapid, direct and "real-time" way to monitor hazards from harmful algal blooms, and a way that would not require sacrificing animals. MBARI's Scholin and his team of engineers and scientists have developed a way to do this, without ever leaving their offices. Basically they have designed and constructed a robotic undersea lab to do the job.

Scholin began by developing DNA probes that recognized the various local species of toxic *Pseudo-nitzschia*. His method broke apart cells vacuumed onto a filter, and then he exposed the cell material to the gene probes that recognize the species of interest.

Scholin's team developed the Environmental Sample Processor (ESP), a seagoing "lab" that is modular and programmable and can test for the presence of a wide array of microorganisms, as well as the toxic algae. Greg Doucette from the Charleston laboratories of the National Oceanic and Atmospheric Administration developed new sensors for the algal toxins, ones that Scholin could add as a module to the ESP.

The robot is now in the testing stage in Monterey Bay, where it is showing its ability to provide information in near real time by relaying information about toxic algae to shore using a radio transmitter at the sea surface. MBARI also has tested the protocols and deployed them on ships off California, in the Gulf of Maine and in the Gulf of Mexico, successfully detecting about 10 different species. Scholin's group is working toward the day, not far off, when we can call up the mooring and ask, "Are there toxic algae in Monterey Bay, and how high is the toxin level out there?"

Global Hazard, Global Triumph

Focused as we are on harnessing the power of technology to protect ourselves, it's easy to forget that people have always found ways to protect themselves from certain hazards. I was reminded of this fact by a discovery, made during a recent sabbatical, of an approach so obvious that I was later embarrassed not to have recognized it.

My sabbatical took me to the Zanzibar Islands, off mainland Tanzania, where I was interested in the likely presence of another genus of toxic algae, *Gambierdiscus toxicus*. (There exist many more harmful algal species than I can mention in this article.) This dinoflagellate grows on seaweeds eaten by reef-visiting fish; the fish ingest ciguatoxin, the poisonous substance made by the alga. *Ciguatera fish poisoning,* characterized by a varied combination of gastrointestinal and neurological symptoms, is a public health problem throughout the tropics, but the scientific literature contained no mention of its having been found in Zanzibar, though I discovered it immediately upon my arrival. Why had it not been reported there before?

Medical professionals in Tanzania told me they'd never heard of ciguatera fish poisoning: "It's not around here," they said. But a fisheries biologist visiting the marine station on Unguja suggested I talk with fishermen. I began visiting remote fishing villages with a Swahili translator, a local islander who knew the fish and the local fishing community well. My translator would introduce me to some of the men after they returned in boats from each morning's fishing and sold their catch to fishmongers. After some persuading by my translator, they consented to be interviewed.

I asked the fishermen whether they had ever experienced symptoms of illness when they ate certain kinds of fish. Commonly they would say no, not *my* fish, but they would suggest that there was such illness down the coast. After I compiled a list of what fish were caught in all the villages, I showed individuals pictures of various fish and asked which fish caused problems. Soon a fuller picture emerged.

Grouper is a fish commonly associated with ciguatera poisoning. People would say no, there's no problem eating grouper. But I thought to ask whether they had a problem eating the liver of grouper. Liver, a rich source of calories, is highly prized in the Zanzibar islands—but when I asked this question, people looked at me as if I were daft. No, they'd say, there's no problem with liver. But I continued: "Well, did you eat the liver?" And they'd say no. Why? "You don't eat the liver. You get sick—vomiting, diarrhea. So there's no problem." And this explained their earlier answer. They avoided the problem by deciding that the liver of grouper was not a food, just as we don't consider the poisonous part of a rhubarb plant to be food.

My time in Africa also reminded me that the mouse is not the only organism that has been widely used as a bioassay for shellfish poisoning. In many parts of the world, the bioassay of choice is not a mouse but a cat. It is common in underdeveloped regions to have plenty of stray animals, including hungry stray cats. So you might say that the oldest bioassay is the "here, kitty" method. A cook calls a stray cat and feeds it a tiny bit of suspect food; if the cat is doing fine in an hour or two, the food is safe to cook.

In the industrialized world, we shy from feeding potential poisons to stray cats and have scarce opportunity to check for elderberry ripening or post sentinels to watch our coastal waters for bioluminescence. We must trust science and technology and insist that health and fisheries officials work to keep us safe against a hazard that will probably always be with us. And yet the more we learn about harmful algae, the more we know that coastal people have always lived with them—and that a characteristic of each age of human history is the ability somehow to enjoy the rich foods that come from the sea, while managing one way or another to protect ourselves from the poisons in the waters that feed us.

References

Bargu, S., and M. W. Silver. 2003. Field evidence of krill grazing on the toxic diatom genus *Pseudo-nitzschia* in Monterey Bay, California. *Bulletin of Marine Science* 72:629–638.

Erlandson, J. M. 2001. The archaeology of aquatic adaptations: paradigms for a new millennium. *Journal of Archaeological Research* 9:287–337

Glibert, P. M., D. M. Anderson, P. Gentien, E. Graneli and K. Sellner. 2005. The global, complex phenomena of harmful algal blooms. *Oceanography* 18(2):132–141.

Landsberg, J. H. 2002. The effects of harmful algal blooms on aquatic organisms. *Reviews in Fisheries Science* 10:113–390.

Lefebvre, K. A., S. Bargu, T. Kieckhefer and M. W. Silver. 2002. From sanddabs to blue whales: The pervasiveness of domoic acid. *Toxicon* 40:971–977.

Price, D. W., K. W. Kizer and K. H. Hansgen. 1991. California's paralytic shellfish poisoning prevention program, 1927–1989. *Journal of Shellfish Research* 10:119–145.

Scholin, C. A., R. Marin III, P. E. Miller, G. J. Doucette, C. L. Powell, P. Haydock, J. Howard and J. Ray. 1999. DNA probes and a receptor binding assay for detection of *Pseudo-nitzschia* (Bacillariophyceae) species and domoic acid activity in cultured and natural samples. *Journal of Phycology* 35:1356–1367.

Shumway, S. E. 1995. Phycotoxin-related shellfish poisonings: bivalve mollusks are not the only vectors. *Reviews in Fisheries Science* 3:1–31.

Sommer, H., W. F. Whedon, C. A. Kofoid and R. Stohler. 1937. Relation of paralytic shell-fish poison to certain plankton organisms of the genus *Gonyaulax*. *Archives of Pathology* 24:537–559.

Wekell, J. C., J. Hurst and K. A. Lefebvre. 2004. The origin of the regulatory limits for PSP and ASP toxins in shellfish. *Journal of Shellfish Research* 23:927–930.

MARY WILCOX SILVER, a native of San Francisco, is a professor in the Ocean Sciences Department at the University of California, Santa Cruz. She has mostly worked on the microbial communities that live on floating "marine snow" and studied how these ubiquitous particles transport materials into the ocean's deep interior. In the past decade she also has become interested in the role of toxic microalgae in coastal oceans, and especially how the toxins may be permeating marine eco-systems, affecting animals living both in the water and on the seafloor. This article is adapted from the Winter 2006 Synergy Lecture at UCSC. Address: Ocean Sciences Department, University of California, Santa Cruz, CA 95064. Internet: msilver@ucsc.edu

Better Safe than Sorry

By being a smart shopper and taking some precautions in the kitchen, you can lower your family's risk of getting sick.

NANCY SHUTE

Andrew Stout's farm in Carnation, Wash., is one of the most successful small organic farms in the country. Each week, Full Circle Farm delivers fresh lettuce, green peas, spring garlic, and spinach to 17 farmers' markets in the Seattle area, as well as to dozens of restaurants and retailers, including Whole Foods Market. Some 2,400 boxes of produce a week go out to families who have bought a share in the farm's riches. His customers are counting on getting freshness and taste—and also on Stout's care when it comes to hygiene. "Bacteria exists everywhere," he says. So he keeps the manure pile away from the packing shed, tests the water used to irrigate and wash vegetables, and keeps an eye on his workers to be sure they wash their hands. "I'm a food provider," he says. "You want to do the absolute best that you can."

The rapidly growing passion for locally grown produce from farmers like Stout and his wife, Wendy Munroe, is one sign of just how nervous Americans have become about the state of food on their plate. Little wonder, given recent headlines: melamine dumped in pet food and fed to millions of chickens and pigs; *E. coli* bacteria killing people who eat spinach; salmonella in peanut butter from Georgia; and just last week, *E. coli* contaminated beef in 15 states. Each year, 76 million Americans get sick from food; more than 300,000 end up in the hospital, and 5,000 die, according to the Centers for Disease Control and Prevention. Confidence in the safety of supermarket food has reached an 18-year low, according to the Food Marketing Institute.

Lately, the melamine scandal has left shoppers wondering whether imported foods can be trusted—a legitimate concern, given that many countries have less-than-sterling safety records, and an increasing percentage of what Americans eat comes from overseas. Clearly, the oversight system has big gaps. The Food and Drug Administration is responsible for monitoring 80 percent of the food supply (everything but meat, which the U.S. Department of Agriculture oversees), but the food inspection program has been underfunded for years, and the agency has little enforcement capability beyond asking companies for voluntary recalls. It is able to inspect less than 1 percent of the $60 billion of food imported each year.

The Food and Drug Administration inspects less than 1 percent of the $60 billion of food imported each year.

Broken. Keeping homegrown food safe, too, requires diligence, from the field through the processing shed and factory, all the way to the supermarket. Farmers are legally bound to produce food that doesn't pose a health risk; so are manufacturers and retailers. Costco, for example, uses its own labs to test food samples for microbes and hires third-party auditors to inspect suppliers' farms and factories. Whole Foods, which increasingly looks abroad for its organic products, requires growers to shun pesticides allowed in their countries but not here. But with little oversight, human error or expediency can cause disease and death. "Our food-safety system is broken," former FDA Commissioner David Kessler told a congressional hearing this month. Last week, the USDA declared that 56,000 pigs fed melamine-tainted feed are safe for human consumption.

So what's a hungry consumer to do? "If you want 100 percent safety, you have to stop eating. You really do," says Marion Nestle, a professor of nutrition at New York University and author of *What to Eat.* "Unless you're growing your own food and you know the quality of your soil, and you know what the animals are eating, you can't be 100 percent safe." Nestle and other food-safety experts would advocate giving the FDA the authority to require food recalls and enough funding to do more inspections. Last month, David Acheson, a longtime food-safety official at the FDA, was named the agency's new "food czar," charged with overhauling the food-safety system. "We're constantly in a situation where, gee, we've got people sick, and we've got to get this food off the market," Acheson says. "What we've got to do as an agency is push that back into prevention." That, clearly, is a long-term project. In the meantime, consumers can make choices that reduce their family's risk.

Self-Defense in the Kitchen

The best defense against food-borne illness, experts say, is mounted at the end of the food chain: the kitchen counter. "The biggest proactive thing you can do, whether you buy organic or conventional food, is wash it and wash your hands," says Stout.

The best defense against food-borne illness is mounted at the end of the chain: the kitchen counter.

The FDA recommends that all produce—organic or not, homegrown or imported—be washed under running water just before it's eaten. Even fruit that will be peeled should be washed, since bacteria from the skin can contaminate the inside during cutting. Firm produce, like melons and cucumbers, should be scrubbed with a clean brush. Drying produce with a clean towel or paper towel may also help remove lingering contamination.

Food should be refrigerated at 40 degrees Fahrenheit to keep nasty bugs from proliferating—a refrigerator thermometer is an essential tool. Produce should be kept separate from uncooked meat, poultry, fish, and eggs; almost all domestic chickens, for example, are contaminated with *campylobacter* bacteria, which causes diarrhea. Cutting boards and other food-prep tools need to be washed with hot soapy water or a diluted bleach solution, especially after they've been used to prepare raw meat.

Salad lovers may lament, but you can't beat cooking food as a way to kill nasty bacteria. That's why milk is pasteurized; it's also why the Department of Agriculture recommends that hamburgers be cooked to 160 degrees, well past pink. Frozen foods are usually cooked in the processing, which makes frozen spinach, raspberries, and strawberries—all nearly impossible to wash thoroughly—a safer bet than fresh.

The Benefit of Buying Organic

In search of safety, many people turn to organic fruits and vegetables, presuming that they are less apt to carry pathogens like *E. coli* than nonorganic produce. That belief has fueled a boom in organic foods at stores as diverse as Whole Foods and Wal-Mart, as well as at regular supermarkets, which now offer organic house brands. Sales of organic food have been growing by 15 percent a year for the past decade, spiking upward after each publicized food contamination incident.

The good news: Organic produce does carry less pesticide and herbicide residue than conventionally grown crops. And under federal law, organic milk, meat, and produce can't contain added growth hormones, antibiotics, or genetically modified organisms. But organic produce and meat can just as easily be contaminated with bacteria, heavy metals, or other pathogens that pose significant health risks. A recent study found that organic chickens were just as likely as their regular brethren to be contaminated with *campylobacter;* some organic birds had more bacteria. And the California spinach contaminated by *E. coli* last year had been grown using organic methods.

Is Local Better?

The new push to buy locally grown food is spurred partly by an interest in supporting environmentally friendly practices that would reduce global warming; food is shipped 1,500 miles on average in the United States, burning fossil fuels and adding shipping costs along the way. But a bigger, if less often expressed, motivation is the belief that domestically grown food is inherently safer than stuff from foreign lands. In a 2002 study, researchers at Colorado State University found that consumers were willing to pay a 19 percent premium for steak labeled: "Guaranteed USA: Born and Raised in the U.S."

Anyone who has traveled in developing countries knows that sanitation standards for food and water are often much less reliable than in this country; people here rarely have to boil water to safely drink it. The same holds true for agriculture: In many countries, human waste is often the only fertilizer available, and pesticides that have been banned in the United States for decades are still used. The melamine-contamination incident, in which Chinese manufacturers adulterated wheat flour with a cheap substance that masquerades as protein, showed only too well how weak links in a global food chain can be exploited by careless or unscrupulous vendors. Given all that, buying from local sources has grown ever more attractive. Proponents count on the people close to home to hold to standards like Stout's.

"Mainly for me, it's knowing the grower, and knowing how they grow the food I'm buying," says Gail Feemstra, a food systems analyst with the University of California–Davis. After the spinach scare, she says, "I went right to the farmers' market and said, 'Give me a big bag of spinach.'" She knew that the spinach she was buying was grown far from the Central California fields implicated in the *E. coli* outbreak.

Beyond the likelihood of getting some quality control, people buying local food may avoid some risks of microbial contamination because the food isn't processed in large facilities that mix products from many farms. But there's still no guarantee; *E. coli* could easily contaminate a spinach patch down the street. Still, farmers' markets have thrived in recent years, supported by rising consumer demand and by state and federal programs designed to help small farmers get their products to market. There are now 4,000 farmers' markets in the United States. Some supermarkets offer a limited selection of locally grown foods in season. And community-supported agriculture programs, in which farmers like Andrew Stout sell shares in their yearly output, provide a two-sided guarantee. Farmers have a guaranteed market, and consumers know that each week they'll get a box of fresh veggies, fruit, or eggs, often delivered right to their doorstep. (CSAS in your area can be found at www.localharvest.org/csa.)

One of Stout's members, Sage Van Wing, 29, created a website when she lived in the San Francisco area, challenging others to eat food grown within 100 miles of home for a month. In the first month, 800 people signed on to the locavores.com challenge. The list has now grown to 2,000 from around the world. Being a

Eat Like a Peasant and Enjoy!

When novelist **Barbara Kingsolver,** 52, moved from Tucson, Ariz., to southwest Virginia with her husband, Steven, and two daughters, they decided that they would spend a year trying to eat local: only food they either raised themselves or bought from nearby growers. They raised turkeys and chickens for meat and eggs, bought milk from a regional dairy, and froze corn and peas for the winter. The results of that experiment are the subject of her new book, *Animal, Vegetable, Miracle: A Year of Food Life.*

Why Did You Decide to Try a Year of Eating Locally?

We talked as a family about undertaking this. It's very important to Steve and me, and very clear to our kids, that food is a subject that matters enormously. Not just what you eat but where it comes from. It's interesting with this new melamine scare, everyone's talking about stepping up inspection. Well, that's going to take money. If the point is to get really cheap ingredients from all over the world to manufacture really cheap food, how cheap is it if you consider inspection and shipping and all the health costs?

You're a Working Mother. How Did You Manage Gardening and Going to Farmers' Markets?

In the beginning, I was worried that it would be a lot of extra work. But I love walking out to the garden at the end of a frustrating day. I look at it as recreation, rather than a chore. We could have gotten a lot more of our vegetables from local farmers if we had wanted to.

I Know Your Husband Was Worried about Going without Coffee, and You Made an Exception for That. What Else Did You Miss?

We all had our separate worries about what we were giving up—Alaskan salmon or Gummi Bears. The really surprising part of this journey is that it was not an exercise in deprivation or slaving over the stove. It really was an adventure. The day the farmers' market opened and it was snowing, nobody wanted to go. I put on the mom pep talk and said, "Our farming friends need us." We had one of the best meals of our year that week: an asparagus morel bread pudding that was amazing.

Isn't It More Expensive to Buy Food at the Farmers' Market?

The most surprising thing was to find that we ate locally and splendidly, organic whole foods for a year, at a cost of about 50 cents per person per meal. It's cheaper than eating out or eating food prepared by other people. If you think about it, it's peasant food, eaten the world over.

—N.S.

Farm of the Future?

The United States, like other countries, is becoming increasingly urban; by 2030, 60 percent of the world's population and 87 percent of North Americans will live in cities. This trend invites a vision of the future that isn't pretty: city dwellers subsisting on tasteless pink tomatoes, flown in from a distant land.

Dickson Despommier has a different vision: fresh, healthful food grown in glittering high-rise city farms. These "vertical farms" would produce crops, poultry, and fish year-round in a controlled environment free of pollutants, parasites, and dangerous microbes. "What would happen to Harlem if vertical farms were a reality?" Despommier asks. "Diets would change; disease would change."

A challenge. Despommier is an unlikely farmer. He's a professor of parasitology at Columbia University who teaches about creatures like hookworm, a parasite that causes anemia and stunts children's development. It remains a major problem in tropical countries, where it spreads through human waste used to fertilize crops. Eight years ago, Despommier challenged the public health students he teaches to come up with a system of growing food for the world that would reduce the risk of disease. Each year, a new class has built on the concept.

Their vertical farm prototype is based on existing technology for hothouse farming and would recycle some of the 1.4 billion gallons of water that New Yorkers pour down the drain every day. All of New York City could be fed with about 150 farms, 30-stories each, Despommier estimates. Pests and late-spring frosts wouldn't be an issue. Fish could live in tanks; poultry in small pens. Plants could be raised hydroponically—without soil, suspended in water or some other solution. Sunlight would be used to generate heat and electricity to optimize growing conditions around the clock. "We shouldn't be at the whims of nature when it comes to providing calories and water," Despommier says.

Several New York companies, including IBM, have expressed interest in developing the concept; Despommier predicts a farm will be in business within 10 years.

—N.S.

locavore wasn't as easy as Van Wing expected; she couldn't find local sources for salt, spices, or bread. Avocados were hard to come by, too. (Exceptions, she decided, are just fine.)

Not labeled. Eating local, or even domestic, is complicated by the fact that most supermarket foods aren't labeled by source. For years, consumer groups and domestic producers have battled for federal legislation to make country-of-origin labeling required for all imported foods. But those requirements have been applied only for seafood. Even if there were universal labeling, most consumers don't have time to bone up on the differences in agricultural practices from country to country so as to make labels of much use.

221

The bottom line: Consumers can take steps to protect themselves, but they can't do it all on their own. "Everybody ought to be screaming bloody murder on the lack of oversight of the food supply and talk to their government representatives," says Nestle. Acheson agrees. "I'd like company X and Y to say, all of this food is safe because we do X and Y," he says. Consumers are now pushing for more information about the safety of the food they eat. "And that's a good thing."

Meanwhile, Stout's reach continues to grow. Demand for the CSA's products is now at the point where it's economically feasible to fly his veggies to 2,000 produce-hungry customers in Alaska, offering practically same-day service. "We are kind of local for southeast Alaska, where there are no farms whatsoever," Stout says. "This is as close as it gets."

With Sarah Baldauf and Adam Voiland

Detecting Mad Cow Disease

New tests can rapidly identify the presence of dangerous prions—the agents responsible for the malady—and several compounds offer hope for treatment.

STANLEY B. PRUSINER

L ast December mad cow disease made its U.S. debut when federal officials announced that a holstein from Mabton, Wash., had been stricken with what is formally known as bovine spongiform encephalopathy (BSE). The news kept scientists, government officials, the cattle industry and the media scrambling for information well past New Year's. Yet the discovery of the sick animal came as no surprise to many of us who study mad cow disease and related fatal disorders that devastate the brain. The strange nature of the prion—the pathogen at the root of these conditions—made us realize long ago that controlling these illnesses and ensuring the safety of the food supply would be difficult.

As researchers learn more about the challenges posed by prions—which can incubate without symptoms for years, even decades—they uncover strategies that could better forestall epidemics. Key among these tools are highly sensitive tests, some available and some under development, that can detect prions even in asymptomatic individuals; currently BSE is diagnosed only after an animal has died naturally or been slaughtered. Researchers have also made some headway in treating a human prion disorder called Creutzfeldt-Jakob disease (CJD), which today is uniformly fatal.

Identifying the Cause

Although mass concern over mad cow disease is new in the U.S., scientific efforts to understand and combat it and related disorders began heating up some time ago. In the early 1980s I proposed that the infectious pathogen causing scrapie (the sheep analogue of BSE) and CJD consists only of a protein, which I termed the prion. The prion theory was greeted with great skepticism in most quarters and with outright disdain in others, as it ran counter to the conventional wisdom that pathogens capable of reproducing must contain DNA or RNA [see "Prions," by Stanley B. Prusiner; *Scientific American,* October 1984]. The doubt I encountered was healthy and important, because most dramatically novel ideas are eventually shown to be incorrect. Nevertheless, the prion concept prevailed.

In the years since my proposal, investigators have made substantial progress in deciphering this fascinating protein. We know that in addition to causing scrapie and CJD, prions cause other spongiform encephalopathies, including BSE and chronic wasting disease in deer and elk [see "The Prion Diseases," by Stanley B. Prusiner; *Scientific American,* January 1995]. But perhaps the most startling finding has been that the prion protein, or PrP, is not always bad. In fact, all animals studied so far have a gene that codes for PrP. The normal form of the protein, now called PrP^C (C for cellular), appears predominantly in nerve cells and may help maintain neuronal functioning. But the protein can twist into an abnormal, disease-causing shape, denoted PrP^{Sc} (Sc for scrapie, the prion disease that until recently was the most studied).

Unlike the normal version, PrP^{Sc} tends to form difficult-to-dissolve clumps that resist heat, radiation and chemicals that would kill other pathogens. A few minutes of boiling wipes out bacteria, viruses and molds, but not PrP^{Sc}. This molecule makes more of itself by converting normal prion proteins into abnormal forms: PrP^{Sc} can induce PrP^C to refold and become PrP^{Sc}.

Cells have the ability to break down and eliminate misfolded proteins, but if they have difficulty clearing PrP^{Sc} faster than it forms, PrP^{Sc} builds up, ruptures cells and creates the characteristic pathology of these diseases—namely, masses of protein and microscopic holes in the brain, which begins to resemble a sponge. Disease symptoms appear as a result.

Prion diseases can afflict people and animals in various ways. Most often the diseases are "sporadic"—that is, they happen spontaneously for no apparent reason. Sporadic CJD is the most common prion disease among humans, striking approximately one in a million, mostly older, people. Prion diseases may also result from a mutation in the gene that codes for the prion protein; many families are known to pass on CJD and two other disorders, Gerstmann-Sträussler-Scheinker disease and fatal insomnia. To date, researchers have uncovered more than 30 different mutations in the PrP gene that lead to the hereditary forms of the sickness—all of which are rare, occurring about once in every 10 million people. Finally, prion disease may result from an infection, through, for instance, the consumption of bovine prions.

Tracing the Mad Cow Epidemic

The world awoke to the dangers of prion disease in cows after the BSE outbreak that began ravaging the British beef industry in the mid-1980s. The truly novel concepts emerging from prion science forced researchers and society to think in unusual ways and made coping with the epidemic difficult. Investigators eventually learned that prions were being transmitted to cattle through meat-and-bone meal, a dietary supplement prepared from the parts of sheep, cattle, pigs and chickens that are processed, or rendered, for industrial use. High heat eliminated conventional pathogens, but PrP^{Sc} survived and went on to infect cattle.

As infected cattle became food for other cattle, BSE began appearing throughout the U.K. cattle population, reaching a high of 37,280 identified cases in 1992. The British authorities instituted some feed bans beginning in 1989, but it was not until 1996 that a strict ban on cannibalistic feeding finally brought BSE under control in the U.K.; the country saw 612 cases last year. Overall the U.K. has identified about 180,000 mad cows, and epidemiological models suggest that another 1.9 million were infected but undetected.

For many people, the regulations came too late. Despite the British government's early assurances to the contrary, mad cow disease proved transmissible to humans. In March 1996 Robert Will, James Ironside and Jeanne Bell, who were working in the National CJD Surveillance Unit in Edinburgh, reported that 11 British teenagers and young adults had died of a variant of Creutzfeldt-Jakob disease (vCJD). In these young patients the patterns of PrP^{Sc} deposition in the brain differed markedly from that found in typical CJD patients.

Many scientists, including myself, were initially dubious of the presumed link between BSE and vCJD. I eventually changed my mind, under the weight of many studies. One of these was conducted by my colleagues at the University of California at San Francisco, Michael Scott and Stephen DeArmond, who collected data in mice genetically engineered to resemble cattle, at least from a prion protein point of view (the PrP gene from cattle was inserted into the mouse genome). These mice became ill approximately nine months after receiving injections of prions from cattle with BSE or people with vCJD, and the resulting disease looked the same whether the prions originated from cows or vCJD patients.

As of February 2004, 146 people have been diagnosed with vCJD in the U.K. and another 10 elsewhere. No one knows exactly how many other people are incubating prions that cause vCJD. Epidemiological models suggest that only a few dozen more individuals will develop vCJD, but these models are based on assumptions that may prove wrong. One assumption, for example, is that vCJD affects only those with a particular genetic makeup. Because prions incubate for so long, it will take some time before we know the ultimate number of vCJD cases and whether they share similar genetics.

In vCJD, PrP^{Sc} builds up, not just in the brain but also in the lymphoid system, such as the tonsils and appendix, suggesting

Overview/Rooting Out Prions

- The recent discoveries of cows afflicted with mad cow disease, or bovine spongiform encephalopathy (BSE), in the U.S., Canada and elsewhere emphasize the need for better tests and policies so that infection does not spread and the public is reassured.
- Several new tests make such identification much more rapid than it has been, and some are able to identify low levels of dangerous prions so officials will not have to wait until cattle are sick before they know there is a problem.
- Invention of a blood test would allow diagnoses to be made for living animals and for people and could be important if some of the therapies now being investigated to treat prion diseases in humans prove to be effective.

that PrP^{Sc} enters the bloodstream at some point. Animal studies have shown that prions can be transmitted to healthy animals through blood transfusions from infected animals. In response to this information, many nations have enacted stricter blood donation rules. In the U.K., people born after 1996, when the tough feed ban came into force, can receive blood only from overseas (those born before are considered already exposed). In the U.S., those who spent three months or more in the U.K. between 1980 and 1996 cannot give blood.

Although such restrictions have contributed to periodic blood shortages, the measures appear justified. Last December the U.K. announced the vCJD death of one of 15 individuals who received transfusions from donors who later developed vCJD. The victim received the transfusion seven and a half years before his death. It is possible that he became infected through prion-tainted food, but his age argues against that: at 69, he was much older than the 29 years of typical vCJD patients. Thus, it seems fairly likely that vCJD is not limited to those who have eaten prion-infected beef.

Since the detection of mad cow disease in the U.K., two dozen other nations have uncovered cases. Canada and the U.S. are the latest entrants. On May 20, 2003, Canadian officials reported BSE in an eight-and-a-half-year-old cow that had spent its life in Alberta and Saskatchewan. (The country's only previous mad cow had arrived as a U.K. import 10 years earlier.) Although the animal had been slaughtered in January 2003, slow processing meant that officials did not test the cow remains until April. By then, the carcass had been turned into pet food exported to the U.S.

Seven months later, on December 23, the U.S. Department of Agriculture announced the country's first case of BSE, in Washington State. The six-and-a-half-year-old dairy cow had entered the U.S. at the age of four. The discovery means that U.S. officials can no longer labor under the misconception that the nation is free of BSE. Like Canada, U.S. agricultural interests want the BSE problem to disappear. Financial woes stem primarily from reduced beef exports: 58 other countries are

A Worldview

Many countries have reported cattle afflicted with BSE, but cases of human infection thought to stem from eating meat from sick cows—variant Creutzfeldt-Jakob disease (vCJD)—remain low, for now at least.

Country	BSE Cases	vCJD Deaths (current cases)
Austria	1	0
Belgium	125	0
Canada	2	1
Czech Republic	9	0
Denmark	13	0
Falkland Islands	1	0
Finland	1	0
France	891	6
Germany	312	0
Greece	1	0
Hong Kong	0	1*
Ireland	1,353	1
Israel	1	0
Italy	117	1
Japan	11	0
Lichtenstein	2	0
Luxembourg	2	0
Netherlands	75	0
Oman	2	0
Poland	14	0
Portugal	875	0
Slovakia	15	0
Slovenia	4	0
Spain	412	0
Switzerland	453	0
U.S.	1	0(1)†
U.K.	183,803	141(5)
Worldwide		151(6)

*Awaiting confirmation
†British subject

keeping their borders shut, and a $3-billion export market has largely evaporated.

Designing Diagnostics

The most straightforward way to provide this assurance—both for foreign nations and at home—appears to be simple: just test the animals being slaughtered for food and then stop the infected ones from entering the food supply, where they could transmit pathogenic prions to humans. But testing is not easy. The USDA uses immunohistochemistry, an old technique that is cumbersome and extremely time-consuming (taking a few days to complete) and so is impractical for universal application.

Accordingly, others and I have been working on alternatives. In the mid-1980s researchers at my lab and elsewhere produced new kinds of antibodies that can help identify dangerous prions in the brain more efficiently. These antibodies, similar to those used in the standard test, recognize any prion—normal or otherwise. To detect PrP^{Sc}, we need to first remove any trace of PrP^{C}, which is done by applying a protease (protein-degrading enzyme) to a brain sample. Because PrP^{Sc} is generally resistant to the actions of proteases, much of it remains intact. Antibodies then added to the sample will reveal the amount of PrP^{Sc} present. Using a similar approach, a handful of companies, including Prionics in Switzerland and Bio-Rad in France, have developed their own antibodies and commercial kits. The results can be obtained in a few hours, which is why such kits are proving useful in the mass screenings under way in Europe and Japan. (Japan discovered its first case of BSE in 2001 and by this past April had reported 11 infected cows in total.)

These rapid tests, however, have limitations. They depend on PrP^{Sc} accumulating to detectable amounts—quite often, relatively high levels—in an animal's brain. Yet because BSE often takes three to five years to develop, most slaughter-age cattle, which tend to be younger than two years, usually do not test positive, even if they are infected. Therefore, these tests are generally most reliable for older bovines, regardless of whether they look healthy or are "downers." At the moment, downer cattle, which cannot stand on their own, are the group most likely to be tested.

Until new regulations came out in January, the U.S. annually sent roughly 200,000 downers to slaughter for human consumption. Of these, only a fraction were tested. Over the next year, however, the USDA will examine at least 200,000 cows for BSE. (Whether milk can be affected remains open; my laboratory is testing milk from BSE-infected cattle.)

Because of the limitations of the existing tests, developing one that is able to detect prions in the bulk of the beef supply—that is, in asymptomatic young animals destined for slaughter—continues to be one of the most important weapons in confronting prion disease outbreaks.

Scientists have pursued several strategies. One tries to boost the amount of PrP^{Sc} in a sample so the prions are less likely to escape detection. If such an amplification system can be created, it might prove useful in developing a blood test for use in place of ones that require an animal's sacrifice (not enough PrP^{Sc} circulates in the blood to be detectable by current methods). Claudio Soto of Serono Pharmaceuticals and his colleagues have attempted to carry this out. They mixed brain preparations from normal and scrapie-infected hamsters and then subjected the mixture to sound pulses to break apart clumps of PrP^{Sc} so it could convert the normal form of the prion protein into the rogue version. The experiment resulted in a 10-fold increase in protease-resistant prion protein. Surachai Supattapone of Dartmouth Medical School has obtained similar results.

Testing for Mad Cow

Four kinds of tests are now used to detect dangerous prions (PrP^{Sc}) in brain tissue from dead cattle. By identifying infected animals, public health officials and farmers can remove them from the food supply. Some of these tests, however, are time-consuming and expensive, so researchers are working to develop the ideal diagnostic: one that could quickly detect even tiny amounts of PrP^{Sc} in blood and urine and thus could work for live animals and people. The hope is to forestall outbreaks by catching infection as early as possible and, eventually, to treat infection before it progresses to disease.

Bioassay

The bioassay can take up to 36 months to provide results and can be very expensive. Its advantage is that it can reveal particular strains of prions as well as how infectious the sample is based on the time it takes for the test animal to become sick.

Immunohistochemistry

The first test used to specifically detect prions, immunohistochemistry is considered the gold standard that other tests must meet. But because technicians must examine each slide, the process is very time-consuming, often taking as many as seven days, and is not useful for mass screenings.

Immunoassay

Many companies produce immunoassays, and these rapid tests are now in widespread use in Europe; they have just been introduced to the U.S. Results can be had within eight hours, and hundreds of samples can be run simultaneously. These tests work well only with high levels of PrP^{Sc}.

Conformation-Dependent Immunoassay (CDI)

This automated test can detect very low levels of PrP^{Sc} and reveal how much of the dangerous prion is present in the sample without first having to degrade PrP^C. It informs experts about the animal's level of infection within five hours or so. It has been approved for testing in Europe. CDI is being tested on tissue from live animals and might one day serve as a blood test.

Spontaneous Prion Disease

Because prion diseases have aspects that resemble those caused by viruses, many people use viral analogies when talking about prions. But these analogies can sow confusion. An example is the presumed origin of the mad cows in Canada and the U.S. Although it is true that bovine spongiform encephalopathy (BSE) first appeared in the U.K. and then spread elsewhere through exported prion-contaminated feed, the idea of a traditional bacterial or viral epidemic is only partly helpful. In those situations, quarantines or bans can curb the spread of disease. But prions can arise spontaneously, which is an extremely important characteristic that distinguishes prions from viruses. In fact, any mammal is capable of producing prions spontaneously.

Spontaneous prion disease, for instance, is thought to have triggered the epidemic of kuru, which decimated a group called the Fore in New Guinea in the past century. The theory is that Creutzfeldt-Jakob disease occurred in an individual, whose brain was then consumed by his or her fellow Fore in a funerary rite involving cannibalism. The continued practice created a kuru epidemic.

Similarly, a feed ban that prevents cows from eating the remains of other animals is crucial in containing BSE. But such bans will not eliminate the presence of mad cows when pathogenic prions arise spontaneously. If every year one out of a million humans spontaneously develops a prion disease, why not the same for cows? Indeed, I suspect that the North American BSE cases could well have arisen spontaneously and that afflicted animals have occasionally appeared unrecognized in herds ever since humans started cattle ranching. We have been extraordinarily lucky that a past spontaneous case did not trigger an American BSE epidemic. Or perhaps small epidemics did happen but went undetected.

Still, many prefer the idea that the two mad cows in North America acquired prions from their feed. Such reasoning allows people to equate prions with viruses—that is, to think of prions only as infectious agents (even though they can also be inherited and occur spontaneously)—and to offer a seemingly plausible plan to eradicate BSE by quarantining herds. But ignoring the revolutionary concepts that govern prion biology can only hamper efforts at developing an effective program to protect the American public from exposure to these deadly agents. We must think beyond quarantine and bans and test for prions even in the absence of an epidemic.

—S.B.P.

Another strategy focuses on the intricacies of the protein shape instead of trying to bolster amounts. A test I developed with my U.C.S.F. colleague Jiri Safar, for example, is based on the ability of some antibodies to react with either PrP^C or PrP^{Sc}, but not both. Specifically, the antibody targets a portion of the prion protein that is accessible in one conformation but that is tucked away in the alternative conformation, much like a fitted corner of a sheet is hidden when folded. This specificity means that test samples do not have to be subjected to proteases. Removing the protease step is important because we now know that a form of PrP^{Sc} is sensitive to the action of proteases, which means that tests that eliminate PrP^C probably also eliminate most or all of the protease-sensitive PrP^{Sc}—and so these tests could well underestimate the amount of PrP^{Sc} present by as much as 90 percent.

Our test—the conformation-dependent immunoassay (CDI)—gained approval for use in Europe in 2003 and might be sensitive

The Significance of Strains

One distinct characteristic that prions do share with viruses is variability; they can come in strains—forms that behave somewhat differently. Laboratory work convincingly shows that prion strains result from different conformations of PrP^{Sc}, but no one has yet figured out exactly how the structure of a given strain influences its particular biological properties. Nevertheless, strains can clearly cause different illnesses. In humans, Creutzfeldt-Jakob disease, kuru, fatal insomnia and Gerstmann-Sträussler-Scheinker disease all result from different strains. Sheep have as many as 20 kinds of scrapie. And BSE may also come in various versions. For instance, a 23-month-old animal in Japan and another cow in Slovakia had a good deal of PrP^{Sc} in the midbrain, whereas in most cases PrP^{Sc} tends to accumulate in the brain stem.

The necessity of recognizing and understanding strains has become clear in studies of sheep exposed to BSE prions. A particular breed of sheep, called the ARR/ARR genotype, resists scrapie, the ovine form of mad cow disease (the letters refer to amino acids on the sheep's prion protein). And so several European countries have created scrapie eradication programs based on breeding flocks with the ARR/ARR genotype. Yet these sheep get sick when inoculated with BSE prions. That these sheep are resistant to scrapie but susceptible to BSE prions has important implications for farming practices. It can be argued that such homogeneous populations will only serve to increase the incidence of BSE prions in sheep. Such a situation could prove to be dangerous to humans who consume lamb and mutton because scrapie strains do not seem to sicken humans, whereas BSE prions do. Experiments with animals should be conducted to see if BSE prions from sheep are as deadly as those from cows before livestock experts continue with selective sheep-breeding programs.

—S.B.P.

enough to detect PrP^{Sc} in blood. The CDI has already shown promise in screening young cows. In the fall of 2003 Japan reported two cases of BSE in cows 21 and 23 months of age. Neither animal showed outward signs of neurological dysfunction. In the case of the 23-month-old cow, two commercially available tests for PrP^{Sc} returned inconclusive, borderline-positive results, but the CDI showed that the brain stem harbored malevolent prions.

Neither of these cases would have been discovered in Europe, where only cattle older than 30 months (24 months in Germany) must be tested if they are destined for human consumption. Initially the Japanese government proposed adopting the European Union's testing protocol. But consumer advocates forced the government to change its policy and test every slaughtered animal. Given that seemingly healthy animals can potentially carry pathogenic prions, I believe that testing all slaughtered animals is the only rational policy. Until now, the tests have been inadequately sensitive. But the advent of rapid, sensitive tests means universal screening can become the norm. (I understand that this statement could seem self-serving because I have a financial interest in the company making the CDI test. But I see no other option for adequately protecting the human food supply.)

Some New Insights

During our work on the CDI, we discovered a surprising fact about the development of prion disease. As alluded to above, we found that the prion is actually a collection of proteins having different degrees of resistance to protease digestion. We also learned that protease-sensitive forms of PrP^{Sc} appear long before the protease-resistant forms appear. Whether protease-sensitive PrP^{Sc} is an intermediate in the formation of protease-resistant PrP^{Sc} remains to be determined. Regardless, a test that could identify protease-sensitive forms should be able to detect infection before symptoms appear, so the food supply can have maximum protection and infected patients can be assisted as early as possible. Fortunately, by using the CDI, my colleagues and I have been able to detect low levels of the protease-sensitive forms of PrP^{Sc} in the blood of rodents and humans.

Hunting for prions in blood led us to another surprise as well. Patrick Bosque, now at the University of Colorado's Health Sciences Center, and I found prions in the hind limb muscles of mice at a level 100,000 times as high as that found in blood; other muscle groups had them, too, but at much lower levels. Michael Beekes and his colleagues at the Robert Koch Institute in Berlin discovered PrP^{Sc} in virtually all muscles after they fed prions to their hamsters, although they report high levels of prions in all muscles, not just in the hind limbs. (We do not know why our results differ or why the hind limbs might be more prone to supporting prions than other muscles are.) These findings were not observed in rodents exclusively but in human patients as well. U.C.S.F. scientists Safar and DeArmond found PrP^{Sc} in the muscles of some CJD patients, and Adriano Aguzzi and his colleagues at the University of Zurich identified PrP^{Sc} in the muscles of 25 percent of the CJD patients they examined.

Of course, the ideal way to test for prions would be a non-invasive method, such as a urine test. Unfortunately, so far the only promising lead—discovery of protease-resistant PrP in urine—could not be confirmed in later studies.

Novel Therapies

Although new diagnostics will improve the safety of the food and blood supply, they will undoubtedly distress people who learn that they have a fatal disease. Therefore, many investigators are looking at ways to block prion formation or to boost a cell's ability to clear existing prions. So far researchers have identified more than 20 compounds that can either inhibit prion formation or enhance prion clearance in cultured cells. Several compounds have been shown to extend the lives of mice or hamsters when administered around the time they were inoculated with prions, but none have been shown to alter the course of disease when administered well after the initial infection occurred. Furthermore, many of these agents require high doses to exert their effects, suggesting that they would be toxic in animals.

Beyond the problem of potential toxicity that high doses might entail lies the challenge of finding drugs that can cross the blood-brain barrier and travel from the bloodstream into brain tissue. Carsten Korth, now at Heinrich Heine University in Dusseldorf, and I—and, independently, Katsumi Doh-ura of Kyushu University in Japan and Byron Caughey of the National Institute of Allergy and Infectious Diseases—have found that certain drugs known to act in the brain, such as thorazine (used in the treatment of schizophrenia), inhibit prion formation in cultured cells. Another compound, quinacrine, an antimalarial drug with a structure that resembles thorazine, is approximately 10 times as powerful.

Quinacrine has shown some efficacy in animals. My co-workers and I administered quinacrine to mice, starting 60 days after we injected prions into their brain, and found that the incubation time (from the moment of infection to the manifestation of disease) was prolonged nearly 20 percent compared with untreated animals. Such an extension might be quite significant for humans with prion diseases if they could be made to tolerate the high levels of quinacrine needed or if more potent relatives of the drug could be made. My U.C.S.F. colleagues Barnaby May and Fred E. Cohen are pursuing the potency problem. In cell cultures, they have boosted the effectiveness of quinacrine 10-fold by joining two of its molecules together.

Another therapeutic approach involves the use of antibodies that inhibit PrP^{Sc} formation in cultured cells. Several teams have had some success using this strategy. In mice inoculated with prions in the gut and then given antibodies directed against prion proteins, the incubation period was prolonged. So far, however, only a few patients have received antiprion drugs. Quinacrine has been administered orally to patients with vCJD and to individuals who have the sporadic or genetic forms of prion disease. It has not cured them, but it may have slowed the progression of disease; we await further evidence.

Physicians have also administered pentosan polysulfate to vCJD patients. Generally prescribed to treat a bladder condition, the molecule is highly charged and is unlikely to cross the blood-brain barrier, so it has been injected directly into a ventricle of the brain. The drug has apparently slowed the progression of vCJD in one young man, but it seems unlikely that it will diffuse throughout the brain because similarly charged drugs—administered in the same way—have not.

A controlled clinical trial is needed before any assessment of efficacy can be made for quinacrine and other antiprion drugs. Even an initial clinical trial may prove to be insufficient because we have no information about how delivery of the drug should be scheduled. For example, many cancer drugs must be given episodically, where the patient alternates periods on and off the drug, to minimize toxicity.

Although the road to a successful treatment seems long, we have promising candidates and strategies that have brought us much further along than we were just five years ago. Investigators are also hopeful that when a successful therapy for prion disease is developed, it will suggest effective therapies for more common neurodegenerative diseases, including Alzheimer's, Parkinson's and amyotrophic lateral sclerosis (ALS). Aberrant, aggregated proteins feature in all these diseases, and so lessons learned from prions may be applicable to them as well.

More to Explore

Prions. Stanley B. Prusiner. 1997 Nobel Prize lecture. Available from the Nobel Foundation site at: www.nobel.se/medicine/laureates/1997/index.html

Prion Biology and Diseases. Second edition. Edited by Stanley B. Prusiner. Cold Spring Harbor Laboratory Press, 2004.

Advancing Prion Science: Guidance for the National Prion Research Program. Edited by Rick Erdtmann and Laura B. Sivitz. National Academy Press, 2004.

STANLEY B. PRUSINER is professor of neurology and biochemistry at the University of California San Francisco School of Medicine. He is a member of the National Academy of Sciences, the Institute of Medicine and the American Philosophical Society. In 1997 he won the Nobel Prize in Physiology or Medicine for his discovery of and research into prions. This is his third article for *Scientific American*. In the spirit of disclosure, Prusiner notes that he founded a company, InPro Biotechnology, which offers several prion tests, some of which are licensed to Beckman Coulter.

Yeasts in Foods and Beverages: Impact on Product Quality and Safety

GRAHAM H. FLEET

Introduction

The impact of yeasts on the production, quality and safety of foods and beverages is intimately linked to their ecology and biological activities. Recent advances in understanding the taxonomy, ecology, physiology, biochemistry and molecular biology of yeasts have stimulated increased interest in their presence and significance in foods and beverages. This has led to a deeper understanding of their roles in the fermentation of established products, such as bread, beer and wine, and greater awareness of their roles in the fermentation processes associated with many other products. As the food industry develops new products and processes, yeasts present new challenges for their control and exploitation. Food safety and the linkage between diet and health are issues of major concern to the modern consumer, and yeasts have emerging consequences in this context. On the positive side, there is increasing interest in using yeasts as novel probiotic and biocontrol agents, and for the nutrient fortification of foods. On the negative side, food-associated yeasts could be an under-estimated source of infections and other adverse health responses in humans.

Two books, entirely devoted to the occurrence and significance of yeasts in foods and beverages, have recently been published[1**,2**] and another includes several chapters on food spoilage yeasts.[3] These publications demonstrate the expanding academic and industrial interest in the field. This article reviews recent developments in understanding the ecology and biology of yeasts in foods and beverages and discusses how these impact on product quality and safety.

New Analytical Tools

The ability to isolate, enumerate and identify yeasts to genus, species and strain levels is fundamental to understanding their occurrence and significance in foods and beverages. Although cultural procedures remain basic to these needs, molecular methods are making the study of yeast ecology much more attractive and convenient than ever before.[4*,5]

Yeast Taxonomy and Species Identification

Whereas the identification of new yeast isolates once required the laborious completion of 80 to 100 morphological, biochemical and physiological analyses, this task is now quickly achieved by DNA sequencing. The DNA sequences of the genes encoding the D1/D2 domain of the large (26S) subunit of ribosomal RNA are known for all yeast species, and the sequence of the ITS1-ITS2 region of rRNA, as well as other genes, is known for many. These sequence–phylogenetic data have led to a complete revision of yeast taxonomy, and the description of many new genera and species.[6*] Although sequencing of ribosomal genes is now the accepted method for yeast identification, restriction fragment length polymorphism (RFLP) analysis of the ITS1-ITS2 region is a less expensive, faster alternative, and databases containing the results of such analyses have been established for food yeasts.[5]

Nucleic acid probes and real-time PCR detection methods have been described for some species, such as *Saccharomyces cerevisiae, Brettanomyces bruxellensis* and *Zygosaccharomyces bailii*,[4*,5,7] and a novel probe-flow cytometric assay has been reported for various *Candida* species.[8]

Strain Differentiation

The distinctive character of many breads, beers and wines can be linked to particular strains of *S. cerevisiae* used in the fermentation.[9] Consequently, differentiation of yeasts at the subspecies level is an important requirement. Molecular methods developed for this purpose include pulsed-field gel electrophoresis (PFGE) of chromosomal DNA and PCR-based methods such as random amplification of polymorphic DNA (RAPD), amplified fragment length polymorphism (AFLP), RFLP, and profiling of microsatellite DNA. A simpler, faster method is based on RFLP analysis of mitochondrial DNA, where no PCR amplification of DNA is required.[4*,5,10] These methods are not only useful for quality assurance typing of yeast starter cultures and spoilage species, but they have been used to reveal the ecological complexity of the yeast flora associated with many food and beverage fermentations. For example, it is now known that the fermentation of wine, cheese, meat sausages and other products not only involves the successional contributions from many different species of yeast, but successional growth of numerous strains within each species also occurs.[11,12*]

Culture-Independent Analysis

Most branches of microbial ecology now accept that viable but non-culturable species occur in many habitats, including foods and beverages. Detection of these organisms requires extraction and analysis of the habitat DNA. One approach that is finding increasing application is PCR in conjunction with denaturing gradient gel electrophoresis (DGGE) or temperature gradient gel electrophoresis (TGGE). Total DNA is extracted from the food, and yeast DNA is specifically amplified using PCR and primers targeting regions of rDNA. The yeast DNA is then resolved into amplicons for individual species by DGGE or TGGE. These amplicons are extracted from the gel and their species identity determined through sequence analysis. PCR-DGGE/TGGE has been applied to analyse the yeast communities associated with grapes, wine, sourdough, cocoa bean, coffee bean and meat sausage fermentations.[4*,5,13*,14,15] There is good agreement in the results obtained by cultural and PCR-DGGE/TGGE methods, although in some cases species that were not identified by agar culture were recovered by PCR-DGGE—suggesting the presence of non-culturable flora. However, the reverse also occurs, where PCR-DGGE has not detected yeasts that were isolated by culture. Many factors affect the performance of PCR-DGGE/TGGE analyses and further research is required to understand and optimize the assay conditions.[4*,13*]

Molecular Understanding of the Yeast Response

As yeasts grow in foods and beverages, they utilize carbon and nitrogen substrates and generate a vast array of volatile and non-volatile metabolites that determine the chemosensory properties of the product and its appeal to the consumer. Some yeasts produce extracellular pro-teases, lipases, amylases and pectinases that also impact on product flavour and texture. The biochemistry of these reactions and their linkage to product quality are generally well known.[16**] Now, genomic studies using sequence, DNA array, and proteomic analyses enable the linkage of these responses to the expression and regulation of individual genes.[17*] Only a few such studies have been performed with food and beverage yeasts, and these have yielded interesting new insights. For example, during wine and beer fermentations, S. cerevisiae exhibits sequential expression and regulation of many genes associated with carbon, nitrogen and sulfur metabolism, as well as other genes required to tolerate stresses such as high sugar concentration, low pH, ethanol and nutrient deficiency.[17*,18,19] Genomic analyses also give molecular explanations of the remarkable tolerance of some yeasts to the extremes of high salt and sugar contents in some foods (e.g. Debaryomyces hansenii in cheese brines, Zygosaccharomyces rouxii in sugar syrups and

fruit juice concentrates), and to organic acid preservatives in other foods (e.g. Z. bailii in salad dressings and soft drinks).[20*]

Beyond Brewing, Baking and Wine Yeasts

Although research on the contribution of S. cerevisiae to beer, bread and wine fermentations continues to be a focus, there is expanding interest in the role of yeasts in other products.[12*]

It is now well recognized that yeasts make important contributions to the process of cheese maturation, where various strains of D. hansenii, Yarrowia lipolytica, Kluyveromyces marxianus and S. cerevisiae frequently grow to high populations. They contribute to the development of cheese flavour and texture through proteolysis, lipolysis, utilization of lactic acid, fermentation of lactose and autolysis of their biomass.[21] In a similar way, D. hansenii, Y. lipolytica and various Candida species affect flavour, texture and colour development in fermented salami style sausages and country cured hams.[15,22] Many breads, especially sour dough varieties, are still produced by traditional fermentation processes where no commercial strains of baker's yeast are added. Although indigenous strains of S. cerevisiae are prominent in many of these fermentations, other yeasts are significant and include Saccharomyces exiguus, Candida milleri, Candida humilis, Candida krusei (Issatchenkia orientalis), Pichia anomala, Pichia membranifaciens and Y. lipolytica. These yeasts grow in cooperation with lactic acid bacteria, giving distinctive flavours to the final product.[23]

High-value cash crops such as cocoa beans and coffee beans also undergo processes that involve the action of yeasts.[24] Coca beans must be fermented to generate the precursors of chocolate flavour, and various species of Saccharomyces, Hanseniaspora, Candida, Issatchenkia and Pichia contribute to the process.[14,25] Coffee beans are processed to remove pulp and other mucilaginous materials that surround the seeds, and species of Candida, Saccharomyces, Kluyveromyces, Saccharomycopsis, Hanseniaspora, Pichia and Arxula have been associated with these fermentations.[26] A vast array of traditional fermented foods and beverages are produced in African, Asian and South American countries from raw materials such as maize, wheat, cassava, rice, soy beans and fruit. Fermentation is essential in contributing to the quality, safety and nutritional value of these products. Aspects of their microbial ecology are just starting to emerge, and demonstrate important contributions from numerous yeast species.[27,28*]

Collectively, the ecological studies of yeasts in products other than beer, bread and wine are providing the knowledge base for developing a new generation of yeast starter cultures, beyond S. cerevisiae.

Microbial Interactions and Biocontrol

Yeasts rarely occur in food and beverage ecosystems as single cultures. Exceptions occur in highly processed products where spoilage outbreaks by single, well-adapted species are known: for example, Z. rouxii in high sugar products.[29]

Generally, most habitats are comprised of a mixture of yeasts, bacteria, filamentous fungi and their viruses, and product quality is determined by the interactive growth and metabolic activity of the total microflora. Even within yeasts themselves, there can be significant species and strain interactions that impact on the population dynamics of the ecosystem. The diversity and complexity of these microbial interactions is just beginning to emerge.[11,30,31]

A network of yeast–yeast interactions occurs in most ecosystems, and is observed in fermentations of wine, cheese, meat, and cocoa beans. These interactions manifest themselves as the successive growth and death of different yeast species and strains within each species, as the fermentation progresses. The mechanisms underlying these ecological shifts are numerous. Explanations include the different rates of nutrient transport and uptake by the different species and strains, their sensitivities to metabolic end products (e.g. ethanol), and responses to killer toxins.[11] Cell–cell interactions might also occur through the production of quorum sensing molecules[32] and unexplained spatial phenomena.[33] Defining the metabolic outcomes of these interactions and their impact on product quality remains a greater challenge, as demonstrated by the interactive effects of *S. cerevisiae* and *Saccharomyces bayanus* strains on the chemical composition and flavour of wines.[34]

Interactions between yeast and bacteria are often seen as the inhibitory effects of yeasts on bacteria through ethanol production; however, the relationships are much broader than this. The death and autolysis of yeast cells releases vitamins and other nutrients that stimulate the growth of important flavour-enhancing bacteria, such as the malolactic bacteria in wine fermentations,[11,31] staphylolcocci, micrococci and brevibacteria during cheese maturation,[21] and lactic acid bacteria during sour dough fermentations.[23] Ethanol, produced by yeasts during cocoa bean fermentations, stimulates the growth of acetic acid bacteria that oxidize the ethanol to acetic acid. This acid is essential for killing the cocoa beans (seeds) and triggering endogenous bean metabolism that generates the precursors of chocolate flavour.[24,25] Some yeasts utilize the organic acids that occur in cheeses, fruit products and salad dressings, causing an increase in product pH and growth of spoilage and pathogenic bacteria.[30] Some bacteria are antagonistic towards yeasts. Excessive growth of lactic acid bacteria and acetic acid bacteria on grapes produces acetic acid and other substances that inhibit the growth of yeasts in grape juice, causing stuck or sluggish wine fermentations and loss of process efficiency.[11,31]

Interactions between yeast and fungi have not been widely studied, except in the context of biocontrol. Fungal growth on wine grapes produces substances that inhibit the growth of yeasts during grape juice fermentation.[11] By contrast, some yeasts improve the growth of *Penicillium* spp. during the maturation of cheeses.[35] Several species within the genera *Candida, Pichia, Metschnikowia, Cryptococcus* and *Pseudozyma* have strong antifungal properties mediated through the production of lytic enzymes, toxic proteins, toxic fatty acids and ethyl acetate, and have potential for the biocontrol of fungi. Commercial preparations of some species are now available for the pre- and post-harvest control of fruit, vegetable and grain spoilage fungi.[36,37]

Yeasts and Food Safety

As part of daily life, humans consume large populations of yeasts without adverse impact on their health. Unlike bacteria and viruses, yeasts are rarely associated with outbreaks of food-borne gastroenteritis, intoxications or other infections. Nevertheless, caution is needed, and further research on this topic is required.[38]

Significant 'lay' literature connects the dietary intake of yeasts with a range of gastrointestinal, respiratory, skin, migraine and even psychiatric disorders. Overgrowth of yeasts in the gastrointestinal tract might contribute to the development of these disorders, but immune reactions to yeast cell wall polysaccharides and responses to yeast-produced amines and sulfur dioxide could also occur. The connection between yeast, the human response and food is largely based on dietary observations. If foods suspected to contain yeasts or their products are removed from the diet, the adverse responses disappear, but return when such foods are reintroduced.[38,39]

Yeasts are not aggressive, infectious organisms, but some species such as *Candida albicans* and *Cryptococcus neoformans* are opportunistic pathogens that cause a range of muco-cutaneous, cutaneous, respiratory, central nervous system and organ infections, as well as general fungemia.[40] Individuals with weakened health and immune systems are at greatest risk, and include cancer, AIDS and hospitalized patients, and those undergoing treatment with immunosuppressive drugs, broad spectrum antibiotics and radio-chemotherapies. The greater frequency of such individuals in the community has led to increased reporting of yeast infections. Moreover, an increasing number of yeast species has been implicated, including many found in foods (e.g. *S. cerevisiae, C. krusei, C. famata, P. anomola, Rhodotorula* spp.[38,41] Infections caused by *S. cerevisiae* are notable because of its extensive use in the food industry, and infections with this yeast have been reported in immunocompetent individuals.[42,43] It is thought that hospitalized patients become exposed to high levels of yeasts through the biofilms they form on catheters and other invasive devices, and that these yeasts probably originate from the hands of hospital workers and the foods brought into the hospital environment.[38] More research is needed to establish stronger linkages between the role of foods in contributing to yeast infections. Information is needed on the survival and growth of yeasts throughout the gastrointestinal system, the potential for yeasts to translocate from the gastrointestinal tract to the blood system, and the general occurrence of yeasts in the hospital and health care environments. The circumstances whereby a nonpathogenic yeast, such as *S. cerevisiae,* becomes pathogenic also require investigation.

Probiotic and Other Health Benefits

Probiotics are viable microorganisms that are beneficial to consumers when ingested in appropriate quantities. Although certain species of lactic acid bacteria are prominent as probiotic organisms, there is increasing interest in yeasts as probiotics.[38,44,45] *S. cerevisiae* var *boulardii* has been used for many years as an oral biotherapeutic agent for treating a range of diarrheal disorders. This species colonizes the intestinal tract where, in a probiotic function, it combats diarrhoea-causing bacteria.[44*,46] Food carrier systems for this yeast need to be developed for its commercial application as a probiotic, but technical obstacles have been encountered. When incorporated into some products, it caused gassy, ethanolic spoilage and off-flavours.[47,48] Of greater concern are reports of fungemia infections caused by *S. boulardii.*[42**,43] Other yeasts mentioned as potential probiotics include *D. hansenii, Kluy. marxianus, Y. lipolytica, I. orientalis, P. farinosa* and *P. anomala,* but further research is required.[38*] Yeasts are increasingly used as probiotics in the livestock and aquaculture industries.[38*]

Yeast products, principally derived from *S. cerevisiae,* have been used for many years as ingredients and additives in food processing. These products include flavourants, enzymes, antioxidants, vitamins, colourants and polysaccharides.[49,50] Three points are worthy of mention. First, many of these products are prepared from yeast cells after they have been processed by autolysis. Despite its commercial significance, molecular understanding of yeast autolysis is still very limited and more research is needed to optimize this process.[51,52] Second, most products are derived from *S. cerevisiae.* The yield and range of products could be increased by screening for their presence in other yeast species and strains, as demonstrated for the vitamin folic acid,[53] cell wall polysaccharides[54] and autolysates.[55] Finally, there remains undiscovered bioactivity and functionality in yeast products. Whereas the glucan polysaccharides from the walls of *S. cerevisiae* were originally valued for their water-binding and rheological functionalities, it is now recognized that they can stimulate the immune system, lower serum cholesterol, exhibit antitumour activity, and adsorb substances such as mycotoxins.[38*,49]

Conclusions

Advances in molecular technologies have provided new analytical tools for studying the diversity and biological activities of yeasts associated with food and beverage production, although more research is still required on the ecology and activities of yeasts in products other than beer, bread and wine. The interactions between yeasts and the ecosystems in which they occur provide another area for future study; yeasts form interactions with other species and strains, along with bacteria, other fungi, protozoans and their viruses, but as yet these relationships remain poorly described and understood. Interest in the public health significance of yeasts in foods and beverages is also increasing, in both positive and negative contexts. Again, we are likely to see future developments in this regard.

Update

Debaryomyces hansenii is one of the most significant yeasts in food and beverage production, and this is highlighted in a recent review of its phylogeny, ecology, physiology, molecular biology and its biotechnological potential.[56] As mentioned in the conclusion, yeast interactions between themselves and with other

organisms have implications for food quality and safety, and further research is needed on these topics. Aspects of yeast cell interactions have been considered in a recent review that discusses their underlying molecular mechanisms, how they impact on growth and survival and how they affect pathogenicity.[57]

References and Recommended Reading

Papers of particular interest, published within the annual period of review, have been highlighted as:

• of special interest

•• of outstanding interest

1. Boekhout T, Robert V (Eds): *Yeasts in Food. Beneficial and*
•• *Detrimental Aspects.* Behr's Verlag; 2003.
  Comprehensive discussions of yeasts in foods and beverages—an emphasis is placed on commodities.

2. Querol A, Fleet GH (Eds): *Yeasts in Food and Beverages.*
•• Springer; 2006.
Comprehensive discussions of yeasts in foods and beverages—emphasis on ecology and biology of yeasts.

3. Blackburn C (Ed): *Food Spoilage Microorganisms.* CRC Press; 2006.

4. Beh AL, Fleet GH, Prakitchaiwattana C, Heard GM: Evaluation
• of molecular methods for the analyses of yeasts in foods and beverages. In *Advances in Food Mycology.* Edited by Hocking AD, Pitt JT, Samson RA, Thrane U. Springer; 2006:69–106.
Reviews and lists recent literature on molecular methods used for the analysis of yeasts in foods and beverages.

5. Fernandez-Espinar JT, Martorell P, de Llanos R, Querol A: Molecular methods to identify and characterise yeasts in foods and beverages. In *Yeasts in Food and Beverages.* Edited by Querol A, Fleet GH. Springer; 2006:55–82.

6. Kurtzman CP, Fell JW: Yeast systematics and phylogeny—
• implications of molecular identification methods for studies in ecology. In *Biodiversity and Ecophysiology of Yeasts.* Edited by Rosa CA, Peter G. Springer; 2006:11–30.
Outlines the most recent changes to yeast classification and taxonomy, based on DNA sequencing and phylogenetic analyses.

7. Rawsthorne H, Phister T: A real-time PCR assay for the enumeration and detection of *Zygosaccharomyces bailii* from wine and fruit juices. *Int J Food Microbiol* 2006, **112:**1–7.

8. Page BT, Kurtzman CP: Rapid identification of *Candida* species and other clinically important yeast species by flow cytometry. *Appl Environ Microbiol* 2005, **43:**4507–4514.

9. Fleet GH: *Saccharomyces* and related genera. In *Food Spoilage Microorganisms.* Edited by Blackbrun C. CRC Press; 2006:306–335.

10. Schuller D, Valero E, Dequin S, Caseal M: Survey of molecular methods for the typing of wine yeast strains. *FEMS Microbiol Lett* 2004, **231:**19–26.

11. Fleet GH: Yeast interactions and wine flavour. *Int J Food Microbiol* 2003, **86:**11–22.

12. Romano P, Capece A, Jespersen L: Taxonomic and ecological
• diversity of food and beverage yeasts. In *Yeasts in Food and Beverages.* Edited by Querol A, Fleet GH. Springer; 2006:13–53.
Good overview of diversity and roles of yeasts in fermented foods and beverages.

13. Prakitchaiwattana J, Fleet GH, Heard GM: Application and
• evaluation of denaturing gradient gel electrophoresis to analyse the yeast ecology of wine grapes. *FEMS Yeast Res* 2004, **4:**865–877.

Provides a critical discussion of merits and limitations of the use of DGGE for analysing yeasts in foods and beverages.

14. Nielsen DS, Hanholt S, Tano-Debrah K, Jespersen L: Yeast populations associated with Ghanaian cocoa fermentation analysed using denaturing gradient gel electrophoresis (DGGE). *Yeast* 2005, **22**:271–284.

15. Cocolin L, Urso R, Rantsiou K, Cantoni C, Comi G: Dynamics and characterisation of yeasts during natural fermentation of Italian sausages. *FEMS Yeast Res* 2006, **6**:692.

16. Swiegers JH, Bartowsky EJ, Henschke PA, Pretorius IS:
•• Yeast and bacterial modulation of wine aroma and flavour. *Aust J Grape Wine Res* 2005, **11**:139–173.

Comprehensive, well illustrated review of the biochemical production of flavour and aroma compounds by microorganisms in foods and beverages.

17. Bond U, Blomerg A: Principles and applications of genomics
• and proteomics in the analysis of industrial yeast strains. In *Yeasts in Food and Beverages*. Edited by Querol A, Fleet GH. Springer; 2006:173–213.

Good, basic introduction to yeast genomics and its applications in food and beverage fermentations.

18. Varela C, Cardenas J, Melo F, Agosin E: Quantitative analysis of wine yeast gene expression profiles under winemaking conditions. *Yeast* 2005, **22**:369–383.

19. Brejning J, Arneborg N, Jespersen L: Identification of genes and proteins induced during the lag and early exponential phase of lager brewing yeasts. *J Appl Microbiol* 2005, **98**:261.

20. Tanghe A, Prior B, Thevelein JM: Yeast responses to stress. In
• *Biodiversity and Ecophysiology of Yeasts*. Edited by Rosa CA, Peter G. Springer; 2006:175–195.

Good, general review of the biology and practical significance of the stress responses in yeasts.

21. Addis E, Fleet GH, Cox JMC, Kolak D, Leung T: The growth, properties and interactions of yeasts and bacteria associated with the maturation of Camembert and blue-veined cheeses. *Int J Food Microbiol* 2001, **69**:25–36.

22. Samelis J, Sofos JN: Yeasts in meat and meat products. In *Yeasts in Food—Beneficial and Detrimental Aspects*. Edited by Boekhout T, Robert V. Behr's Verlag; 2003:239–265.

23. De Vuyst LD, Neysens P: The sourdough microflora: biodiversity and metabolic interactions. *Trends Food Sci Technol* 2005, **16**:43–56.

24. Schwan R, Wheals AE: Mixed microbial fermentations of chocolate and coffee. In *Yeasts in Food—Beneficial and Detrimental Aspects*. Edited by Boekhout T, Robert V. Behr's-Verlag; 2003:429–449.

25. Ardhana M, Fleet GH: The microbial ecology of cocoa bean fermentations in Indonesia. *Int J Food Microbiol* 2003, **86**:87–99.

26. Masoud W, Cesar LB, Jespersen L, Jakobsen M: Yeasts involved in fermentation of Coffee arabica in East Africa, determined by genotyping and by direct denaturing gradient gel electrophoresis (DGGE). *Yeast* 2004, **21**:549–556.

27. Aidoo KE, Nout MJR, Sarkar PK: Occurrence and function of yeasts in Asian indigenous fermented foods. *FEMS Yeast Res* 2006, **6**:30–39.

28. Nout MJR: Traditional fermented products from Africa,
• Latin Amercia and Asia. In *Yeasts in Food—Beneficial and Detrimental Aspects*. Edited by Boekhout T, Robert V. Behr's-Verlag; 2003:451–473.

Demonstrates the diversity and significance of yeasts in many products little known to western consumers.

29. Stratford M: Food and beverage spoilage yeasts. In *Yeasts in Food and Beverages*. Edited by Querol A, Fleet GH. Springer; 2006:335–380.

30. Viljoen B: Yeast ecological interactions. Yeast-yeast, yeast-bacteria, yeast-fungi interactions and yeasts as biocontrol agents. In *Yeasts in Food and Beverages*. Edited by Querol A, Fleet GH. Springer; 2006:83–110.

31. Alexandre H, Costello PJ, Remize F, Guzzo J, Guilloux-Benatier M: *Saccharomyces cerevisiae–Oenococcus oeni* interactions in wine: current knowledge and perspectives. *Int J Food Microbiol* 2004, **93**:141–154.

32. Hogan DA: Quorum sensing: alcohols in a social situation.
• *Curr Biol* 2006, **16**:R457–R458.

Novel discussion of the concept of quorum sensing and its mechanisms in yeast biology.

33. Arneborg N, Siegumfeldt H, Andersen GH, Nissen P, Daria VR,
• Rodrigo PJ, Gluckstad J: Interactive optical trapping shows that confinement is a determinant of growth in a mixed yeast culture. *FEMS Microbiol Lett* 2005, **245**:155–159.

Novel use of laser optical technology to demonstrate that spatial phenomena might affect yeast cell–cell interactions.

34. Howell K, Cozzolino D, Bartowsky E, Fleet GH, Henschke PA: Metabolic profiling as a tool for revealing *Saccharomyces* interactions during wine making. *FEMS Yeast Res* 2006, **9**:91–101.

35. Hansen TK, van der Tempel T, Cantor MD, Jakobsen M: *Saccharomyces cerevesiae* as a starter culture in mycelia. *Int J Food Microbiol* 2001, **69**:101–111.

36. Fleet GH: Yeasts in fruit and fruit products. In *Yeasts in Food—Beneficial and Detrimental Aspects*. Edited by Boekhout T, Robert V. Behr's-Verlag; 2003:267–287.

37. Passoth V, Fredlund E, Druvefors UA, Schnurer J: Biotechnology, physiology and genetics of the yeast *Pichia anomala*. *FEMS Yeast Res* 2006, **6**:3–13.

38. Fleet GH, Balia R: The public health and probiotic signficance of
• yeasts in foods and beverages. In *Yeasts in Food and Beverages*. Edited by Querol A, Fleet GH. Springer; 2006:381–398.

First major review of positive and negative public health issues relating to yeasts in foods and beverages.

39. Eaton TK: Moulds, yeasts, ascospores, basidiospores, algae and lichens: toxic and allergic reactions. *J Nutrit Environ Med* 2004, **14**:187–201.

40. Hazen KC, Howell SA: *Candida, Cryptococcus* and other yeasts of medical importance. In *Manual of Clinical Microbiology* 8th *edition*. Edited by Murray PR. American Society for Microbiology; 2003:1693–1711.

41. Hobson RP: The global epidemiology of invasive *Candida* infections—is the tide turning? *J Hosp Infect* 2003, **55**:159–168.

42. Enache-Angoulvant A, Hennequin C: Invasive *Saccharomyces*
•• infections: a comprehensive review. *Clin Inf Dis* 2005, **41**:1559–1568.

A thorough review and discussion of human infections caused by *S. cerevisiae*—an industrial yeast not normally considered to be a risk to human health.

43. de Llanos R, Querol A, Peman J, Gobernado M, Fernandez-Espinar MT: Food and probiotic strains from the *Saccharomyces cerevisiae* species as a possible origin of human systemic infections. *Int J Food Microbiol* 2006, **110**:286–290.

44. van der Aa Kuhle A, Skovgaard K, Jespersen L: *In vitro*
• screening of probiotic properties of *Saccharomyces cerevisae* var *boulardii* and foodborne *Saccharomyces cerevisiae* strains. *Int J Food Microbiol* 2005, **101**:29–40.

Provides a good discussion of issues related to the use of yeasts as probiotic organisms.

45. Sullivan A, Nord CE: The place of human probiotics in human intestinal infections. *Int J Antimicrob Agents* 2003, **20**:313–319.

46. Czervoka D, Rampal P: Experimental effects of *Saccharomyces boulardii* on diarrheal pathogens. *Microbes Infect* 2002, **4**:733–739.

47. Lourens-Hattingh A, Viljoen BC: Growth and survival of probiotic yeast in dairy products. *Food Res Int* 2001, **34:**791–796.

48. Heenan CN, Adams MC, Hosken RW, Fleet GH: Survival and sensory acceptability of probiotic microorganisms in a non-fermented frozen, vegetarian dessert. *Lebensm Wiss Technol* 2004, **37:**461–466.

49. Dawson KA: Not just bread or beer: new applications for yeast and yeast products in human health and nutrition. In *Nutritional Biotechnology in the Feed and Food Industry.* Edited by Lyons TP, Jaques FA. Nottingham University Press; 2002:225–232.

50. Abbas CA: Production of antioxidants, aromas, colours, flavours and vitamins by yeasts. In *Yeasts in Food and Beverages.* Edited by Querol A, Fleet GH. Springer; 2006:285–334.

51. Zhao J, Fleet GH: Degradation of RNA during the autolysis of *Saccharomyces cerevisiae* produces predominantly ribonucleotides. *J Ind Microbiol Biotechnol* 2005, **32:**415–423.

52. Alexandre H, Guilloux-Benatier M: Yeast autolysis in sparkling wine—a review. *Aust J Grape Wine Res* 2006, **12:**119–217.

53. Hjortmo S, Patring J, Jastrebova J, Andlid T: Inherent biodiversity of folate content and composition in yeasts. *Trends Food Sci Technol* 2005, **16:**311–316.

54. Nguyen TH, Fleet GH, Rogers PL: Composition of the cell wall of several yeast species. *Appl Microbiol Biotechnol* 1998, **50:**206–212.

55. Lukondeh T, Ashbolt NJ, Rogers PL: Evaluation of *Kluyveromyces marxianus* as a source of yeast autolysates. *J Ind Microbiol Biotechnol* 2003, **30:**52–56.

56. Breuer U, Harms H: *Debaryomyces hansenii*— an extremophilic yeast with biotechnological potential. *Yeast* 2006, **23:**415–437.

57. Palkova Z, Vachova L: Life within a community: benefit to yeast long term survival. *FEMS Microbiol Rev* 2006, **30:**806–824.

The Microbiology of Cocoa Fermentation and Its Role in Chocolate Quality

ROSANE F. SCHWAN AND ALAN E. WHEALS

1. Introduction

1.1. Cocoa and Chocolate

Probably originating in Mesoamerica,[1] chocolate or cacao had already been used as a food, a beverage, and as medicine for over 2,000 years before Hernando Cortés brought it to Europe in 1528.[2,3] Its special status in human culture is reflected in its Latin name with genus *Theobroma,* meaning food of the gods. The specific name, *cacao,* probably originated as an Olmec word from Mexico.[2] The principal varieties are Criollo, now rarely grown because of its disease susceptibility, Forastero from the Amazonas region, and a hybrid, Trinitario, the latter two forming most of the "bulk" market. The Arriba type, with a "fine" flavor, is grown in Ecuador. World annual production is approximately 2.5M tonnes and the major producers are the Ivory Coast, Ghana, Indonesia, Brazil, Nigeria, Cameroon, Malaysia, and Ecuador, but there are many other smaller producers, particularly of "fine" cocoa, which constitutes about 5% of world trade. Firms that make chocolate almost exclusively are Mars, Hershey, and Rowntree-Mackintosh, but other important companies are the beverage conglomerate, Jacobs-Suchard, and several multi-nationals such as Nestlé, Cadbury-Schweppes, Philip Morris, Unilever, and Zareena.

Trade in cocoa is complex: farmers produce fermented beans, warehouses store beans, processors turn this into cocoa products, traders ship to mainly North America and Europe, and manufacturers convert this into consumable products. The "first" world dominates the commodities market that determines the price of cocoa for the "third" world farmers.

After reaching a peak of well over US$3,000/tonne in 1977 the price of roasted beans has fallen to an average about US $1,000/tonne during the last decade. There has been a long battle in Europe to prevent chocolate products that contain only approximately 20% (w/w) cocoa solids being called chocolate but a compromise has been reached with terms such as "Family milk chocolate" being legally permitted (EU Directive 2000/36/ EC). This revision of the 1973 Council Directive (73/241/EEC) permits up to 5% non-cocoa vegetable fat to be used in the manufacture of chocolate throughout the EU. It would probably result in a loss in demand for cocoa beans exceeding 184,000 tonnes. If there were worldwide adoption, the loss of revenue to cocoa producers could be more than US$1.5bn.

1.2. Fermentation

Mature fruits (pods) rise directly from the stem of the cocoa tree and are thick walled and contain 30–40 beans (seeds). Each bean consists of two cotyledons and an embryo (radicle) surrounded by a seed coat (testa) and is enveloped in a sweet, white, mucilaginous pulp that comprises approximately 40% of seed fresh weight. A microbial fermentation and drying process is required to initiate the formation of the precursors of cocoa flavor.[4] Harvested seeds are immediately allowed to undergo a natural fermentation during which microbial action on the mucilaginous pulp produces ethanol and acids as well as liberating heat. A schematic of a microbial succession summarizes the key events during the process that occurs during cocoa fermentations in Bahia, Brazil. Diffusion of these metabolites triggers complex biochemical reactions to occur in the cotyledons. The testa provides a barrier to acid penetration into the bean and diffusion out of undesirable theobromine, caffeine, and polyphenols. The seed embryo is killed and the fruit tissues degrade which makes it much easier to dry the beans. This can be done in the sun (using movable roofs to protect from tropical showers) with regular turning until the water content is less than 8%, which takes from one to four weeks. Alternatively, artificial dryers are used but it is important to keep the temperature not exceeding 60°C and to dry slowly (at least 48 hours) during which time some excess acids may volatilize and some oxidation will occur, both of which are beneficial. The beans can then be stored for up to a year but staling will eventually occur. At this stage the cut beans show a purple color due to the presence of anthocyanins.

1.3. Processing

The next stage is to roast the beans from 5 to 120 minutes and from 120°C to 150°C depending on nature of the beans and the required product. There may also be a pre-roast and thermal shock (to loosen the husk). During this process the cracked husks are air-separated (winnowing) from the entire separated cotyledons (nibs), which undergo a further series of chemical reactions leading to the development of full chocolate flavor. The roasted beans are then processed into chocolate. The nibs are ground several times at elevated temperatures to make a fluid paste (cocoa liquor) that on cooling yields cocoa mass, a dark bitter material with astringent flavors from the polyphenols and tannins. Typically 2/3 of this material is then pressed to separate cocoa butter, a pale yellow, fatty liquid without any cocoa flavor, and cocoa (press) cake, a dark brown residue (58% of the total). The cocoa cake will then be ground to cocoa powder for use by the confectionery and other industries. Cocoa cake is a strongly flavored but inedible material that needs further processing to become palatable. To make finished chocolate products, including confectionery, most of the cocoa butter is mixed back with the cocoa mass (liquor) together with sugar, sweeteners, milk products, emulsifiers, and cocoa butter substitutes depending on the requirements of the final product. For the finest chocolate, "conching" is performed in order to get fine crystallization. The chocolate is typically heated to between 50 and 60°C for several hours, although it can be up to 5 days for specialist chocolate, while lecithin is added followed by repeated milder heating and cooling cycles before filling moulds.

Cocoa butter, like all fats, is composed of a mixture of fatty acids and is typically the saturated fatty acids palmitic acid (25%) and stearic acid (35%), the monounsaturated fatty acid oleic acid (35%), and the polyunsaturated fatty acid linoleic acid (3%) with some others (2%). The melting point of cocoa butter is around 35°C with softening around 30–32°C and it becomes brittle fracture below 20°C.

1.4. Health and Nutrition

Most of the health problems associated with high chocolate consumption stem from the high concentration of carbohydrates in processed chocolate rather than the chocolate itself. The basis of its "addictive" properties for chocoholics has not been identified although cannabinoids are found in chocolate at low levels.[5] In Colombia the nutritional role of chocolate is emphasized because natural cane sugar and chocolate are combined into a nutritious beverage with an excellent balance of carbohydrates, lipids, and proteins. Possible medicinal/health benefits of chocolate have been reported for many years but it is only recently that some of these claims are being more clearly identified and studied.[3] Research shows that the cocoa bean and its derived products are rich in specific antioxidants, including catechins and epicatechin, and especially the polymers procyanidins and polyphenols similar to those found in vegetables and tea. Metabolic and epidemiological studies indicate that regular intake of such products increases the plasma level of antioxidants, a desirable attribute as a defense against reactive oxygen species (ROS). The antioxidants in cocoa can prevent the oxidation of LDL-cholesterol, related to the mechanism of protection in heart disease. Likewise, a few studies show that ROS associated with carcinogenic processes is also inhibited.[6] The fats from cocoa butter are mainly stearic triglycerides (C18:0) that are less well absorbed than other fats and tend to be excreted in the feces. Thus, cocoa butter is less bioavailable and has minimal effect on serum cholesterol.[6]

Since the starting material is sterile, the fermentation process creates hot, acid conditions, and the beans are roasted at over 100°C, it is not surprising that there has never been a single report of *Escherichia coli* or *Salmonella* spp. contamination in cocoa mass although some bacilli may survive.[7] Food poisoning organisms rarely have been reported in the final processed chocolate, presumably arising from contamination at a late stage in the factory.[8] The technology of chocolate production effectively limits mycotoxin contamination by moulds that might have occurred in the period at the end of fermentation, during drying or if allowed to get wet during transport and storage. Mycotoxins have been found on shells but never in cotyledons, perhaps because of the inhibitory presence of methylxanthines.

1.5. Chemistry

Chocolate flavors and aromas have been the subject of extensive research. Unfermented cocoa seeds do not produce cocoa flavor on roasting so an understanding of the development of cocoa flavor precursors during fermentation is required. Bitter and astringent flavors are due to polyhydroxyphenols such as catechins, flavan-3-ols, anthocyanins, and proanthocyanadins. Polyphenols tend to diffuse out of the bean during the fermentation and also are oxidized by polyphenol oxidazes to produce mostly insoluble tannins. There is also a loss during drying and roasting.[9] Since there are abundant health claims for polyphenols[10–14] efforts are being made to maintain their levels while avoiding taste problems.[15] Theobromine and caffeine and their complexes are major components of cocoa's bitter taste but they also tend to diffuse out of the bean during fermentation. Endogenous acids (malic, tartaric, oxalic, phosphoric, citric) are probably less important because it is the diffusion of lactic and acetic acids into the bean that dominate bean acidity. In turn they depend on the sugars in the pulp and availability of oxygen for their production by bacteria. Some lactic acid is lost during drying but most of the acetic acid remains. Therefore it is important to ensure that neither the initial conditions nor fermentation and drying produce excess acid.

The source of hundreds of volatiles found in roasted beans (both fermented and unfermented) are the reducing sugars, free amino acids, and oligopeptides. The sugars come from sucrose and its hydrolysis products, glucose and fructose, in addition to being released from glycosides. Most amino acids and oligopeptides are produced during acid hydrolysis that occurs during fermentation. These compounds undergo non-enzymatic browning reactions during drying and roasting. These Maillard reactions are condensations between the α-amino group of amino acids, proteins, or amines and the carbonyl group of reducing sugars. They are quite distinct from caramelization of sugars, which does not involve amino acids. Typical "hammy" off-flavors are produced by over-fermentation when bacilli and filamentous fungi grow on cocoa husks and nibs to produce short chain fatty acids. Smoky off-flavors from wood fires used for drying are now less of a problem.

Chocolate shares with wine the distinction of being an ancient fermented product with a combination of nutritional, medicinal, and mystical properties. The global improvement in wine quality over the last 25 years has been significantly due to better control of the fermentation process itself. The purpose of this review is to describe research over the last 15 years into the fermentation process and discoveries on how cocoa fermentation is involved in production of chocolate flavor precursors. If implemented, this knowledge will enable high quality natural chocolate to be routinely produced and, perhaps yield better financial returns for farmers.

2. The Fermentation Process
2.1. Cocoa Pulp: The Fermentation Substrate

Cocoa pulp is a rich medium for microbial growth. It consists of 82–87% water, 10–15% sugar, 2–3% pentosans, 1–3% citric acid, and 1–1.5% pectin.[16] Proteins, amino acids, vitamins (mainly vitamin C), and minerals are also present. The concentration of glucose, sucrose, and fructose is a function of fruit age.[17] More glucose and fructose and a slight increase in total sugar concentration were observed in samples 6 days after harvest than in freshly harvested (ripe) pods.[18] In a comparative analysis of pulp from beans collected in the Ivory Coast, Nigeria, and Malaysia, differences were found in the amounts of water, citrate, hemicellulose, lignin, and pectin.[19] Pectin content, approximately 1% on a fresh weight basis, was found to 37.5 and 66.1 g kg^{-1} dry weight pulp.

Seeds within the ripe pod are microbiologically sterile. When the pod is opened with a knife, the pulp becomes contaminated with a variety of microorganisms many of which contribute to the subsequent fermentation. Organisms come mainly from the hands of workers, knives, unwashed baskets used for transport of seeds, and dried mucilage left on the walls of boxes from previous fermentations.

2.2. Microbial Fermentation

On small-holdings, fermentations are often done in heaps of beans from about 25 kg to 2000 kg enclosed by banana or plantain leaves with some turning to assist aeration. Baskets, lined and covered with leaves, are also used. In larger farms fermentations are performed in large, perforated wooden boxes allowing pulp to drain away and air to enter. Although they can hold up to 2000 kg of beans the depth does not exceed 50 cm to ensure good aeration. The beans are covered with banana leaves or sacking to conserve the heat generated during fermentation. To ensure uniform fermentation and increase aeration, beans are manually turned up to once per day. Some are tiered on slopes that facilitates transfer of beans from one box to a lower one with simultaneous aeration. Plantations usually ferment for a longer period than small-holders and 6 to 7 days is usual.

Changes in the local climatic conditions influence the sequence of microorganisms involved in cocoa fermentation but a similar succession of groups of organisms has often been reported.[20,21,22] The microbial succession in the fermentation process has been clearly established.[16,20–24] Early on in the fermentation, several species of yeasts proliferate, leading to production of ethanol and secretion of pectinolytic enzymes.

This is followed by a phase in which bacteria appear, principally lactic-acid bacteria and acetic-acid bacteria, which is followed by growth of aerobic spore-forming bacteria. Finally, some filamentous fungi may appear on the surface.

The initial acidity of the pulp (pH 3.6), due to citric acid, together with low oxygen levels, favor colonization by yeasts[25] that are able to utilize pulp carbohydrates under both aerobic and anaerobic conditions. The size of the yeast population increases from 10^7 CFU/g of pulp to 10^8 CFU/g of pulp during the first 12 h, then remains almost constant for the next 12 h after which there is a dramatic decline of four orders of magnitude over the next day followed by a slower decrease leading to a final population of only 10 viable cells per gram of pulp.[22]

The amended conditions favor the development of lactic-acid bacteria. The number of these organisms reaches a peak around 36 hours after the fermentation process begins and the bacterial population reached 6.4×10^7 CFU/g of pulp. This period of time is coincident with the decline of the yeast population.[22,24] The lactic acid bacteria exhibit the fastest growth rate during the 16–48 h period of fermentation and are present in greater numbers, but not necessarily in biomass, than yeasts for a short period of time.[22] As aeration of the fermenting mass increases and the temperature rises above 37°C, acetic acid bacteria became the dominant organisms, and the population reached a peak at 88 hours with 1.2×10^7 CFU/g of pulp.[22] This stage in the microbial succession is reflected in a decline in the concentration of ethanol and lactic acid, and increase in acetic acid. The exothermic reactions of acetic-acid bacteria raise the temperature of the fermenting mass even further up to 50°C or more. The decrease in the number of acetic-acid bacteria from three days onwards is probably due to their inhibition by the high temperature in the cocoa mass. The strong odor of acetic acid, evident from 48 to 112 h, decreases progressively towards the end of the fermentation. After 120 hours of fermentation acetic acid bacteria were not found. There is a minor increase in the number of yeasts to 3.5×10^3 CFU/g of pulp[22] around 132–160 hours. This is due to growth of thermotolerant yeasts utilizing some of the acids coinciding with an increase in the oxygen content in the fermenting mass[22] as well as survivors in the cooler external layers of the fermentation.

Aerobic, spore-forming bacteria can be isolated during the first three days of fermentation with populations around 10^4 CFU/g of pulp but their numbers remain virtually unchanged. Thereafter they start to dominate the microbial population to such an extent that they form over 80% of the microflora,[22,25,26] reaching 5.5×10^7 CFU/g of pulp.[22] This phase in the succession coincides with increases in oxygen tension, temperature, and pH of the fermenting mass. Filamentous fungi are found in small numbers throughout the fermentation, most commonly in the aerated and cooler, superficial areas of the fermenting mass. At the end of the fermentation the beans are usually transferred to platforms and sun-dried. During this process, commencing after 156 hours, there is a sharp decrease in the total microbial population. During sun drying cocoa beans are often humidified to help the workers remove the rest of the mucilage with their feet but eventually only microorganisms that are able to form spores, bacilli, and filamentous fungi can survive.

Table 1   Yeasts Isolated from Cocoa Fermentations in Four Countries

Brazil[22]	Ghana[32]	Malaysia[32]	Belize[100]
Candida bombi, Candida pelliculosa, Candida rugopelliculosa, Candida rugosa, Kloeckera apiculata, Kluyveromyces marxianus, Kluyveromyces thermotolerans, Lodderomyces elongisporus, Pichia fermentans, S. cerevisiae var. chevalieri, Saccharomyces cerevisiae, Torulaspora pretoriensis	Candida spp., Hansenula spp., Kloeckera spp., Pichia spp., Saccharomyces spp., Saccharomycopsis spp., Schizosaccharomyces spp., Torulopsis spp.	Candida spp., Debaryomyces spp., Hanseniaspora spp., Hansenula spp., Kloeckera spp., Rhodotorula spp., Saccharomyces spp., Torulopsis spp.	Brettanomyces clausenii., Candida spp., C. boidinii, C. cacoai, C. guilliermondii, C. intermedia, C. krusei, C. reukaufii, Kloeckera apis, Pichia membranaefaciens, Saccharomyces cerevisiae, Saccharomyces chevalieri, Saccharomycopsis spp., Schizosaccharomyces malidevorans, Schizosaccharomyces spp.

Table 2   Lactic Acid Bacteria Isolated from Cocoa Fermentations in Four Countries

Brazil[29]	Ghana[32]	Malaysia[32]	Belize[100]
Lactobacillus. Acidophilus, Lb. brevis, Lb. casei, Lb. Delbrueckii, Lb. fermentum Lb. Lactis, Lb. Plantarum Lactococcus lactis, Leuconostoc mesenteroides, Pediococcus acidilactici, P. dextrinicus	Lb. collinoides Lb. fermentum Lb. mali Lb. plantarum	Lb. collinoides, Lb. plantarum	Lb. brevis, Lb. buchneri, Lb. casei, Lb. Casei pseudoplantarum, Lb. cellobiosus, Lb. delbrueckii, Lb. fermentum, Lb. fructivorans, Lb. gasseri, Lb. kandleri, Lb. plantarum, Leuconostoc mesenteroides, Ln. oenos, Ln. paramesenteroides

2.3. Yeasts

Yeasts have been isolated from cocoa fermentations by many groups[23] but only four studies have simultaneously identified yeasts and bacteria (Table 1). To avoid confusion the names used in the original literature have been retained but current nomenclature is given in the appendix. Other studies have identified isolates of the genera *Candida, Pichia, Saccharomyces, Kloeckera, Trichosporon,* and *Schizosaccharomyces* in Java;[16] *Kloeckera apis, Candida pelliculosa, Candida tropicalis,* and *Saccharomyces cerevisiae* in Indonesia;[27] and *Pichia membranaefaciens, Saccharomyces cerevisiae, Candida zeylanoides, Torulopsis candida, T. castelli,* and *T. holmii* in the Ivory Coast.[28] It is not possible to determine whether these differences in the yeast flora were due to geography or to fermentation practices. In the most comprehensive study[22] (Table 1), frequency of species with time was also monitored in detail. *Saccharomyces cerevisiae* was the dominant yeast in the cocoa beans taken from boxes immediately after filling. *Kloeckera apiculata* grew during the early phase of fermentation but declined rapidly such that it could not be isolated after 24 h of fermentation which probably reflects its intolerance of ethanol at concentrations above 4% (v/v).[22] *Kluyveromyces marxianus* grew slowly at the outset of fermentation and then declined gradually. Two different strains of *S. cerevisiae* dominated the alcoholic fermentation phase and survived throughout the fermentation process. Small numbers of *Pichia fermentans* and *Lodderomyces ellongisporus* were isolated but only during the first few hours of fermentation. *Candida* spp. increased in numbers after 24 h. *Candida rugosa* was present up to the end

of fermentation when the temperature was approximately 50°C. *Torulospora pretoriensis* and *Kluyveromyces thermotolerans* were found also when the temperature of the fermenting mass was approximately 50°C. The yeast flora was abundant and varied, which is not surprising since cocoa bean pulp contains, on average, 14% of sugars. Of these, 60% is sucrose and 39% a mixture of glucose and fructose.[18] All these sugars are fermented by the above species, but even so, *S. cerevisiae* was the most common species of yeast identified in the study probably because of its rapid growth and ethanol-tolerance. It was also found in high numbers during the first 24 h of cocoa fermentation in Trinidad.[21] *Kluyveromyces marxianus, K. thermotolerans, Candida* spp, and *Torulospora pretoriensis,* which were present in considerable numbers in the Brazilian study, have not been reported from cocoa bean fermentations in other countries.[21,23]

2.4. Bacteria
A. Lactic-Acid Bacteria

Lactic-acid bacteria increased in numbers when part of the pulp and "sweatings" had largely drained away, and the yeast population was declining. Yeast metabolism favors the growth of acidoduric, lactic-acid bacteria. Of the lactic acid bacteria isolated from cocoa fermentations[21] (Table 2), *Lactobacillus fermentum, Lb. plantarum, Leuconostoc mesenteroides,* and *Lactococcus (Streptococcus) lactis* were the most abundant species in the first 24 h of fermentation. In Bahia (Brazil), six *Lactobacillus* spp. and two species of the genus *Pediococcus* together with *Lactococcus lactis* and *Leuconostoc mesenteroides* were isolated[29]

Table 3   Acetic Acid Bacteria Isolated from Cocoa Fermentations in Four Countries

Brazil[31]	Ghana[32]	Malaysia[32]	Belize[100]
Acetobacter aceti subsp. liquefaciens, A. pasteurianus, A. peroxydans, Gluconobacter oxydans subsp. suboxydans	*Acetobacter ascendens, A. rancens, A. xylinum, Gluconobacter oxydans*	*Acetobacter lovaniensis, A. rancens, A. xylinum, Gluconobacter oxydans*	*Acetobacter spp., Gluconobacter oxydans*

(Table 2). In general, the *Lactobacillus* spp. were present at the early stages whereas *Lactococcus* spp. occurred during the final stages of fermentation. Lactic acid bacteria were isolated in cocoa fermentation in Indonesia and *Lactobacillus plantarum* and *Lactobacillus cellobiosus* were the principal species.[27]

B. Acetic-Acid Bacteria

After the decline in the populations of yeasts and lactic-acid bacteria, the fermenting mass becomes more aerated. This creates conditions suitable for the development of acetic-acid bacteria. These bacteria are responsible for the oxidation of ethanol to acetic acid and further oxidation of the latter to carbon dioxide and water. The acidulation of cocoa beans and the high temperature in the fermenting mass, which causes diffusion and hydrolysis of proteins in the cotyledons, has been attributed to the metabolism of these organisms. Thus the acetic acid bacteria play a key role in the formation of the precursors of chocolate flavor.[30] In general, the members of genus *Acetobacter* were found more frequently than those of *Gluconobacter* (Table 3).[31] Species of *Acetobacter aceti* and *Acetobacter pasteurianus* were isolated in Indonesia but the populations were only approximately 10^5 to 10^6 CFU/g.[27]

C. Aerobic Spore-Forming Bacteria

Increased aeration, increased pH value (3.5 to 5.0) of cocoa pulp, and a rise in temperature to about 45°C in the cocoa mass in the later stages of fermentation are associated with the development of aerobic spore-forming bacteria of the genus *Bacillus*[21,26,32] (Table 4). Many *Bacillus* spp. are thermotolerant and others grow well at elevated temperatures. *B. stearothermophilus, B. coagulans,* and *B. circulans* were isolated from cocoa beans that had been subjected to drying and roasting (150°C) temperatures.[7]

Aerobic spore-forming bacteria produce a variety of chemical compounds under fermentative conditions. These may contribute to the acidity and perhaps at times to the off-flavors of fermented cocoa beans. Indeed it has been suggested that C_3–C_5 free fatty acids found during the aerobic phase of fermentation and considered to be responsible for off-flavors of chocolate[34] are produced by *B. subtilis, B. cereus,* and *B. megaterium.* Other substances such as acetic and lactic acids, 2,3-butanediol, and tetramethylpyrazine, all of which are deleterious to the flavor of chocolate, are also produced by *Bacillus* spp.[34,35]

2.5. Filamentous Fungi

Filamentous fungi are not considered to be an important part of the microbial succession of cocoa fermentation.[16] They have been found quite often, however, in the well-aerated parts of the fermenting mass and during the drying process.[36,37] It is likely

Table 4   Aerobic Spore-Forming Bacteria Isolated from Cocoa Fermentations in Four Countries

Brazil[26]	Trinidad[21]	Ghana and Malaysia[101]
Bacillus brevis, B. cereus, B. circulans, B. coagulans, B. firmus, B. laterosporus, B. licheniformis, B. macerans, B. megaterium, B. pasteurii, B. polymyxa, B. pumilus, B. stearothermophilus, B. subtilis	*Bacillus cereus, B. cereus var. mycoides, B. coagulans, B. licheniformis, B. megaterium, B. pumilus, B. stearothermophilus, B. subtilis*	*Bacillus licheniformis, B. subtilis*

that they may cause hydrolysis of some of the pulp and even the testa of the seeds; they may also produce acids or impart off-flavors to the beans.[37] Filamentous fungi isolated from fermenting cocoa in Bahia were *Aspergillus fumigatus, A. niger, Fusarium moniliforme, F. oxysporum, Lasiodiplodia theobromae, Mucor racemosus, Mucor* sp., *Paecilomyces varioti, Penicillium citrinum, P. implicatus, P. spinosum, Thielaviopsis ethaceticus, Trichoderma viridae,* and three different isolates of *Mycelia sterilia.*[37] Although the numbers were small, a great diversity of species was seen in the first 44 h of fermentation. Thereafter *Aspergillus fumigatus* and *Mucor racemous* dominated the fungal population up to the end of fermentation. Most of these fungi are reported to be unable to grow at temperatures higher than 45°C, but they have been isolated when the temperature of the fermenting mass was around 50°C. It is not uncommon for yeast species isolated from Brazil to show higher maximum growth temperatures than the corresponding species isolated from temperate sources.[38]

3. Roles of Microorganisms During Cocoa Fermentation

The great majority of flavor compounds (*ca.* 400) are formed due to biochemical and enzymatic reactions that occur within the cotyledon. The major role of microorganisms is to produce acids and alcohols that will penetrate the testa and start the chemical reactions that will form the precursors of chocolate flavor. There is no evidence that enzymes from the microorganisms penetrate the testa and create flavor compounds but hydrolytic enzymes inside the beans are activated by microbial metabolites such as acetic acid.[39,40,41] Many different species of microorganisms have been characterized and the microbial succession has been defined. So far the roles of all these microorganisms have not been explicitly described particularly in their relative contribution to the overall quality of the final product. The first step in understanding this is to determine the physiology of the microorganisms and what they contribute to the dynamics of the fermentation process. Then it is possible to define the potential ecological roles of these microorganisms.

3.1. Roles of Yeasts
A. Ethanol Production

The sugar-rich, acidic pulp presents ideal conditions for rapid yeast growth. Conversion of sucrose, glucose, and fructose to ethanol and CO_2 is the primary activity of the fermentative yeasts. Measurements of ethanol show clearly how, after rising in concentration in the pulp, it penetrates the cotyledons of the beans. However, it is reputedly the acetic acid that kills the beans.[30]

B. Breakdown of Citric Acid

Some of the yeasts, including *Candida* spp. and *Pichia* spp., metabolize citric acid causing the pH value to increase in the pulp which allows growth of bacteria. The loss of citric acid both in the "sweatings" and by microbial metabolism causes an alkaline drift in pH. This, together with the increasing levels of alcohol and aeration, inhibits the yeasts and their activity wanes.

C. Production of Organic Acids

Several of the yeast isolates produce organic acids including acetic, oxalic, phosphoric, succinic, and malic acids. These weak organic acids will have a buffering capacity and will tend to reduce fluctuations in pH.

D. Production of Volatiles

Yeasts produce a large array of aroma compounds, principally fusel alcohols, fatty acids, and fatty acid esters[42] and different species produce different aromas.[42,43] It is known that volatile compounds are important in the development of full chocolate flavor.[44,45] The five major yeasts that produce these volatiles (*Kloeckera apiculata, S. cerevisiae, S. cerevisiae* var. *chevalieri, Candida* sp., and *Kluyveromyces marxianus*) have been studied individually. *Kloeckera apiculata* and *S. cerevisiae* var. *chevalieri* were the major producers of volatiles such as isopropyl acetate, ethyl acetate, methanol, 1-propanol, isoamyl alcohol, 2,3 butanediol, diethyl succinate, and 2-phenylethanol. Among the yeasts with high fermentative power *S. cerevisiae* var. *chevalieri* produced large amounts of aroma compounds suggesting that these strains might be collaborating in the elaboration of aroma and flavor characteristics (Schwan, R. F. unpublished

observations). Although this study was done in pure culture, it does give clues as to which species might be added. Studies on wine fermentations have shown the importance of the range of volatiles that can be produced by different strains and different species.[43,46,47]

E. Production of Pectinolytic Enzymes

Some strains produce pectinolytic enzymes[20,28,48–50] that break down the cement between the walls of the pulp cells and the resultant juice (or "cacao honey") drains away as "sweatings." The collapse of the parenchyma cells in the pulp between the beans results in the formation of void spaces into which air-percolates. Only 4 out of 12 yeast species showed pectinolytic activity (*K. marxianus, S. cerevisiae* var. *chevalieri, Candida rugopelliculosa,* and *K. thermotolerans*). Only the first two showed substantial activity and only *K. marxianus* produced large quantities of heat stable endopolygalacturonase (PG). It had strong maceration activity that reduced cocoa pulp viscosity during the first 36 hr of fermentation when *K. marxianus* was the most abundant pectinolytic yeast. PG of *K. marxianus* has been studied in more detail.[49,50] None of the bacterial species present in the early stages of the fermentation have been shown to have pectinolytic activity. This enzyme activity is crucial during the first 24 hours because it breaks down the pulp and allows penetration of oxygen into the fermenting cocoa mass enabling aerobic acetic acid bacteria to grow.

F. Yeast Varieties

It is likely that all of these biochemical transformations are necessary for a normal fermentation and species that perform some or all of them are probably essential but the other yeast species are probably unimportant. Indeed, some of them could be defined as transients that only show spasmodic appearance and it is possible that the different species found in different countries or in different types of fermentation are not important in respect of the fermentation process per se. It may be both necessary and sufficient for there to be representatives of each physiological/ecological group to provide the appropriate transformations during the fermentation (see section 5).

3.2. Roles of Bacteria
A. Lactic-Acid Bacteria

- **Production of lactic acid.** The great majority of lactic-acid bacteria isolated during cocoa fermentation utilize glucose via the Embden-Meyerhof pathway yielding more than 85% lactic acid. However, some species utilize glucose via the hexose monophosphate pathway producing 50% lactic acid, and ethanol, acetic acid, glycerol, mannitol, and CO_2. Their relative proportion will thus change the composition of the pulp substrate and thus may consequently change the microbial succession.
- **Production of citric acid.** Lactic-acid bacteria first contribute to an increase in acidity by producing citric acid and then lower the pH by metabolizing it and liberating non-acid by-products.[51] All lactic acid

bacteria isolated from cocoa fermentations were able to metabolize malic and citric acids.[29,51] Dissimilation of these acids leads to an overall drop in acidity and rise in pH value. Lactic acid bacteria are virtually non-proteolytic and their ability to ferment amino acids is also restricted with only two, serine and arginine, that are extensively attacked by some of these organisms.[52]

B. Acetic-Acid Bacteria

- **Production of acetic acid.** These bacteria are responsible for the oxidation of ethanol to acetic acid and further oxidation of the latter to carbon dioxide and water. The exothermic reactions of the acetic-acid bacteria raise the temperature of the fermenting mass, sometimes to 50°C or more. The acidity of cocoa beans, the high temperature in the fermenting mass, and the diffusion and hydrolysis of protein in the cotyledons has been attributed to the metabolism of these microorganisms. Concentrations of a maximum of 6 g/L of pulp of acetic acid was found in cocoa pulp after 88 hours of fermentation.[24] However, it disappeared quickly from the pulp when the mass temperature rose above 50°C.[22,24] Part of this acid is volatilized and part penetrates the testa (approximately 2%) and is responsible for killing the embryo.[53] Ethanol, acids, and water diffusing into the cotyledon act as solvents so that cellular components are transported to sites of enzyme activity and vice versa. The detailed levels of chemical reactions inside the bean are still unknown. It is clear that excess acid will interfere with chocolate flavor[22,54] even though most of the acetic acid will eventually be volatilized.

C. Aerobic Spore-Forming Bacteria

The *Bacillus* species that were isolated produced a variety of chemical compounds under fermentative conditions such as 2,3 butanediol, pyrazines, acetic, and lactic acid. These bacteria may contribute to the acidity and perhaps at times to off-flavors of fermented cocoa beans. Oxygen is one of the factors that determines the microbial succession. Facultatively anaerobic yeasts are metabolically active at the beginning of fermentation when oxygen is not available because of its occlusion by the mucilaginous pulp surrounding the seed. Lactic acid bacteria are the next group in the succession and are microaerophilic. When the pulp has been degraded, oxygen becomes more plentiful and then the strictly aerobic acetic acid bacteria develop.

3.3. Conclusions

Growth in relation to sugar and oxygen are the key parameters that establish and change the microbial succession. Ethanol tolerant yeasts ferment the sugars at low pH (pH 3.5 and 4.2) and pectinolytic enzymes open the structure of the pulp for the ingress of air. Lactic acid bacteria are micro-aerophilic and members of the homolactic group are able to ferment sugars and tolerate this acidity. The acetic acid bacteria are

aerobic and can grow at high concentrations of ethanol and tolerate temperatures around 45°C. They produce acetic acid from sugars and also can oxidize ethanol to acetic acid and then to CO_2 and water. These conclusions about the physiological roles of the major groups were experimentally tested (see section 5).

4. Cocoa Pulp

Cocoa pulp is the raw material on which the fermentation proceeds and this section will describe how it seems to be a key determinant of both quality and financial viability of the process. Not only is the quantity of pulp crucial in affecting the efficiency and nature of the fermentation, but excess pulp can also be sold as a high value commodity.

4.1. Quantity of Pulp Surrounding the Cacao Seed

Not all pulp is necessary for a successful fermentation of cocoa beans. Loss of pulp occurs naturally during a fermentation because the 'sweatings' drain out through the holes in the fermentation box. This liquid is almost transparent and is rich in fermentable sugars, pectin, and acids. In Brazil, it has been used traditionally to make jelly. Today the juice for commercial jelly production is pressed from the seeds before fermentation. The economy of cocoa-producing areas in Brazil is very dependent on the acceptance of its products in the market. Processing of post-harvest residues and by-products of cacao (eg. cacao juice, cacao jam, vinegar and liquor of cacao juice) may offer opportunities for diversification on farms, especially where cocoa production is the major enterprise.[55] Revenue generated by these products exceeds that obtained from selling cocoa beans to processors. Ghana and Malaysia are also developing these industries.[56,57,58]

4.2. Mechanical Removal of Cocoa Pulp

Brazilian and Malaysian cocoas tend to be extremely acidic (cotyledon pH about 4.2) and this has adversely affected the development of their international markets. Removal of some of the pulp before fermentation reduces acidity and this presents a possible solution to the acidity problem. It was reported that at least 10% by weight can be removed by pressing the beans prior to fermentation without measurable consequences.[59] A normal bean fermentation occurred when up to 20% of total fresh weight of beans (including pulp) was removed.[59] This produced a less acidic cocoa in Brazil[60] although the acidity of beans was not reduced when some of the pulp was removed prior to fermentation in Malaysia.[61] A decrease in volume, water, and sugar content in cocoa pulp occurred when beans were spread out in a thin layer before fermentation in Malaysia and this method produced cocoa with less acidity.[62] Genetic differences in material cultivated in Malaysia and Brazil may be responsible for these differences since cacao cultivars in Malaysia have about three times more pulp sugars than the Brazilian *comum* cultivar.[51,60] Using a modified domestic washing machine it was shown[63] that partial (20%) removal of cocoa pulp gave an accelerated fermentation. There was a more rapid progression in the microbial succession, temperature increase, and rise in pH value of the cotyledon from 4.8 (as in traditional cocoa fermentation) to 5.5. Unfortunately these results could not be reproduced using a commercial depulping machine. This was probably due to differences in the technology: centrifugation is the basis of separation in a washing machine while gentle scraping is the principle of depulpers, but they also tend to remove the tightly adhering mucilaginous layer immediately surrounding the bean.[59] Pulp extraction on a larger scale for the cacao juice industry has been done with commercially available depulpers.[64] Such depulpers remove from 17 to 20% of pulp in terms of the fresh weight of the seed. Some depulpers leave loose mucilage, but little sugar on the seeds. This mucilage blocks the void spaces in the cocoa mass, impairs aeration, causes under-fermentation, and extends the fermentation period. If this occurs, there is no reduction of acidity of the cocoa compared to traditional box fermentations. The viscosity of the pulp still needs to be reduced in addition to reduction of pulp quantity.

Washing of the seeds has been used to produce a product suitable for wine production.[65] This also yielded a pulp-depleted bean that when fermented gave rise to fermented cocoa beans that were less acidic. Such a process does require a very high level of water quality and worker hygiene.

4.3. Enzymatic Removal of Cocoa Pulp

The addition of pectinolytic enzymes improves the efficacy of mechanical pulp extractors. One liter of a 0.2% (w/w) solution of pectinase (Ultrazym 100G, Novo Nordisk Ferment) sprayed over the seeds and allowing a reaction time of 30 minutes, increases the quantity of pulp extracted to approximately 23% compared to the batch-type depulper.[66,67] This value represents an increase of about 5% of total weight over the machine. Based on an assessment of both external and internal ("cut-test") color of beans, total fermentation time was reduced from seven to four days and the acidity of the final product was reduced.[55] As the pectin chains were broken by the added enzymes, the pulp had a lower viscosity. This change also helps pulp processing for pasteurized juice as well as for cacao soft drink production that is bottled and stored at ambient temperatures. Some laboratory experiments have suggested that the yield could be improved by inoculation with pectinolytic yeasts.[67] However, the experiments are difficult to evaluate since only 1 kg samples of beans were used, no monitoring of the microbial population was done, and the fermented beans were only analyzed by anthocyanin content. Since good color and flavor are also due to bacterial activity, the results could have been partly due to other microbiological activity.

Addition of commercial enzymes is costly and prohibitive on a large scale. Two alternative approaches are being explored to provide a better quality of fermented beans by speeding up this process; (1) to increase microbial pectinolytic activity at the onset of fermentation, and (2) making a source of enzyme obtainable from yeast cultures themselves.

4.4. Pectinases Produced by Yeasts

Pectins give pulp its sticky, viscous, and cohesive properties. Pectin and pectic acid, the natural substrates of pectic enzymes, are branched heteropolysaccharides in which the backbone contains L-rhamnose residues and αa-D-(1,4)-linked residues of D-galactopyranosiduronic acid.[68,69] The neutral sugars, D-galactose and L-arabinose and sometimes D-xylose and L-fucose, form the side-chains of the pectin molecule. The carboxyl groups of the D-galactopyranosiduronic acid residues are partially esterified with methanol. Some of the secondary alcohol groups at C-2 and C-3 are acetylated.[69] The degree of esterification, the proportion of neutral saccharides, and degree of polymerization are the principal elements of heterogeneity in pectic compounds of diverse origins.[69] Enzymes that attack pectin can be assigned to two main groups: (1) de-esterifying enzymes (pectinmethylesterases, PME) that remove the methoxyl groups from the esterified acid, and (2) chain-splitting enzymes (depolymerases) that split the βb-(1,4)-glycosidic bond, either by hydrolysis (polygalacturonases, PG) or by trans-elimination (pectin and pectate lyases, PL). An increasing number of yeast species have been discovered to have pectinolytic activity.[70] In Java pectinolytic yeasts belonging to the genera *Candida, Pichia, Saccharomyces,* and *Zygosaccharomyces* were found.[71] Yeasts from cocoa fermentations produced various pectinolytic enzymes that aided the maceration of cocoa pulp and the drainage of "sweatings."[28] They claimed that *Saccharomyces chevalieri* (now classified as *S. cerevisiae*[72,73]), *Torulopsis candida,* and *T. holmii* produced PME and that *S. chevalieri* and *Candida zeylanoides* secreted PG. Genome sequencing of *Saccharomyces cerevisiae* (http://genome-www.stanford/saccharomyces) has revealed a PG gene but not a PME gene suggesting either erroneous assays or mis-identification of species in at least one case. *S. chevalieri, Candida norvegensis,* and *Torulopsis candida* were the only pectinolytic yeasts isolated from cocoa fermentations in another study.[48,74] In trials with pure "starter" cultures of yeasts, including *Kluyveromyces marxianus,* among isolates from cocoa fermentations, the pH value did not rise during the early stages of fermentation.[74] The researchers suggested that *K. marxianus* interfered with the development of the wild yeast flora. Among the other strains studied, *C. norvegensis* produced the greatest amount of extracellular enzyme. They found that the yeast enzymes had the same optimum pH value of activity (5.0) but differed from each other in their optimum temperature and thermal stability. The enzymes of *T. candida* and *K.fragilis* had the highest optimum temperature (60°C).

Of the 12 yeast species isolated from cocoa fermentations in Brazil, *K. marxianus, S. cerevisiae* var. *chevalieri, C. rugopelliculosa,* and *K. thermotolerans* produced extracellular endopolygalactoronase (endoPG).[50] Neither PME nor PL was detected in culture filtrates. The amounts and properties of each PG differed but all were relatively unstable compared with that of *K. marxianus,* which was also found to be the most active producer of PG. This strain fermented the major pulp sugars as well as degrading pectin. High yields of PGs were obtained with self-induced anaerobic batch fermentations of *K. marxianus* with 100 g l^{-1} glucose as the sole carbon source[75] but production is inhibited by oxygen.[76,77] Addition of pectin or polygalacturonic acid to the growth medium did not increase enzyme secretion, indicating that PG production is constitutive under these conditions but it was unable to grow on pectin or galacturonic acid as the sole carbon source like *Rhodotorula spp.*[78] PG secreted by *K. marxianus* could macerate potato and cucumber slices and decrease the viscosity of cocoa pulp by 50% within 18 min.[49,50] These data suggest that the anaerobic conditions that rapidly predominate after initiation of natural cocoa fermentations are ideal for the appearance of the enzyme but that its production becomes self-limiting as the pulp drains away and air percolated through the fermenter.

The PG secreted from *K. marxianus* was characterized and showed activity from pH 4 to 6, with an optimum at pH 5 typical of endoPG secreted by yeasts. Unlike some pectinases, *K. marxianus* endoPG activity was not affected by buffers used across the pH range studied. The effect of temperature on endoPG activity from *K. marxianus* was similar to that reported for PGs from other yeasts. From concentrated culture supernatant of *K. marxianus,* gel filtration resolved four peaks containing PG activities. The relative molecular masses were calculated and the four PG forms had apparent Mr of 47, 41, 35, and 33 kDa. According to analysis of all bands by densitometry, about 85% of total protein secreted into culture medium by *K. marxianus* consisted of PG.[50] A study of the kinetics of appearance of the enzyme using sub-cellular fractionation showed it was secreted by the classical yeast secretory pathway. Since high endoPG activity in early stages of fermentation speeds the fermentation process and leads to better quality of chocolate, overproducer strains might be useful in improving quality. An attempt was made to do this by conventional chemical mutagenesis using nitrosoguanidine. However, with a constitutive (deregulated) gene, substantial improvements in productivity were never likely and in a screen of 18,000 mutagenized cells only a few strains produced enhanced levels of the enzyme and the best was only 25% above wild type levels.[50] It was concluded that a more directed approach might be more profitable.

4.5. Genetics of EndoPG Production

The genes for endoPG have recently been cloned from strains of *Saccharomyces cerevisiae*,[79–82] *Saccharomyces bayanus*,[83] and *Kluyveromyces marxianus.*[82,84] One of the most interesting observations with respect to *Saccharomyces cerevisiae* was that the non-pectinolytic laboratory strains used in genetic studies contain a PG homologue.[85] By isolating the structural gene and putting it on an expression vector it has been shown that it is secreted and functional.[79,82] Transcription analysis has shown that the gene can be induced under special conditions of nitrogen starvation with induction of the pseudohyphal development pathway.[86] Only one copy of the gene is present on the genome in both of these yeasts in contrast to filamentous fungi where at least four are present in *Aspergillus niger.*[87] Overexpression strains should facilitate the process of producing a pure enzyme. In the absence of contaminating enzymes and undesirable by-products, such as methanol from pectin methylesterase, PG from *K. marxianus* could be used directly on cocoa beans to speed up the process and enhance the quality of the final product.

As an alternative to over-producing strains, the possibility of local production of the enzyme has been investigated. This would have the additional benefit of being suitable for use on any other pectinaceous fruit. To this end the endoPG from a number of these strains has been taken from the original cloning strain (*S. cerevisiae*) and transferred in turn to both *K. lactis* and *K. marxianus* to create new expression systems (Jia and Wheals, unpublished data). The advantage of these hosts is that (1) the plasmid carrying the endoPG is stable and requires no selection using conventional systems and (2) that the strains can be grown on either cheese whey (an industrial waste product) or on sugar cane juice, a widely available and cheap commodity in tropical countries where cacao is grown. The enzyme output from these sources is at least 50% higher than wild strains and constitutes about 90% of secreted protein. The medium in which the cells are grown is thus suitable for direct use without purification or concentration.

Although the addition of an enzyme would be useful, even better would be the inoculation of fermentations with overproducer strains, particularly if they were stable enough to continually re-infect fresh batches of cocoa beans. Indeed, *K. marxianus* has the status of an organism that is generally regarded as safe (GRAS). However, over-expressing strains are constructed with heterologous DNA sequences and are therefore classified as genetically modified organisms (GMOs). To use such a GMO would require satisfying the safety aspects for the regulatory authorities of the country concerned. This is a lengthy and costly process. Even more important is that GMOs are currently under public scrutiny and are often perceived as abnormal and undesirable. Chocolate produced with such a strain would never find a market!

Alternative strategies have been tried (increase in chromosomal copy number; site-directed mutagenesis of the active site of endoPG) but without success (Jia and Wheals, unpublished data). It therefore seems likely that the best approach would be to screen additional strains for the desired enhanced activity.

5. Chocolate Quality

One of the reasons that chocolate quality has not been a priority for farmers is that there is no financial incentive to produce high quality fermented cocoa beans. Poor practices are widely reported and even led Ecuador, with special status for quality, to have its rating downgraded in 1994. However, as increasing numbers of farmers and countries are attempting to take control of all the processing that occurs in the tropical countries and the global market reaches saturation, quality will become even more important. This section is directed to pointing towards general improvement in practice that can be achieved by using appropriate procedures.

5.1. Starter Cultures

From knowledge of the microorganisms responsible for spontaneous cocoa fermentations and their physiological roles during the process, an attempt was made to manipulate the fermentation.[24] From 12 yeast species, and 30 bacterial species that had been identified, a defined microbial cocktail was selected for use as an inoculum. It consisted of one pectinolytic yeast species, two lactic acid bacterial species, and two acetic acid bacterial species.

The yeast *Saccharomyces cerevisiae* var. *chevalieri* produces pectinase, can ferment all pulp sugars at pH 3.5–4.2, is ethanol tolerant, and was present at the beginning of natural fermentation. The lactic-acid bacteria were selected after observing production of lactic acid at acidic pH, oxygen requirement, and temperature tolerance. Two species of *Lactobacillus* were selected—*L. lactis* and *L. plantarum*. The best producers of acetic acid that were also tolerant to temperatures of 45°C were isolates of *Acetobacter aceti*. *Gluconobacter oxydans* was also added to oxidize the ethanol to acetic acid and to CO_2 and water.

A cocktail of these microorganisms was inoculated on cocoa beans immediately after the pod was broken open and left to ferment in sterilized 200 kg wooden boxes for 7 days. The three key metabolites in the pulp, ethanol, lactic acid, and acetic acid showed similar sequential rises and falls to that found in spontaneous fermentation. Contamination from extraneous microorganisms was kept to a minimum. The beans were then dried and roasted and chocolate was produced by the usual means. A taste panel found the product as good as a "natural fermentation." However, natural fermentations are subject to random fluctuations in the inoculum and the fermentation does not always proceed correctly. Defined cocktails should always be more reliable if only because of the lack of spoilage organisms and the fermentation time was less than the normal fermentation.

These were encouraging results since the fermentation occurred normally and the product was more than acceptable. It also suggests that the physiological analysis presented in section 3 is correct and that the consortium of microorganisms was correctly chosen with respect to their physiological roles. Clearly there is still room for improvement since there was no guarantee that the best species had been selected or that they had been inoculated at the most appropriate rates. For example, subsequent enzymological work suggested that *Kluyveromyces marxianus* alone or *K. marxianus* together with *S. cerevisiae* might be a better choice of yeast(s).

One way to explore the generality of this result is to examine the species found in other countries (Tables 1–3). In the only other study where yeast identification went to species level, the pectinase-producing variety of *Saccharomyces cerevisiae* (var chevalieri) was found. In the other studies both *Saccharomyces* and *Candida* genera were found, both of which include pectinolytic yeasts. It is likely therefore that all cocoa fermentations contained both strongly fermentative and pectinolytic yeasts. With respect to the bacteria, there is little change between countries and the ones used in the cocktail were always present. These results show that representatives of the three major groups may be sufficient to complete the complex fermentation and that the basic features of the microbiology of cocoa fermentation are understood.

Since there is no quick way in which raw materials and tree varieties are going to change, inoculation with a defined starter culture may be an important way in which reliable fermentations may be achieved relatively quickly. On a global scale, a large number of companies already supply fresh pressed, dried, or "instant" yeast cultures for baking, brewing, or wine making. Some of these companies also have the capacity to produce batches of special yeasts and other, non-yeast microorganisms to order. Once a suitable strain or, more likely, consortium of strains has been defined it will be possible to create cultures for direct addition to initiate fermentations. Provided the product is of high quality and the chocolate producers are prepared to pay a premium for this enhanced quality then it will be economic for farmers to purchase and continue to use starter culture strains. Education and training for the farmers may also encourage them to let the fermentation take its full course.

A potential problem with using starter cultures is that the resident microbial flora will compete with, and may even outgrow, the starter culture inoculum. Clearly, starting with a high density will help the new population to establish itself but it will certainly be necessary to reduce the host microbiota. This will not be easy without a change in some traditional practices. For example, washing the pods before breakage does help considerably to reduce contamination but at present farmers break open the pods in the fields with contaminated machetés, transport them in unwashed containers, and pour them straight into wooden fermentation boxes containing the residues of the previous batch. Bringing unopened pods close to the boxes for washing would only help if there were a source of washing water that was free of fecal contamination. It would also require disposal of the husks that are currently left to rot in the fields. Knives, containers, and fermentation boxes would all need decontaminating, perhaps with a disinfectant. Further ahead we can envisage the need for proper pasteurization of the materials. This would certainly enable control of the inoculum to be achieved but there could be adverse effects on the chemistry of the pectin and the beans leading to deleterious changes in overall quality. Furthermore it implies bringing the pods to a more centrally located "factory" with appropriate facilities and economies of scale. Field trials are now in progress (RF Schwan, unpublished data).

5.2. Manipulation of the Fermentation

Depulping is an approach that is advantageous to the course of the fermentation because excess pulp can lead to an over-acid fermented bean but the precise design of the depulping machine is important. Pulp removal by shearing alone in commercial depulpers was not as effective as a combination of centrifugation and shearing given by a domestic washing machine. New designs may be needed and certainly the cost needs to be kept as low as possible.

The natural course of a fermentation takes about seven days and the microbial flora evolves in a more or less predictable manner. It has already shown that supplementing natural endopolygalacturonase (PG) with commercial pectinase led to a faster fermentation and a higher quality product. If non-GMO yeast strains can be produced that can secrete enhanced amounts of endoPG, these should have the same effect as adding pectinase. Aeration is known to be important in the fermentation but adequate supplies of oxygen are dependent on reduced viscosity of the pulp and regular turning of the bean mass. Mechanical turning is not a realistic alternative and forced aeration requires careful control.[65] Redesign of fermenters could be possible and this is being investigated.

It is likely that the events of the first 24 hours of the fermentation entrain the subsequent microbial succession (see section 3.1.E) but a detailed study of the lactic acid and acetic acid bacteria may also be useful. Reducing the amount of free carbohydrate will undoubtedly affect the growth and biochemistry of the bacteria but it is unclear what effect this will have with respect to quality and flavor.

Bacterial spore-formers and filamentous fungi that appear during the latter stages of the fermentation are usually associated with the appearance of off-flavors and spoilage. A more precise investigation needs to be done on this in order to establish a clear endpoint to the fermentation when harvesting is optimal. It is rather difficult to assess this stage since the state of the beans throughout the box is not uniform but delay in termination could lead to loss of quality.

5.3. Flavor

Under-fermented beans have an astringent and bitter taste due partly to the presence of high levels of polyphenols. Attempts have been made to stimulate polyphenol oxidase activity that decreases during fermentation and drying, by acidic incubation. This has had some success in ameliorating the effect in some poorer quality Indonesian beans.[88] Alternatively Mars Inc. (Hackettstown, NJ) has patented a method to maintain high levels of putatively beneficial polyphenols.[15] Chocolate flavor chemistry is very complex and is determined by both the cocoa plant variety and the fermentation and roasting process[89,90] and it would be premature to say that it is fully understood, but work over the last thirty years, particularly by the Technical University of Braunschweig, has resulted in the determination of the more important chocolate flavor precursors.[39,40,41,44] In essence, it is proposed to be the combination of two proteases, aspartic endopeptidase and serine carboxy-(exo)peptidase, on vicilin (7S)-class globulin (VCG) storage protein that produces the cocoa-specific precursors. Experiments with alternative, related storage proteins or alternative peptidases both fail to produce appropriate flavor precursors. The aspartic endopeptidase splits VCG at hydrophobic amino acid residues. The products of this hydrolysis are substrates for the serine exopeptidase that removes the hydrophobic amino acid residues at the carboxyl terminus of the hydrophobic oligopeptides. Roasting of these precursors in the presence of reducing sugars produced significant cocoa aroma. However, which of the hydrophobic amino acid residues and hydrophilic oligopeptides are responsible for the cocoa aroma is not yet known.[91]

Both of these enzymes are very pH-dependent in their activity. When the pH during proteolysis approaches pH 3.8 (the optimum for aspartic endopeptidase), more hydrophobic oligopeptides and less free amino acids are produced. On the other hand pH values close to 5.8 (the optimum for serine exopeptidase), lead to an increase in hydrophilic oligopeptides and hydrophobic amino acids. If the pH becomes too acid too soon (pH < 4.5)

there will be both a final reduction in flavor precursors and an over-acid product. Thus, with respect to the organic acids that diffuse slowly into the cotyledons, timing of initial entry, duration of the period of optimum pH and final pH are crucial for optimum flavor. In other words, fermentation may still be the key to cocoa quality.[44] This is further emphasized by an analysis of VCG proteins and their proteolytic degradation products amongst five widely used cacao genotypes (Forastero, Criollo, Trinitario, SCA 12, and UIT1). Although they can give rise to different quality chocolate, all had similar potential for producing raw cocoa with high aroma potential.[92] Two aspartic proteinase (EC 3.4.23) genes that are expressed during seed development have been cloned from *T. cacao*[93] and thus prospects for the identification of the flavor precursors are now good.

Although final acidity can be reduced by slow (sun) drying to allow the acids to volatilize[94] and produce a less acid product, it is the acidity during the fermentation that is more crucial for flavor development.[95] Although acetic and lactic acids produced during fermentation are the key determinants, oxalic acid, one of the endogenous acids, has been reported to be a significant contributor to flavor.[96]

5.4. Fermenter Design

Further work along these lines will produce a more precise understanding of the entire process and elucidate how the various microbiological factors determine the final outcome of the fermentation. Both the biology and the chemistry are complex and it will need a comprehensive, simultaneous, and dynamic analysis of all aspects in controlled conditions if a really good understanding of the process is to be achieved such that defined changes will have predictable and quantitative outcomes. To achieve some of these aims a sterilizable stainless steel vessel of novel design capable of turning a 50 kg load of beans has been constructed. Inoculum, aeration, and turn rate can be controlled, temperature monitored and samples taken at intervals. Early results show that it can mimic the natural conditions of fermentation boxes and produce fermented beans for making good quality chocolate in five days (Freire, Schwan and Serodio, unpublished data). Combined with defined inocula there is the prospect of producing the best quality chocolate reliably and in less time. Another approach has been to modify a rotary drier enabling it to ferment up to 9 tons of beans.[97] Initial results also show that it performed well in comparison with traditional fermentations although the process was stopped after four days. Research with such fermenters will accelerate the number of variables that can be studied and allow earlier use of this information by farmers. These are the first reports of making largescale, controllable, and mechanized fermenters in an otherwise very traditional industry.

It is clear that the research done by major manufacturers on roasting and processing has enabled them to produce good quality products from inferior sources. However in a consumer oriented world where less processing and additives are desirable, attention to the primary aspects of good quality plant material and well controlled fermentations will lead to both improved final products and the need for less processing.

6. Conclusions

We propose that reliable fermentations giving consistently high quality products can be achieved by a combination of three procedures:

- Control of the amount of free pulp at the start of the fermentation such that the final pH is not too acid,
- Use of a defined starter culture so that there is a well ordered succession and timely production of acids diffusing into the cotyledons, and
- Improved fermenter design to optimize the physical aspects of the process especially aeration

Another approach to improving flavor would be to make chocolate from *Theobroma grandiflorum* (cupuaçu). This has a different and more fruity flavor but can be processed in much the same way[98] to produce "cupulate." Although there could be potential competition with the chocolate market, evidence from other commodities strongly suggests that the total market will increase.

Relatively little attention has been paid to the microbiology of cocoa fermentation over the last 20 years and yet it remains at the heart of the question of quality. The clearest demonstration that the basic physiology and ecology of cocoa fermentation is understood is that biotechnological manipulation of the component parts (microorganisms, amount of pulp, etc) can lead to understandable and reasonably predictable effects—but this is only a start. Where does this research lead in the longer term? The historical evolution of other natural fermentation systems provides clues to the direction cocoa fermentation technology might take. Beer fermentations were originally all conducted locally because of difficulties in transporting the beer and relied on contamination by natural microbiota to start the fermentation. Later, breweries developed their own strains for use as starter cultures, optimized the design and operation of the fermentation vessels, and also paid more attention to the quality of the raw materials. Now large centralized breweries that send their pasteurized or sterile products nationally and internationally have substantially replaced local breweries and the whole process, from raw materials to microorganisms to fermentation vessels to post-fermentation processing, is very closely controlled. The development of wine and cider fermentations shows similar trends but they are not as far advanced. The essential trend in all three industries is for better control at all stages of the process.

Cocoa fermentations are still at the first stage. There is relatively little control over the raw materials and perhaps not even the best varieties of *Theobroma cacao* have been planted—quality may have taken second place to quantity in choice of tree, and the monocultures have laid themselves open to devastating fungal pathogens such as witches broom (*Crinipellis perniciosa*) in Brazil and pod rot (*Phytophthora* spp.) in Africa.[99] Natural microflora initiate the fermentation and the sometimes poor and certainly variable quality of the product may reflect the vagaries of chance contamination. Not enough is known about the detailed relationship between the microorganisms and the quality of the product. The fermentation vessels are open wooden boxes designed to allow both aeration and drainage of sweatings and development work needs to be done on them in respect of shape, size and ease of use in turning the beans. All aspects of the process need attention.

In the last 20 years, multinational companies have put most effort into two areas: encouraging farmers to maximize production and into improving processing of the fermented beans. The fermentation process itself has been largely neglected. This review emphasizes the need for more work on this aspect of the process and how it can pay dividends in improving the quality of the final product.

References

1. De La Cruz, M., Whitkus, R., Gomez-Pompa, A., and Mota-Bravo, L. 1995. Origins of cacao cultivation. *Science,* **375:**542–543.
2. Coe, S.D. and Coe, M.D. 1996. *The true history of chocolate,* London: Thames and Hudson.
3. Erdman, J.W., Wills, J., and Finley, D'A. (eds). 2000. *Chocolate: Modern science investigates an ancient medicine. J. Nutrition,* **130:** Supplement 8S.
4. Lopes, A.S. and Dimick, P.S. 1995. Cocoa fermentation. In *Biotechnology: A comprehensive treatise,* vol. 9, pp. 563–577. Reed, G. and Nagodawithana, T.W., Eds. (2nd ed.), Enzymes, Food and Feed. Weinheim: VCH.
5. Di Tomaso, E., Beltramo, M., and Piomelli, D. 1996. Brain cannabinoids in chocolate, *Nature,* **382:**677–678.
6. Weisburger, J.H. 2001. Chemopreventive effects of cocoa polyphenols on chronic diseases. *Exp. Biol. Med.,* **226:**891–897.
7. Barrile, J.C., Ostovar, K., and Keeney, P.G. 1971. Microflora of cocoa beans before and after roasting at 150°C. *J. Milk Food Technol.,* **34:**369–371.
8. Cordier, J.L. 1994. HACCP in the chocolate industry. *Food Control,* **5:**171–175.
9. Wollgast, J. and Anklam, E. 2000. Review of polyphenols in *Theobroma cacao:* Changes in composition during the manufacture of chocolate and methodology for identification and quantification. *Food Res. Intern.,* **33:**42–447.
10. Tapiero, H., Tew, K.D., Nguyen Bal, G., and Mathé, G. 2002. Polyphenols: Do they play a role in the prevention of human pathologies? *Biomed. Pharmacother.,* **56:**200–207.
11. Hatano, T., Miyatake, H., Natsume, M., Osakabe, N., Takizawa, T., Ito, H., and Yoshida, T. 2002. Proanthocyanidin glycosides and related polyphenols from cacao liquor and their antioxidant effects. *Phytochem.,* **59:**749–758.
12. Yamagishi, M., Osakabe, N., Natsume, M., Adachi, T., Takizawa, T., Kumon, H., and Osawa, T. 2001. Anticlastogenic activity of cacao: Inhibitory effect of cacao liquor polyphenols against mitomycin C-induced DNA damage. *Food Chem. Toxicol.,* **39:**1279–1283.
13. Carnésecchia, S., Schneidcra, Y., Lazarusb, S.A., Coehloc, D., Gosséa, F., and Raul, F. 2002. Flavanols and procyanidins of cocoa and chocolate inhibit growth and polyamine biosynthesis of human colonic cancer cells. *Cancer Letts.,* **175:**147–155.
14. Kris-Etherton, P.M. and Keen, C.L. 2002. Evidence that the antioxidant flavonoids in tea and cocoa are beneficial for cardiovascular health. *Current Opinion in Lipodology,* **13:**41–49.
15. Kealey, K.S., Snyder, R.M., Romanczyk, L.J., Geyer, H.M., Myers, M.E., Withcare, E.J., Hammerstone, J.F., and Schmitz, H.H. 2001. Cocoa components, edible products having enhanced polyphenol content, methods of making same and medical uses, United States Patent 6,312,753, Mars Incorporated, USA.

16. Roelofsen, P.A. 1958. Fermentation, drying, and storage of cocoa beans. *Adv. Food Res.,* **8:**225–296.

17. Saposhnikova, K. 1952. Changes in the acidity and carbohydrates during growth and ripening of the cocoa fruit: Variations of acidity and weight of seeds during fermentation of cacao in Venezuela. *Agro. Tropical,* **42:**185–195.

18. Berbet, P.R.F. 1979. Contribuiçao para o conhecimento dos açúcares componentes da amêndoa e do mel de cacau. *Rev. Theobroma,* **9:**55–61.

19. Pettipher, G.L. 1986. Analysis of cocoa pulp and the formulation of a standardised artificial cocoa pulp medium. *J. Sci. Food Agricult.,* **37:**297–309.

20. Rombouts, J.E. 1952. Observations on the microflora of fermenting cocoa beans in Trinidad. *Proc. Soc. Appl. Bacterial.,* **15:**103–111.

21. Ostovar, K. and Keeney, P.G. 1973. Isolation and characterization of microorganisms involved in the fermentation of Trinidad's cacao beans. *J. Food Sci.,* **38:**611–617.

22. Schwan, R.F., Rose, A.H., and Board, R.G. 1995. Microbial fermentation of cocoa beans, with emphasis on enzymatic degradation of the pulp. *J. Appl. Bacteriol. Symp. Supp.,* **79:**96S–107S.

23. Lehrian, D.W. and Patterson, G.R. 1983. Cocoa fermentation. In *Biotechnology, a Comprehensive Treatise,* vol. 5, pp. 529–575. Reed, G. Ed. Basel: Verlag Chemie.

24. Schwan, R.F. 1998. Cocoa fermentations conducted with a defined microbial cocktail inoculum. *Appl. Environ. Microbiol.,* **64:**1477–1483.

25. Schwan, R.F., Lopez, A., Silva, D.O., and Vanetti, M.C.D. 1990. Influência da frequência e intervalos de revolvimentos sobre a fermentaçao de cacau e qualidade do chocolate. *Rev. Agrotróp.,* **2:**22–31.

26. Schwan, R.F., Vanetti, M.C.D., Suva, D.O., Lopez, A., and Moraes, C.A. de. 1986. Characterization and distribution of aerobic, spore-forming bacteria from cacao fermentations in Bahia. *J. Food Sci.,* **51:**1583–1584.

27. Ardhana, M.M. 1990. Microbial ecology and biochemistry of cocoa bean fermentations, PhD thesis, University of New South Wales, Australia.

28. Gauthier, B., Guiraud, J., Vincent, J.C., Porvais, J.P., and Galzy, P. 1977. Comments on yeast flora from traditional fermentation of cocoa in the Ivory Coast. *Rev. Ferment. Ind. Aliment.,* **32:**160–163.

29. Passos, F.M.L., Silva, D.O., Lopez, A., Ferreira, C.L.L.F., and Guimares, W. V. 1984. Characterization and distribution of lactic acid bacteria from traditional cocoa bean fermentations in Bahia. *J. Food Sci.,* **49:**205–208.

30. Forsyth, W.G.C. and Quesnel, V.C. 1963. Mechanisms of cocoa curing. *Adv. Enzymol.,* **25:**457–492.

31. Passos, F.M.L. and Passos, F.J.V. 1985. Descrição e classificação de bactérias acéticas isoladas da fermentação do cacau, com base em uma analise numérica. *Rev. Microbiol.,* **16:**290–298.

32. Carr, J.G., Davies, P.A., and Dougan, J. 1979. *Cocoa fermentation in Ghana and Malaysia II.* University of Bristol Research Station, Long Ashton, Bristol and Tropical Products Institute, Gray's Inn Road, London.

33. Lopez, A. and Quesnel, V.C. 1973. Volatile fatty acid production in cacao fermentation and the effect on chocolate flavour. *J. Sci. Food Agricult.,* **24:**319–326.

34. Lopez, A. and Quesnel, V.C. 1971. An assessment of some claims relating to the production and composition of chocolate aroma. *Intern. Chocolate Rev.,* **26:**19–24.

35. Zak, D.L., Ostovar, K., and Keeney, P.G. 1972. Implication of *Bacillus subtilis* in the synthesis of tetramethylpyrazine during fermentation of cacao beans. *J. Food Sci.,* **37:**967–968.

36. Maravalhas, N. 1966. Mycological deterioration of cocoa beans during fermentation and storage in Bahia. *Intern. Chocolate Rev.,* **21:**375–378.

37. Ribeiro, N.C.A., Bezerra, J.L., and Lopez, A. 1986. Micobiota na fermentaçao do cacau no estado da Bahia, Brazil, *Rev. Theobroma,* **16:**47–55.

38. Pataro, C., Santos, A., Correa, S.R., Morais, P.B., Linardi, V.R. Rosa, C.A. 1998. Physiological characterization of yeasts isolated from artisanal fermentations in an *aguardente* distillery. *Rev. Microbiol.,* **29:**104–108.

39. Biehl, B., Heinrichs, J., Voigt, G., Bytof, G., and Serrano, P. 1996. Nature of proteases and their action on storage proteins in cocoa seeds during germination as compared with fermentation. In *12th Cocoa Research Conference, Salvador,* pp. 18–23. Lagos, Nigeria: Cocoa Producers Alliance.

40. Biehl, B., Heinrichs, J., Ziegeler-Berghausen, H., Srivastava, S., Xiong, Q., Passern, D., Senyuk., V.I., and Hammoor, M. 1993. The proteases of ungerminated cocoa seeds and their role in the fermentation process. *Angew. Bot.,* **67:**59–65.

41. Voigt, J., Biehl, B., Heinrichs, H., Kamaruddin, S., Gaim Marsoner, G., and Hugi, A. 1994. *In vitro* formation of cocoa specific aroma precursors: Aroma-related peptides generated from cocoa seed protein by cooperation of an aspartic endoprotease and a coarboxypeptidase. *Food Chem.,* **49:**173–180.

42. Suomalainen, H. and Lehtonen, M. 1979. The production of aroma compounds by yeast. *J. Inst. Brew.,* **85:**149–156.

43. Mateo, J.J., Jimenez, M., Huerta, T., and Pastor, A. 1991. Contribution of different yeasts isolated from of Monastrell grapes to the aroma of wine. *Int. J. Food Microbiol.,* **14:**153–160.

44. Biehl, B. and Voigt, J. 1999. Biochemistry of cocoa flavour precursors. In *Proceedings of the 12th International Cocoa Research Conference, Salvador, Brazil, 1996,* pp. 929–938. Lagos, Nigeria: Cocoa Producers Alliance.

45. Lopez, A.S. 1974. The contribution of volatile compounds to the flavour of chocolate and their development during processing. PhD Thesis, University of the West Indies/Cacao Research Unit.

46. Romano, P. 1997. Metabolic characteristics of wine strains during spontaneous and inoculated fermentation. *Food Technol. Biotechnol.,* **35:**255–260.

47. Mauricio, J.C., Moreno, J., Zea, L., Ortega, J.M., and Medina, M. 1997. The effects of grape must fermentation conditions on volatile alcohols and esters formed by *Saccharomyces cerevisiae.* *J. Sci. Food Agric.,* **75:**155–160.

48. Sanchez, J., Guiraud, J.P., and Galzy, P. 1984. A study of the polygalacturonase activity of several yeast strains isolated from cocoa. *Appl. Microbiol. Biotechnol.,* **20:**262–267.

49. Schwan, R.F., Cooper, R.M., and Wheals, A.E. 1996. Endopolygalacturonase of the yeast *Kluyveromyces marxianus* is constitutive, highly active on native pectin and is the main extracelleular protein. In *Pectins and Pectinases,* pp. 861–868. Visser, J. and Voragen, A.G.J., Eds. Amsterdam: Elsevier Press.

50. Schwan, R.F., Cooper, R.M., and Wheals, A.E. 1997. Endopolygalacturonase secretion by *Kluyveromyces marxianus* and other cocoa pulp-degrading yeasts. *Enz. Microb. Technol.,* **21:**234–244.

51. Carr, J.G. Cocoa. 1982. In *Economic Microbiology,* pp. 275–292. Rose, A.H., Ed. London: Academic Press.

52. Cogan, TM. 1995. Flavour production by dairy starter cultures. *J. Appl. Bacterial. Symp. Supp.,* **79:**49S–64S.

53. Quesnel, V.C. 1965. Agents inducing the death of the cacao seeds during fermentation. *J. Sci. Food Agricult.,* **16:**441–447.

54. Lopez, A.S. and Dimick, P.S. 1991. Enzymes involved in cocoa curing. In *Food Enzymology,* pp. 211–236. Fox, P.F., Ed. Amsterdam: Elsevier.

55. Freire, E.S., Romeu, A.P., Passos, F.J. Mororo, R.C., Schwan, R.F., Collado, A.L., Chepote, R.E., and Ferreira, H.I. 1990. *Aproveitamento de resíduos e subprodutos da póscolheita do cacau.* Boletim Técnico, CEPLAC/ Cocoa Research Centre/ CEPEC, Bahia, Brazil.

56. Selamat, J., Yusof, S., Jimbun, M., Awang, M., and Abdullah, R. 1996. Development of juice from cocoa pulp. In *Proceedings of the Malaysian International Cocoa Conference 1994,* pp. 351–357. Kota Kinabalu, Sabah: Malaysian Cocoa Board.

57. Buamah, R., Dzogbefia, V.P., Oldham, J.H. 1997. Pure yeasts culture fermentation of cocoa (*Theobroma cacao L*): Effect on yield of sweatings and cocoa bean quality. *World J. Microbiol. Biotech.,* **13:**457–462.

58. Adomako, D. and Takrama, J.F. 1999. Large scale collection of cocoa bean pulp juice (sweatings). In *Proceedings of the 12th International Cocoa Research Conference, Salvador, Brazil, 1996,* pp. 929–938. Lagos, Nigeria: Cocoa Producers Alliance.

59. Lopez, A.S. 1984. The cacao pulp soft-drink industry in Brazil and its influence on bean fermentation. In *Proceedings of 9th International Conference on Cocoa Research, Lome, Toga,* pp. 701–704. Lagos, Nigeria: Cocoa Producers Alliance.

60. Lopez, A. 1979. Fermentation and organoleptic quality of cacao as affected by partial removal of pulp juices from the beans prior to curing. *Rev. Theobroma, 9:*25–37.

61. Chong, C.F., Shepherd, R., and Poon, Y.C. 1978. Mitigation of cocoa bean acidity: Fermentary investigations. In *Proceedings of International Conference on Cocoa and Coconuts, Kuala Lumpur, Malaysia,* pp. 22–27. Lagos, Nigeria: Cocoa Producers Alliance.

62. Biehl, B., Meyer, B., Said, M., and Samarakoddy, R.J. 1990. Bean spreading: A method for pulp preconditioning to impair strong nib acidification during cocoa fermentation in Malaysia. *J. Sci. Food Agricult., 51:*35–45.

63. Schwan, R.F. and Lopez, A. 1987. Mudança no perfil da fermentaçao de cacau ocasionada pela retirada parcial da polpa da semente. In *Proceedings of 10th International Conference on Cocoa Research, Dominican Republic,* pp. 227–282. Lagos, Nigeria: Cocoa Producers Alliance.

64. Passos, F.J.V., Freire, E.S, and Romeu, A.P. 1989. *Desempenho de extratores semi-contínuos para polpa de cacau.* Boletim Técnico no 163, CEPLAC/Cocoa Research Centre (CEPEC), Ilhéus, BA, Brazil.

65. Akinwale, T.O. 2000. Extraction of pulp from fresh cocoa beans for wine production: Physico-chemical and sensory evaluation of the wine, and effect of pulp removal on the quality of cured cocoa beans. *Abstracts of 13th International cocoa Research Conference,* Kota Kinabalu, Malaysia, 9–14 October, 12.

66. Schwan, R.F. and Mororó, R. 1989. Despolpamento enzimático de sementes de cacau. In *Congresso de Ciencia e Tecnologia de Alimentos (Abstracts),* p. 25. Rio de Janeiro, Brazil: Brazil. Soc. Sci. Food Technol.

67. Freire, E.S., Mororó, R.C., and Schwan, R.F. 1999. The cocoa-pulp agroindustry and the uses of its residues in Bahia: Progress achieved in the last ten years. In *Proceedings of the 12th International Cocoa Research Conference, Salvador, Bahia, Brasil, 1996,* pp. 1013–1020. Lagos, Nigeria: Cocoa Producers Alliance.

68. Fogarty, W.M. and Ward, O.P. 1996. Pectinases and pectic polysaccharides, *Prog. Ind. Microbiol., 13:*59–119.

69. Rexová-Benková, L. and Markovic, O. 1976. Pectic Enzymes. *Adv. Carbohydrate Chem. Biochem., 33:*323–385.

70. Blanco, P., Sieiro, C., and Villa, T.G. 1999. Production of pectic enzymes in yeasts. *FEMS Microbiol. Letts., 175:*1–9.

71. Roelofsen, P.A. 1953. Polygalacturonase activity in yeast. *Neurospora* and tomato extract. *Biochim. Biophys. Ada* **10:**410–413.

72. Barnett, J.A., Payne, R.W., and Yarrow, D. 2000. *Yeasts: Classification and identification,* 3rd ed., Cambridge, Cambridge University Press.

73. Kurtzman, C.P. and Fell, J.W. 1998. *The Yeasts: A taxonomic study.* 4th ed., Amsterdam: Elsevier.

74. Sanchez, J., Daguenet, G., Guiraud, J.P., Vincent, J.C., and Galzy, P. 1985. A study of the yeast flora and the effect of pure culture seeding during the fermentation of cocoa beans. *Lebens.-wissen. Technol.,* **18:**69–76.

75. Schwan, R.F. and Rose, A.H. 1994. Polygalacturonase secretion by *Kluyveromyces marxianus:* Effect of medium composition. *J. Appl. Bacterial.* **76:**62–67.

76. Cruz-Guerrero, A., Bárzana, E., García-Garibay, M., and Gómez-Ruiz, L. 1999. Dissolved oxygen threshold or the repression of endo-polygalacturonase production by *Kluyveromyces marxianus. Proc. Biochem.,* **34:**621–624.

77. Wimborne, M.P. and Rickard, P.A.D. 1978. Pectinolytic activity of *Saccharomyces fragilis* cultured in controlled environments. *Biotechnol. Bioeng.,* **20:**231–242.

78. Vaughn, R.H., Jakubczyk, T., Macmillan, J.D., Higgins, T.E., Dave, B.A., and Crampton, V.M. 1969. Some pink yeasts associated with softening of olives. *Appl. Microbiol.,* **18:**771–775.

79. Blanco, P., Sieiro, C., Reboredo, N.M., and Villa, T.G. 1998. Cloning, molecular characterisation, and expression of an endo-polygalacturonase-encoding gene from *Saccharomyces cerevisiae* IM1-8b. *FEMS Microbiol. Letts.,* **164:**249–255.

80. Hirose, N., Kishida, M., Kawasaki, H., and Sakai, T. 1998. Molecular cloning and expression of a polygalacturonase gene in *Saccharomyces cerevisiae. J. Ferment. Bioeng.,* **86:**332–334.

81. Gognies, S., Gainvors, A., Aigle, M., and Belarbi, A. 1999. Cloning, sequencing analysis and over-expression of a *Saccharomyces cerevisiae* endopolygalacturonase-encoding gene *(PGLl). Yeast,* **15:**11–22.

82. Jia, J. and Wheals, A.E. 2000. Endopolygalacturonase genes and enzymes from *Saccharomyces cerevisiae* and *Kluyveromyces marxianus. Current Genet,* **38:**264–270.

83. Naumov, G.I., Naumova, E.S., Aigle, M., Masneuf, I., and Belarbi, A. 2001. Genetic reidentification of the pectinolytic yeast strain SCPP as *Saccharomyces bayanus* var. *uvarum. Appl. Microbiol. Biotechnol.,* **55:**108–111.

84. Šiekštele, R., Bartkeviciute, D., and Sasnauskas, K. 1999. Cloning, targeted disruption and heterologous expression of the *Kluyveromyces marxianus* endopolygalacturonase gene *(EPG1). Yeast,* **15:**311–322.

85. Galibert, F., et al. 1996. Complete nucleotide sequence of *Saccharomyces cerevisiae* chromosome X. *EMBO. J.,* **15:**2031–2049.

86. Madhani, H.D., Galitski, T., Lander, E.S., and Fink, G.R. 1999. Effects of a developmental mitogen-activated protein kinase cascade revealed by expression signatures of signaling mutants. *Proc. Natl. Acad. Sci. USA.,* **96:**12530–12535.

87. Prenicová, L., Benen, J.A.E., Kester, H.C.M, and Visser, J. 1998. *pgaE* encodes a fourth member of the endopolygalacturonase gene family from *Aspergillus niger. Eur. J. Biochem.,* **251:**72–80.

88. Misnawa, S.J., Jamilah, B., and Nazamid, P. 2000. Polyphenol oxidase refeneration of under-fermented dried cocoa beans and its possibility to improve cocoa flavor quality. *Abstracts of 13th International Cocoa Research Conference,* Kota Kinabalu, Malaysia, 128.

89. Luna, F., Crouzillat, D., Cirou, L., and Bucheli, P. 2002. Chemical composition and flavor of Ecuadorian cocoa liquor. *J. Agric. Food. Chem.,* **50:**3527–3532.

90. Rohan, T.A. 1964. The precursors of chocolate aroma: A comparative study of fermented and unfermented beans. *J. Sci. Food Agric., 29:*456–459.

91. Buyukpamukcu, E., Goodall, D.M., Hansen, C.E., Keely, B.J., Kochhar, S., and Wille, H. 2001. Characterization of peptides formed during fermentation of cocoa bean. *J. Agricult. Food Chem.*, **49:**5822–5827.

92. Amin, I., Jinap, S., Jamilah. B., Harikrisna, K., and Biehl, B. 2002. Analysis of vicilin (7S)-class globulin in cocoa cotyledons from various genetic origins. *J. Sci. Food Agricult.*, **82:**728–732.

93. Laloi, M., McCarthy, J., Morandi, O., Gysler, C., and Bucheli, P. 2002. Molecular and biochemical characterisation of two aspartic proteinases TcAP1 and TcAP2 from *Theobroma cacao* seeds. *Planta*, **215:**754–762.

94. Dias, J.C. and Avila, M.G.M. 1993. Influence of the drying process on the acidity of cocoa beans. *Agrotrópica*, **5:**19–24.

95. Clapperton, J.F. 1994. A review of research to identify the origins of cocoa flavour characteristics. *Cocoa Growers' Bulletin*, **48:**7–16.

96. Holm, C.S., Aston, J.W., and Douglas, K. 1993. The effects of the organic acids in cocoa on flavour of chocolate. *J. Sci. Food Agricult.*, **61:**65–71.

97. Zaibunnisa, A.H., Russly, A.R., Jinap, S., Jamilah, B., and Wahyudi, T. 2000. Quality characteristics of cocoa beans from mechanical fermentation. *Abstracts of 13th International Cocoa Research Conference*, Kota Kinabalu, Malaysia, 131.

98. Venturieiri, G.A. and Aguiar, J.P.L. 1988. Composição do chocolate caseiro de amêndoas de cupuaçu *(Theobroma grandiflorum* (Willd ex Spreng) Schum). *Acta Amazonica*, **18:**3–8.

99. Thorold, C. A. 1975. *Diseases of cocoa*. Oxford: Clarendon Press.

100. Thompson, S.S., Miller, K.B., and Lopez, A. 1997. Cocoa and Coffee. In *Food Microbiology Fundamentals and Frontiers*, pp. 649–661. Doyle, M.P., Beuchat, L.R. and Montville, T.J., Eds. Washington, DC: ASM Press.

101. Carr, J.G. and Davies, P.A. 1980. *Cocoa fermentation in Ghana and Malaysia: Further microbial methods and results*. Bristol: University of Bristol.

Acknowledgments—Much of the work described in this paper was done in SETEA at CEPLAC/CEPEC in Itabuna and RFS thanks her colleagues for their consistent support and help over many years, in particular the excellent technical staff who did much of the work. The authors thank CNPq, CAPES and the EC (INCO-DC IC18 CT97 0182) for financial support.

Appendix: Revised Names of Genera and Species

Former Name	Current Name
Yeasts	
Brettanomyces clausenii	*Dekkera anomala*
Candida guilliermondii	*Candida guilliermondii* var. *guilliermondii* and *Candida guilliermondii* var. *membranifaciens*
Candida krusei	*Issatchenkia orientalis*
Candida norvegensis	*Pichia norvegensis*
Candida pelliculosa	*Pichia anomala*
Candida reukaufii	*Metschnikowia reukaufii*
Candida. Cacoai	*Pichia farinosa*
Kloeckera apiculata	*Hanseniaspora uvarum*
Kloeckera apis	*Hanseniaspora guilliermondii*
Kluyveromyces fragilis	*Kluyveromyces marxianus*
Saccharomyces chevalieri	*Saccharomyces cerevisiae*
Saccharomyces fragilis	*Kluyveromyces marxianus*
Schizosaccharomyces malidevorans	*Schizosaccharomyces pombe*
Torulopsis candida	*Candida saitoana*
Torulopsis castelli	*Candida castelli*
Torulopsis holmii	*Saccharomyces exiguus*
Bacteria	
Gluconobacter oxydans subsp. *suboxydans*	*Gluconobacter oxydans*
Lactobacillus acidophilus	*Thiobacillus acidophilus*
Lactobacillus casei pseudoplantarum	*Lactobacillus paracasei* subsp. *paracasei*
Lactobacillus cellobiosus	*Lactobacillus fermentum*
Lactobacillus kandleri	*Weissella kandleri*
Lactobacillus plantarum	*Lactococcus plantarum*

Test-Your-Knowledge Form

We encourage you to photocopy and use this page as a tool to assess how the articles in *Annual Editions* expand on the information in your textbook. By reflecting on the articles you will gain enhanced text information. You can also access this useful form on a product's book support website at *http://www.mhcls.com*.

NAME: DATE:

_____

TITLE AND NUMBER OF ARTICLE:

_____

BRIEFLY STATE THE MAIN IDEA OF THIS ARTICLE:

_____

LIST THREE IMPORTANT FACTS THAT THE AUTHOR USES TO SUPPORT THE MAIN IDEA:

_____

WHAT INFORMATION OR IDEAS DISCUSSED IN THIS ARTICLE ARE ALSO DISCUSSED IN YOUR TEXTBOOK OR OTHER READINGS THAT YOU HAVE DONE? LIST THE TEXTBOOK CHAPTERS AND PAGE NUMBERS:

_____

LIST ANY EXAMPLES OF BIAS OR FAULTY REASONING THAT YOU FOUND IN THE ARTICLE:

_____

LIST ANY NEW TERMS/CONCEPTS THAT WERE DISCUSSED IN THE ARTICLE, AND WRITE A SHORT DEFINITION:

We Want Your Advice

ANNUAL EDITIONS revisions depend on two major opinion sources: one is our Advisory Board, listed in the front of this volume, which works with us in scanning the thousands of articles published in the public press each year; the other is you—the person actually using the book. Please help us and the users of the next edition by completing the prepaid article rating form on this page and returning it to us. Thank you for your help!

ANNUAL EDITIONS: Microbiology 10/11

ARTICLE RATING FORM

Here is an opportunity for you to have direct input into the next revision of this volume.
We would like you to rate each of the articles listed below, using the following scale:

1. **Excellent: should definitely be retained**
2. **Above average: should probably be retained**
3. **Below average: should probably be deleted**
4. **Poor: should definitely be deleted**

Your ratings will play a vital part in the next revision.
Please mail this prepaid form to us as soon as possible.
Thanks for your help!

RATING	ARTICLE	RATING	ARTICLE
	1. Microbial Diversity Unbound		27. Sponge's Secret Weapon Revealed
	2. Extreme Microbes		28. Immunity's Early-Warning System
	3. Pondering a Parasite		29. How Cells Clean House
	4. Say Hello to the Bugs in Your Gut		30. Start Early to Prevent Genital HPV Infection—And Cervical Cancer
	5. Nurturing Our Microbes		31. How Safe Are Vaccines?
	6. An Endangered Species in the Stomach		32. Caution: Killing Germs May Be Hazardous to Your Health
	7. Bacteria Are Picky about Their Homes on Human Skin		33. Why We're Sicker
	8. Biofilms		34. Novel Anti-Infectives: Is Host Defence the Answer?
	9. Understanding Fungi through Their Genomes		35. Malaria Vaccines and Their Potential Role in the Elimination of Malaria
	10. Going with His Gut Bacteria		36. Preventing the Next Pandemic
	11. New Tactics against Tuberculosis		37. An Ill Wind, Bringing Meningitis
	12. Fighting Killer Worms		38. The Flu Hunter
	13. Bacterial Therapies: Completing the Cancer Treatment Toolbox		39. Climate, Environment, and Infectious Disease: A Report from the AIBS 2008 Annual Meeting
	14. Microbial Moxie		40. Mosquito Modifications: New Approaches to Controlling Malaria
	15. Eyeing Oil, Synthetic Biologists Mine Microbes for Black Gold		41. The White Plague
	16. The Bacteria Fight Back		42. Rising Incidence of Valley Fever and Norovirus
	17. Germ Warfare		43. They Came from Above
	18. Routes of Resistance		44. Toxic Salad
	19. Is Your Patient Taking the Right Antimicrobial?		45. Fear of Fresh: How to Avoid Foodborne Illness from Fruits & Vegetables
	20. Constant Struggle to Conquer Bacteria		46. Protecting Ourselves from Shellfish Poisoning
	21. Super Bugged		47. Better Safe than Sorry
	22. New Antimicrobial Agents for the Treatment of Gram-Positive Bacterial Infections		48. Detecting Mad Cow Disease
	23. Bacteriophage Lysins as Effective Antibacterials		49. Yeasts in Foods and Beverages: Impact on Product Quality and Safety
	24. HIV Integrase Inhibitors—Out of the Pipeline and into the Clinic		50. The Microbiology of Cocoa Fermentation and Its Role in Chocolate Quality
	25. Removing the Golden Coat of *Staphylococcus aureus*		
	26. Antibodies for the Treatment of Bacterial Infections: Current Experience and Future Prospects		

BUSINESS REPLY MAIL
FIRST CLASS MAIL PERMIT NO. 551 DUBUQUE IA

POSTAGE WILL BE PAID BY ADDRESSEE

McGraw-Hill Contemporary Learning Series
501 BELL STREET
DUBUQUE, IA 52001

NO POSTAGE
NECESSARY
IF MAILED
IN THE
UNITED STATES

ABOUT YOU

Name   Date

Are you a teacher? ☐ A student? ☐
Your school's name

Department

Address   City   State   Zip

School telephone #

YOUR COMMENTS ARE IMPORTANT TO US!

Please fill in the following information:
For which course did you use this book?

Did you use a text with this ANNUAL EDITION? ☐ yes ☐ no
What was the title of the text?

What are your general reactions to the Annual Editions concept?

Have you read any pertinent articles recently that you think should be included in the next edition? Explain.

Are there any articles that you feel should be replaced in the next edition? Why?

Are there any World Wide Websites that you feel should be included in the next edition? Please annotate.

May we contact you for editorial input? ☐ yes ☐ no
May we quote your comments? ☐ yes ☐ no